Acoustical Imaging

Volume 24

Acoustical Imaging

Recent Volumes in This Series:

Volume 9	Proceedings of the Ninth International Symposium, December 3–6, 1979, edited by Keith Y. Wang
Volume 10	Proceedings of the Tenth International Symposium, October 12–16, 1980, edited by Pierre Alais and Alexander F. Metherell
Volume 11	Proceedings of the Eleventh International Symposium, May 4–7, 1981, edited by John P. Powers
Volume 12	Proceedings of the Twelfth International Symposium, July 19–22, 1982, edited by Eric A. Ash and C. R. Hill
Volume 13	Proceedings of the Thirteenth International Symposium, October 26–28, 1983, edited by M. Kavch, R. K. Mueller, and J. F. Greenleaf
Volume 14	Proceedings of the Fourteenth International Symposium, April 22–25, 1985, edited by A. J. Berkhout, J. Ridder, and L. F. van der Wal
Volume 15	Proceedings of the Fifteenth International Symposium, July 14–16, 1986, edited by Hugh W. Jones
Volume 16	Proceedings of the Sixteenth International Symposium, June 10–12, 1987, edited by Lawrence W. Kessler
Volume 17	Proceedings of the Seventeenth International Symposium, May 31–June 2, 1988, edited by Hiroshi Shimizu, Noriyoshi Chubachi, and Jun-ichi Kusibiki
Volume 18	Proceedings of the Eighteenth International Symposium, September 18–20, 1989, edited by Hua Lee and Glen Wade
Volume 19	Proceedings of the Nineteenth International Symposium, April 3–5, 1991, edited by Helmut Ermert and Hans-Peter Harjes
Volume 20	Proceedings of the Twentieth International Symposium, September 12–14, 1992, edited by Yu Wei and Benli Gu
Volume 21	Proceedings of the Twenty-First International Symposium, March 28–30, 1994, edited by Joie Pierce Jones
Volume 22	Proceedings of the Twenty-Second International Symposium, September 3–7, 1995, edited by Piero Tortoli and Leonardo Masotti
Volume 23	Proceedings of the Twenty-Third International Symposium, April 13–16, 1997, edited by Sidney Lees and Leonard A. Ferrari
Volume 24	Proceedings of the Twenty-Fourth International Symposium, September 23–25, 1998, edited by Hua Lee

A Continuation Order Plan is available for this series. A continuation order will bring delivery of each new volume immediately upon publication. Volumes are billed only upon actual shipment. For further information please contact the publisher.

Acoustical Imaging
Volume 24

Edited by

Hua Lee

University of California, Santa Barbara
Santa Barbara, California

Springer Science+Business Media, LLC

The Library of Congress cataloged the first volume of this series as follows:

International Symposium on Acoustical Holograpy
Acoustical holography; proceedings, v. 1–

v. illus. (part col.), ports. 24 cm.
Editors, 1967– , A. F. Metherell and L. Larmore (1967 with H. M. A. el-Sum)
Symposium for 1967 held at the Douglas Advanced Research Laboratories, Huntington Beach, Calif.

1. Acoustic holography—Congresses—Collected works. I. Metherell, Alexander A., ed. II. Larmore, Lewis, ed. III. el-Sum, Hussein Mohammed Amin, ed. IV. Douglas Advanced Research Laboratories. V. Title.

QC244.5.1.5 69-12533

Proceedings of 24th International Symposium on Acoustical Imaging, held September 23–25, 1998, in Santa Barbara, California.

ISSN 0270-5117

ISBN 978-1-4757-8796-2 ISBN 978-0-306-47108-7 (eBook)
DOI 10.1007/978-0-306-47108-7

Originally published by Kluwer Academic / Plenum Publishers in 2000
Softcover reprint of the hardcover 1st edition 2000

http://www.wkap.nl/

10 9 8 7 6 5 4 3 2 1

A C.I.P. record for this book is available from the Library of Congress

24th International Symposium on Acoustical Imaging

PREFACE

The International Symposium of Acoustical Imaging has been widely recognized as the premier forum for presentations of advanced research results in both theoretical and experimental development. Held regularly since 1968, the symposium brings together leading international researchers in the area of acoustical imaging. The 24th meeting is the third time Santa Barbara hosted this international conference and it is the first time the meeting was held on the campus of the University of California, Santa Barbara.

As many regular participants noticed over the years, this symposium has grown significantly in size due to the quality of the presentations as well as the organization itself. A few years ago multiple and poster sessions were introduced in order to accommodate this growth. In addition, the length of the presentations was shortened so more papers could be included in the sessions. During recent meetings there were discussions regarding the possibility of returning to the wonderful years when the symposium was organized in one single session with sufficient time to allow for in-depth presentation as well as discussions of each paper. And the size of the meeting was small enough that people were able to engage in serious technical interactions and all attendees would fit into one photograph. In light of the constraints of the limited budget with respect to the escalating costs it was not considered feasible. Yet, with the support of the members of the Executive Council, we were very pleased to be able to organize the 24th meeting in the old-fashion manner.

The technical program of the meeting was organized into three components: Advanced Systems and Techniques; Microscopy and Nondestructive Evaluation; and Biomedical Applications. Out of the pool of submissions fifty-seven presentations were included in the three-day meeting and fifty-four papers were selected to be included in the proceedings. Because of the selectivity of the technical program the papers in the proceedings are of the highest quality.

I wish to thank Vice Chancellor France Cordova of University of California, Santa Barbara for her encouragement and support. Most sincere thanks to the session chairs, the members of the International Advisory Board, and especially, the members of the Executive Council for their contributions to the formation of technical program, as well as the overall organization of the conference. We also wish to thank the staff of UC Santa Barbara's Conference Services for their assistance in managing the operations of the symposium.

Hua Lee
University of California, Santa Barbara

CONTENTS

PART I: ADVANCED SYSTEMS AND TECHNIQUES

PART II: MICROSCOPY AND NONDESTRUCTIVE EVALUATION

PART III: BIOMEDICAL APPLICATIONS

NON COHERENT SYNTHETIC APERTURE IMAGING

Pierre Alais, Pierre Cervenka, Pascal Challande, Valérie Lesec

Laboratoire de Mécanique Physique
Université Paris 6, CNRS UPRES A 7068
2 Place de la Gare de Ceinture, F-78210 Saint-Cyr-l'Ecole

INTRODUCTION

A first realization of a synthetic aperture sonar has been related at the third Symposium of Acoustical Holography by Castella [1]. Like in synthetic aperture radar imaging, the acoustical echographic information is added coherently in terms of the complex amplitude obtained from a quadrature synchronous detection at different positions of the emitter/receiver system. The synthetic aperture is limited by the directivity of the emitter and the angular acceptance of the receiver. This technique leads to a transverse or azimutal spatial resolution length Δx which remains independent of the slant range and, in general, much smaller than with a real aperture technique. A classical paper by Cutrona [2] in 1976 discusses the performance of such a technique compared with the classical one. Among the encountered difficulties are : The motion compensation which must be performed within a fraction of a wavelength in terms of the acoustical path ; The spatial sampling which must not create blurring grating lobes and limits severely the speed of exploration. A lot of work has been devoted these last years to this very promising technique [3, 4, 5]. It seems that through new self-focusing algorithms, the perturbations of the vehicle motion can be compensated adequately, but for the yaw which must be known very accurately. Nevertheless, the speed of exploration remains limited and this technique seems adapted to high resolution surveys at the expense of low coverage rates.

Another synthetic aperture technique consists of adding the echographic information received in terms of energy rather than in terms of complex amplitude. The synthetic aperture is limited exactly as in the preceding case, but the phase information is not taken into account : This operation can be defined as a non coherent synthetic aperture imaging technique. In fact, the first author to use this technique was probably Kossoff. In the 1970's [6], he obtained the first good obstetrical ultrasonic images by using, with a manual echographic system, a compound scanning permitting to look at the same target from several angles of view. The compound scanning technique was not used in later systems, including real time electronically scanned systems, because of tissue inhomogeneities which limit the good superposition of images obtained from different azimuths.

Acoustical Imaging, Volume 24.
Edited by Hua Lee, Kluwer Academic/Plenum Publishers, 2000.

The situation is much more favorable in the ocean and the object of this paper is to show that the non coherent synthetic aperture technique, also called "multi-look", if less ambitious than the coherent technique in resolution capability, offers many practical advantages. This technique has been already discussed, at least theoretically, by several authors [7, 8]. H. Lee [9] gives experimental results that compare the multi-look sonar technique with the classical one and also with the coherent synthetic aperture technique. It is well recognized that the non coherent addition of echographic information does not permit to improve substantially the resolution. The focusing effect created in such a way is related to the temporal resolution or to the corresponding range resolution length Δy. A crude model, based on a rectangular profile for an echographic information obtained without any other focusing technique, yields a focusing effect characterized by the following approximate resolution length :

$$\Delta x_{3dB} = 4\Delta y/\Delta\theta, \tag{1}$$

where $\Delta\theta$ is the acceptance angle permitted by the echographic system for operating the synthetic addition (Figure 1).

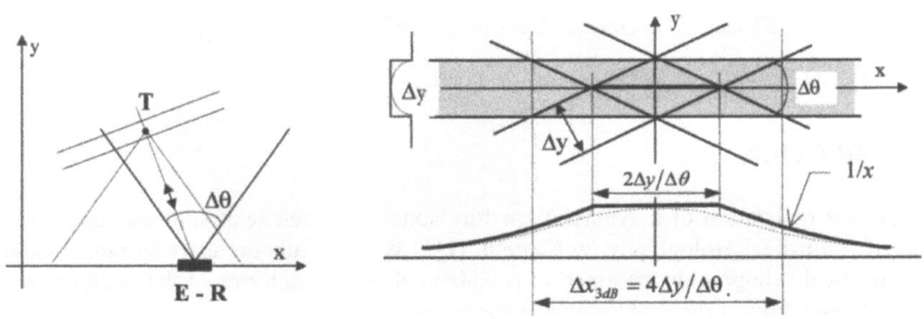

Figure 1 : Focusing by a non coherent synthetic aperture addition of energy

If the operation is performed with a receiving array of real aperture $L = pa$ (p adjacent elements of size a), the focusing capability of the physical antenna (coherent real aperture) at a slant range R is expressed by $\Delta x_{3dB} = R\lambda/L$. The receiving angle of view is classically limited by $\Delta\theta_{max} = \lambda/(2a)$ to keep grating lobes at a reasonably low level. Let us denote $N = L/\lambda$ and $n = a/\lambda$, respectively the length of the array and the width of the each array element measured in wavelengths λ. The focusing effects by non coherent synthetic aperture and by coherent real aperture become equal for the slant range R given by :

$$R = 8\,Nn\,\Delta y \tag{2}$$

It is easy to check that for practical cases, the non coherent synthetic aperture effect is not efficient in terms of resolution as it will be seen in one particular experiment. Hence, it remains important to use a high resolution real aperture system : The synthetic operation affords a small improving effect on the resolution Δx, and at large ranges only.

In fact, our paper aims to show that the non coherent synthetic addition has many other advantages than increasing the focusing effect. The greatest improvement occurs in the image quality because of the contrast enhancement : The non coherent summation cancels the speckle effect induced by the coherent process performed on the physical antenna. Another great advantage is the robustness of this technique with respect to motion perturbations. Errors in the estimation of the motion may be accepted with one or two orders of magnitude larger than in the coherent synthetic aperture process. Spatial under-sampling of pings caused by the vehicle speed is also well tolerated and allows large coverage rates.

THE EXPERIMENTAL SYSTEM

We have built a single sided prototype sonar. The transmitted signal is a linear FM chirp ($F_o = 100$ kHz, $B = 3$ kHz) whose amplitude is modulated with a truncated Gaussian window. The range resolution obtained after pulse compression is about 30 cm. The aperture in site is classical for both transmitter and receivers, but the transmitted beam is rather large in azimuth ≅ 10°, obtained by means of a transmitting antenna whose geometry is an arc of circle (radius ≅ 6 m, curvilinear length ≅ 1.3 m). Two linear arrays are used at receive for interferometric purpose and bathymetric measurements. Each array (L = 1.44 m ≅ 96 λ) is composed of 32 transducers (a = 45 mm ≅ 3 λ). The resulting angular resolution is about 0.7°. In these conditions, the maximum angle of view maintaining the grating lobes at reasonable levels is $\Delta\theta = \lambda/(2a) = 1/6$ rad., i.e. the 10° adopted for the transmitted beam.

According to Eq. (2), the range for which the synthetic addition may be effective is 800 m, so that we can expect only a small improvement for the spatial along track resolution Δx. The following simulation will confirm this situation, but will show also how the speckle is washed out in comparison with conventional imaging technique.

SIMULATIONS

A single point target is assumed to be located at 600 m range. Figure 2 is the image of this target obtained with a single ping. The receiving array is not tapered, so that the first side lobe level is at –13 dB. Figure 3 is the image of the same target obtained after non coherent processing of simulated pings covering a tapered angular aperture of 10° @ –6 dB (length of the simulated trajectory = 105 m). The combination of the high range resolution (thanks to the pulse compression technique) with the rotation that the multi-look angles, results in the scattering of the side lobes whose levels are therefore reduced (–5 dB for the first lobe, –7 dB for the second lobe). A slight gain for the angular resolution (14% @ –6 dB) can be also observed. The larger the target range and the synthetic angular aperture, the most effective is the process with respect to the reduction of the side lobes level and the increase of the resolution.

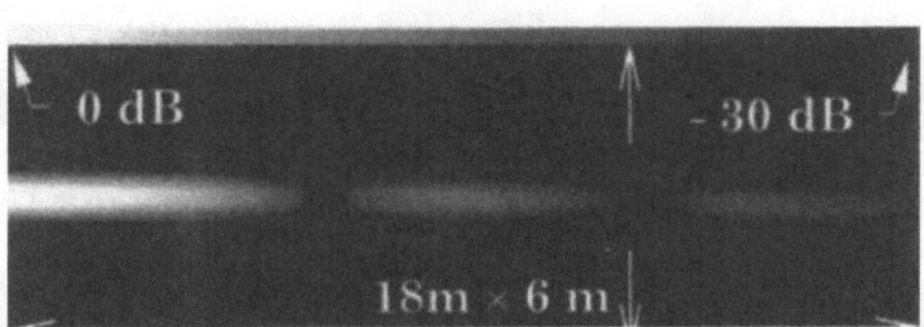

Figure 2 : (half) Point Spread Function - Single ping (target @ 600 m)

Figure 3 : Non coherent synthesis (10° aperture @ –6 dB ≡ 105 m baseline)

A simple configuration is used in Figures 4-6 to show the mechanism of the speckle reduction. We assume two closely spaced point targets (40 cm apart) at 200 m range. In Figure 4, the targets are perfectly symmetrical to the bore-sight direction of the single ping used to build the image. The targets cannot be discriminated by the receiving antenna. They produce a single central spot together with side lobes. In Figure 5, the targets are slightly rotated, so that the range difference is λ/4, i.e. only 3.25 mm ! Now, the antenna lies in between two petals of the backscattered diagram. Because the size of the antenna is large enough to cover a significant part of a petal, the image is not completely black. The central spot is divided in two parts, resulting of an interference process. These two spots are not actual images of the targets which are much closer. Figure 6 shows the image obtained after a non coherent summation of pings such as for building Figure 3. Both unresolved targets are seen as a single spot. A single image is displayed here, because there is absolutely no difference whether the targets are perfectly symmetrical or not. The cause is that the base-line of the non coherent process encompasses in both cases several petals of the diagram. The only noticeable effect is a constant loss of about 5 dB which corresponds roughly to the normalized mean energy returning from a petal.

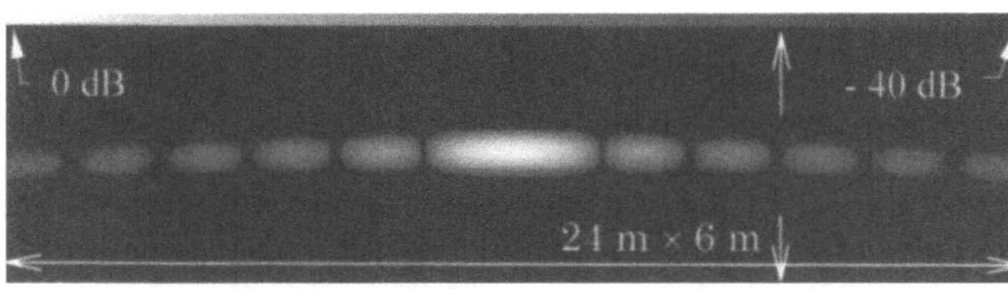

Figure 4 : Central ping - Symmetrical targets (40 cm apart) @ 200 m range

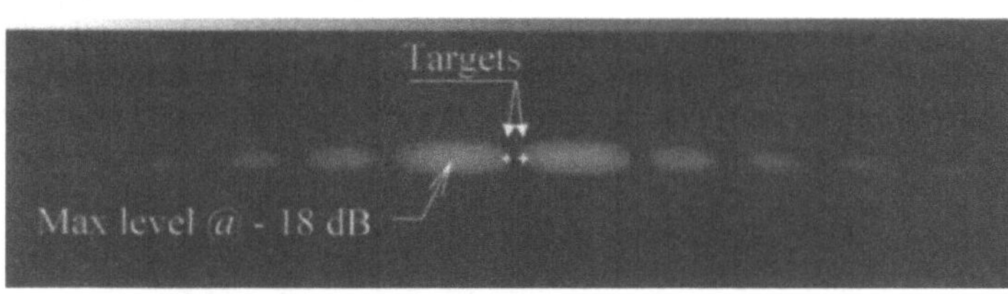

Figure 5 : Central ping - Targets range difference = λ/4

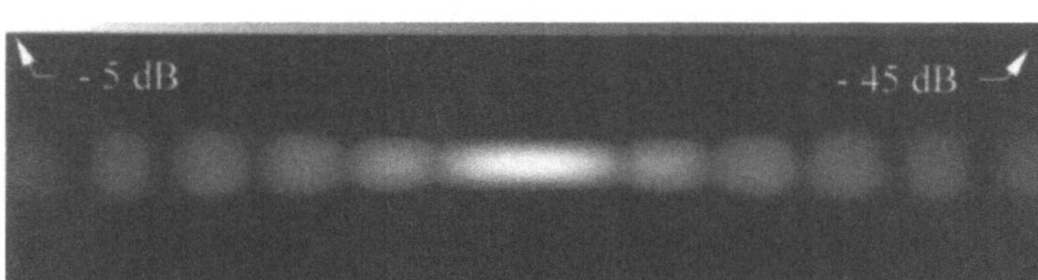

Figure 6 : Non coherent synthesis (10° aperture) - Distance between targets = 40 cm

EXPERIMENTAL RESULTS

Our experimental results have been obtained at the GESMA (Groupe d'Etudes Sous-Marines de l'Atlantique) facility in the bay of Brest. It consists of an horizontal rail that carries a motorized sliding platform. The translation capability is 12 m. The rail is mounted on a dock open to the sea, and may be immersed at different depths. The antenna was about 3 m below the sea surface and 9 m above the bottom. The bore-sight direction aimed at 40° below the horizontal. In this area, the sea bottom is nearly flat. Several objects were set on the bottom, and particularly a cylinder (70 cm diameter, 1,9 m long) at 30 m range and a sphere (1 m diameter) at 50 m range. This range is very small compared to what can be expected at a frequency of 100 kHz, but is convenient with respect to the practical limitation (< 12 m) of the synthetic aperture and the value retained for the angle of view (10°).

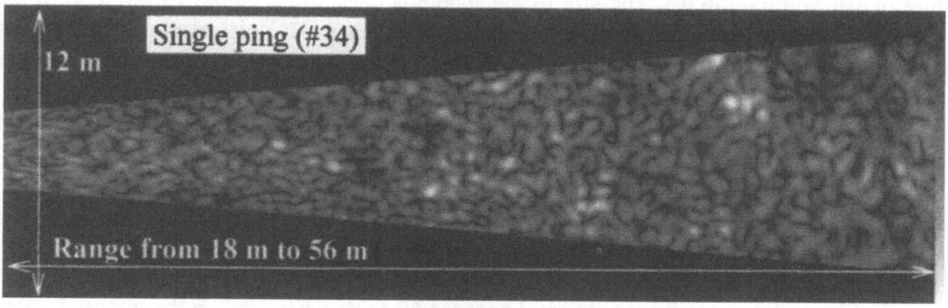

Figure 7

Figure 8

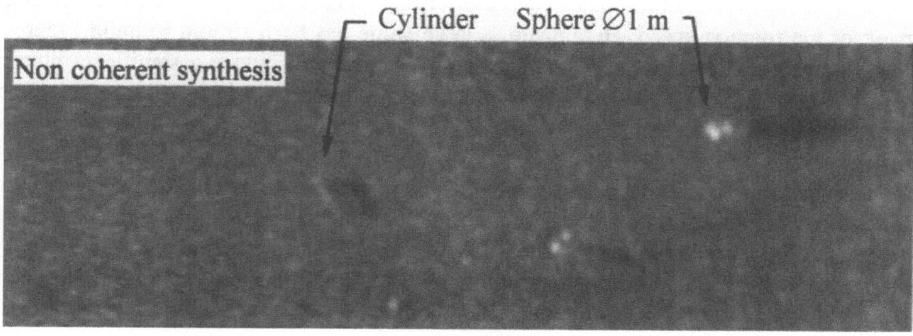

Figure 9

With a first mode of data acquisition, base-band digitized signals received by each of the 32 transducers of a single array were registered. The presented images are computed with these data sets. The representation is in slant range, i.e. there is no radiometric correction. The pixel dimension is 5×5 cm^2. For comparison purpose, the median gray level is adjusted at 35%. No histogram equalization, nor other fancy image enhancement, are performed.

Figure 7 exhibits the image computed with a single ping (shot from the center of the rail). The speckle noise hides most of the features : The echoes from the sphere can be only guessed. Figure 8 is built as a classical side-scan sonar image, i.e. signals received from each ping at bore-sight are properly merged and smoothed. The shadow of the sphere can be seen, and the presence of the cylinder can be guessed. However, the presence of the speckle noise is still very perturbing. Figure 9 is obtained after non coherent processing of all pings, i.e. adding the energy of images such as shown in Figure 7, using a tapered aperture of 10° width @ -6 dB. The size of the sphere is close to the limit of resolution of the receiving antenna. However, the drastic reduction of the speckle noise exemplifies the clear advantage afforded by this technique. The shadows of the cylinder and of the sphere appear clearly.

Let us define the redundancy factor, r, as the number of pings that can be used to build a pixel at a given range $R \leq R_{max}$, i.e. :

$$r = (R/R_{max})\,(c\Delta\theta/2V) \qquad (3)$$

where R_{max} is the maximal slant range, V is the speed of the platform and $\Delta\theta$ is the synthetic angular aperture.

In Figure 9, the conditions are ideal because 60 shots were used on the imposed linear trajectory of 12 m run at low speed (20 cm/s). However, it must be noted that the redundancy factor can be drastically reduced, down to 5. We checked that Figure 9 is only slightly degraded if we use only 6 pings instead of 60. Only zones at very close range are possibly under-sampled. For example, the condition $r > 5$ is satisfied for a vehicle that runs at 6 knots, 100 m above the sea bottom, looking up to $R_{max} = 1000$ m, with a view angle $\Delta\theta = 5°$. In these conditions, the Doppler shift would remain less than 12 Hz, and would afford an equivalent perturbation of 40 mm in the acoustical path with the used chirp. This perturbation is much smaller than the range resolution, so that the Doppler effect would not affect the image quality.

Besides, as mentioned in the introduction, the non coherent technique is very robust with respect to the perturbations induced by the vehicle motion. Figure 10 is built with a bias in the lateral speed of the antenna, i.e. by taking a value that is not null. Such a bias is almost equivalent to introducing a bias in the azimuth. A very large value (16 cm/s @ 2 knots ≡ 9°) has been chosen in order to see some significant difference with the reference image (Figure 11). Another major source of artifact occurs from the bias in the measurement of the rotation speed. Here again, a large value has been chosen to build Figure 12 (1.2° between extreme pings, 0.1°/s @ 2 knots) in order to induce a significant blurring effect. It must be noticed that the image obtained by coherent synthesis (Figure 13 - coherent synthetic aperture over 4° @ -6 dB) is completely destroyed if such biases are introduced.

On the other hand, our system is able to record on line the information corresponding to 15 directions in azimuth, with an angular pitch of .66° in order to cover the entire view angle $\Delta\theta$. These preformed beams use 8 focal zones to keep the best azimuth resolution in the range 10 m – 1000 m. This information permits to build images with a very low degradation compared to Figure 9, with much less computation and much higher speed.

Figure 10

Figure 11

Figure 12

Figure 13

CONCLUSION

The non coherent imaging technique does not aim at the same goal as the coherent synthesis. With the non coherent process, the overall resolution is not much increased with respect to the theoretical performance of the physical array. However, the image quality is definitively improved at a low computational expense, and without stringent requirements about the accuracy of the antennae trajectory and attitude.

The main interests for the non coherent synthesis are : 1) Reduction of the side lobe levels, together with a slight improvement of the resolution in azimuth ; 2) Drastic reduction of the speckle effect so that objects whose size is close to the theoretical longitudinal resolution of the physical antennae can be detected ; 3) Robustness of the image quality with respect to the trajectory and attitude artefacts ; 4) Robustness of the image quality with respect to the vehicle speed as long as the redundancy factor remains larger than 5, which authorizes surveys at high coverage rate ; 5) Improvement of bathymetric measurements through the redundancy factor. This effect will be shown in an other paper. We are looking now for an experiment carried from a vessel, in order to verify and to quantify the capabilities of this technique at larger ranges, i.e. in actual operational conditions.

ACKNOWLEDGMENTS

We thank M. Brussieux, V. Tonard and A. Salaun from GESMA for their very efficient cooperation for carrying the experiments in Brest.

REFERENCES

1. F.R. Castella, *Application of one-dimensional holographic techniques to a mapping sonar system*, Acoustical Holography, Vol. 3, 1970, pp. 247-271.
2. L.J. Cutrona, *Comparison of sonar system performance achievable using synthetic aperture techniques with the performance achievable by more conventional means*, J.A.S.A., Vol. 58, n° 2, August 1975, pp. 336-348.
3. J. Chatillon, M.E. Zakharia, M.E. Bouhier, *Self-focusing of synthetic aperture sonar : validation from sea data*, Proc. 2nd European Conference on Underwater Acoustics, Lyngby, Denmark, July 1994, pp. 727-731.
4. V. Tonard, J. Chatillon, *Acoustical imaging of extended targets by means of synthetic aperture sonar technique*, Acustica, Acta Acoustica, Vol. 83, 1997, pp. 992-999.
5. D. Billon, F. Fohanno, *Theoretical performance and experimental results for synthetic aperture sonar self-calibration*, Oceans'98, Septembre 1998, Nice, France.
6. G. Kossoff, *Progress in pulse echo techniques*, Proc. 2nd World Congress Ultrasonics in Medicine, Rotterdam 1974, pp. 37-48.
7. M.P. Hayes, P.T. Gough, *Broad band synthetic aperture sonar*, IEEE Journal of Oceanic Engineering, Vol. 17, n° 1, Janvier 1992.
8. G. Shippey, T. Nordkvist, *Phased array acoustic imaging in ground coordinates, with extension to synthetic aperture processing*, IEEE Proc. Radar Sonar Navigation, Vol. 143, n° 3, Juin 1996.
9. B.L. Douglas, H. Lee, C.D. Loggins, *A multiple receiver synthetic aperture active sonar imaging system*, Oceans'92, Conf. Records MTS, IEEE, 1992, pp. 300-305.

IMAGING WITH A 2D TRANSDUCER HYBRID ARRAY

Ken Erikson, Jason Stockwell, Allen Hairston, Gary Rich,
John Marciniec, Lee Walter, Kristin Clark & Tim White

Lockheed Martin IR Imaging Systems
Lexington, MA 02421

ABSTRACT

Imaging with fully populated 2D arrays using acoustical lenses in the low MHz frequency range offers the potential for high resolution, real-time, 3D volume imaging together with low power and low cost. A 2D composite piezoelectric receiver array bonded directly to a large custom integrated circuit was discussed[1] at the 23rd International Symposium on Acoustical Imaging. This 128 x 128 (16,384 total) element Transducer Hybrid Array (THA) uses massively parallel, on-chip signal processing and is intended for medical and underwater imaging applications. The system under development, which is a direct analog of a video camera, will be discussed in this paper.

INTRODUCTION

Real-time imaging has been one of the key factors in the widespread acceptance of medical ultrasound and it will certainly be no less important in underwater imaging. 3D Ultrasound (3DUS) volumetric imaging has been gaining acceptance in medical imaging due to the additional applications it enables. Similar interest in 3DUS is growing in underwater imaging.

Real-time-3DUS imaging, however, imposes stringent requirements on an acoustic array and signal processing system. Due to the relatively low velocity of sound, real-time 3DUS at ranges greater than a few cm in tissue or water requires either multiple parallel B-scan beam forming or C-scan parallel data acquisition.

A large two-dimensional aperture must be available continuously to maintain resolution throughout the volume. "Conventional" volume images assembled in post processing from multiple B-scan images use the smaller footprint of beam-formed linear phased arrays at the expense of out of plane resolution and are clearly not real-time.

Traditional electronic beamforming, either for medical or underwater imaging applications requires a large amount of fast signal processing which in turn, usually requires high electrical power consumption. Parallel beamforming, if used, compounds the power

Acoustical Imaging, Volume 24.
Edited by Hua Lee, Kluwer Academic/Plenum Publishers, 2000.

demands. In contrast, an acoustical lens is a two-dimensional "beamformer" which uses zero electrical power and can be fabricated at very low cost. Portability requirements for underwater imaging systems place an upper limit on system size and aperture for SCUBA divers, however, underwater vehicles (UUV's/AUV's) typically can support somewhat larger sizes.

In 1994, our group at Lockheed Martin IR Imaging began development of a 2D ultrasonic array based on infrared (IR) focal plane array technology developed for military applications over the last two decades. Using an integrated circuit previously developed for an IR imager, and a new transducer array, a 42x64 (2688 total) element 2D Transducer Hybrid Array (THA) was developed. Together with a simple two element acoustical lens, this system demonstrated real-time imaging, albeit with very limited dynamic range. Subsequently, we developed a 128x128 (16,384 total) element THA, with a custom integrated circuit specifically designed for 3DUS which was discussed at the 23rd International Symposium on Acoustical Imaging.

This paper first reviews real-time 3DUS system requirements and our basic system approach. Tradeoffs at the system design level are then discussed, followed by a more detailed discussion of the acoustical aspects of the THA itself.

SYSTEM REQUIREMENTS

Table I. Real-time 3D Imaging System Requirements

Bold = primary, * = less important

	Medical Camera	Diver Camera
Targets to be imaged:	**Abdomen** **Breast**	**Mines**
Range:	**0 to 100 mm**	**1 to > 5 meters**
3D Resolution: (voxel dimensions)	**< 1 mm**	**< 10 mm**
Maximum Camera Dimensions		
Volume^2:	*	**< 800 in^3**
Aperture:	**< 100 mm**	**<200 mm**
Acoustical Frame Rate: (defined here as "real-time")	**≥15 Hz**	**≥15 Hz**

SONOCAM™ ACOUSTICAL CAMERA SYSTEM

The SonoCam™ concept is an acoustical analog of an electronic optical camera (Fig. 1). At present, the system is bistatic (Fig. 1a), i.e. it uses a separate transmitter transducer. Development of a monostatic capability (Fig. 1b) is underway. The transmitter pulse-insonifies a volume. Reflected sound energy is collected by the acoustical lens (Fig. 1c) to form a focused image on the Transducer Hybrid Array (THA) (Fig. 1d). The signals from individual elements of the THA's piezocomposite transducer array are electronically read out through a custom integrated circuit intimately attached to the backside of the array by small solder bumps (Fig. 2). Additional standard digital electronics behind the THA (Fig.

1e) provide control and signal processing. A PC-based system, not shown, provides image processing and display.

With the added advantage of range (time) gating multiple planes, the third dimension of data is gathered (Fig. 3). Unlike most other acoustical imaging systems, only one pulse of sound is required for multiple complete planes of data because all 16,384 receiver transducers and preamplifiers receive data simultaneously. The acoustical waveform is sampled in quadrature and stored in the integrated circuit. After reception of all the planes for a single transmitted pulse, the data is multiplexed out through 16 ports, digitized to 12 bits and processed at an aggregate rate of 160 megasamples per second. On the next pulse the range gates are delayed to sample the next volume and this is repeated until the full volume of interest has been sampled. In the present system, with five planes of storage per acoustical pulse, only 16 pulses per acoustical frame are required for a volume 128x128x80 planes deep (1.3 million voxels).

A simple calculation reveals that real-time acoustical frame rates can be maintained well past 100 mm in tissue with such a "plane-at-a-time" system. In contrast, for 128x128x80 voxels, a "conventional", electronically beamformed "line-at-a-time" system such as a 2D phased array would require 32 parallel beamformers to achieve a 100 mm range.

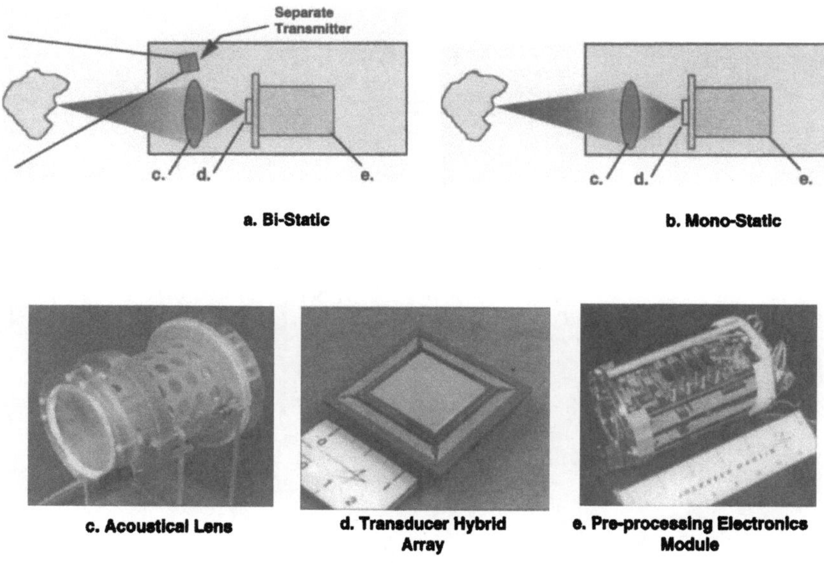

Fig. 1. Acoustical Camera Concept & Components

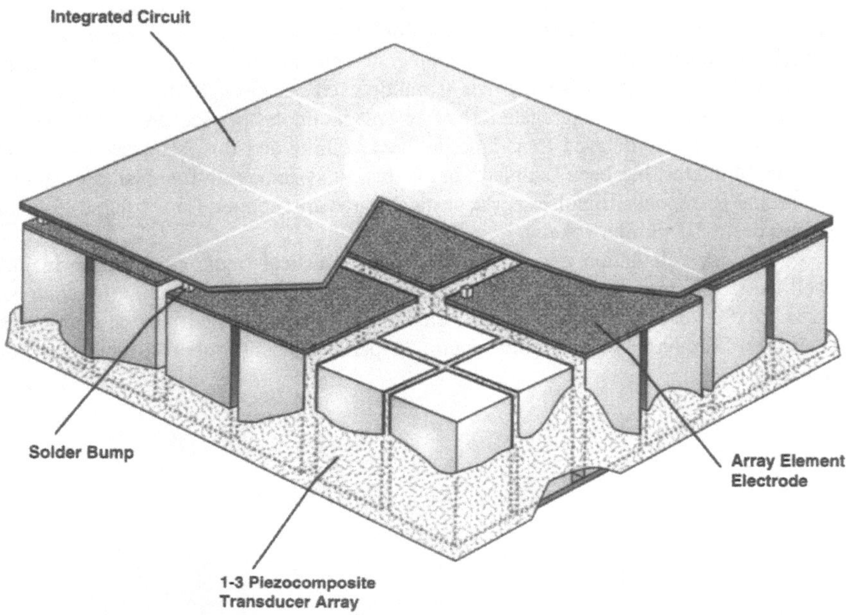

Fig. 2. Transducer Hybrid Array (THA)

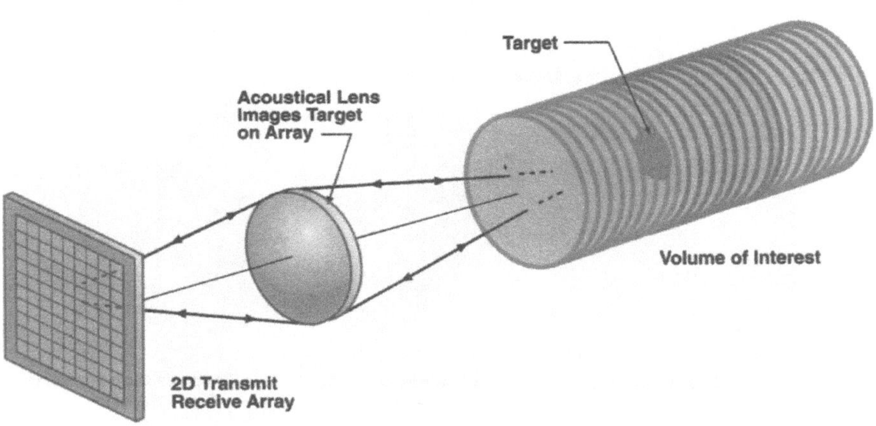

Fig. 3. C-Scan Planar Image Formation (Bistatic)

SYSTEM DESIGN PHILOSOPHY/TRADEOFFS

Using the requirements of Table I, system tradeoffs began with the sonar design equation[3]. Shown graphically in Fig. 4 for a typical scenario at 3 MHz, this equation relates the RF voltage applied to the transmitter transducer, transmitter output pressure, losses in the medium, target characteristics, receiver transducer sensitivity and electronic noise level to range and system performance. The signal level is referred to the input of the preamplifier at each element.

It may be observed in Fig. 4 that large specular or nearly specular targets are at or above the dynamic range of the preamplifier (≈60 dB) over much of the range of interest, whereas small "point-like" targets with lower reflectivity and increased geometrical spreading disappear below the electronics noise after two meters. The full dynamic range of the system may then be used to display target structures, although at longer ranges only the larger ones will be visible.

Large targets at close range may easily be brought into the dynamic range of the system by simply decreasing the transmitter voltage. Signal levels from small targets at longer ranges may be increased somewhat by increasing this transmitter voltage, however, non-linear effects in water (or tissue) set a fundamental limit on what can be achieved.

The Range/Resolution/Wavelength Tradeoff

Wavelength is the single most important parameter of any imaging system, especially in ultrasound where attenuation in the medium is strongly frequency dependent. Choice of operating wavelength involves balancing the imaging requirements with the performance predictions of the sonar equation. For the underwater diver's camera, attenuation in seawater is 3 to 4 dB/meter at 3 MHz. Geometrical spreading losses, however, are greater. A simple calculation reveals that a lens diameter of 200 mm is required for 10 mm resolution at 4 to 5 meters. For the medical camera, a similar calculation shows that with a diffraction limited lens of 100 mm diameter, 1 mm resolution in a plane can easily be obtained at 5 MHz at a range of 100 mm.

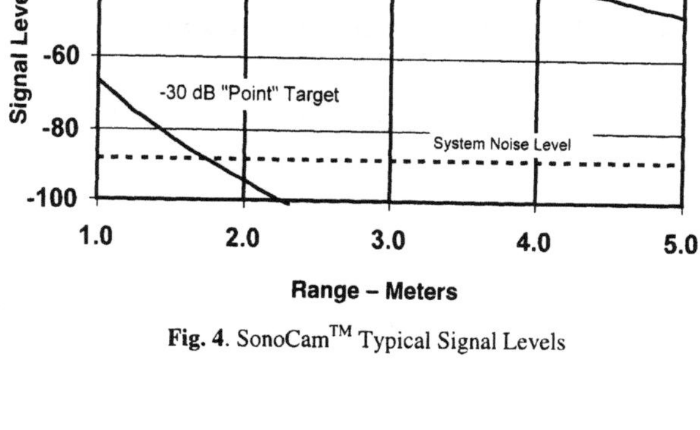

Fig. 4. SonoCam™ Typical Signal Levels

Acoustical Lens

Lens tradeoffs are well understood[4-8] although only rudimentary single element lenses have been used in most previous ultrasound systems. Several wide angle multi-element lenses have been designed and tested at Lockheed Martin using computer-aided lens design software conventionally used for optical lens design. Figure 5 shows a scanned image of a low contrast plastic target rendered in perspective, made with the lens shown in Fig. 1. One millimeter resolution is clearly achieved. This four element f/1.2, 65 mm focal length lens for the medical camera is diffraction limited.

Figure 6 is a plot of the one way point spread function (PSF) of this lens. Note that this PSF is cylindrically symmetric, unlike the asymmetric PSF of typical linear arrays. For example, the left side of Fig. 7 shows a reconstructed C-scan made from a series of B-scan images from a typical phased linear array medical ultrasound system.[9] The target was a 0.5 mm diameter tungsten carbide sphere embedded in a tissue equivalent medium. On the right is a C-scan image using a lens similar to that in Fig. 1. A significant asymmetry is revealed in the linear array PSF due to the small aperture in the plane orthogonal to the B-scan plane.

While this asymmetry artifact is generally ignored in a B-scan system, it becomes a serious limitation for any 3DUS system using a linear array. The out-of-plane resolution loss is easily observed in many commercial 3DUS images by rotating the image reconstruction plane 90 degrees from the B-scan plane.

One criticism of bistatic systems (or any system that does not focus the transmitted beam) is the relatively high sidelobe levels in the one-way PSF. Figure 7 demonstrates that compared to a conventional linear phased array, these sidelobes are confined to a smaller region around the acoustical axis and most importantly are symmetric about this axis.

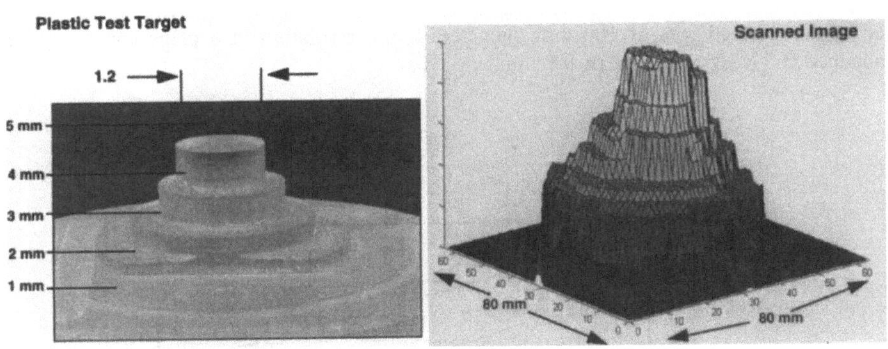

Fig. 5. Ultrasonic Image of Plastic Test Object Made with the Lens of Fig. 1

Magnitude Relative to Central Maximum	- 10 dB	- 20 dB	- 30 dB
Beam Width (mm)			
On Lens Axis	0.64	1.28	5.04
Edge of Field (40 mm)	0.64	1.28	3.63
Corner of Field (57 mm)	0.64	1.28	3.63

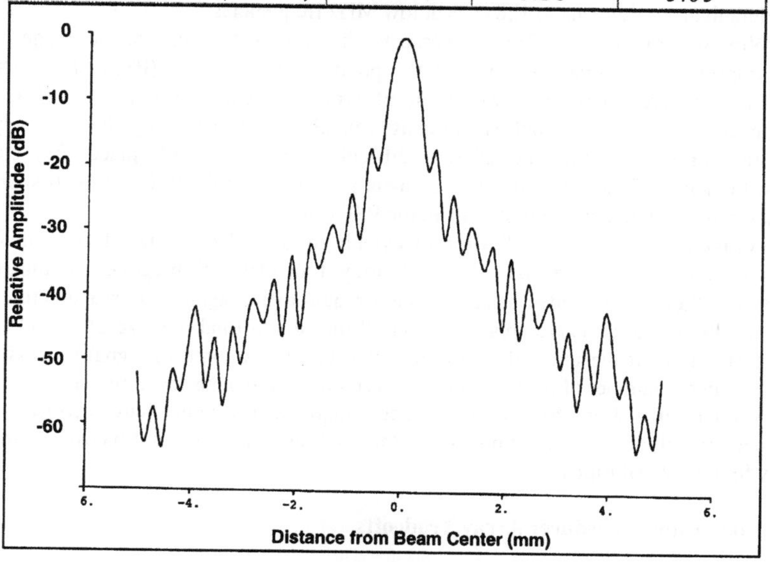

Fig. 6. Acoustical Lens Point Spread Function

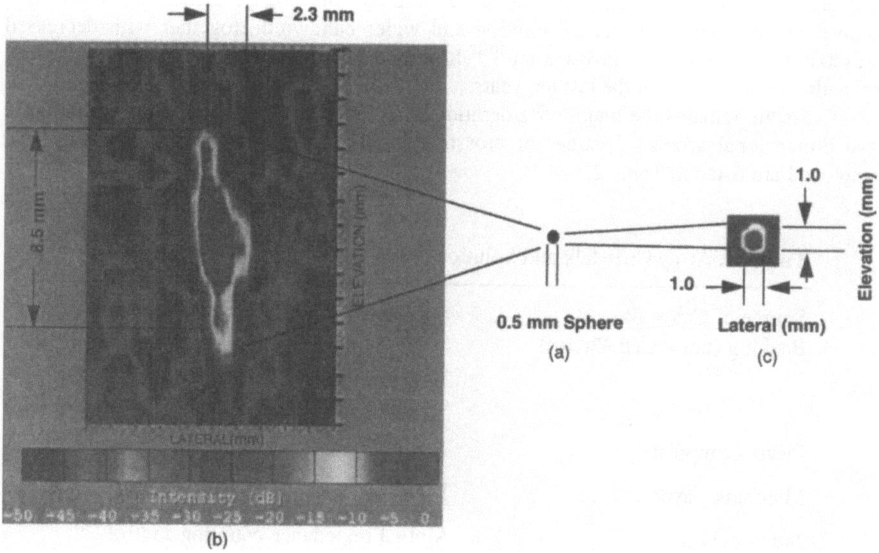

Fig. 7. 2-D Eliminates Out-of-Plane Artifacts

Array/Integrated Circuit Tradeoffs

Tradeoffs are more complex in this area. Clearly maximizing the number of elements in the matrix is always desirable, however, the maximum physical dimensions of the complete array are primarily a function of cost. Integrated circuit dimensions are also constrained by semiconductor photolithography, limiting practical array sizes with one-wavelength array elements to 128x128 at low MHz frequencies.

With an array of only 128x128 elements, the optimal sampling of the image plane is very important. Optimal sampling of the point spread function (PSF) of a diffraction limited lens determines the lower limit of array element size required. The partial coherence of the acoustical pulses and phase sensitive transducers, together with any array acoustical crosstalk makes this analysis difficult. An *ad hoc* rule places five elements across the lens PSF, such that the third element is centered on the PSF and the first and fifth elements are centered on the first nulls in the PSF (Fig. 6).

For the medical camera at 5 MHz, this yields an ideal element size close to the 0.2x0.2 mm unit cell chosen. To realize a 128x128 array, four identical integrated circuits, closely butted together on two sides balance system requirements against IC manufactureability. This results in a total active area of 12.8x12.8 mm and a final die size of 15.5x15.5 mm. The dimensions of the unit cell in the IC determine the amount of signal processing and particularly the number of planes of data storage that can then be accommodated.

Similar calculations for the underwater camera, with a larger lens aperture, yield an optimal unit cell of 0.4 mm. In both cases the final integrated circuit dimensions are within many foundry capabilities.

Piezocomposite Transducer Array Tradeoffs

Transducer tradeoffs are generally well understood, especially for linear arrays. Piezocomposite arrays consist of piezoelectric ceramic posts embedded in an epoxy matrix and are virtually universal in medical ultrasound systems today. Their lower effective acoustical impedance, increased coupling and wider bandwidth, together with decreased crosstalk from Lamb wave propagation[10-12] have played a major role in improved image B-scan ultrasound quality in the last ten years.

Crosstalk remains the major consideration in system design and is a major limitation in two dimensional arrays. Sources of crosstalk in 2D arrays and the solutions we have employed are listed in Table 2.

Table II. Array Crosstalk and Solutions Employed

Source of Crosstalk	Solutions Employed
Backing (Integrated Circuit)	Solder bumps much less than a wavelength minimize energy coupling into the backing.
Piezo-Composite	Array geometry optimization
Matching Layer	Dice at element level[13].
Medium (water)	Mutual impedance coupling control [14,15]

Array Modeling

A one dimensional KLM model[16] together with the equivalent medium approach[11] is the start of our array design process. Subsequently, two dimensional characteristics have been modeled using PZFlex[17]. Figure 8 shows an initial result of a 2D, five element model of the 3 MHz piezocomposite array which demonstrates some of the crosstalk issues and the solutions we have employed. Note that the calculated pressure fields in Fig. 8 are plotted logarithmically, i.e. an 80 dB pressure range is plotted. Figure 8b shows the reverberation one-half cycle after driving the center element with a single cycle sine wave. The sound pulse is just emerging into the water and a relatively small amount of energy has propagated through the solder bumps into the silicon. Some energy has also propagated horizontally in the piezocomposite.

One-half cycle later (Fig. 8c), energy from the medium has coupled back into the nearest neighbor elements, however, the two posts in the element are out of phase, resulting in cancellation of the signal from the element itself. In Fig. 8d, one-half cycle later, the posts have reversed phase and cancellation is still present. Although this 2 dimensional model is illustrative, a full 3D model together with experimental confirmation is underway.

Array Measurements

We have designed and built a number of 128 x 128 5 MHz arrays with 0.2 x 0.2 mm elements for medical imaging applications and 64x64 3 MHz arrays with 0.4 x 0.4 mm elements for underwater imaging applications. These receiver arrays are essentially air-backed due to the very small coupling from the array to the silicon through the metal bumps. Figure 9 is a graph of measured 3 MHz array performance compared with a KLM[16] model. The bandwidth in such an array is primarily determined by the matching of the piezoelectric composite to the water load. While the bandwidth is somewhat lower than that required for B-scan applications, it is adequate for the range resolution needed in the underwater camera. Array sensitivity in this air-backed configuration is excellent.

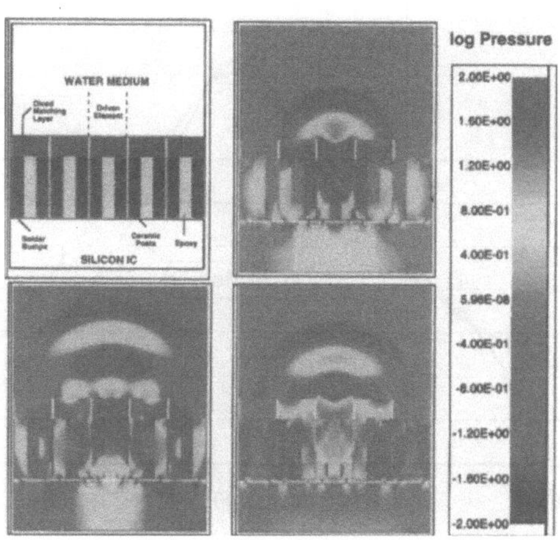

Fig. 8. Acoustical Propagation Model of 5 Elements (Clockwise, from upper left) a: THA Cross-section, b: Acoustical pulse Emerging into water, ½ cycle after single cycle drive on center element; c: ½ cycle later than b.; d: ½ cycle later than c.

SUMMARY/FUTURE WORK

The SonoCam™ acoustical imaging system is currently under development at Lockheed Martin. Figure 10 is a concept drawing of the underwater camera. It is intended to be hand-held by a SCUBA diver operating in relatively shallow water which may have near zero optical visibility. The compact camera has a display together with focusing controls, menu selection and other controls similar to a video camera.

Future work includes development of a zoom acoustical lens and further optimization of the piezocomposite array through 3D modeling as well as a high voltage capability in the integrated circuit which will permit monostatic operation. Additional MEMS transducer technologies are also being investigated in conjunction with Prof. Khuri-Yakub and his colleagues at Stanford University[18] and J. Bernstein and his colleagues at Draper Laboratory[14] as part of the DARPA sponsored Sonoelectronics Program[20].

ACKNOWLEDGMENTS

This material is based upon work supported by the Indian Head Division, Naval Surface Warfare Center under contract No. N00174-98-C-0019. This work has been further supported by DARPA, US Navy (NAVEODTECHDIV and ONR) contracts as well as Lockheed Martin internal funding. The continuing support of Elliott Brown and Heather Dussault, DARPA, Bruce Johnson, NAVEODTECHDIV and Wallace Smith, DARPA/ONR is sincerely appreciated. We also thank Najib Abboud for his assistance with PZ Flex™ and Heidi Burnham, Heidi Savino and Bill Willis for their help in production of this paper.

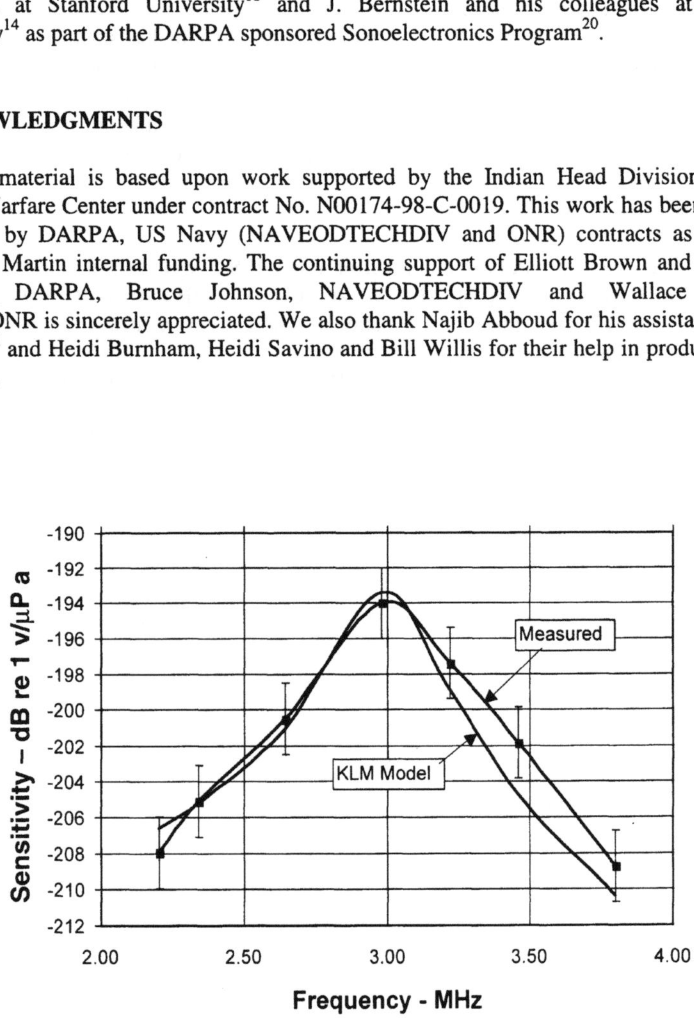

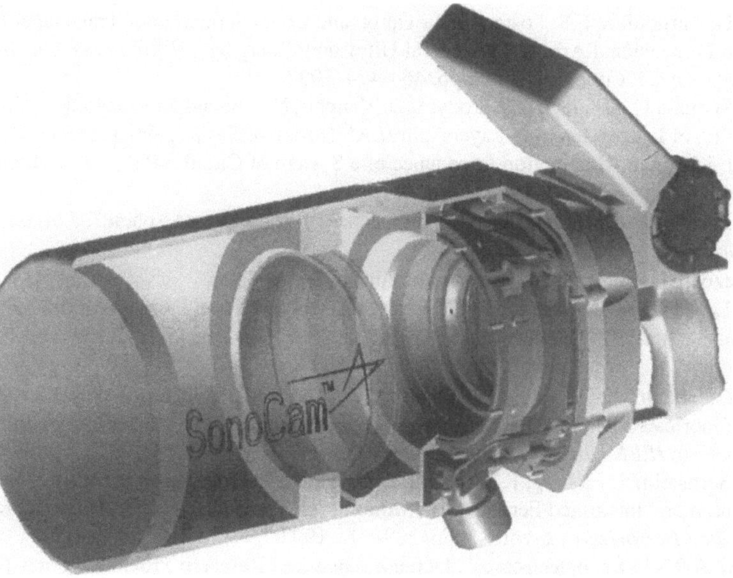

Figure 10. SonoCam™ (Artist's Concept)

REFERENCES

[1] K. Erikson et al, "A 128 x 128 (16k) Ultrasonic Transducer Hybrid Array", in *Acoustical Imaging, Vol. 23*, ed. Lees & Ferrari, Plenum Press, New York, pp. 485-494, 1997.
[2] Bruce Johnson, NAVEODTECHDIV, Indian Head, MD, Personal Communication.
[3] R.J. Urick, *Principles of Underwater Sound*, McGraw-Hill, New York, 1983, Ch.2.
[4] D. Sette, "Ultrasonic Lenses of Plastic Materials", *J. Acoust. Soc. Am.*, vol. 21, pp. 375-381, 1949.
[5] D.L. Folds, "Focusing Properties of Solid Ultrasonic Cylindrical Lenses", *J. Acoust. Soc. Am.*, vol. 53, pp. 826-834, 1973.
[6] Y. Tannaka & T. Koshikawa, "Solid-Liquid Compound Hydroacoustic Lens of Low Aberration", *J. Acoust. Soc. Am.*, vol. 53, pp. 590-595, 1973.
[7] H.W. Jones & C.J. Williams, "Lenses and Ultrasonic Imaging", in *Acoustical Holography, Vol. 7*, ed. L.W. Kessler, Plenum Press, NY, pp. 133-153, 1977.
[8] B. Kamgar-Parsi, B. Johnson, D.L. Folds & E. Belcher, "High-Resolution Underwater Acoustic Imaging with Lens-Based Systems", *Int. J. Imaging Syst. Technol.*, vol. 8, pp. 377-385, 1997.
[9] D. Phillips, X. Chen, C. Raeman, K. Parker, "Acoustic Lens Characterization in a Scattering Medium - Summary Report", Univ. of Rochester, Rochester, NY, 18 March, 1996.
[10] T.R. Gururaja, W.A. Schulze, L.E. Cross, R.E. Newham, B.A. Auld & Y.J. Wang, "Piezoelectric composite materials for ultrasonic transducer applications. Part I: Resonant modes of vibration of PZT rod-polymer composites", *IEEE Trans. Ultrason. Ferroelec. Freq. Control*, vol.32, pp. 481-498, 1985.
[11] W. A. Smith & B. A. Auld, "Modeling 1-3 Composite Piezoelectrics: Thickness-Mode Oscillations", *IEEE Trans. Ultrason. Ferroelec. Freq. Control*, vol.38, pp. 40-47, 1991.

[12] D.H. Turnbull & F.S. Foster, "Fabrication and Characterization of Transducer Elements in Two-Dimensional Arrays for Medical Ultrasound Imaging", *IEEE Trans. Ultrason. Ferroelec. Freq. Control*, vol. 39, pp. 464-474, 1992.
[13] G. Wojcik, C. Desilets, L. Nikodym, D. Vaughn, N. Abboud and J. Mould, "Computer Modeling of Diced Matching Layers", in *IEEE Ultrason. Symp.*,1996, pp.1503 - 1508.
[14] S.J. Klapman, "Interaction Impedance of a System of Circular Pistons", *J. Acoust. Soc. Am.*, vol. 11, pp. 289-295, 1940.
[15] C.H. Sherman, "Mutual Radiation Impedance of Sources on a Sphere", *J. Acoust. Soc. Am.*, vol. 31, pp. 947-952, 1959.
[16] PiezoCAD, Sonic Concepts, Woodinville, WA, 98072.
[17] G.L. Wojcik, D.K. Vaughn, V. Murray & J. Mould, "Time-Domain Modeling of Composite Arrays for Underwater Imaging", in *IEEE Ultrason. Symp.*, 1994, pp.1027-1032.
[18] I. Ladabaum, X. Jin, H.T. Soh, F. Pierre, A. Atalar & B.T. Khuri-Yakub, "Microfabricated Ultrasonic Transducers: Towards Robust Models and Immersion Devices", in *IEEE Ultras. Symp.*, 1996, pp. 335-338.
[19] J. Bernstein, S. Finberg, K. Houston, L. Niles, H. Chen, L. Cross, K. Li, K. Udayakumar, "Integrated Ferroelectric Monomorph Transducers for Acoustic Imaging", *Integrated Ferroelectrics*, vol. 15, pp. 289-307, 1997.
[20] "BAA 97-33 Sonoelectronics", Defense Advanced Research Projects Agency (DARPA), Arlington, VA 22203, July 1997.

Inferring 3-dimensional animal motions from a set of 1-dimensional multibeam returns

Jules S. Jaffe

Marine Physical Lab, Scripps Institution of Oceanography
La Jolla, CA 92093-0238

Abstract

With the development of the FTV underwater multibeam sonar imaging system (Jaffe,1995) and the more recent development of our combined optical and acoustical imaging system: OASIS (Jaffe,1998), we now have the ability for both tracking and identifying animals (primarily zooplankton) in the sea. Past uses of the system have been mainly for abundance estimation, however, more recently we have become interested in attempting to infer the characteristics of animal behavior from the sonar returns. In the context of this problem, we sought to develop methods for processing the data which would be sensitive to extremely small motions of the animals, as it was suspected that the animals were mostly quiescent during some times of observation (daytime). As in most sonar systems, our system has much better range resolution that lateral or azimuth resolution. Tracking animals in three dimensions with this anisotropic resolution results in a coarser estimate of animal trajectory than desired. On the other hand, under the assumption of an isotropic distribution of animal motions the three dimensional probability density function for animal displacements: $pdf_{3d}(\Delta\vec{r})$ should be derivable from a measurement of the one dimensional probability density function $pdf_{1d}(\Delta\rho)$ which measures displacements only in range. Since the one dimensional probability density function can be estimated from the measured variations in animal range with higher resolution than azimuth, using the methodology developed in this paper, one can obtain a higher accuracy version of the 3-dimensional pdf by using only the range information. Here, the methodology of the transformation is presented.

Introduction

Certainly, at the present time, our understanding of the oceans continues to be primarily limited by the lack of technology for measurement. In particular, understanding the functioning of certain aspects of the ecosystem are hindered by the lack

Acoustical Imaging, Volume 24.
Edited by Hua Lee, Kluwer Academic/Plenum Publishers, 2000.

of observational tools for the observation of animal behavior. In pursuit of this goal, over the last decade, our group in underwater imaging has been developing both optical and acoustical imaging tools for obtaining information about animal activity. As a result of one of our expeditions, an interesting data set was collected over several summers in a fjord in British Columbia, Canada: Saainch Inlet. Here, the intermediate size animals (.5 cm - 2.0 cm) are almost completely dominated by the zooplankter *Euphausia pacifica*, an animal which is important in the food chain as it eats plants and is consumed by fish. These animals perform on a daily cycle what is arguably the largest commute on the planet, ascending to surface waters during dusk and returning to their deeper residence at dawn. One interesting question concerns their metabolic rate during their daytime residence at depth. Do the animals "shut down" their activity and thus conserve their energy expenditure? Since metabolic rate has been shown to be related to swimming behavior (Torres and Childress, 1983), the observation of swimming activity can be related to animal metabolism. In order to examine this hypothesis we deployed our 3-dimensional imaging system, FishTV so that we could track the animals and determine what their swimming behavior was during daylight hours.

FishTV is a multibeam echosounder (Jaffe et al, 1995) which operates at a frequency of 445 kHz so that it can track animals at ranges of 3 - 10 m whose sizes range from 1 cm and up. The system uses eight rectangular apertures for transmitting and a complementary set of eight transducers for receiving. The transducers are arranged in a spiral fashion so that the entire field of view of the system (16 degrees by 16 degrees) is observed. Animal echoes from targets can be recorded at frame rates up to 4 Hz with angular resolution of 2 degrees by 2 degrees and range resolution 1 cm. Since a two degree beam subtends 18 cms at a distance of 5 meters the system possesses much better range than azimuth resolution at these distances. Tracking animals in three dimensions is thus limited to the more coarse resolution associated with the azimuthal characteristics. Alternatively, if animal motions can be considered isotropic then, in principle, the observed changes in range can be used to infer the three dimensional characteristics of the animal motion. This paper presents the theory behind this idea and introduces the idea of performing tomographic inversion on observed one dimensional probability density functions in order to compute the "true" three dimensional probability density function associated with this behavior.

Experimental Methodology

Data were collected in Saainch Inlet with the system deployed at a depth of 70 meters. The system was mounted on a current vane which pointed it steadily into the current in moderate current velocities (5 cm/s). The system was thus downstream from the animals and observations of the animals trajectories did not seem to be affected by the presence of the system. During the daytime, animals were located in a layer of approximately 20 meters thickness which spanned the depths from 80 to 100 meters. The sonar was angled downward at approximately 27 degrees and thus peered down into the animal layer. Records of animal reflectivity with respect to range showed the existence of many targets that were drifting toward the sonar system at a rate which was due to current flow. In addition, animal activity was noted, exemplified by targets that were not traveling at the same speed as the vast majority of (drifting) targets.

The sonar records the echo intensity of the individual targets which are located in each of the eight by eight or 64 composite beams. Data consists of 3-dimensional matrices of dimensions 8 x 8 (beams) x 512 (range bins) which are collected as a function of time at rates of up to 4 Hz. Algorithms developed in our group are then used to identify the 3-dimensional positions and to also estimate the target strengths of the targets. Observed target strengths are based on a calibration of the system. At frame rates of up to 4 Hz, several sets of thousands of frames were recorded over several weeks. Here, we consider the formulation of the algorithm that was applied to data that were recorded in the daytime over several sessions. Our working hypothesis in implementing this algorithm on these data sets is that the isotropy assumption does apply.

For each frame, data relating to the set of 10 strongest target reflections were retained for further processing. We considered only the range direction as, noted above, since this dimension presented the opportunity for the most accurate determination of target displacement. Next, considering pairs of frames as a function of time delay, the set of 10 x 10 target displacements were computed for all of the 10 targets in each of the frames. Taking only the 10 strongest target reflections minimized the chances of mistaking the successive positions of two different targets as a translation because the 10 targets were distributed sparsely in the measured volumes. Moreover, only target displacements were retained if the target stayed in the same of each of the 64 beams. At the end of this stage of the processing a set of M target displacements, $\{\Delta\rho_i|\Delta T\}, i = 1, M$, were in hand for a given time delay (ΔT) for each of the data collection episodes. We use the notation $\Delta\rho$ to indicate that only displacements along the axes of the beams were measured, corresponding to this one dimensional motion. Additional details about the sonar system and the data processing needed in order to get to this stage are elaborated in (Jaffe et al., 1995 and Jaffe et al., 1999).

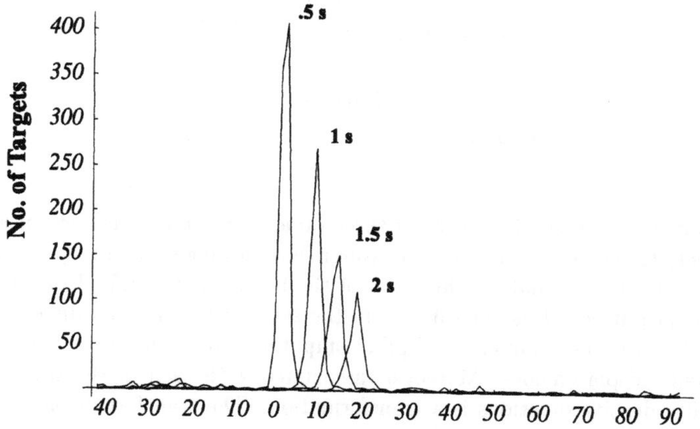

Figure 1. A plot of the number of observed animal displacement versus actual displacement. The graphs are for time delays of .5, 1.0, 1.5 and 2 seconds.

The set of displacement data were then histogrammed in order to compute an estimate of the probability density function for animal range displacement as a function of time. Figure 1 shows a set of histograms which were derived from one of the observation sessions. Shown in the Figure is the number of targets as a function of target displacement for four time intervals, .5 s, 1 s, 1.5 s, and 2 s. Two interesting features of the set of curves are that the peak is both displaced and broadened. The displacement is due to the current ($\sim 15cm/s$) and the broadening is due to the movement of the animals. The curves indicate that the animals' positions can be thought of as a diffusion process with longer time intervals leading to increased displacements of the animals. However, as mentioned above, the animals' positions were only measured in range. Regarding this as the true 3-dimensional pdf for animal displacements would be in error. Considering the set of true 3-dimensional displacements $\{\Delta\vec{r}_i|\Delta T\}, i = 1, M$, the system data when analyzed in this mode only measures the projection of these sets of vectors onto a single axis: $M_{3d\to 1d} : \{\Delta\vec{r}_i|\Delta T\} \Rightarrow \{\Delta\rho_i|\Delta T\}$. Considering the probability density function for $\Delta\vec{r}$, $pdf(\Delta\vec{r}|\Delta T)$ this distribution is mapped by a different operator to the distribution for $\Delta\rho$: $Pdf_{3d\to 1d} : pdf_{3d}(\Delta\vec{r}_i|\Delta T) \Rightarrow pdf_{1d}(\Delta\rho_i|\Delta T)$.

Theory

The question addressed in this article is the uniqueness and the computational feasibility of inverting for the 3-dimensional probability distribution $pdf_{3d}(\Delta\vec{r})$ from the one dimensional probability density function $pdf_{1d}(\Delta\rho)$. Of course, this is only possible if the 3-dimensional probability distribution function is isotropic. In this case, the 3-dimensional pdf is independent of angle and can be completely specified by a radial line at any angle. As an aid in the following discussion we consider the following diagram:

$$
\begin{array}{ccc}
 & M_{3d\to 1d} & \\
\{\Delta\vec{r}_i|\Delta T\} & \Rightarrow & \{\Delta\rho_i|\Delta T\} \\
\Downarrow & & \Downarrow \\
 & Pdf_{3d\to 1d} & \\
pdf_{3d}(\Delta\vec{r}|\Delta T) & \Rightarrow & pdf_{1d}(\Delta\rho|\Delta T)
\end{array}
$$

As noted, $\{\Delta\vec{r}_i|\Delta T\}$ is the set of three dimensional measurements of animal displacement, $\{\Delta\rho_i|\Delta T\}$ is the set of measurements in range, essentially a projection of the three dimensional data within each of the beams, $pdf_{3d}(\Delta\vec{r}|\Delta T)$ is the "true" 3-dimensional pdf, which is derivable from the set of 3-dimensional displacements and $pdf_{1d}(\Delta\rho|\Delta T)$ is the 1-dimensional pdf, computable from the set of 1-dimensional or "projected" displacements. Moreover, associated with each of the sets of data and probability density functions is a transformation. The transformation $M_{3d\to 1d}$ is the transformation that takes the 3-dimensional set of data and transforms it into the set of 1-dimensional data. So, for example if a displacement of $\Delta\vec{r}$ occurs in one of the beams, without loss of generality take $\Delta\vec{x}$ to be the beam direction. Then, measurement of only the radial displacement maps $\Delta\vec{r} = \{\Delta x, \Delta y, \Delta z\} \to \{\Delta x\}$. The other transformation, $Pdf_{3d\to 1d}$, maps the 3-dimensional pdf to the 1-dimensional pdf. It is easy to imagine what this transformation is, based on the theory of the projection of functions, that is, the 1-dimensional pdf is a double projection of the 3-dimensional pdf. So, for example, again taking the x axis as the beam direction

$$pdf_{1d}(\Delta x) = \int\int pdf_{3d}(\Delta x, \Delta y, \Delta z)dydz$$

Given the fact that the 3-dimensional pdf is centrosymmetric, a radial slice through the 'true' 3-dimensional pdf can be computed from the 1-dimensional pdf via an inversion of this relationship using some standard theorems from tomography (Kak and Slaney, 1987). As one example of how this can be accomplished, a one dimensional Fourier Transform can be taken of the 1-dimensional pdf which would then be followed by an inverse 3-dimensional Hankel transform, yielding a central slice through the 3-dimensional pdf, our desired result.

Therefore, given the 1-dimensional pdf one can obtain a 3-dimensional pdf under the assumption that the 3-dimensional pdf is centrosymmetric. Is this pdf the correct one? With reference to the flow diagram, we ask whether it is possible to invert $Pdf_{3d\rightarrow 1d}$ to obtain $Pdf_{1d\rightarrow 3d}$ so that the 3-dimensional pdf can be obtained from the 1-dimensional pdf. Here, we sketch a proof of this conjecture by exploiting the invertability of the relationships.

Assume that it is not true that we have obtained the correct 3-dimensional pdf from the given 1-dimensional pdf. Given the 1-dimensional pdf we can obtain a data set $\{\Delta\rho_i|\Delta T\}$ that corresponds to it. This data set is in fact unique, except for some scaling factor for the number of data values. Moreover, assume that the transformation $M_{3d\rightarrow 1d}$ is in fact uniquely invertible, allowing us to compute a set of data $\{\Delta\vec{r_i}|\Delta T\}$. Since the 3-dimensional pdf can be uniquely computed from this data, our statement is in contradiction, because the projection of this 3-dimensional pdf should be the 1-dimensional pdf. So, assuming that we can invert $M_{3d\rightarrow 1d}$ we have a unique answer inversion for the pdf. Here, we conjecture that this is true and leave the exercise for a future publication.

Conclusions

In this article we have proposed a processing algorithm for obtaining a 3-dimensional probability density function from a corresponding 1-dimensional probability density function. Based on the experimental configuration of our sonar system and the physical state of the environment, under the assumption of isotropy, the 1-dimensional probability density function can be inverted to obtain the 3-dimensional one. This method confers considerable advantages upon our data analysis as our sonar system obtains much higher resolution in range information than azimuth. Thus, with the correct choices of axes, the 1-dimensional probability density function can confer the higher range resolution to the 3-dimensional probability density function resulting in a more accurate estimate of animal displacement.

In the example presented here, that of the acquisition of sonar data from a fjord in British Columbia, the methodology was applied to the data, resulting in a small degree of narrowing of the distributions. As an interesting example of the method, if the probability density function is Gaussian distributed the inversion will have no effect on the form of the pdf. This is, again, a well known phenomena in computed tomography as the Gaussian is invariant under Fourier Transform and thus under the projection operators considered here. Of course, one does not know a-priori what the degree of difference the technique will make as it depends on the data. It thus makes sense to apply the transformation in any case.

In order to accomplish this goal, this paper introduces the idea of performing tomography on probability density functions. This seems to be a new use of these methods which might have other applications in cases where different sonar systems are being used, or the observation of 1-dimensional probability density functions is much easier than the 3-dimensional one. An open question for further study concerns the invertability and uniqueness of the projection data under the centrosymmetric constraint.

Acknowledgments

The author would like to thank his colleagues, Mark D. Ohman and Alex DeRobertis for helpful discussions. We also gratefully acknowledge the support of the National Science Foundation.

References

Jaffe, J. S., E. Reuss, D. McGehee and G. Chandran, "FTV, a sonar for tracking macrozooplankton in three-dimensions", Deep Sea Research, Vol 42, No. 8, 1995, pp 1495-1512.

Jaffe, J. S., Ohman, M. D., and De Robertis, A. 1998. OASIS in the sea: measurement of the acoustic reflectivity of zooplankton with concurrent optical imaging. Deep Sea Res. **45**, 7: 1239-1253.

Jaffe, J. S., M. D. Ohman and A. De Robertis, "Sonar estimates of daytime activity levels of *Euphausia pacifica* in Saanich Inlet ", Can. Journal of Fisheries & Aquatic Sciences,(to appear) Sept, 1999.

Kak, A.C. and Slaney., M. 1987. Principles of Computerized Tomographic Imaging, IEEE Press, New York.

Torres, J.J., and J.J. Childress. 1983. Relationship of oxygen consumption to swimming speed in *Euphausia pacifica* 1. Effects of temperature and pressure. Mar. Biol. **74**: 79-86.

DEVELOPMENT OF AN ULTRASONIC FOCUSING SYSTEM BASED ON THE SYNTHETIC APERTURE FOCUSING TECHNIQUE

P. Acevedo, J. Juárez and S. Rodríguez

DISCA, IIMAS-UNAM. Apartado Postal 20-726, 01000, México, D.F. México

INTRODUCTION

The Synthetic Aperture Focusing Technique (SAFT) is an ultrasonic image technique designed to enhance the performance of conventional ultrasonic testing procedures. SAFT is used for performing two basic functions: i) detection of defects within structural and functional components, and ii) classification and/or characterization of these detected defects in terms of size, shape, orientation, location, and composition.

This paper describes very briefly the system`s interconnection and in more detail the signal processing preceding the use of SAFT.

SYSTEM'S INTERCONNECTION

The system was implemented using the following hardware and devices: A pulse-echo board, an ultrasonic transducer, a HC11 microcontroller board, a step motor control board, an IBM compatible personal computer, a digital oscilloscope, a positioning mechanic system and a phantom.

The signals to excite the transducer are obtained fron the pulse-echo board and the HC11 board, both signal are constantly generated ("allow" and "shot" signals). The HC11 board also generates the signals to control the step motors. One of the oscilloscope's test probes is connected to the amplifier's output in the pulse-echo board and the other is connected to the sincronization signal, with the help of these two signals echos are viewed (A scans). The Oscilloscope's serial port 1 is connected directly to to the PC's series port using DB-9 connectors. Finally for communication the PC's serial port 2 is connected to the microcontroller via the RS-232 interface. Figure 1 shows a general diagram of the system's interconnection.

Acoustical Imaging, Volume 24.
Edited by Hua Lee, Kluwer Academic/Plenum Publishers, 2000.

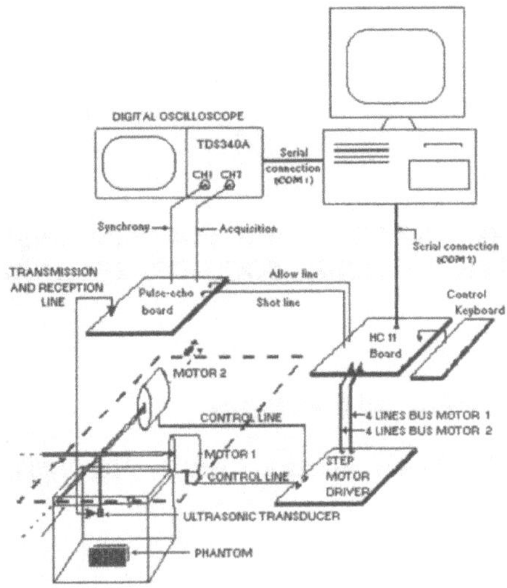

Figure 1. General Diagram of the System's Interconnection.

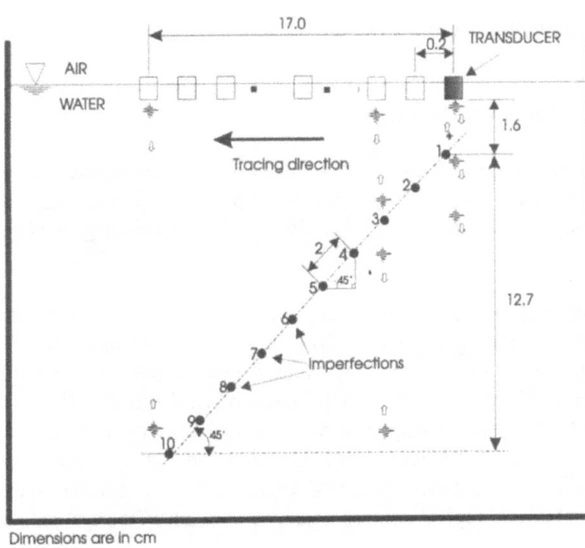

Figure 2. Water tank with transducer and phantom.

SIGNAL ACQUISITION

For a better understanding of the signal acquisition process it is useful to see figure 2. Here the ultrasonic transducer is shown moving horizontaly and linerly at regular intervals, under this transducer the phantom (constructed using acrylic and 10 equidistant strings) is located.The transducer is constantly emitting acoustic waves and receiving echoes generated by the strings simulating imperfections. A signal is acquired when is convenient, this is based in the signal's characteristics displayed at the oscilloscope. Figure 3 shown an A scan, here it is possible to observe the echoes from the strings showing the phenomeno of attenuation according to the relative possition of the transducer in relation to the string.

SIGNAL PROCESSING

As the images generated using ultrasonic techniques were not obtained directly from the signals received by the transducer, it was necessary to process each one of them in order to extract useful information and then to build the final image. As the useful information is in the signal amplitude and echo flying time it was necessary to rectify, apply gain compensation and filter each one of the captured ultrasonic signals obtained using a phantom before using SAFT [1].

Rectification

In order to ease the rectification of the ultrasonic signals MATLAB signal processing toolbox [2] was used. Data from the Oscilloscope was stored in matrix form, then to rectify them the ABS ( ) instruction was used, this instruction yield the absolute value of the respective matrix, if we take into account that all data is in matrix form when we apply the ABS ( ) instruction we are performing a full wave rectification.

Gain Compensation

Due to the fact that the signal processing is digital, compensation is not apply to the echoes' amplifier but this compensation is based on the characteristic curve of the transducer, all captured echoes are analized and the ones with higher amplitude are plotted [3]. Matlab software joins each one of these points by straight lines, giving as a result a non characteristic curve of the transducer. Therefore, it is necessary to smooth this curve, then based on this curve all remaining echoes are compensated.

Filtering

Most of the electronic systems used in the signal processing field employ filters, sometimes are used to select signals with especial features from others which are simply random. Filters are also used to improve signal-noise ratio. Filtering process is used to supress high frequency components from rectified ultrasonic signals, practically leaving the envelope of each one of these signals. This envelope is the representation of the received echo's amplitud, this is very important since using the envelope's information is possible to generate the image or images either using gray scale levelsor just a black and white representation.

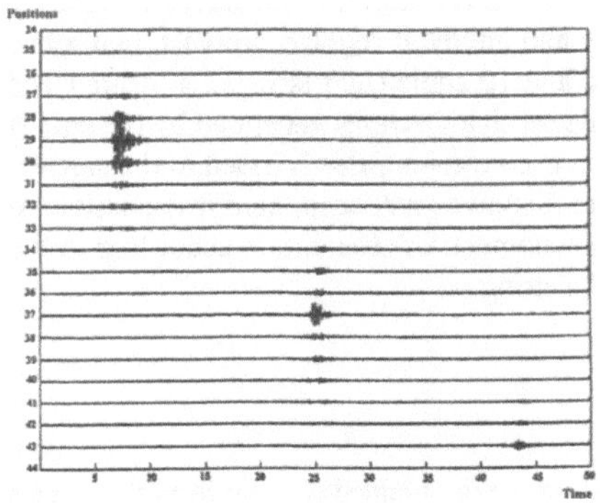

Figure 3. A scan array.

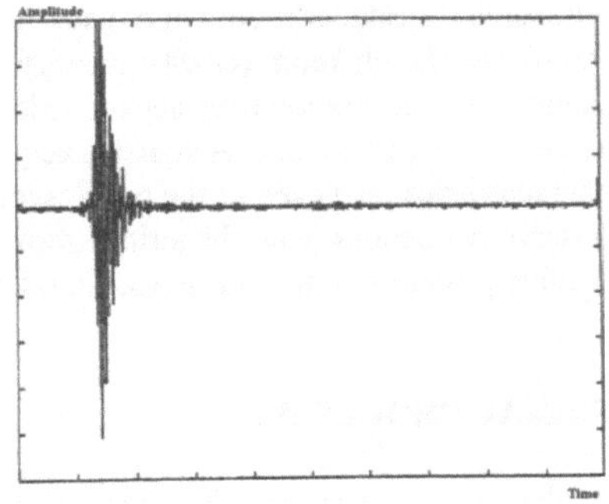

Figure 4. Typical A scan signal.

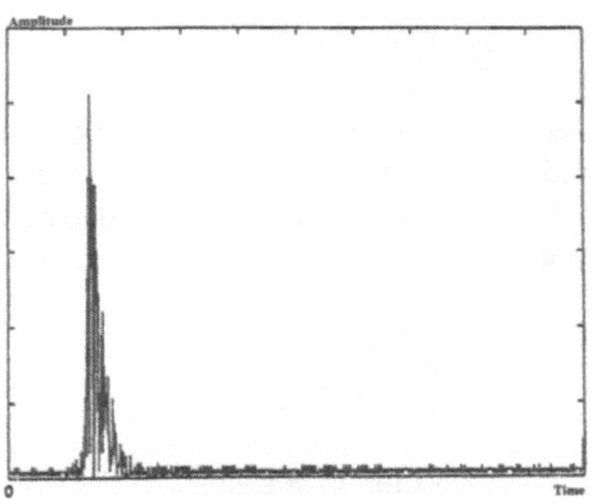

Figure 5. Rectified signal after using the ABS instruction.

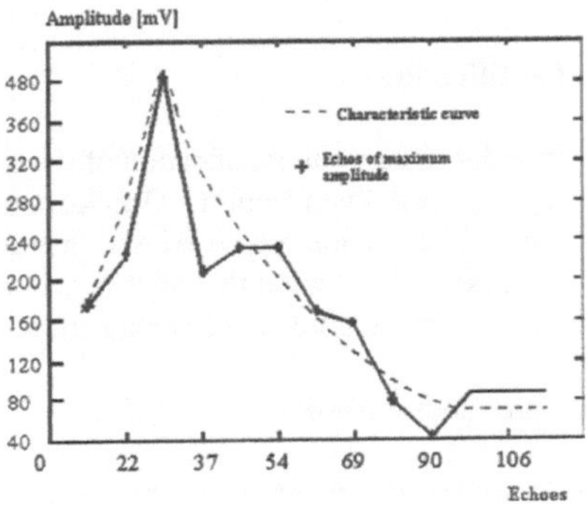

Figure 6. Transducer's characteristic gain curve.

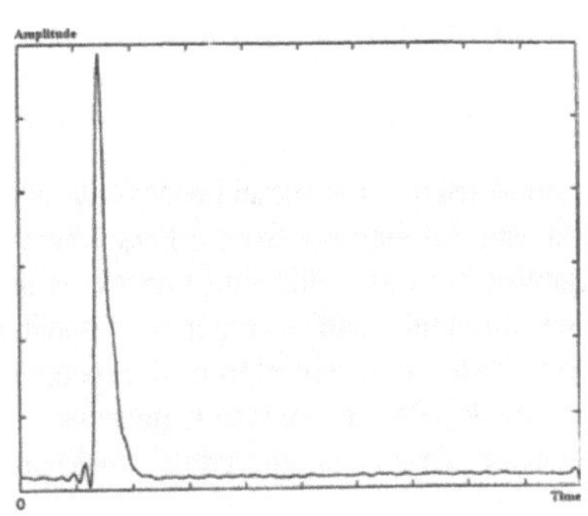

Figure 7. Filtered signal using a Butterworth filter.

EXPERIMENTS

All signal processing was achieved using MATLAB [4] which is a high-performance interactive software package for scientific and engineering numeric computation. Matlab provided a great flexibility to the system since it was possible to modify several parameters, mainly during filtering, in fact this is one of the main innovative features of the system.

As the number of acquired signals is large (85 signals), to exemplify the rectification process we will choose one of the most representative echo signals, the original signal (without rectification) is stored in the matrix named E29, this signal is shown in figure 4.

After using the Matlab function ABS ( ) we have:

R29 = ABS (E29);

the rectified signal is stored in the new matrix R29 which is shown in figure 5. Figure 6 shows the gain compensation curve.

As mentioned before Matlab allows filter simulation (Chebycheb, Butterworth, bandpass, high pass and low pass filters). In our case a lowpass Butterworth filter was used, first it is necessary to define the filter:

[B,A] = BUTTER (N,W_n);

where N is the filter order and W_n is the cut frequency, this frequency may be within a range of $0.0 < W_n < 1.0$. The values selected in our case are $N = 9$ and $W_n = 0.1$, then:

[B,A] = BUTTER (9,0.1).

Once the filter has been selected, it is necesary to define the following function:

Y = FILTFILT (B,A,X);

this instruction allows to apply the defined filter by the matrixes A and B to the matrix X, and the filtered signal is stored in the matrix Y. Applying this instruction to our echo signal:

F29 = FILTFILT (B,A,R29);

the filtered signal is stored in the matrix F29, this is shown in figure 7.

As the Oscilloscope only captures what is on its screen, information captured by the PC is plotted in a 0 to 1000 scale corresponding to the time of the signal. Due to this, the initial flying time of each one of the echoes is not considered and it is necessary some kind of compensantion. To achieve this it is necessary to consider the initial flying time which may be obtained from the oscilloscope, knowing these time values we can mark the echo boundaries for the first and last phantom's strings.

In this case it is necessary to consider the elapsed time after the echoes either for the first and the penultimate string since this information is necessary to generate the final image. As before and after the pulse-echoes originated by the strings ther is no signal and considering that the ultrasonic signal are stored in vectors, zeros are added to compensate the required time, using for this the following instruction:

$$C_i = [\text{zeros } (cz_l,1);\ F_i;\ \text{zeros } (cz_r,1)],$$

where C_i is the matrix where the compensated signal is stored, cz_l is the number of added zeros to the left, cz_r is the number of added zeros to the right and F_i is the filtered signal. Figure 8 shows the number of zeros that must be added to each one of the vectors that store echoes and also the value of the zeros expressed in time.

FINAL CONSTRUCTION OF THE IMAGE

Once the A scan signals are acquired, these are processed step by step and independently following the process previously described. This process gives enough information to graph "points" of the total image, these "points" represent defects located under the transducer and at certain depth. "Points " are black on black background (no defect) for depths along all the vertical line where defects are not detected. Plotted "points" are displayed in gray scale levels according to the amplitud of the echo reflected.

Resuming, "points" are black if there is not energy reflected, they are gray (gray scale level) if their amplitude is less than the maximum value and they are white if their amplitude is maximum. To achieve this, the display software requires that the values to be disolayed are between zero (black) and one (white), to allot a gray level color to values between cero and one. To solve this minor problem it was necessary to normalize all the values stored in the vectors for each A scan signal. This can be easily done using the following Matlab instruction:

$$Y = 1 - (Ymax - K_i)/Ymax;$$

where Y is the name of the normalized vector, K_i is the Kth A scan value and Ymax is the maximus A scan value.

Each transducer's position along the tracing length forms a column of the image created, for each position the transducer detects whatever there is under from the top to the botom of the specimen. The Image is created by columns which have a length proportional to the maximum depth of the specimen and along this length the echoes are represented and set according to their real depth in the specimen. This can be expressed in simple words as follows; columns are black but having white or gray dots where there is a defect.

APPLICATION OF SAFT

It was decided to apply SAFT to A scans obtained from strings one, five and ten (see figure 2). These strings were selected as representatives since their position is at the beggining, at the middle, and at the end.

The effective aperture of the selected sum for SAFT was of five positions. That is, for a total aperture of ten positions (numbered from 1 to 10) we have the following sum: 1+2+3+4+5, 2+3+4+5+6;...... ,6+7+8+9+10.

The display process for SAFT images is the same as the one described in the preceding secction [5].

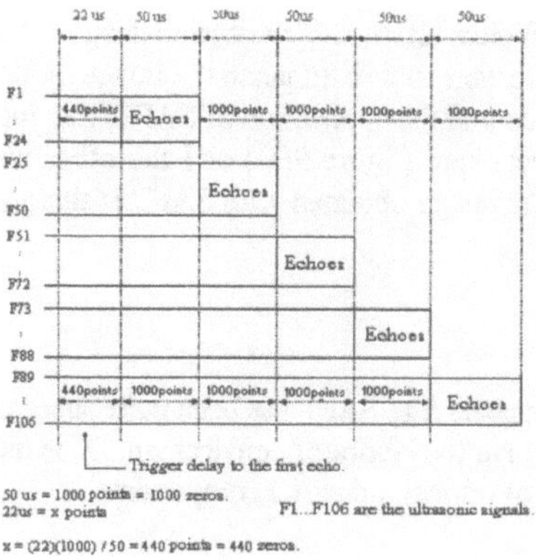

Figure 8. Time compensation diagram of ultrasonic signals.

(a)

(b)

Figure 9. Image of string number 10 obtained a) using B scan technique, b) using SAFT.

RESULTS

The system developed is an efficient alternative to obtaine images of good quality. Application of SAFT proved to be an important tool to enhance the image quality. In order to prove the validity of this system and of course the application of SAFT, two images are shown, one was obtained using the B scan technique (figure 9(a)) and the other was obtained using SAFT (figure 9(b)). It is clear that the image obtained with SAFT is sharper and clearer.

CONCLUSIONS

From the results obtained, it appears that SAFT has been quite successful at providing a flexible high-resolution imaging system in the laboratory enviroment. This method of imaging appears to be ideal for the inspection of critical industrial components.

The application of SAFT imaging methods to field testing appears to be a feasible and highly desirable extension of the basic research that has been carried out so far. With the development ofa special-purpose, real-time SAFT processor, this transfer of defect detection/imaging technology from the laboratory to the field appears to be both practical and feasible.

ACKNOWLEDGMENT

The authors would like to thank Mr. M. Fuentes Cruz for his practical assistance in the construction of the experimental system.

REFERENCES

1. P. Fish. *Physics and Instruments of Diagnostic Medical Ultrasound*, John Wiley & Sons Ltd., 1990.
2. T.P. Krauss and L. Shure. *Signal Processing Toolbox, User's Guide*. The Math Works Inc. 1995.
3. R.L. Powis and W.J. Powis. *A Thinker's Guide to Ultrasonic Imaging*, Urban & Schwarzenberg. Baltimore-Munich, 1984.
4. MATLAB User's Guide, High-Performance Numeric Computation and Visualization Software for Microsoft Windows, The Math Works Inc. 1992.
5. C.M. Thomson and L. Shure. *Image Processing Toolbox, User's Guide*. The Math Works Inc. 1995.

HIGH-RESOLUTION PROCESS IN ULTRASONIC REFLECTION TOMOGRAPHY

P. Lasaygues, J.P. Lefebvre, M. Bouvat-Merlin

CNRS Laboratoire de Mécanique et d'Acoustique
31 Chemin Joseph Aiguier
13402 Marseille cedex 20
France
email : lasaygues@lma.cnrs-mrs.fr

INTRODUCTION

Non Destructive Testing of materials is the main application of Ultrasonic Reflection Tomography (URT). This method results from a linearization of the Inverse Acoustic Scattering Problem, named Inverse Born Approximation (IBA). URT allows perturbations (theoretically small) of a reference medium to be visualized. For media with weak inhomogeneities, one chooses the reference medium to be homogeneous : the mean medium. This leads to a "Constant Background" IBA method, whose practical solution results in regular angular scanning with broad-band pulses, allowing one to cover slice-by-slice the spatial frequency spectrum of the imaged object. This leads to "Reconstruction-From-Projections" algorithms like those used for X-ray Computed Tomography. For media with strong heterogeneities, the problem is quite non-linear and there is in general no single solution. However, for example, one is generally concerned only by flaws, which appear to be strong (but small and localized so that the result is a small disturbance) inhomogeneities in well known media, the part of component to be inspected. In this case, one can use a "Variable Background" IBA method - the reference background being the water-specimen set - to reconstruct the perturbation.

URT fails when strong multiple scattering occurs (strong contrast and large object with respect to wavelength). In this case, one would guess that low frequency (less than 1 MHz) tomography will have a larger domain of validity than the classical one. But, the usual algorithm leads to poor resolution images, inappropriate for material imagery.

To improve URT, we used a deconvolution technique. Our enhancement procedure is based on Papoulis deconvolution i.e. on an extension of the generalized inversion in the complementary bandwidth of the electro-acoustic set-up. The procedure was tested on a square aluminum rod and a triangular PVC rod, smaller than the wavelength.

FOUNDAMENTAL BASIS OF URT

Basic principles of URT are now well established for weakly varying media as biological structures [1].

If we suppose the medium is composed of a known part (the background) identified by the density ρ_0 and the celerity c_0, and an unknown part (the perturbation) identified by ρ and c, the Equation that describes acoustic propagation/diffusion phenomena in the medium (including the boundary and Sommerfeld conditions) result from the Pekeris Equation and is given, for weak-scattering ($\rho c << \rho_0 c_0$), by :

$$\left[-\frac{1}{c_0^2}\frac{\partial^2 .}{\partial t^2}+\Delta .\right]p=-\frac{\alpha}{c_0^2}\frac{\partial^2 p}{\partial t^2}+\vec{grad}\, p.\vec{grad}\, \xi-\frac{1}{2}\vec{grad}\, p.\vec{grad}\, \alpha \qquad (1)$$

where p is the acoustic pressure. $\alpha=\frac{c^2-c_0^2}{c^2}$ is the quadratic fluctuation of celerity and

$\xi=\mathrm{Log}\left(Z/Z_0\right)$ is the Logarithmic Fluctuation of Acoustic Impedance ($Z=\rho c$).

If we know the Green function of the non-perturbed problem, we can found a solution by using the Lippman-Schwinger integral equation and one can calculate the far field solution of this equation in back-scattering, within the Born approximation (mono-scattering) [2] :

$$p_{d(1)}^{\infty}(\vec{x},t)=\frac{1}{4\pi|\vec{x}|}h_{d(1)}^{\infty}\left(\vec{x},t-\frac{|\vec{x}|}{c_0}\right) \text{ with } h_{d(1)}^{\infty}(\vec{x},t)=\frac{c_0}{4}\frac{\partial}{\partial x_3^2}\left\{\int_{R^2}\xi(\vec{x}_3')\,d^2\vec{x}\right\}_{x_3=\frac{c_0 t}{2}} \qquad (2)$$

where $h_{d(1)}^{\infty}$ is the first order far-field response of the medium for incidence $\vec{n}_0$ (colinear with $\vec{x}_3$) and observation $\vec{n}$.

In space-frequency domain, the back-scattering transfer function is given by $H_{d(1)}^{\infty}(\vec{n}_0,\omega)=-2k^2\hat{\xi}(\vec{K})$ with $\hat{\xi}(\vec{K})=\int_{R^3}\xi(\vec{x})e^{-i\vec{K}.\vec{x}}d^3\vec{x}$ and $\vec{K}=-2k\vec{n}_0$ where $\hat{\xi}(\vec{K})$ is the Fourier transform of the logarithmic fluctuation of acoustics impedance.

And if one takes in account only the amplitude of the echos (echography), one gets $a_{d(1)}^{\infty}(\vec{n}_0,\omega)=\hat{r}(\vec{K})$ where $\hat{r}(\vec{K})$ is the spatial Fourier transform of the reflectivity of the medium $r(\vec{x})$.

The goal of URT is to get reflectivity images from back-scattered measurements. Rotating of the emitter-receiver transducer around the "object" and emitting broad-band pulses at each position leads to the same situation as in X-ray tomography ; slice-by-slice spectral coverage of the object spectrum $\hat{r}(\vec{K})$.

So reconstruction can be made by any tomographic reconstruction algorithm. We chose the classical algorithm of summation of back-projections of filtered projections [3]

ELECTRONIC AND MECHANICAL APPARATUS

The general architecture of the mechanical system is that of a first generation X-ray tomograph : a main symmetric arm holds two transverse arms allowing the parallel translation of two transducers [4]. Angular scanning is allowed by the rotation of either the main arm or the object holder. The transducers can also be positioned and oriented with high precision, allowing linear and sectorial scanning. All motions are insured by six stepping motors sequentially driven by a programmable translator-indexer device fitted with a power multiplexer.

The electro-acoustic device consists at a Panametric® pulse-receiver and Imasonic® piezo-composite transducers (central frequency 250 kHz and wavelength 6 mm in water). In practice, only a limited number of views are available : the scanned sector is typically 90° or 180°, and the angular increment 2°. Moreover the frequency band-width of the employed pulses is very limited, typically one octave. The resolving power of the system is then limited. A typical numerical signal is composed of 1024 samples at a sampling period of 50 nsec.

Low and high frequency can be restored by use of a deconvolution algorithm that enhances the resolution. We operate an improvement of the spectral bandwidth by Papoulis deconvolution based essentially on a non-linear adaptive extrapolation of the Fourier domain.

PAPOULIS DECONVOLUTION

Position of the problem

In the case of our linear, stationary and causal device, input and output are linked by the convolution integral :

$$y(t) = \int_{-\infty}^{t} x(t)h(t-z)\,dz = (h * x)(t) \tag{3}$$

where h is the impulse response of the device. When the input x(t) can not reasonably be approached by a Dirac pulse, h(t) must be restored from the output signal y(t). The deconvolution is the numerical solution of this convolution integral. The theory of the inverse problem that we exposed in the previous paragraph shows an idealistic character because it doesn't integrate the frequency restrictions introduced by the electro-acoustic set-up and the mechanical system. To attenuate the effect of filtering, we must deconvolve the emitted signal and received signal.

The function h(t) to be restored is the impulse response of the medium ; x(t) is the transmitted pulse measured by reflection on a perfect plane reflector, for example the interface between air and water ; and y(t) is the observed signal.

If x, y and h are functions with Fourier transforms X, Y and H (real problem), we can write equation (3) in the frequency domain and the solution by inverse filtering is :

$$H(\upsilon) = Y(\upsilon)X^{-1}(\upsilon) \tag{4}$$

so that $h(t) = \int_{-\infty}^{\infty} H(\upsilon)\, e^{2i\pi t\upsilon}\, d\upsilon$.

The transform $H(\upsilon)$ fully determines h(t), and the sole knowledge of $Y(\upsilon)$ and $X(\upsilon)$ is sufficient. But equation (4) is formal and has a meaning only if the function X^{-1} exists and is never nul for any value of υ. In practice, since x(t) is a frequency band-limited signal, equation (4) shows that $H(\upsilon)$ is known only on the finite interval wherein $X(\upsilon)\neq 0$. There are also problems when the input signal is small, reduced to noise.

We try to estimate the function $H(\upsilon)$, noted $\tilde{H}$, by minimization of the quadratic residual error :

$$J = |Y - \tilde{H}X|^2 = (Y - \tilde{H}X)^*(Y - \tilde{H}X) \tag{5}$$

where $(\ )^*$ represents the conjugate complex. If we define a function G by $G(\upsilon) = \frac{X^*(\upsilon)}{X^*(\upsilon)X(\upsilon)} = \begin{cases} X^{-1} \text{ if } \upsilon \in \text{support of } (X) \\ 0 \quad \text{elsewhere} \end{cases}$ then we have :

$$\tilde{H}(\upsilon) = G(\upsilon)Y(\upsilon) \tag{6}$$

and the regularized version $G(\upsilon) = \frac{X^*(\upsilon)}{X^*(\upsilon)X(\upsilon) + \alpha}$, where α can be the input noise, allows to overcome the difficulties of small input signals. But the bandwidth of the restored impulse response is still that of the input signal, i.e. typically the bandwidth of the transducers. A solution to this problem was given by Papoulis.

Papoulis extrapolation

The method proposed by Papoulis [5] to determine h(t) as a function of its Fourier transform within a band, is a non-linear adaptive modification of a extrapolation method [6]. It takes advantage of the finite width of impulse responses in both time and frequency. We consider a function h(t) and its Fourier transform $H(\upsilon)$. If we define the frequency band $B = [\upsilon_1, \upsilon_2]$ of X, we assume that $H(\upsilon)$ is known only in the segment B, i.e. that only the function

$$W_0(\upsilon) = \tilde{H}(\upsilon)P(\upsilon) = G(\upsilon)Y(\upsilon)\,P(\upsilon) \text{ where } P(\upsilon) = \begin{bmatrix} 1 \text{ if } \upsilon \in B \\ 0 \text{ if } \upsilon \notin B \end{bmatrix} \tag{7}$$

is known. The problem is to compute $H(\upsilon)$ for all frequencies, i.e. to compute $H(\upsilon)$ from the knowledge of $W_0(\upsilon)$. First, one computes the inverse transform of $W_0(\upsilon)$:

$$w_0(t) = \frac{1}{2\pi}\int_B W_0(\upsilon)e^{j\upsilon t}d\upsilon \tag{8}$$

since h(t) is a time-limited signal (finite impulse response) of support T, h(t) = 0 if $t \notin T$. Given (estimating) T, we then define the new function $h_0(t)$:

$$h_0(t) = \begin{cases} w_0(t) & \text{if } t \in T \\ 0 & \text{if } t \notin T \end{cases} \quad \text{and} \quad H_0(\upsilon) = \int_T h_0(t)e^{-j\upsilon t}dt \tag{9}$$

and we define the new function W_1 :

$$W_1(\upsilon) = \begin{cases} W_0(\upsilon) = G(\upsilon)Y(\upsilon) & \text{if } \upsilon \in B \\ H_0(\upsilon) & \text{if } \upsilon \notin B \end{cases} \tag{10}$$

$$\Rightarrow W_1(\upsilon) = P(\upsilon)G(\upsilon)Y(\upsilon) + [1 - P(\upsilon)]H_0(\upsilon) \tag{11}$$

and by inverse Fourier transform, we compute the first order impulse response of the medium.
And so on, we can computed the nth step of the algorithm by Fourier transform of the $(n-1)^{th}$ iteration.

The main problem is the estimation of B and T ; i.e. the level under which the signal is considered as noise. We, now, illustrate result given by Papoulis deconvolution, reduced to its first iterate.

DECONVOLUTION IN LOW FREQUENCY TOMOGRAPHY

The method is now applied to an object "c" immersed in a tank and fixed at the supposed center of the bench. Two objects are tested. The first is a square aluminum rod and the second is an triangular PVC rod. The section of rods being smaller than the central wavelength (6 mm in water ; the central frequency of the transducer is 250 kHz). The size of the section is 4 mm for the two rods. The back scattering response of the object without correction is given by :

$$y_c(\phi, t) = h_a(\phi, t) * h_c(\phi, t) \tag{12}$$

where $h_a(\phi,t)$ is the impulse response of the whole apparatus and $h_c(\phi,t)$ is the true unknown impulse response of the object. In fact, the response $h_a(\phi,t)$ is the convolution product of the mechanical response $h_m(\phi,t)$ and the electro-acoustic response of the transducer $h_t(t)$ (Figure 1). The function $h_c(\phi,t)$ we want to image is determined after two deconvolutions :

$$h_c(\phi,t) = y_c(\phi,t) \overset{*}{\underset{*}{}} h_m(\phi,t) \overset{*}{\underset{*}{}} h_t(t) \tag{13}$$

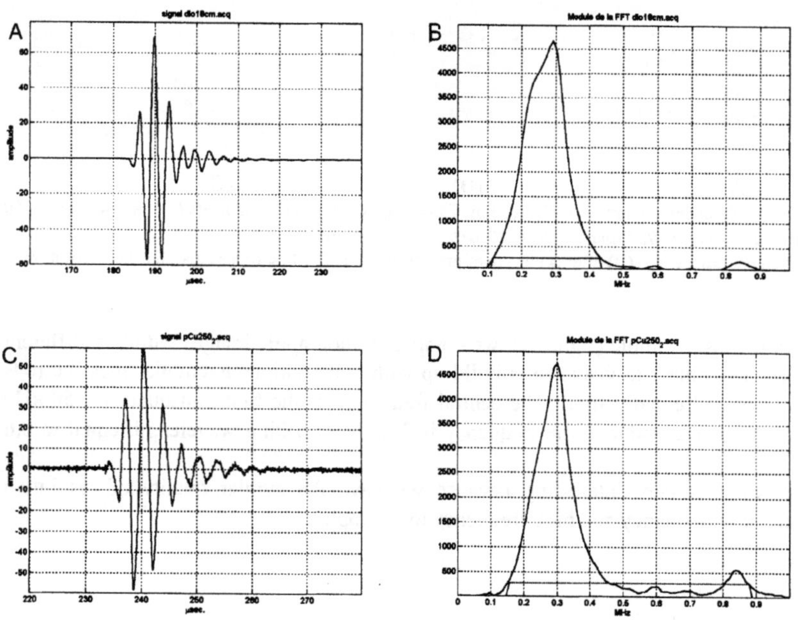

Figure 1 : Impulse response and transfert function of the transducer and of the mechanical system ; (A-B) $h_t(t)$ is the response to a water-air interface ; (C-D) $h_m(\phi,t)$ is a response to a copper wire

The input signal $X(\upsilon)$ takes significant values in the band at -20 dB : B = [120, 590] kHz. So, $H_c(\phi,\upsilon)$ can be determined reliably in this band and by inverse FFT, we compute

the function $w_1(t)$. Some details can be found in [8]. If the signal $y_c(\phi,t)$ is only significative (>-10 dB) on a portion T, the time-limited first order estimate $h_1(\phi,t)$, is a good approximation of the idealized medium.
Images without treatment (Figures 2-a, 3-a) should be compared with the deconvolved images of rods (Figures 2-c, 3-c) after centering all projections (Figures 2-b, 3-b). The dimension of rods is well restored and the contour is reinforced by the signal processing.

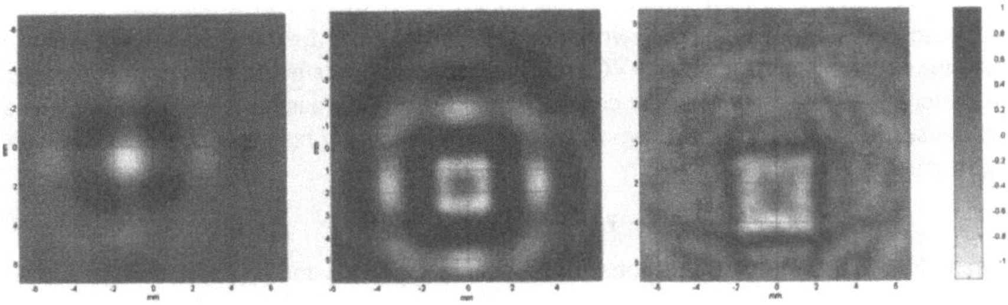

(A) (B) (C)

Figure 2. Low Frequency Ultrasonic Reflection Tomography of a Square Aluminum Rod (4 mm) ; *180 projections ($\delta\alpha = 2°$) and 1024 samples, $F_e = 20$ MHz ; 255 x 255 pixels.*

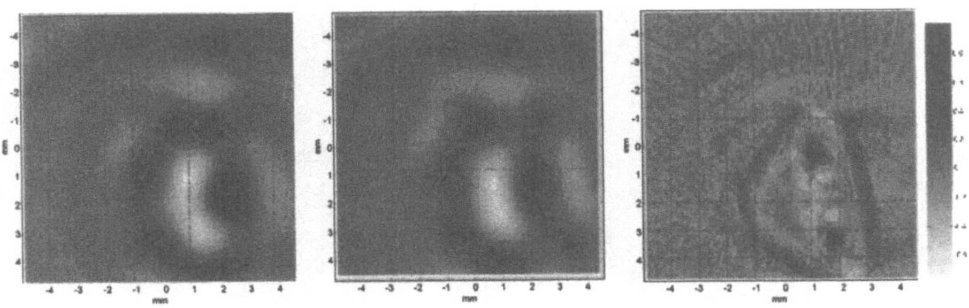

(A) (B) (C)

Figure 3. Low Frequency Ultrasonic Reflection Tomography of a Triangular PVC Rod (4 mm) ; *90 projections ($\delta\alpha = 4°$) and 1024 samples, $F_e = 20$ MHz ; 255 x 255 pixels.*
(A) without signal processing, (B) after centering of projections (C) after first order Papoulis Deconvolution

With the last example (figure 4), we compare tomographic images with two different frequencies. The object to visualize is build up with two contiguous circular metallic rods (section of one rod equal 2 mm). The central frequency of the first transducer is 250 kHz and for the second transducer, the frequency is 2,25 MHz with a wavelength equal to 0,6 mm.

The result is very interesting because we have the similar quality of resolution between the low frequency and high frequency tomography.

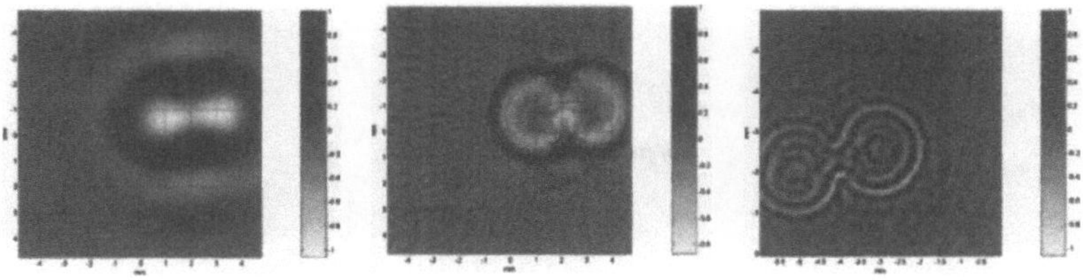

(A) (B) (C)

Figure 4. Comparison between low and high URT of two contiguous circular metallic rods (2 mm) ; *90 projections ($\delta\alpha$ = 4°) and 1024 samples, F_e = 20 MHz ; 255 x 255 pixels.*
(A) F = 250 kHz (before signal processing), (B) F =250 kHz (after deconvolution) (C) F = 2,25 MHz

CONCLUSION

In this paper, we have exposed a solution to improve the resolution in Low Frequency Ultrasonic Tomography. Since the basic principle of ultrasonic reflection tomography prohibits the inspection of objects with strong contrast and large extension, we turn down the frequency of the transducer, in order to increase the penetration length of the wave and the validity of the method. But this is done at the expense of resolution.

To restore resolution, we proposed a signal processing method based on Papoulis deconvolution. We implemented this algorithm and tried to operate an improvement from an aluminum rod smaller than the wavelength.

The result is an imagery with the same quality as that given by a more high frequency method.

REFERENCES

1. D. Hiller and H. Ermert, System analysis of ultrasound reflection mode computerized tomography, *IEEE Trans. Sonic Ultrasonic SU-31*, p 240, (1984).
2. S. Mensah and J.P. Lefebvre, Ultrasonic reflection tomography : specific problems and adapted solutions, in *Mathematical Methods in Medical Imaging*, FL Bookstein and al. (eds), SPIE Philadelphia, vol. 2299, p.264, (1994)
3. S. Mensah and J.P. Lefebvre, Enhanced Diffraction Tomography, *IEEE Trans. Ultrason. Ferroelec. Freq. Control*, Vol 44, Number 6, pp 1245,1252 (1997).
4. Internet address of the Laboratory (http://alphalma.cnrs-mrs.fr/pi/api.html)
5. A. Papoulis and C. Chamzas, Improvement of range resolution by spectral extrapolation, in *Ultrasonic Imaging 1*, p. 121 (1979)
6. A. Papoulis, A new algorithm in spectral analysis and band-limited extrapolation, in *IEEE Trans. Circuits Syst.*, vol. Cas-22, 9, p. 735, (1975).
7. P. Lasaygues, J.P. Lefebvre and S. Mensah, High-resolution low frequency ultrasonic tomography, in *Ultrasonic Imaging*, vol. 19, p 278, (1997).

SPATIAL COHERENCE AND BEAMFORMER GAIN

Jeffrey C. Bamber[1], Ronald A. Mucci, Donald P. Orofino[2]

Hewlett Packard Company, Medical Products Group, Imaging Systems
Andover, MA, 01810-1099, USA
[1]present address: Physics Department, Institute of Cancer Research and Royal Marsden NHS Trust, Downs Road, Sutton, Surrey, SM2 5PT, U.K.
[2]present address: The Mathworks, Natick, MA 01760-9889, USA

INTRODUCTION

Many acoustical imaging systems rely upon the assumption that received backscattered waves are perfectly coherent. Poor performance of such systems results as a consequence of poor spatial coherence. The spatial coherence of the backscattered field is influenced by system-, target-, and propagation-dependent factors, such as the transmit beam, scattering properties, and aberration and attenuation, respectively (Goodman, 1985; Liu and Waag, 1995; Mallart and Fink, 1991). This paper demonstrates the relationship between imaging performance and spatial coherence in medical pulse-echo imaging. Performance is quantified here in terms of the gain attained by a conventional delay and sum beamformer. The results are intended for applications such as aperture design, beamformer development, and aberration compensation.

THEORY AND ANALYSIS METHODS

Spatial coherence was used as a measure of received acoustic field similarity as a function of the spatial separation of the observations of RF sensor data, time aligned to compensate for geometrical path-length differences. The spatial coherence function $R(l)$ is defined as the normalized correlation between the pressure fields received by two sensors separated by a distance l. $R(l)$ is estimated from RF sensor signals, $x_n(k)$, by computing

$$\hat{R}(m) = \frac{N}{N-m} \frac{\sum_{k=k_1}^{k_2} \sum_{n=0}^{N-m-1} x_n(k) x_{n+m}(k)}{\sum_{k=k_1}^{k_2} \sum_{n=0}^{N-1} x_n^2(k)}, \tag{1}$$

over pairs of sensors, n, separated by lag m, within a total aperture of N elements. Averaging occurs over a time window, from sample k_1 to sample k_2, usually several wavelengths.

In a companion paper in this volume (Bamber, 1998), the application of the van Cittert-Zernike theorem, a classical theorem of statistical optics, is discussed. According to this

Acoustical Imaging, Volume 24.
Edited by Hua Lee, Kluwer Academic/Plenum Publishers, 2000.

theorem, if the insonified volume for a single imaging pulse can be treated as an incoherent source then the spatial coherence of the echoes from a fixed depth is determined entirely by the transmit beamwidth. The broader the beam, the shorter the spatial coherence length, defined as the distance over which the spatial coherence function remains above some predefined level, such as 0.5. As shown by Mallart and Fink (1991), this theorem also predicts that incoherent scattering from *within the transmit focal plane* gives rise to a spatial coherence function that is the autocorrelation function of the transmit aperture. For a rectangular aperture this is triangular, running from a value of one at zero lag, to zero when the lag equals the aperture width. The triangular coherence function is independent of both frequency and depth of the transmit focus.

The gain of a conventional beamformer was defined as the ratio of its output power to the total power of the sensor signals, estimated for an M-element receive aperture as

$$\hat{G}_M = \frac{1}{N-M+1}\sum_{j=1}^{N-M+1}\frac{\sum_{k=k_1}^{k_2}\left[\sum_{n=j}^{j+M-1}x_n(k)\right]^2}{\sum_{k=k_1}^{k_2}\left[\sum_{n=j}^{j+M-1}x_n^{\,2}(k)\right]}, \tag{2}$$

which includes averaging over all subapertures of size M in an N-element receive array.

It may be shown that this expression for beamformer gain can be re-written as

$$\hat{G}_M = 1 + 2\sum_{n=1}^{M-1}\frac{(M-n)}{M}R(n), \tag{3}$$

from which a number of important statements may be made: a) the gain of the conventional beamformer is equal to the weighted area of the spatial coherence function, b) the area under the spatial coherence function represents an upper bound on realizable gain, c) the gain depends more strongly on lower order lags, and d) negative coherence values will degrade beamformer gain.

EXPERIMENT AND RESULTS

In vitro (phantom)

A uniform diffusely scattering foam target was imaged with an HP SONOS 1000™ scanner, using a 128 element, 3.5 MHz sector probe, maintaining the transmit focus at 8 cm (~ 10 cm fixed elevation focus) with no transmit apodization. The transmit aperture size was varied in order to achieve different levels of spatial coherence at each of several analysis depths. Transmit aperture sizes of 128, 96, 64 and 48 elements were employed.

To infer the relative sizes of insonified volumes for subsequent analysis, transmit beam patterns at each depth of interest were estimated by an unusual technique of reconstructing 90° sector images of the foam using RF data for a single (0°) transmit beam direction only. A summary of the trends noted from this analysis is given in Fig. 1, where the -6 dB beamwidths have been plotted as a function of transmit aperture for each of the three depths of interest: the transmit focal depth, within the near field and beyond the focal depth. Corresponding spatial coherence functions, computed using Eq. 1, are shown in Fig. 2.

At the focal depth, as anticipated, the widest aperture (128 transmit elements) produced the greatest coherence length and each decrease in transmit aperture size produced a corresponding decrease in spatial coherence. At 4 cm depth, the spatial coherence associated with the largest transmit aperture produced the least amount of spatial coherence. At 14 cm the amount of spatial coherence was found to be quite similar for all four transmit apertures. These results are consistent with the transmit beamwidths shown in Fig. 1.

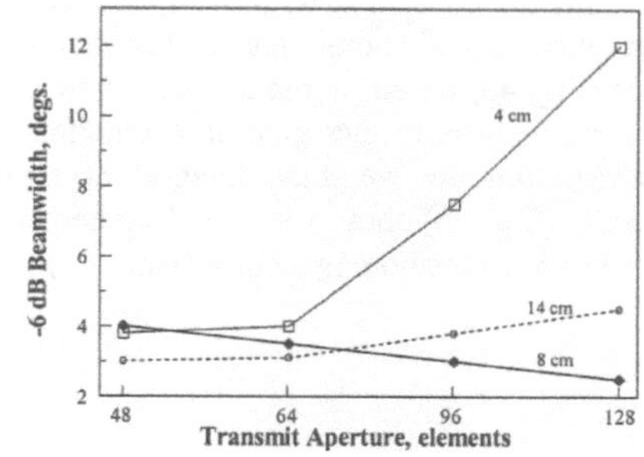

Figure 1. Beamwidths at -6 dB, as a function of transmit aperture, for analysis depths 4 cm, 8 cm and 14 cm.

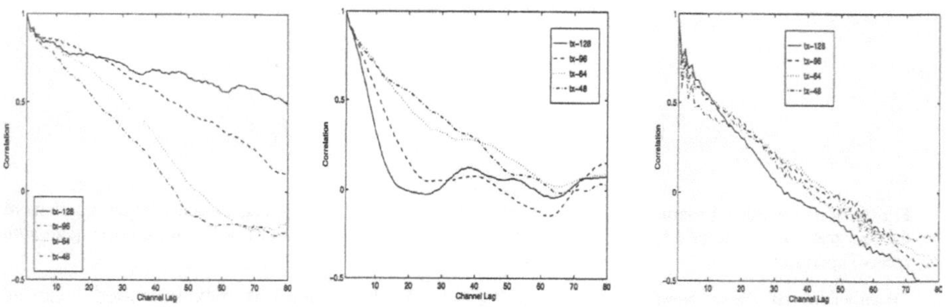

(a) (b) (c)

Figure 2. Measured spatial coherence functions for diffuse backscattering from foam, for each of the four transmit aperture sizes show in Fig.1; (a) at the focal depth of 8 cm, (b) at 4 cm and (c) at 14 cm.

Beamformer performance (gain) was estimated directly from the RF sensor signals for the foam datasets using Eq. 2. The results, plotted again as a function of receive aperture size, are shown in Fig. 3. At the focal depth, the largest aperture produced the greatest

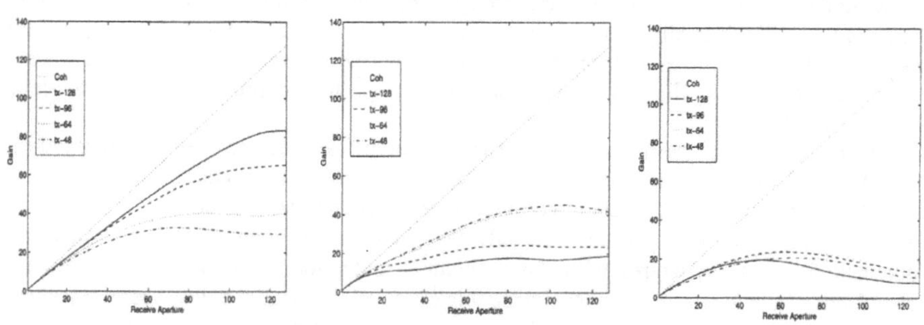

(a) (b) (c)

Figure 3. Measured beamformer gain as a function of receive aperture size, for diffuse backscattering from foam, for each of the four transmit aperture sizes show in Fig.1; (a) at the focal depth of 8 cm, (b) at 4 cm and (c) at 14 cm. The dotted straight-lines show the results expected for conditions of perfect spatial coherence.

beamformer gain, with each decrease in transmit aperture producing a corresponding decrease in gain. Overall beamformer gain is lower for the 4 cm and 14 cm depths than for the focal depth and, at 4 cm depth, beamformer performance is poorest for the largest transmit aperture. These results are consistent with the spatial coherence curves in Fig. 2. The consistency with theoretical expectations was further demonstrated by synthesizing triangular coherence curves (Fig. 4a) based on the autocorrelation functions of each of the transmit apertures. Predictions of beamformer gain as a function of receive aperture size were then obtained by computing the weighted integral of these triangular functions, according to Eq.3. The results (Fig. 4b) show very good agreement with the experimental observations for backscattering from the homogeneous foam.

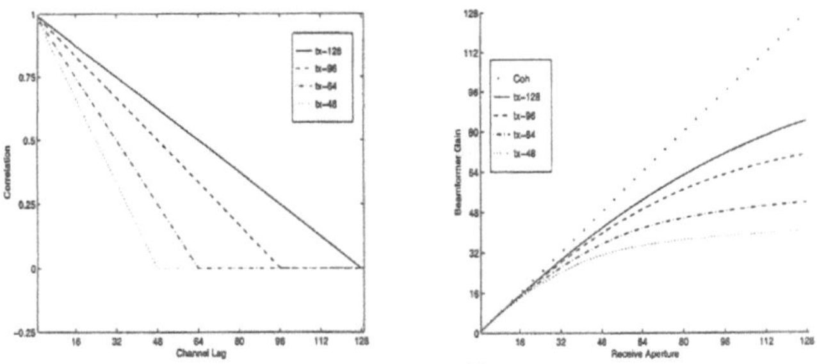

(a) (b)

Figure 4. (a) Predicted spatial coherence curves synthesized from the triangular autocorrelation functions of the four transmit apertures employed and (b) corresponding predicted variations of the beamformer gain with receive aperture size.

Discussion of these results appears at the end of the paper but they provided baseline confirmation that when the target is diffusely scattering and non-aberrating, the experimental behaviour for spatial coherence and beamformer gain is consistent with theoretical predictions for conditions of incoherent scattering.

In vivo

Using the RF channel sensor data corresponding to two apical four-chamber views of the human heart of two volunteers, the spatial coherence of the received acoustic field and the beamformer gain were computed, as above, for four anatomical features selected when viewing the conventional B-mode sector image. One volunteer provided what was generally regarded as a "high quality" image, the other a relatively "poor quality" image containing considerable image clutter and apparent aberration. The structures selected for study were:

(a) blood pool in the left ventricular cavity, expected to provide diffuse low-level backscatter,
(b) myocardium, expected to provide variable but diffuse medium-strength backscatter,
(c) mitral valve leaflet, expected to provide strong specular backscatter, and
(d) pericardium, also expected to provide strong specular backscatter.

The motivation for this part of the study was to compare spatial coherence functions obtained for these structures with those obtained under the ideal the diffuse scattering conditions of the *in vitro* experiment, since it is expected that spatial coherence is associated with both target structure and overall image quality. Table 1 shows the depths of the selected structures for each of the two images. Figure 5 shows the results for spatial coherence and beamformer gain for the "good" image and those for the "poor" image are shown in Fig. 6.

Table 1. Depths (in cm) of selected echo structures for each of the two images

	Pericardium	Valve	Septal Myocardium	Lateral Myocardium	Ventricular Cavity
Good Image	9.0	8.8	5.7	7.8	6.9
Poor Image	7.3	7.4	7.3	7.5	7.3

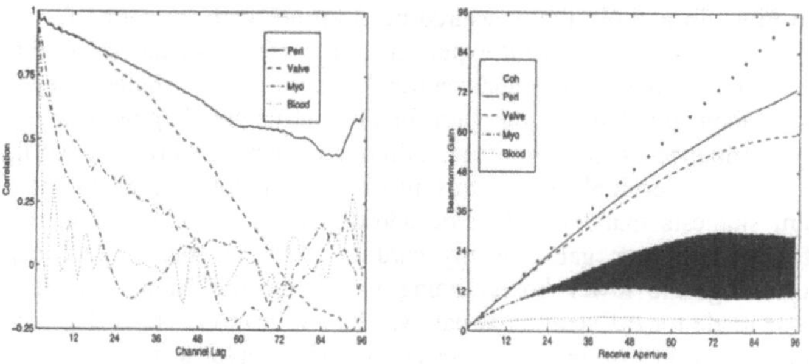

(a) (b)

Figure 5. *In vivo* results for the "good" image: (a) measured spatial coherence functions for the selected anatomical structures shown in Table 1 and described in the text, and (b) corresponding measured beamformer gain as a function of receive aperture; key: Peri = pericardium, Valve = mitral valve, Myo = septal myocardium (upper curve) and lateral myocardium (lower curve), Blood = ventricular cavity, Coh = expected result for coherent received wave. Image acquired with a 96 element, 210 µm pitch, 3.5 MHz centre frequency, array.

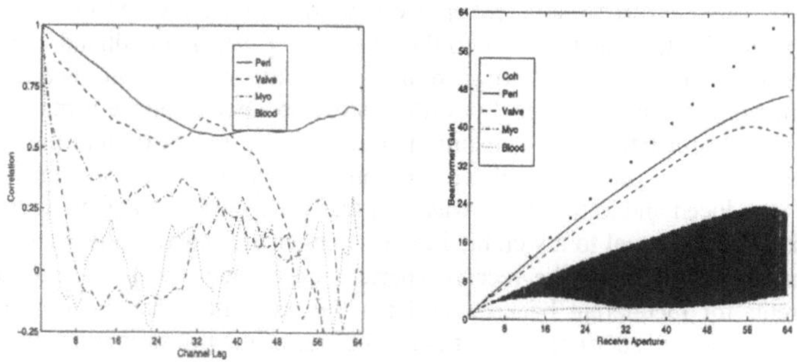

(a) (b)

Figure 6. *In vivo* results for the "poor" image: (a) measured spatial coherence functions for the selected anatomical structures shown in Table 1 and described in the text, and (b) corresponding measured beamformer gain as a function of receive aperture (key as for Fig. 5). Image acquired with a 64 element, 320 µm pitch, 2.5 MHz centre frequency, array.

With reference to Figs. 5a and 6a, relatively high spatial coherence was observed for waves scattered from the pericardium and the mitral valve leaflet, which is consistent with the anticipated specular nature of these two structures and with their associated high image brightness. Two curves have been used to depict the range of spatial coherence from myocardium. The higher curve corresponds to septal myocardium, which was easily identifiable in the image. The lower curve corresponds to myocardium in the anterior lateral wall, which was much less clearly imaged than septal myocardium. Lastly, the curve labelled blood is for echoes apparently arising in the ventricular cavity. However, since true backscatter from blood is substantially less than for soft tissue the echoes used for this analysis were both week and probably represented clutter of some kind. Indeed the measured spatial coherence function for this region has an impulse-like shape with an oscillatory nature that we have been able to show is consistent with an additive combination of random noise and backscatter from off-axis tissues. In the "poor" quality image this type of result was seen both for the ventricular cavity and for the anterior lateral wall, for which myocardium which was not clearly identifiable. It appeared that insufficient beamformer gain necessary to discern myocardium was a consequence of the impulse-like nature of the spatial coherence function, probably resulting from aberration broadening of the transmit beam and increased noise content of the received signal.

Refering to Figs. 5b and 6b, the measured beamformer gain functions for pericardium and valve are much closer to the result expected for a perfect specular target than the gain functions observed for myocardium and ventricular cavity. Nevertheless, beyond receive aperture sizes of about 60 elements the gain for these structures begins to level off as the beamformer performance is compromised. Although such scattering is unlikely to be perfectly specular, the resemblance of this result to that obtained from the incoherent scattering foam suggests that there may be additional loss of spatial coherence due to aberration. The beamformer gain for myocardium falls far below that expected for incoherent scattering, the lower bound being worse for the "poor" image. A similar statement can be made for the ventricular cavity. The implication here is that increasing the receive aperture will do little to improve detectability of the myocardium.

SUMMARY AND CONCLUDING DISCUSSION

In the space available only some of the main implications of the results can be discussed.

Firstly, spatial coherence can be used to quantify the performance degradation caused by aberration and attenuation, using the area under the spatial coherence function as a first order indicator. The measured spatial coherence functions were shown to be consistent with predictions based on the van Cittert Zernike theorem, given a broadband model of the intensity distribution across the scattering volume, and additive contributions due to random noise and coherent off-axis structures. For diffuse structures performance is maximum at the transmit focus and proportional to transmit aperture size. In the near field performance is greatly reduced and is inversely related to transmit aperture size. Beyond the focus performance is reduced and in this study was not greatly affected by transmit aperture size.

Beamformer gain is equal to a weighted integral under the spatial coherence function. It increases linearly with size of the receive aperture only when the backscattered field is highly coherent; for incoherent backscatter from the transmit focus the limiting gain is reached when the receive aperture is about twice the size of the transmit aperture. However, two-thirds of the realizable gain is achieved with a receive aperture equal in size to the transmit aperture, a result consistent with the value obtained from the ultrasonic focusing criterion of Mallart and Fink (1994). Receive apertures larger than about twice the measured coherence length offer little improvement in performance. Near-field beamformer performance may be degraded by using too large a transmit aperture. Beyond the focus, transmit and receive aperture sizes could generally be reduced without loss of performance.

Spatial coherence is a significant factor limiting the performance of conventional beamformers. It is related, via beamformer gain, to the varying degrees of difficulty experienced with imaging different cardiac structures and may account for a substantial amount of the apparent variation of target strength in practical imaging situations.

REFERENCES

J.C. Bamber, R.A. Mucci, D.P. Orofino, K. Thiele, B-mode speckle texture: the effect of spatial coherence, *companion paper in this volume* (1998).

J.W. Goodman, *Statistical Optics*, Wiley, New York (1985).

D.L. Liu, R.C. Waag, About the application of the van Cittert-Zernike theorem in ultrasonic imaging, *IEEE Trans Ultrason Ferroelec and Frequ Control.* **42**:590-601 (1995).

R. Mallart, M. Fink M, The van Cittert-Zernike theorem in pulse echo measurements, *J Acoust Soc Am.* **90**:2718-2727 (1991).

R. Mallart, M. Fink, Adaptive focusing in scattering media through sound speed inhomogeneities - the van Cittert-Zernike approach and focusing criterion, *J Acoust Soc Am.* **96**:3721-3732 (1994).

A NEW APPROACH FOR CALCULATING WIDEBAND FIELDS

S. Leeman, A. J. Healey* and J. P. Weight*

King's College School of Medicine and Dentistry
London U.K.
*City University, London U.K.

INTRODUCTION

Ultrasonic techniques are extensively used in medicine and engineering. In many applications short 'bursts' or pulses of ultrasound are employed. It is often highly desirable, if not essential, to have an accurate knowledge of the structure of such transient fields, as they propagate (in realistic media). Unfortunately, such a knowledge is difficult to obtain as the structure of the field may be very complex, due to effects such as diffraction. The influences of diffraction are considerable in many applications due to the relative dimensions of typical spatial frequencies and source apertures used, coupled with the fact that relevant interactions and measurements are performed relatively 'close' to the source, and that the fields employed may be considered to be 'highly coherent'. There is, of course, also the problem of designing and tailoring appropriate fields and transducers. Hence there is an obvious need for pulsed field simulation and prediction tools.

There are a number of approaches towards the simulation problem including the Angular - Spectrum method; finite element and finite difference techniques; Tupholm/Stephanishen formalism, etceteras. This paper presents an alternative method that we have called the Directivity Spectrum approach. It may be regarded as a generalisation of the Angular Spectrum method as the Angular Spectrum may be obtained directly from the Directivity Spectrum, but not vice-versa. Due to the different boundary conditions used, the problematic evanescent waves which appear in the Angular Spectrum are not present in the Directivity Spectrum formalism. It is particularly suited to dealing with very short pulsed transient fields, and possesses efficient numerical time complexities. Propagation in lossy media is easily incorporated and there is also the potential to include non-linear propagation effects efficiently (although these are not presented here). The proposed method is also ideally suited to be used in conjunction with other techniques. For example a finite element approach may be used to generate the boundary conditions as

Acoustical Imaging, Volume 24,
Edited by Hua Lee, Kluwer Academic/Plenum Publishers, 2000.

input to the Directivity Spectrum calculation. The field may then be propagated to any desired location (both forward and backward) in a lossy media, with no accumulation of numerical errors with propagation distance. Emphasis is placed on the display of simulation results from simple geometries as it allows direct comparison of the quality of the simulation output to well known theoretical results.

THEORY

Consider a (four-dimensional) ultrasound (pressure) wave, $p(\mathbf{r},t)$, propagating linearly in a loss-less medium according to the canonical wave equation, $\nabla^2 p(\boldsymbol{r},t) - \frac{1}{c^2}\frac{\partial^2 p(\boldsymbol{r},t)}{\partial t^2} = 0$, where c denotes the velocity of ultrasound. The pressure field may be expressed as,

$$p(\boldsymbol{r},t) = \frac{1}{(2\pi)^3}\int\int\int_{-\infty}^{\infty} d^3k\, F(\boldsymbol{k})\exp(i\{\boldsymbol{k}\bullet\boldsymbol{r} - \omega t\}) \qquad (1)$$

where $F(\mathbf{k})$ is the spatial Fourier transform of the field at some time, say t=0.

$$F(\boldsymbol{k}) = \int\int\int_{-\infty}^{\infty} d^3r\;\; p(\boldsymbol{r},0)\exp(-i\{\boldsymbol{k}\bullet\boldsymbol{r}\}) \qquad (2)$$

$F(\mathbf{k})$ is termed the 'Directivity Spectrum'. The terminology results from the interpretation of these Fourier components ($F(\mathbf{k})$) to the far field directivity function[1].

Equation (2) has a ready interpretation. Any solution of the canonical wave equation (propagating forward into an unbounded medium) may be expressed as a set of travelling plane waves. For some specific point in time, say t=0, the pressure field in three dimensional space, may be decomposed into it's constituent plane wave components by the application of a three dimensional spatial Fourier transformation. As long as the field is propagating linearly in an 'ideal' loss-less medium (i.e. according to the canonical wave equation) the field at some other time instant may be determined by propagating each of the plane wave components for the appropriate time interval. Note that this operation corresponds to a simple phase factor in the Fourier domain. Hence $F(\mathbf{k})$, which is determined from the boundary conditions, $p(\mathbf{r},t)$, provides all the information for a full four dimensional knowledge of the pressure field, $p(\mathbf{r},t)$, and completely characterises the field. If the field is propagating in an homogenous lossy medium, $p_\alpha(\mathbf{r},t)$ then dispersive attenuation and velocity may be expressed via,

$$p_a(\boldsymbol{r},t) = \frac{1}{(2\pi)^3}\int\int\int_{-\infty}^{\infty} d^3k\, F_a(\boldsymbol{k})\exp(i\{\boldsymbol{k}'\bullet\boldsymbol{r}' - \omega t\}) \qquad (3)$$

with complex k formalism, $\boldsymbol{k}' = \boldsymbol{k} + ia(k)\hat{\boldsymbol{k}}$, and $\omega' = \omega(\boldsymbol{k})$.

SIMULATION DETAILS

The boundary conditions for the Directivity Spectrum approach are a 'snapshot' of the spatial pressure distribution at some fixed time instance (for convenience say t=0), i.e. $p(\mathbf{r},0)$. In contrast the Angular Spectrum which requires a knowledge of the time dependence of the pressure field in some plane (for convenience the x-y plane located at z=0), i.e. $p(x,y,0,t)$. It is the prudent selection of boundary condition in the Directivity

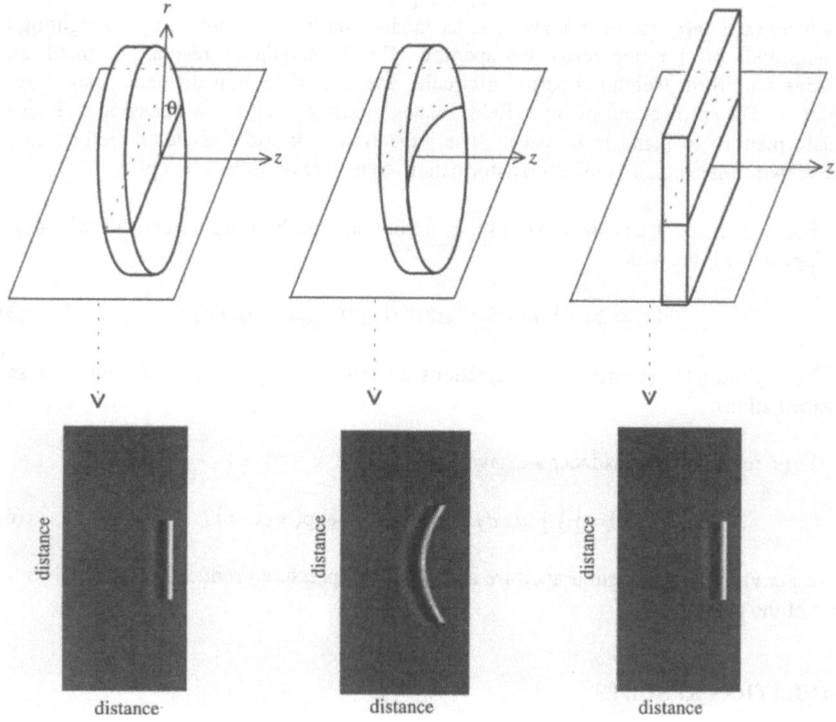

Figure 1: Schematic showing the simulation geometry. The pressure field is shown in the spatial domain as a 'snapshot' taken at time zero, just as the pulse 'departs' the source. These are the simplified boundary conditions used as input to the Directivity Spectrum calculation (more exact boundary conditions would account for the radiation pressure experienced across the source aperture generated via the 'edges' of the source). The upper graphics attempt to represent the 'envelope' of the spatial pulse i.e. the three-dimensional spatial surface 'encapsulating' the pulse. The grey scale image represents the (raw rf) pressure in the two-dimensional plane indicated. Lighter shades depict positive (above ambient) pressures, and darker shades indicate negative (below ambient) pressures, and are presented on a linear grey scale. Left - the circular planar transducer; middle - the focused circular transducer and, right - the rectangular transducer.

Spectrum approach which leads to a much more natural description of transient pulsed fields, and the absence of evanescent waves in the formalism.

Consider the simplified simulation model for a 'snapshot' in time (at t=0), of the (simplified) spatial distribution of pressure of a pulsed field from such a planar circular aperture transducer, just as it emerges from the aperture. The simplified geometry is shown in Figure 1. The simplified approach, used here for convenience only, neglects the radiation pressure experienced across the source due to the source 'edges' which are considered as minor in these examples for the case of very short transient fields. However the incorporation of such effects are easily included. The functional form of the pressure field in cylindrical co-ordinates is best expressed via, $p(r,\theta,z) = f(z)\, g(r)$, with

$$F(\boldsymbol{k}) = 2\pi \int\int dr\, dz\, r f(z)\, g(r)\, \exp(-i\{\boldsymbol{k} \bullet \boldsymbol{r}\}) \tag{4}$$

The function $f(z)$, may be interpreted as the pressure produced by the excitation of the transducer face. $g(r)$ can be thought of as an apodisation function, providing a weighting to the amplitude of vibration across the aperture. For the simulation results presented, $g(r)$ provides a uniform weighting across a circular aperture of 13 mm diameter, and is zero outside. The projection of this field relates directly (via two-dimensional Fourier Transformation) to plane in **k** space. The projection in **k** space relates directly (via an inverse two-dimensional Fourier Transformation) to the pressure field in a plane[1].

For a focused transducer we use a similar approach in spherical co-ordinates, $p(r,\theta,\varphi) = f(r)\, g(\theta)$, with

$$F(k) = 2\pi \int\int dr\, d\theta\, r^2 \sin\theta\, f(r)\, g(\theta) \exp(-i\{k \bullet r\}) \tag{5}$$

For cylindrical symmetry the projections we require are provided via Fourier Bessel transformations.

For a rectangular transducer we have,

$$F(k) = \int\int\int dx\, dy\, dz\, f(z)\, g(x,y) \exp(-i\{k \bullet r\}) \tag{6}$$

where $g(x,y)$ is the rectangular aperture and $f(z)$ is the pressure produced orthogonal to the plane of the aperture.

SIMULATION RESULTS

The four-dimensional nature of transient fields presents problems of visualisation. With the angular spectrum approach it is usual (as this is what the technique directly calculates) to present representations of the pressure distribution in a two dimensional plane, orthogonal to the field propagation direction, and indicate how this varies with time. The temporal behaviour of the pressure field is then calculated for arbitrary location of this plane. With the Directivity Spectrum approach it is more natural to present 'snapshot' images of the spatial distribution of the pressure field at a particular instant in time. The format of such an image is shown in Figure 1.

Figure 2 shows the simulation output for the well known example of a circular planar transducer, with 13 *mm* diameter aperture, and a centre frequency of 3.5 MHz. For all three transducer geometries $f(z)$, (and $f(r)$ in the case of the circular transducer) was modelled by the function $z\exp(-az)$. The direct and edge wave components are clearly visible, as is the phase change in the edge wave across the edge of the geometric 'shadow' of the transducer aperture. Figure 3 gives the result for a 19 *mm* diameter aperture circular focused transducer of 90 *mm* nominal focal length (with the same 3.5 MHz centre frequency as per Figure 2). Figure 4 shows the result for a square transducer of 13 *mm* sides (again with the same 3.5 MHz centre frequency). The 'direct', 'edge' and 'corner' waves are clearly visible in the simulation output. The approach provides a rapid computational method to determine effects on pulse shape such as attenuation and velocity dispersion. Note also that it is trivial to extend the approach to include multi-element transducers, such as two- and three- dimensional phased arrays.

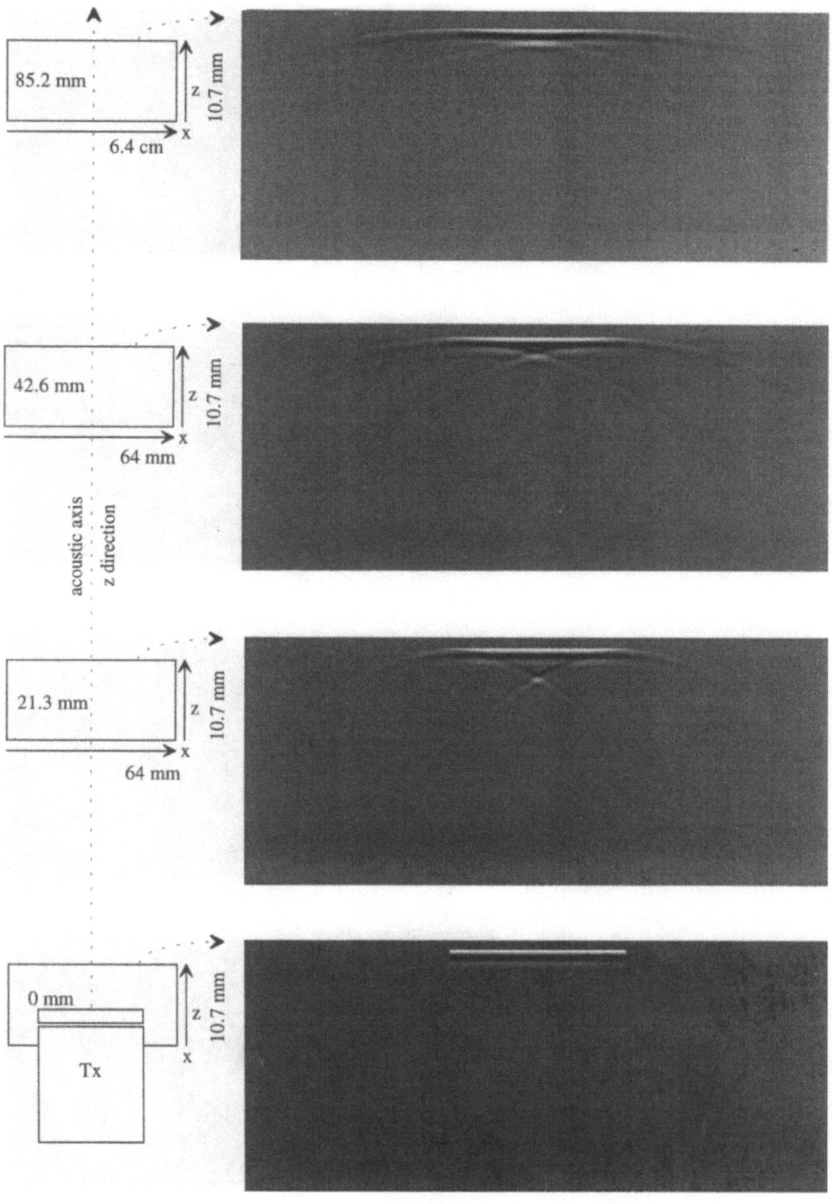

Figure 2: simulation output for the well known example of a circular planar transducer, with 13 *mm* diameter aperture. The Two-dimensional grey scale images on the right hand side represent the ultrasound pressure pulse on a linear grey scale. Light shades represent positive pressures (above ambient) and darker shades represent negative (below ambient) pressures. Image dimensions are 64 *mm* by 10.7 *mm*, the images are not displayed proportionately in order to more clearly present the structure of the field. The direct and edge wave components are clearly visible, as is the phase change of the edge wave, associated with the geometric 'shadow' of the transducer aperture. Images are 'snapshots', from bottom to top at 0, 13.8, 27.6, and 55.2 μs, corresponding to an axial distance of 0, 21.3, 42.6, and 85.2 *mm* from the transducer face.

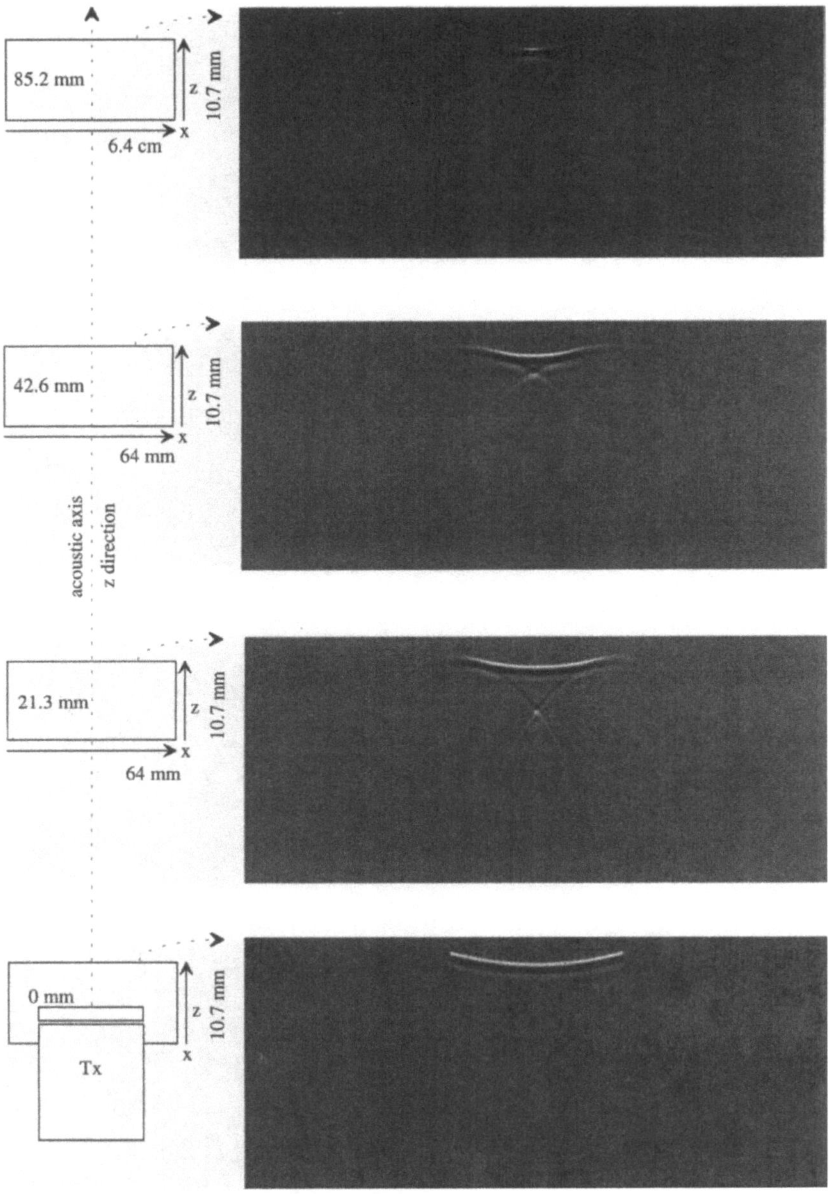

Figure 3: simulation output for a circular aperture focused transducer, with diameter aperture, nominal focus at 90 *mm*. The Two-dimensional grey scale images on the right hand side represent the ultrasound pressure pulse on a linear grey scale. Light shades represent positive pressures (above ambient) and darker shades represent negative (below ambient) pressures. Image dimensions are 64 *mm* by 10.7 *mm*, the images are not displayed proportionately in order to more clearly present the structure of the field. Images are 'snapshots', from bottom to top at 0, 13.8, 27.6, and 55.2 μs, corresponding to an axial distance of 0, 21.3, 42.6, and 85.2 *mm* from the transducer face.

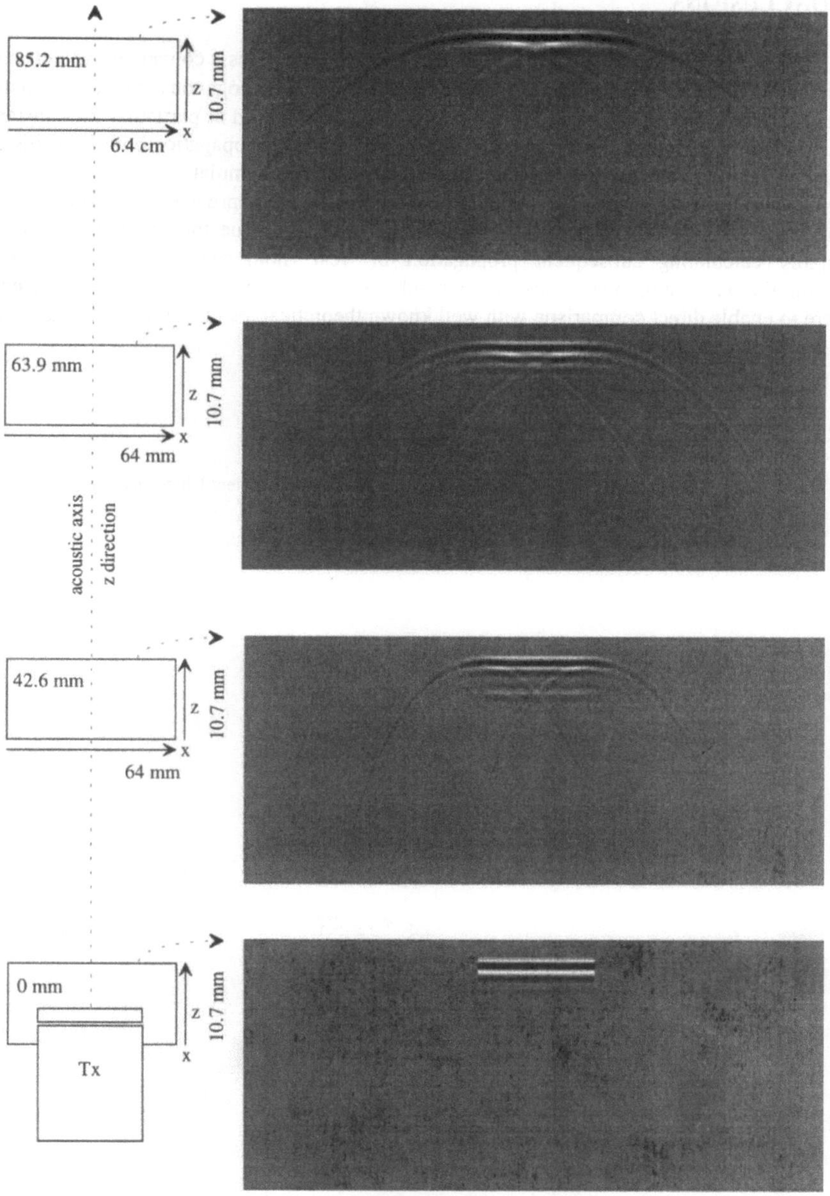

Figure 4: simulation output for a square aperture unfocused transducer. The Two-dimensional grey scale images on the right hand side represent the ultrasound pressure pulse on a linear grey scale. Light shades represent positive pressures (above ambient) and darker shades represent negative (below ambient) pressures. Image dimensions are 64 *mm* by 10.7 *mm*, the images are not displayed proportionately in order to more clearly present the structure of the field. Images are 'snapshots', from bottom to top at 0, 27.6, 41.4, and 55.2 μs, corresponding to an axial distance of 0, 42.6, 63.9, and 85.2 *mm* from the transducer face.

CONCLUSIONS

The Directivity Spectrum approach presented here provides a convenient and efficient method to simulate transient acoustic fields propagating in homogenous media. Simulation output directly provides spatial 'snapshots' of the acoustic field at particular time instants. Attenuation and velocity dispersion are easily incorporated. Propagation both forwards and backwards to any location is easily accomplished without accumulation of numerical errors with propagation distance. The boundary conditions for the approach may be provided by other methods (such as finite element models). The technique then provides a tool for rapidly calculating subsequent propagation in 'real' homogenous media, free from accumulation of numerical errors. Only results of simple geometries have been presented here to enable direct comparison with well known theoretical results, although the approach is applicable to arbitrary geometries.

REFERENCES

1. A. J. Healey, S. Leeman, and J. P. Weight, Space-Time Imaging of Transient Ultrasound Fields, Int J Imaging Syst Technol, **8** 45-51, 1997.

HIGH-RESOLUTION ACOUSTIC ARRAYS USING OPTIMUM SYMMETRICAL-NUMBER-SYSTEM PROCESSING

David Jenn, Phillip Pace, and John P. Powers

Department of Electrical and Computer Engineering
Naval Postgraduate School
Monterey CA 93943-5121
email: jpowers@nps.navy.mil

Introduction

Acoustic arrays can be used for imaging or direction finding. A critical measure of the array's capability is its resolution; here, we will be concerned with the angular resolution of the array, that is, its ability to measure the angular bearing of the target from the array axis. The optimum symmetrical number system (OSNS) can be used to design a simple interferometric acoustic array and to process the data from that array to achieve high resolution angle-of-arrival information. Small angular resolutions may be obtained from a few elements that are spaced a few to several wavelengths apart from each other. The OSNS scheme is based on decomposing the acoustic spatial filtering operation into parallel sub-operations that are each simpler to perform. The results from the parallel operations are combined into an overall high-resolution result. The approach that we use is similar to the approach used to design an RF direction-finding system based on the same concepts [1–5].

Figure 1 indicates the array geometry. Three array elements are indicated (i.e., there are two interferometers); their spacing is determined (from the technique described below) to be 1.50λ and 2.75λ from the end element, where λ is the nominal wavelength of the source. (More array elements may be added, if desired, and different spacings can be prescribed if the element spacings cannot be physically realized.) The angle of arrival, ϕ, is measured from the perpendicular to the array, as shown, and spans a range between $-90°$and $+90°$. The array elements are receivers that are omnidirectional in the right half-plane. The spacing of the elements and the processing of the signals from them depends on the principles of the optimum symmetrical number system; we now offer a short review of those principles before describing the system design.

Review of Optimum Symmetrical Number System (OSNS)

We begin by choosing N pairwise relatively prime integers (i.e., numbers taken in pairs that contain no common divisor other than 1); these N integers constitute the

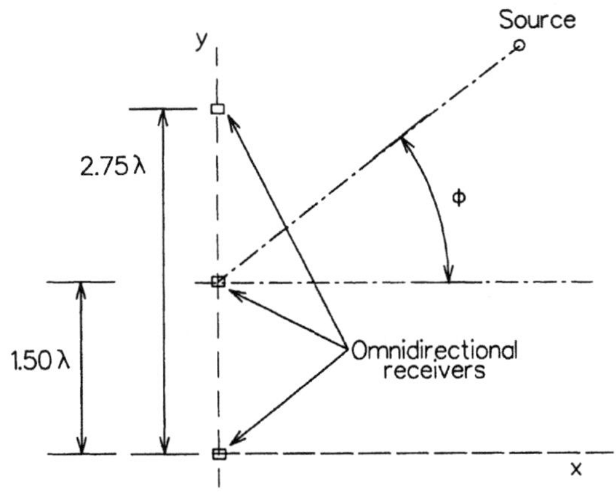

Figure 1. Array geometry.

Table 1. Sequences formed from members of two-moduli OSNS ($m_1 = 3$ and $m_2 = 4$).

Index	0	1	2	3	4	5	6	7	8	9	10	11	12
$m_1 = 3$	0	1	2	2	1	0	0	1	2	2	1	0	0
$m_2 = 4$	0	1	2	3	3	2	1	0	0	1	2	3	3

moduli of system, $m_1, m_2, \ldots, m_N$. (For a simple illustration, we choose $N = 2$ with $m_1 = 3$ and $m_2 = 4$.) For each modulus, m_i, we can generate the sequence,

$$x_m = [0, 1, 2, \ldots, m_i - 1, m_i - 1, \ldots, 2, 1, 0, 0, 1, \ldots] . \tag{1}$$

Table 1 shows the two sequences for our example.

We can consider the index value (i.e., the top row of the table) as an ever-increasing input signal. We note that the values for any particular modulus "fold" as the index continuously increases. In the $m_1 = 3$ row of Table 1, the sequence rise to a maximum of two, then falls back to zero, and then rises to two again. This pattern repeats every six values as the index (the top row of the table) continues to increase. The other row follows the pattern of rising and falling but with different maximum values and, hence, different periods. The $m_2 = 4$ row reaches a maximum of three and has a period of eight. From this we deduce that the sequence period is $2m_i$ and reaches a maximum value of $m_i - 1$. (This folding is reminiscent of the folding of a continuously varying phase, every 2π radians.)

As we count up, we note that no column of values is duplicated until we get to an index of 12 (that duplicates the index 11 column). Hence, the unambiguous dynamic range for our example is 12 (i.e., there are 12 unambiguous states before repetition begins. This value of *dynamic range*, M, is found from $M = m_1 \cdot m_2 \cdots m_N = \prod_{i=1}^{N} m_i$. For our example, this would be $M = 3 \cdot 4 = 12$ levels.

Figure 2a illustrates the transfer function of the folding circuit that is used to convert the wide dynamic range of the input signal into a small dynamic range output signal. Parallel folding circuits with differing periodicities remove the ambiguity of the single circuit. The folding transfer functions do not need to be linear; they can be nonlinear. Similarly they can have both positive and/or negative values and can extend into negative values of the input function. Examples of folding circuits include phase detectors (mixer followed by a lowpass filter) and Mach-Zehnder electro-optic modulators.

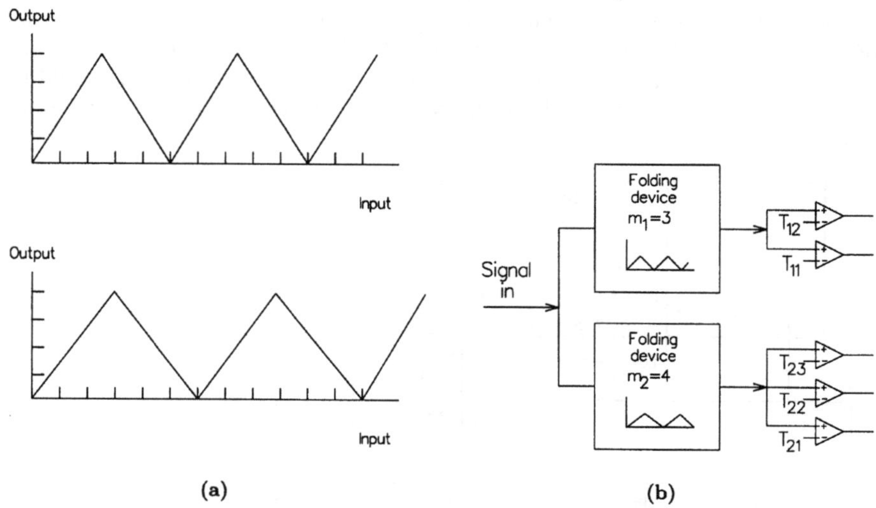

Figure 2. (a) Transfer function of typical folding circuits and (b) folding circuits followed by array of comparators functioning as counting circuits.

Figure 2b shows the folding circuits followed by a bank of comparators. The comparators indicate the range of the corresponding folding-circuit output voltage as a "thermometer" code (i.e., the comparators successively turn on and then off as the input signal increases). Combining the comparator outputs in a logic-gate array allows the user to determine a digital representation of the input value. There are $m_i - 1$ comparators for each channel, each with a different threshold voltage. The threshold voltage T_{ij} for the j-th comparator of the i-th channel is found by dividing the input signal dynamic range into M equal segments and using the transfer function to find the corresponding output value, which is the desired threshold value.

Direction-finding System Concept

Our simple two-modulus example produced only 12 quantization levels, too few for practical use. If we choose our moduli as $m_1 = 6$ and $m_2 = 11$, we will obtain 66 levels, a value with more utility. For each modulus, the periodic sequence described above can be constructed. These sequences have periods of $2m_i$ (i.e., 12 and 22, for our example) and the combination of all sequences extends unambiguously for M values where $M = m_1 m_2 \ldots m_N = 6 \cdot 11 = 66$, in our example).

The spacings of the N array elements (measured from a base element) are determined by $d_i = M\lambda/4m_i$ where λ is wavelength of the source [5]. For our conceptual system, the element spacings from the base element are 2.75λ and 1.5λ.

The signal processing of the system (shown in Fig. 3) measures the phase of the i-th element relative to the base element. The phase detectors consist of a mixer and a low-pass filter and produce a (normalized) output voltage of $v_i = \cos(2d_i \sin\phi/\lambda)$ where ϕ is the angle of arrival with values between -90°and +90°. The phase-detector voltage is a periodic waveform when plotted against $\sin\phi$. While the voltage out of a *single* phase detector has too many ambiguities to measure the exact phase, the voltages out of *multiple* phase detectors, each with a different period when plotted against $\sin\phi$, can be used to resolve the ambiguities and produce a more exact measurement. In particular, the range of $\sin\phi$ from -1 to $+1$ (corresponding to the range of ϕ values of $-\pi/2 \geq \phi \leq +\pi/2$) can be subdivided into M different resolution cells. Hence, the resolution of $\sin\phi$ is $2/M = 2/(m_1 m_2 \cdots m_N)$. For our example, this value would be $2/66 = 0.030$. Each resolution cell for $\sin\phi$ can then be transformed into a (nonuniform) resolution cell for ϕ.

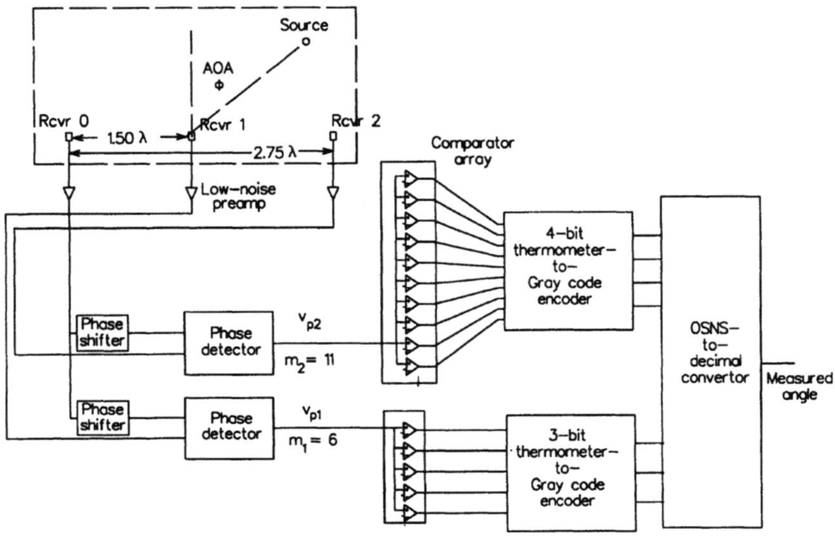

Figure 3. Block diagram of signal processing.

The signal processing required to perform the calculation of the angle of arrival consists of phase detectors feeding a bank of $m_i - 1$ comparators. (For our example, there are 5 and 10 comparators in the two comparator banks.) The threshold voltages of each comparator are set in accordance with the OSNS theory with output of each comparator bank being a thermometer code representing the value of the input in terms of the corresponding OSNS sequence. A digital processing circuit, consisting of an EPROM or an ASIC, converts the thermometer-code output of each comparator bank into a Gray-code output and then combines all of the Gray-code outputs into the high-resolution decimal representation of the angle of arrival (AoA).

The threshold voltages of the comparators are found as follows. We know that there are $m_i - 1$ thresholds to find for a given modulus, m_i. In addition, the range of $\sin\phi$ is divided up into M even increments of size, $\Delta(\sin\phi) = (1-(-1))/M = 2/M$ (where we recognize that the extreme values of $\sin\phi$ are 1 and -1). For our example, we are seeking five threshold values for $m_1 = 6$ and ten threshold values for $m_2 = 11$. The value of the increments of $\sin\phi$ is 0.0303, as described, and, so, the values of $(\sin\phi)_j$ at each increment are $-1+1(0.0303), -1+2(0.0303), -1+3(0.0303), \ldots, -1+M(0.0303) = +1$.

For $m_1 = 6$, we know that the phase-detector output voltage, v_{p1}, will be

$$v_{p1} = \cos\left(\frac{2\pi d_1}{\lambda}\sin\phi\right) = \cos(5.5\pi\sin\phi) \,, \tag{2}$$

where $d_1 = M\lambda/4m_1 = 2.75\lambda$.

Similarly, for $m_2 = 11$, we find that the phase-detector output voltage, v_{p2}, will be

$$v_{p2} = \cos\left(\frac{2\pi d_2}{\lambda}\sin\phi\right) = \cos(3\pi\sin\phi) \,, \tag{3}$$

where $d_2 = M\lambda/4m_2 = 1.50\lambda$. Table 2 shows the values of $(\sin\phi)_j$ for the first eleven increments and the corresponding values of v_{p1} ($m_1 = 6$) and v_{p2} ($m_2 = 11$). Ignoring the repeated values and the values of +1 and −1, we find the five required threshold values for $m_1 = 6$ to be −0.5, −0.866, 0, 0.5, and 0.866. Similarly, we find the ten threshold values for $m_2 = 11$ as −0.959, −0.841 −0.655, −0.415, −0.1423, 0.1423, 0.415, 0.655, 0.8141, and 0.959. .The computed threshold values are also valid if one converts each value of $\sin\phi$ into the corresponding values of ϕ.

Figures 4a and b show the phase detector voltage (normalized) for the $m_1 = 6$ case plotted against $\sin\phi$ and ϕ, respectively. Superimposed on the plots are the five thresholds of the corresponding comparators. Similarly, Figures 4c and d show the normalized voltages for the $m_2 = 11$ case. The corresponding threshold voltages are also superimposed on the plots.

The predicted resolution of the angle-of-arrival is shown in Fig. 5a. The quantity, $\sin\phi$, is evenly divided into M resolution cells, resulting in a nonuniform division of ϕ that is evident in part (b) of the figure. The errors in sampling the phase (deviations from the ideal straight-line plot) are also shown in the figure. The errors in the measurement of $\sin\phi$ are uniform at 0.0303 over the entire range of values. The errors in ϕ are 1.73°(0.0303 radians) for small values of ϕ (broadside) and increase to a maximum error of 14.25°(0.249 radians) for ϕ equal to ±90°(end-fire direction).

Summary

We have presented the concept of designing an acoustic array and its signal processing based on the properties of the Optimum Symmetrical Number System (OSNS). The

Table 2. Computed values of $(\sin\phi)_j$ and the corresponding values of phase-detector voltage, v_{p1} for $m_1 = 6$ and v_{p2} for $m_2 = 6$. These voltage values represent the thresholds of the comparators

j	$(\sin\phi)_j$	v_{p1}	v_{p2}
1	−0.970	0.5	−0.959
2	−0.939	−0.866	−0.841
3	−0.909	−1	−0.655
4	−0.879	−0.866	−0.415
5	−0.848	−0.5	−0.1423
6	−0.818	0	0.1423
7	−0.788	0.5	0.415
8	−0.757	0.866	0.655
9	−0.727	1.0	0.8141
10	−0.697	0.866	0.959
11	−0.667	0.50	1

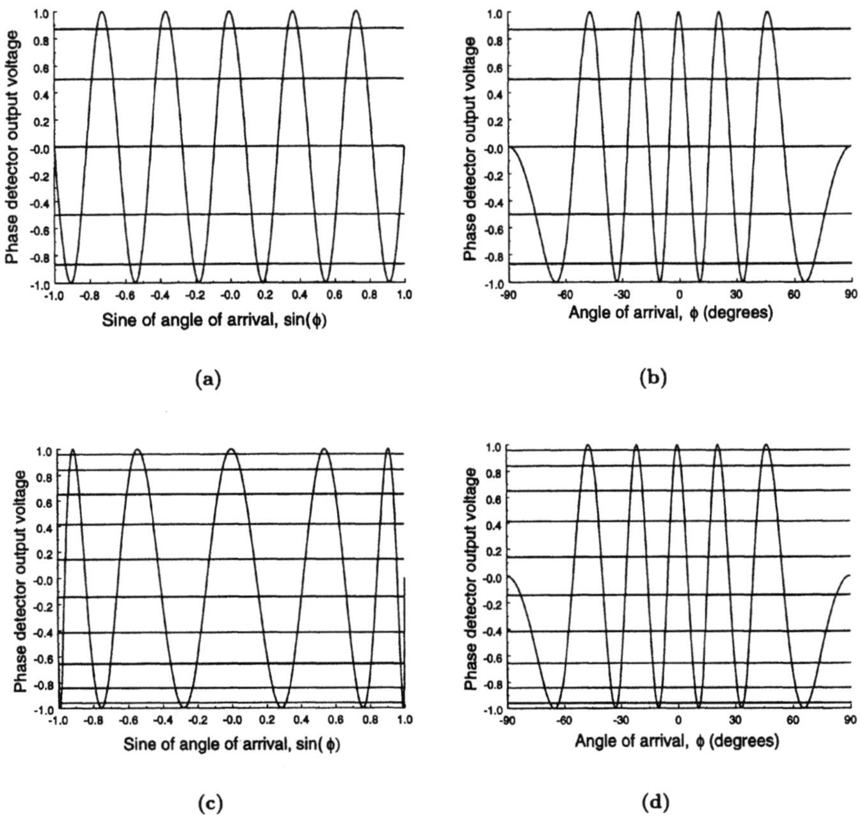

Figure 4. (a) Normalized response of phase detector vs. the sine of the angle of arrival for $m = 6$, (b) normalized response of phase detector vs. the angle of arrival for $m = 6$, (c) normalized response of phase detector vs. the angle of arrival for $m = 11$, and (d) normalized response of phase detector vs. the angle of arrival for $m = 11$. Threshold voltages for the corresponding comparator banks are shown as horizontal lines.

folding properties of the OSNS can be applied to the folding properties of the electronic phase detectors in the array processing system. Table 3 summarizes the relations between the OSNS assumptions and the array system properties. The number of moduli, N, determines the number of elements $(N + 1)$and the number of processing channels. It also contributes to the total number of comparators required. The product of the moduli determines the dynamic range of the system, M. This property determines the element spacing and the angular resolution. Generally, for a given resolution, one can tradeoff the overall length of the array against the number of channels (and the number of comparators).

Acknowledgments

The authors wish to thank Prof. David Styer of the University of Cincinnati for his insight into the symmetrical number system and our thesis students for their excellent work on this project. The authors also wish to acknowledge funding by the Depart-

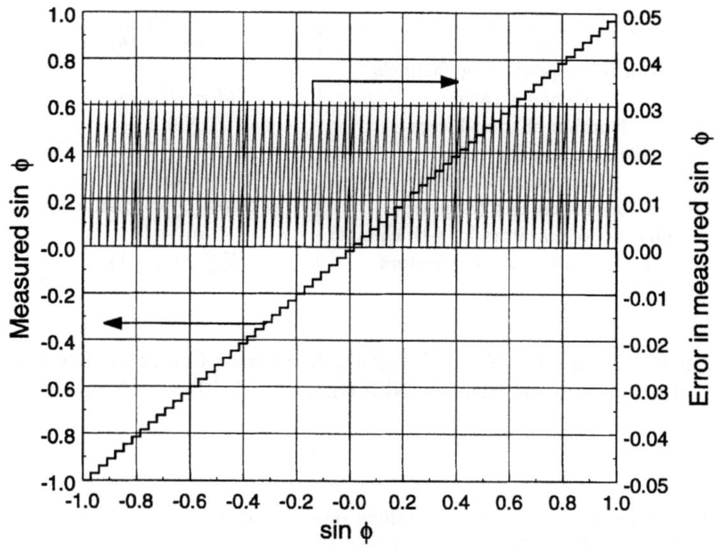

(a)

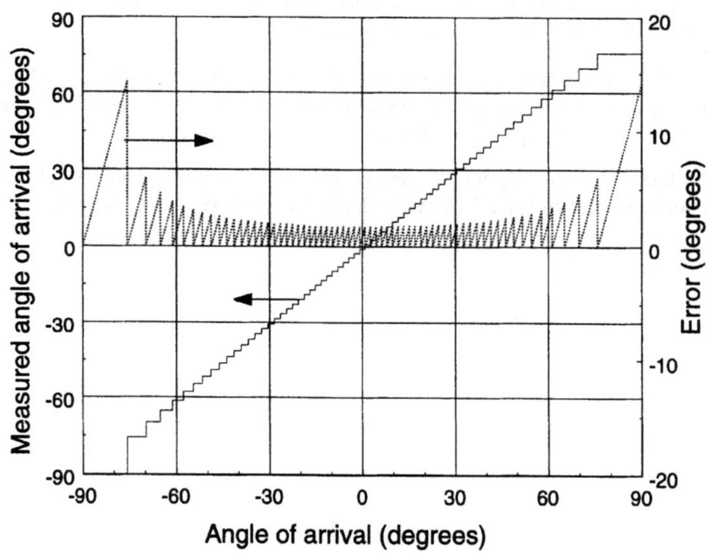

(b)

Figure 5. Simulated angle-of-arrival results and error for (a) $\sin\phi$ and (b) ϕ.

Table 3. Summary of system properties

Property	Value
Moduli	$m_1, m_2, \ldots, m_N$
Number of processing channels	N
Dynamic range	$M = \prod_{i=1}^{N} m_i$
Number of transducers	$N+1$
Element spacing (from end)	$d_i = M\lambda/4m_i \quad (i = 1 \ldots N)$
Resolution of $\sin\phi$	$2/M$
Resolution of ϕ (degrees)	$\sim 180/M$ (for small ϕ)
Number of comparators for i-th channel	$m_i - 1$
Total number of comparators	$\sum_{i=1}^{N}(m_i - 1)$

ment of Defense through the Naval Postgraduate School's Center for Reconnaissance Research and partial support from the NPS Direct-Funded Research Program.

References

[1] L. E. M. Rodrigues, "High-resolution residue antenna architectures for wideband direction finding," Master's thesis, Naval Postgraduate School, June 1996.

[2] P. Papandreau, "Design and prototype development of an optimum symmetrical number system direction finding array," Master's thesis, Naval Postgraduate School, March 1997.

[3] D. Jenn, P. Pace, T. Hatziathanansiou, and R. Vitale, "Symmetrical number system phase sampled DF antenaa architectures," in *1998 IEEE AP-S International Symposium and URSI North American Radio Science Meeting*, 1998.

[4] D. Jenn, P. Pace, T. Hatziathanasiou, and R. Vitale, "High resolution wideband direction finding arrays based on optimum symmetrical number system encoding," *Electronics Letters*, vol. 34, no. 11, pp. 1062–1064, 1998.

[5] T. N. Hatziathanasiou, "Optimum symmetrical number system phase sampled direction finding antenna architectures," Master's thesis, Naval Postgraduate School, June 1998.

FREQUENCY WEIGHTING OF DISTRIBUTED FILTERED BACKWARD PROPAGATION IN ACOUSTIC TOMOGRAPHY

Stephanie Lockwood and Hua Lee

Department of Electrical and Computer Engineering
University of California at Santa Barbara
Santa Barbara, CA 93106

ABSTRACT

The objective of this paper is to provide the formulation of the spatial-frequency weighting of distributed filtered backward propagation in acoustic tomography. This formulation gives the exact frequency-weighting characteristics for various data acquisition configurations.

INTRODUCTION

Similar to the classical X-ray tomography, acoustic tomography requires a spatial-frequency weighting procedure to compensate for the nonuniform density of frequency samples during tomographic superposition [1]. Typically for mathematical analysis and modeling, this frequency-domain compensation procedure is performed after the tomographic superposition as a centralized process.

Frequently in advanced imaging schemes, the image reconstruction operation is instead conducted sequentially, due to constraints of computation and algorithm structures. As a result, the frequency weighting process needs to be conducted in a distributed manner within each individual projection prior to the tomographic superposition. It can be achieved by using a one-dimensional frequency weighting degenerated from the centralized two-dimensional version.

The degeneration is straightforward for the case of X-ray tomography because of the simplicity associated with the linear feature of a projection's spectral content in the overall frequency coverage [2]. However, for the case of diffraction tomography in acoustic imaging, the degeneration becomes more complex due to the unique characteristics of the spectral distribution which is associated with diffraction and angular offset in bistatic schemes. Therefore, identifying the exact formulation of the frequency weighting of distributed filtered backward propagation becomes a subject of significant importance.

Acoustical Imaging, Volume 24.
Edited by Hua Lee, Kluwer Academic/Plenum Publishers, 2000.

In this paper, we first establish the link between the frequency index of the detected wavefield and the corresponding location in the final spectral distribution of the tomographic image. With this relationship, the exact formulation of the spatial-frequency content can then be structured in a simple and closed form. Subsequently, this weighting function is incorporated into the backward propagation filter as the magnitude component of the transfer function. This allows us to perform filtered backward propagation for diffraction tomography in a distributed manner with high degree of efficiency, accuracy, and flexibility for various imaging data acquisition configurations.

ENHANCEMENT FILTER

The spectral coverage of the detected wavefield by a transmission-mode diffraction tomographic imaging system can be illustrated as Fig. (1). It is limited to the lowpass component of the overall spectrum. The loss of high frequency components is caused by the attenuation of evanescent waves and the cutoff frequency is $\sqrt{2}/\lambda$. The circular symmetry is due to the nature of rotational scan.

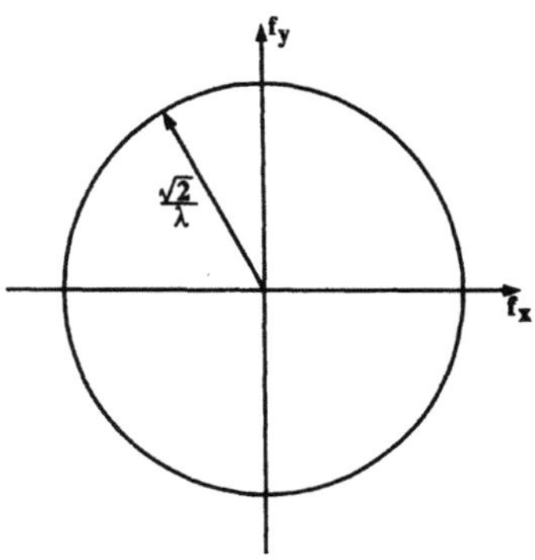

Figure 1. Spectral coverage of the detected wavefield of a transmission-mode diffraction tomographic imaging system.

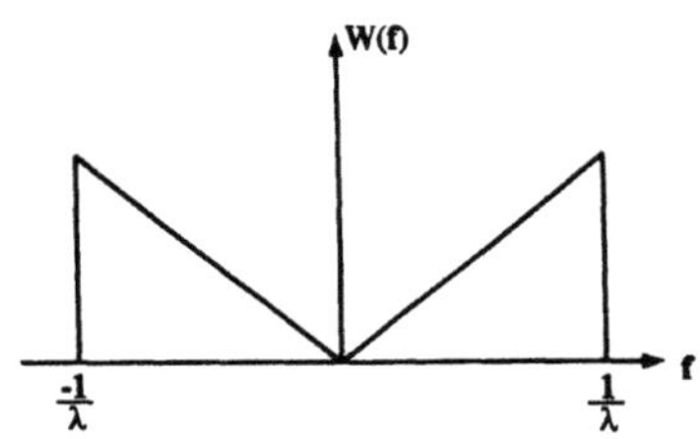

Figure 2. Cross-section of the filter's transfer function.

Because of the nonuniform density distribution of frequency samples, a two-dimensional filter is commonly used to compensate for the lower density of high-frequency samples, as given by

$$\begin{aligned} W(\mathbf{f}) = W(f_x, f_y) &= \|\mathbf{f}\| \\ &= \left(f_x^2 + f_y^2\right)^{1/2} \end{aligned} \tag{1}$$

where $\mathbf{f} = [f_x, f_y]$, and f_x and f_y are the spatial frequencies in the x and y directions, respectively. The cross-sectional variation of the filter's transfer function is shown as Fig. (2).

Because of the similar problem of nonuniform frequency sample density, we encounter the same problem and therefore the same filtering procedure is also applied in the classical case of X-ray tomography.

Typically, the procedure is performed as an enhancement filtering process subsequent to the tomographic superposition of all sub-image components generated from various projections. When the tomographic image formation is conducted sequentially as it is often preferred, this filtering process needs to be performed within each individual projection in a decentralized manner before the sub-images are combined into the final result. Accordingly, the frequency weighting corresponding to the enhancement filter will be converted and distributed into the reconstruction procedure of the sub-images. In the case of X-ray tomography, this decentralized version is known as .ul convolution back projection where the filtering is conducted prior to the backprojection process instead [2]. To do so, the frequency weighting is degenerated into a one-dimensional version, given as

$$W(f) = |f|. \tag{2}$$

The frequency weighting of the degenerated one-dimensional enhancement filter has frequency characteristics similar to the original multi-dimensional version. This is because the spectrum of an X-ray projection is a line segment cutting through the origin, according to the *Central Projection Theorem*, which allows a straightforward conversion for the degeneration [2]. Eq. (2) is, in fact, identical to the cross-sectional profile of the 2-D frequency weighting.

DIFFRACTION TOMOGRAPHY

In the case of diffraction tomography, the conversion is more complex due to the fact that the spectrum of a wavefield projection is a semicircle instead of a linear segment in the case of X-ray tomography. The spectrum of a single backward propagated sub-image is a semicircle, the half of the circle toward the direction of receiving aperture, as illustrated by Fig. (3). The one-to-one correspondence between the spectrum of the received wavefield and the semicircle spectrum of the backward propagated image is illustrated as Fig. (4).

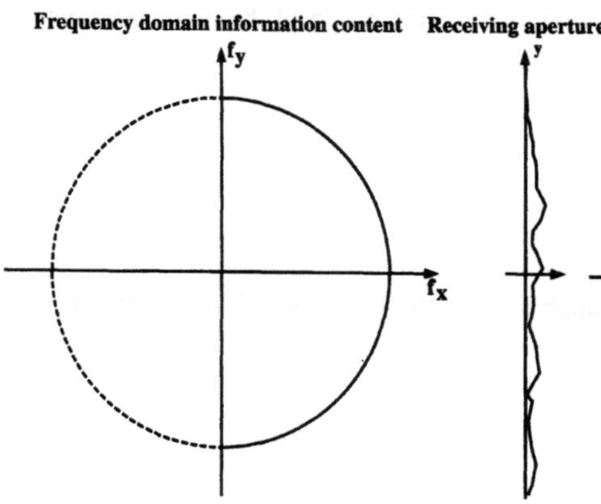

Figure 3. Spectrum of a single backpropagated detected wavefield.

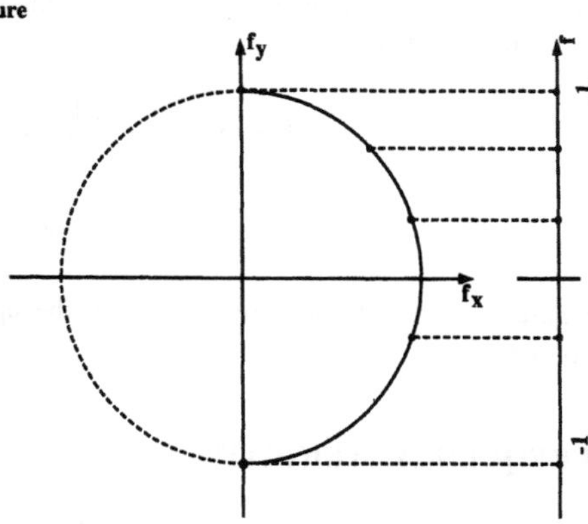

Figure 4. One-to-one correspondence of the received wavefield and the semicircle spectrum of the backward propagated image.

Commonly in active diffraction tomography, we illuminate the object distribution with a known wavefield, and model the object as a secondary source. As a result, to produce the sub-images correctly, we need to remove the contribution of the illumination wavefield to obtain the signal content directly related to the object variation. When the illumination wavefield is a plane wave, which is most widely utilized, the process for the removal of illumination wavefield becomes a simple demodulation operation which can be achieved by shifting the spectrum toward the direction of the illuminating source by $1/\lambda$.

To better describe the demodulation effect, we utilize three simple cases of transmission-mode, reflection-mode, and a more general bistatic mode with arbitrary angular offset, as illustrated in Fig. (5). Fig. (5a) shows the spectral content of the backward propagated image formed from the transmission-mode received wavefield. The semicircle spectrum is shifted away from the aperture direction and toward the direction of illumination source as previously described. Fig. (5b) is the spectrum of the sub-image corresponding to the reflection-mode imaging. The semicircle spectrum is now shifted toward the direction of the aperture since the illumination source and aperture are on the same side for this operating mode. Consequently, the spectral content is shifted away from the origin which explains the nature of high-frequency content in reflection-mode imaging.

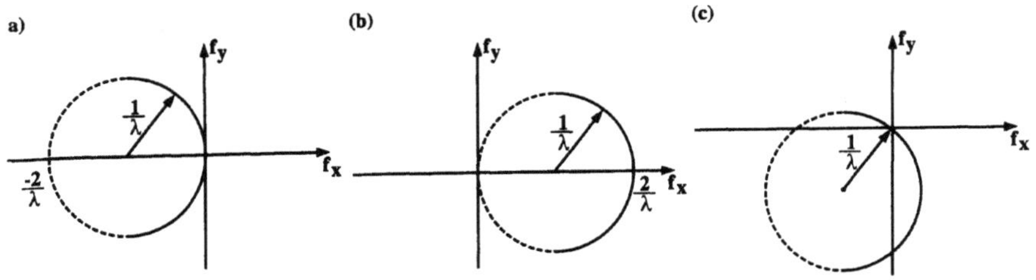

Figure 5. Demodulation effect on image spectrums. Shift shown for (a) transmission mode, (b) reflection mode, and (c) general bistatic mode imaging systems.

Fig. (5c) illustrates a more general case of bistatic imaging configuration. After the frequency shift associated with the demodulation process, the origin of the new coordinate system becomes,

$$\begin{aligned} \mathbf{f} = [f_x, f_y] &= \left[\frac{1}{\lambda}\cos(\theta+\pi), \frac{1}{\lambda}\sin(\theta+\pi)\right] \\ &= -\frac{1}{\lambda}[\cos\theta, \sin\theta] \end{aligned} \tag{3}$$

where θ is the angular offset between the aperture and the direction of the illumination source.

FREQUENCY WEIGHTING

To formulate the frequency weighting of the enhancement filter, we need to establish the correspondence between the one-dimensional frequency index of the received wavefield and the final position of that particular frequency component in the two-dimensional spatial frequency domain. To do so, we need to combine the frequency relocation processes of backward propagation in forming the preliminary sub-images and frequency shift associated with the demodulation process.

It is known that the received wavefield at the aperture is a lowpass signal and the frequency content is bounded by the cutoff frequency of $1/\lambda$,

$$\begin{aligned} \mathbf{P}(f) &= F\{p(y)\} \\ &= \begin{cases} \mathbf{P}(f), & |f| < 1/\lambda \\ 0, & otherwise \end{cases} \end{aligned} \tag{4}$$

According to the point-by-point mapping associated with backward propagation, as illustrated in Fig. (4), the spatial frequency index of the received wavefield is mapped to a unique location along the semicircle at,

$$
\begin{aligned}
f \leftrightarrow \mathbf{f} &= [f_x, f_y] \\
&= \left[\left(1/\lambda^2 - f^2\right)^{1/2}, f\right].
\end{aligned} \tag{5}
$$

Combining with the frequency shift associated with the demodulation of the illumination plane wave, the frequency component is now mapped to

$$
\begin{aligned}
\mathbf{f}' &= \left[\left(\frac{1}{\lambda^2} - f^2\right)^{1/2}, f\right] - \left[\frac{1}{\lambda}\cos(\theta+\pi), \frac{1}{\lambda}\sin(\theta+\pi)\right] \\
&= \left[\left(\frac{1}{\lambda^2} - f^2\right)^{1/2} + \frac{1}{\lambda}\cos\theta,\ f + \frac{1}{\lambda}\sin\theta\right]
\end{aligned} \tag{6}
$$

Then, within the spectrum of each individual projection, the weighting of frequency f can be obtained by evaluating the norm of the corresponding frequency vector in the final image

$$
\begin{aligned}
W(f) &= \|\mathbf{f}'\| \\
&= \left\{\left[\left(\frac{1}{\lambda^2} - f^2\right)^{1/2} + \frac{1}{\lambda}\cos\theta\right]^2 + \left[f + \frac{1}{\lambda}\sin\theta\right]^2\right\}^{\frac{1}{2}} \\
&= \left\{\frac{2}{\lambda^2} + \frac{2}{\lambda}f\sin\theta + \frac{2}{\lambda}\left(\left(\frac{1}{\lambda^2} - f^2\right)^{1/2}\right)\cos\theta\right\}^{\frac{1}{2}} \\
&= \frac{\sqrt{2}}{\lambda}\left\{1 + f\lambda\sin\theta + \left(\frac{1}{\lambda^2} - f^2\right)^{1/2}\cos\theta\right\}^{\frac{1}{2}}.
\end{aligned} \tag{7}
$$

We can now calculate the weighting of the special case of transmission-mode by setting $\theta = \pi$, and Eq. (7) then becomes

$$
W(f) = \frac{\sqrt{2}}{\lambda}\left\{1 - \left(1 - f^2\lambda^2\right)^{1/2}\right\}^{1/2}. \tag{8}
$$

For the special case of reflection-mode operation, we let $\theta = 0$ and Eq. (7) is reduced to

$$
W(f) = \frac{\sqrt{2}}{\lambda}\left\{1 + \left(1 - f^2\lambda^2\right)^{1/2}\right\}^{1/2}. \tag{9}
$$

Figs. (6) and (7) show the frequency weighting profiles of the transmission and reflection modes respectively.

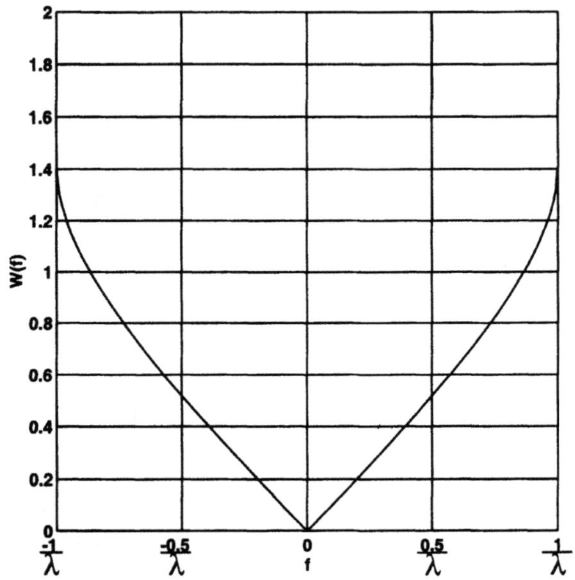

Figure 6. Weighting profile for the transmission case.

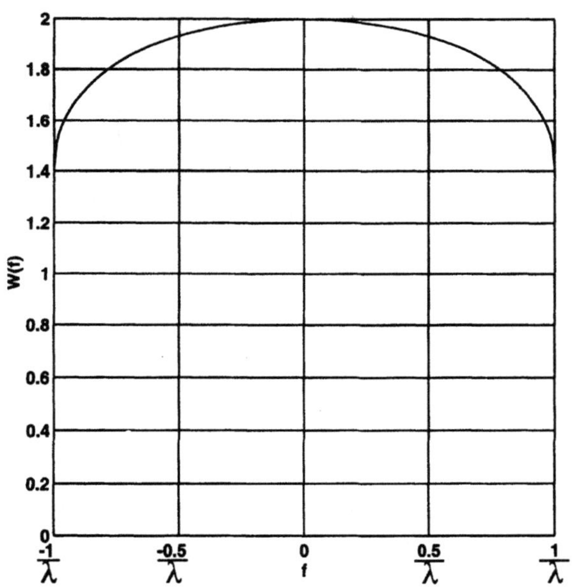

Figure 7. Weighting profile for the reflection case.

Fig. (8) illustrates the change of weightings as the angular offset between the illumination source and aperture position varies. This also describes the change from a highpass weighting in the transmission mode to the lowpass of the reflection mode. Because of the significant difference in frequency weighting characteristics as the angular offset varies, this addresses the importance of accurate use of the filter in decentralized filtered backward propagation.

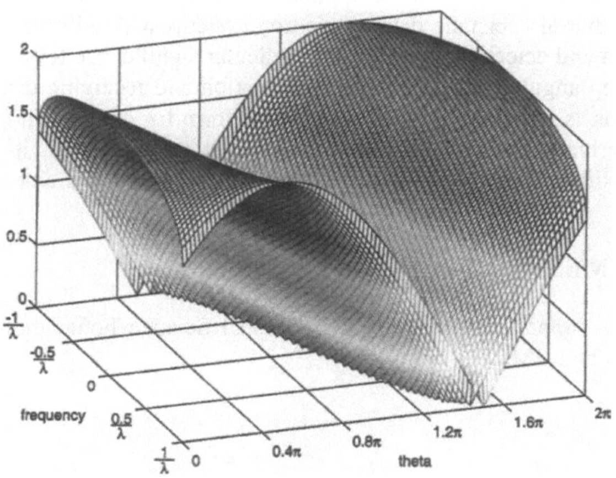

Figure 8. Weighting profile for the general case when the angle between the illumination source and aperture position varies.

It is worth noting that if we simplify Eq. (8) by taking the first-order approximation of the frequency-weighting characteristics, it reduces to Eq. (2). This explains the common adoption of the frequency weighting in classical X-ray case for diffraction tomography without serious errors, especially for systems with narrower effective spatial-frequency bandwidth coverage due to limited aperture size. On the other hand, however, we also need to point out that the approximation is valid only for the case of transmission-mode. The quality of the final reconstruction can be compromised by the inaccurate use of weighting in filtered backward propagation when the system is operating under other data-acquisition configurations.

Mathematically, the frequency weighting process is a separate step in the image reconstruction operation. Because it is performed sequentially with the backward propagation image formation process, these two steps can be combined by adopting $W(f)$ as the magnitude component, and the overall transfer function associated with distributed filtered backward propagation now becomes

$$H(f) = \begin{cases} W(f) \exp\left(j2\pi r\sqrt{\frac{1}{\lambda^2} - f^2}\right), & |f| < \frac{1}{\lambda} \\ 0, & otherwise \end{cases} \tag{10}$$

where the range direction is denoted as r so that the formulation can be coordinate independent.

With this modification, the frequency weighting in diffraction tomography can be conducted accurately when it is performed in the decentralized format. Yet, there is no increase in terms of computation complexity since the modified backward propagation procedure remains a single-step filtering process. This also points out that the magnitude variation of the filtering process of filtered backward propagation is governed by the angular offset of the data acquisition configuration while the phase components remain the same.

CONCLUSION

In this paper, we provided the exact formulation of the transfer function for distributed filtered backward propagation for acoustic tomography. This is achieved by identifying the frequency enhancement weighting and then building it into the backward propagation filter. This allows us to perform filtered backward propagation in a distributed format for various imaging configurations. The transfer function is in a simple and closed form with a high level of flexibility and stability. It can be applied to not only the traditional transmission or reflection modes, but also bistatic data acquisition systems with arbitrary angular offset between illumination and detection. It will be of particular significance for circular-scanned schemes with varying angular offset between illumination and receiving directions.

This formulation is presented in two-dimensional form for coherent imaging systems for simplicity in derivation and illustration. With slight modifications, it can be easily extended into three-dimensional cases with wideband illumination and detection.

ACKNOWLEDGEMENTS

This research is supported in part by the National Science Foundation under Grant No. CMS-9309775 and 3M.

REFERENCES

1. A. J. Devaney. A filtered backpropagation algorithm for diffraction tomography. *Ultrasonic Imaging*, 4:336–350, 1982.
2. H. J. Scudder. Introduction to computer aided tomography. *Proceedings of the IEEE*, 66:628–637, June 1978.

EXACT SOLUTION OF TWO-DIMENSIONAL MONOCHROMATIC INVERSE SCATTERING PROBLEM AND SECONDARY SOURCES SPACE SPECTRUM

V.A.Burov [1], S.A.Morozov [1], O.D.Rumiantseva [1], E.G.Sukhov [2], S.N.Vecherin [1] and A.Yu.Zhucovets [1]

[1] Moscow State University, Faculty of Physics, Dept. of Acoustics
Moscow, 119899, Russia
[2] Institute of Control Sciences of Russian Academy of Sciences

2-D Algorithm of Scatterer Reconstruction

The aim of the present report is to discuss results of programming realization of a method concerning solution of acoustical inverse scattering problem in the two-dimensional monochromatic case. The method allows to reconstruct a scatterer function $v(\vec{r}) = k_0^2 - k^2(\vec{r})$ on the base of measurement of the scattering amplitude $f(\vec{k},\vec{l})$; $k_0 = \omega/c_0$, $k(\vec{r}) = \omega/c(\vec{r})$ - wave numbers for the background homogeneous medium and for inhomogeneous one; c_0, $c(\vec{r})$ - phase velocities, correspondingly. Wave vectors $\vec{k}$ and $\vec{l}$ characterize the direction of falling of a plane wave $\exp(i\vec{k}\vec{r})$ and the scattering direction, respectively: $\vec{k}, \vec{l} \in R^2$, $\vec{k}^2 = \vec{l}^2 = k_0^2$. With practical point of view, measurement of $f(\vec{k},\vec{l})$ is not very convenient, and these values can be recalculated from the near field scattering data. The method was proposed for quantum mechanics scattering and was formulated in terms of functional analysis [1,2]. An application of the method to acoustical problems is discussed in [3]. An 'intrinsic' mechanism of its action, limits of its validity and physical sense of reconstructing operations are revealed.

The main moment of the method is a formal extension of the values $f(\vec{k},\vec{l})$ to 'non physical' domain of complex wave vectors with an imaginary part orthogonal to a real part:

$$\vec{k} = \vec{k}_R + i\vec{k}_I, \ \vec{k}_R \perp \vec{k}_I, \ \vec{k}^2 = k_0^2; \ \ \vec{k}_R, \vec{k}_I \in R^2.$$

The extension is realized by the generalized Green function introduced by L.D. Faddeev [4]. In the algorithm considered here the infinitesimal orthogonal addition $\vec{k}_I \to 0$ is used. In the 2-D problem such addition has two orientations – 'left' and 'right', which two types of the generalized Green functions G^+ and G^- correspond to. The 'classical' outgoing (G_0^+) and incoming (G_0^-) Green functions are analogies for $G^\pm$. Indeed, parallel ($\vec{k}_I \uparrow\uparrow \vec{k}_R$) and anti-parallel ($\vec{k}_I \uparrow\downarrow \vec{k}_R$) infinitesimal additions $\vec{k}_I \to 0$ correspond to G_0^+ and G_0^-, respectively. The generalized Lippmann-Schwinger equations with appropriate functions $G^\pm$ give generalized wave fields $\Psi^\pm(\vec{r},\vec{k})$.

The reconstruction algorithm has several stages. Firstly, extended 'left' and 'right' scattering amplitudes $h^\pm(\vec{k},\vec{l})$ are formed by the data $f(\vec{k},\vec{l})$. The extension is connected with a solution of linear integral (algebraic – in case of discrete computer model) equations. The data are included both in an operator kernel (a matrix) and in the right part (a free term). Secondly, a new operator $\rho(\vec{k},\vec{l})$ is build from $h^{\pm}(\vec{k},\vec{l})$ by a solution of equations similar to ones described above. It is the operator of connection between the solutions $\Psi^{+}(\vec{r},\vec{k})$ and $\Psi^{-}(\vec{r},\vec{k})$ of the initial Helmholtz equation for inhomogeneous medium. As a result, the data $f(\vec{k},\vec{l})$ are included in $\rho(\vec{k},\vec{l})$ by essentially nonlinear manner. Thirdly, the obtained operator $\rho(\vec{k},\vec{l})$ allows to find values of a difference field

$$K(\vec{r},\vec{k}) = \exp(-i\vec{k}\vec{r}) \cdot [\Psi^{+}(\vec{r},\vec{k}) - \Psi^{-}(\vec{r},\vec{k})].$$

In its turn, the function $K(\vec{r},\vec{k})$ is connected with the sought scatterer function by the linear relationship:

$$v(\vec{r}) = -\frac{k_0}{2\pi}\left(i\frac{\partial}{\partial x} + \frac{\partial}{\partial y}\right)\int_0^{2\pi} K(\vec{r},\varphi)e^{i\varphi}\,d\varphi, \quad \vec{r} = \{x,y\}, \quad \vec{k} = \{k_0,\varphi\}.$$

Such a 'linearization' of the problem is reached thanks to both symmetry properties of the generalized Green functions $G^{\pm}(\vec{r},\vec{k})$ with respect to the direction of the vector $\vec{k}$ and the integration over angle φ of the falling field. Such the peculiarities lead to a compensation of effects of the multiple scattering for all orders of rescatterings.

It is possible to point out a number of characteristic peculiarities of the algorithm.

a) Effects of the multiple scattering of waves are taken into consideration, and therefore the algorithm is nonlinear with respect to the scattering data. Nevertheless, it is linear with respect to the sought scatterer function.

b) The method is rigorous and valid for scatterers with an adsorption and without it. However the algorithm is applied only for a class of the scatterers with a restricted complication of their space spectrum and which are not very contrasting as related to the given background. The class is defined by conditions of providing the solution uniqueness and stability. The uniqueness takes place for the mean power scatterers, i.e. there are not areas of a strong focusing of wave fields within the scattering domain. The stability is achieved for the scatterers, which create such a small back scattering that it may be ignored. The back scattering is absent if a space spectrum of the secondary sources has not space frequencies with magnitude more than $2k_0$.

c) A frequency range of the probing waves, in which an acoustical scatterer belongs to the class described above is estimated. Within the framework of this range the reconstruction is the most reliable. The usable frequency range for medical applications extends from tens of kilohertz to several megahertz.

d) The scatterer function can be found at any fixed space point $\vec{r}$ independently of its values at other ones. Such a locality of the problem is convenient for practical applications, in particular, for medical problems of acoustical tomography.

e) The number of the computing operations needed for numerical scatterer reconstruction is noticeably less than in traditional iterative methods. Actually, let the data $f(\vec{k},\vec{l})$ be measured for N orientations of the vector $\vec{l}$ at each of N fixed orientations of the vector $\vec{k}$. Then the scatterer reconstruction at N^2 space points supposes the order of N^4 operations, whereas iterative algorithms for the mean power scatterer reconstruction need at least N^5 operations.

The algorithm has some disadvantages. Firstly, it does not allow a simple generalization to the 3-D problem. Secondly, an essential condition is that the background medium (without a scatterer) has not an absorption. Thirdly, a generalization of the algorithm to pulse regime leads to a very large increase of the number of operations.

Programming realization of the method allows investigating a number of model problems. The high computing efficiency of the algorithm conditioned by a relatively small volume of operations is confirmed. Simultaneously, a possibility appears to study a character and a degree of solution destruction as passing the limits of the guarantee validity of the algorithm. Such the investigations are first used for the considered method. Thus, the algorithm has been examined by the model data for scatterers with different power and width of space spectrum. The data $f(\vec{k},\vec{l})$ were simulated with the help of the Lippmann-Schwinger equation solved by the iterative method. To simplify calculations the central symmetry of the scatterer was proposed.

The scatterers of the first type have *the Gauss shape* $v(\vec{r})$. The e^{-1} half-width of the function $v(\vec{r})$ is $d \approx 0.82\lambda_0$. Such a value d provides that the space scatterer spectrum $\tilde{v}(\vec{\xi})$ is very well localized inside the $2k_0$ - radius circle. Consequently, the back scattering is absent in practice, and the reconstruction algorithm remains rigorous. A maximal contrast of the scatterer $\Delta c / c_0 = \max_{\{\vec{r}\}} c(\vec{r}) / c_0 - 1$ regulates a power of the scatterer. By such a manner, the described model of the scatterers allows to investigate how the reconstruction depends on the scatterer power. A conditionality of the algebraic equations systems appears to be deteriorated respectively smoothly as the power increases (the most critical step of the algorithm is finding $K(\vec{r},\vec{k})$ by $\rho(\vec{k},\vec{l})$). Simultaneously, a norm of the data - $\| f(\vec{k},\vec{l}) \|$ - is controlled, because the condition

$$\| f(\vec{k},\vec{l}) \| < 1/(3\pi) \tag{1}$$

is refereed to [1, 2] as the sufficient condition of the solution uniqueness.

Results of the reconstruction of the Gauss shape scatterer without absorption (i.e. $v(\vec{r})$ is real-valued) are shown on fig.1. The scatterer is strong: $\Delta c / c_0 = 0.6$ and an additional phase shift within the scattering domain is $\Delta\psi \approx 1.23\pi$. In spite of the fact that the norm $\| f(\vec{k},\vec{l}) \| \approx 16/(3\pi)$ is noticeably more than the guarantee norm (1), the reconstructed scatterer estimation $\hat{v}(\vec{r})$ is ideal (fig.1a). At the same time, the solution turns out to be sufficiently stable to accidental noise interference, though the inverse conditionality of matrices is just $0.2 \div 0.03$. So, the interference with magnitude $\approx 3\%$ of the maximal magnitude of the scattering data does not break down the solution (fig.1b). To demonstrate strong rescattering effects, the scatterer reconstruction in the Born (a single scattering) approximation has been carried out (fig.1c): the estimation $\hat{v}_{born}(\vec{r})$ is far from truth.

The scatterers of the second type have *the shape* $v(\vec{r})$ *of a hollow cylinder* (fig. 2a). External and internal diameters are $(25/8)\lambda_0$ and $(19/8)\lambda_0$. Such a model allows investigating an influence of the back scattering, which can be considered with respect to the algorithm as certain interference. For this purpose the scatterer of the mean power is examined: $\Delta c / c_0 = 0.1$, $\Delta\psi \approx 0.14\pi$; the back scattering amplitude is $\approx 5\%$ of the forward scattering one. Since the space spectrum $\tilde{v}(\vec{\xi})$ is restricted by a circle of the $(11/4)k_0$ - radius for simulation of the data $f(\vec{k},\vec{l})$, the scatterer $v^{cut}(\vec{r})$ (fig. 2a) is expected after restoring. Its reconstruction without noise interference is shown in fig. 2b: $\| f(\vec{k},\vec{l}) \| \approx 8/(3\pi)$, the inverse conditionality ≈ 0.4. Reconstruction errors have the only cause – the back scattering, an influence of which on the reconstruction quality is comparable to that of the accidental noise interference. The reconstruction in a single scattering approximation (fig. 2c) confirms the strong rescattering effects.

Estimations of Secondary Sources Space Spectrum

A further important problem connecting with the numerical solution of the inverse scattering problem is the general problem independently of the concrete method of the solution. The matter is the necessity of an adequate going from continuous functional values to samples for the model scattering data or the data obtained in a real experiment. This raises such a question as a connection between a volume of the scattering data samples and a possibility to provide the uniqueness and stability of the inverse problem solution [5]. It turns out that when the multiple scattering effects are taken into consideration the volume of the data samples needs to be essentially larger than the volume of independent parameters characterizing the sought scatterer. From this point of view, the scattering data are redundant. The solution uniqueness is provided when the redundancy coefficient RC overcomes a critical value: $RC \approx s^2 / s_{born}^2$. Here s_{born} and s - width of the secondary sources space spectrum in the Born (a single scattering) approximation and when multiple scattering effects are taken into account. If, simultaneously, this critical volume can be obtained by the sampling interval not less than the Raleigh's limit [5], then the solution is both unique and stable. It should be noted that the solution stability is provided (independently of the uniqueness question) if the secondary sources space spectrum is localized within the $2k_0$-radius circle for the two-dimensional monochromatic

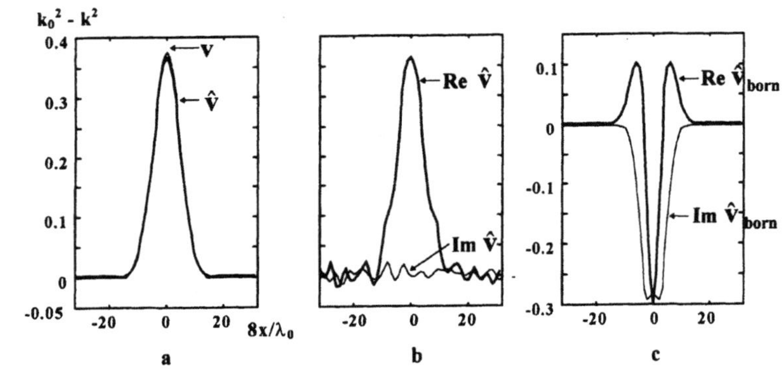

Figure 1. Central section of Gauss scatterer: a) true scatterer and one reconstructed without noise interference; b) the scatterer reconstructed by the data with 3% noise level; c) the scatterer reconstructed in Born approximation.

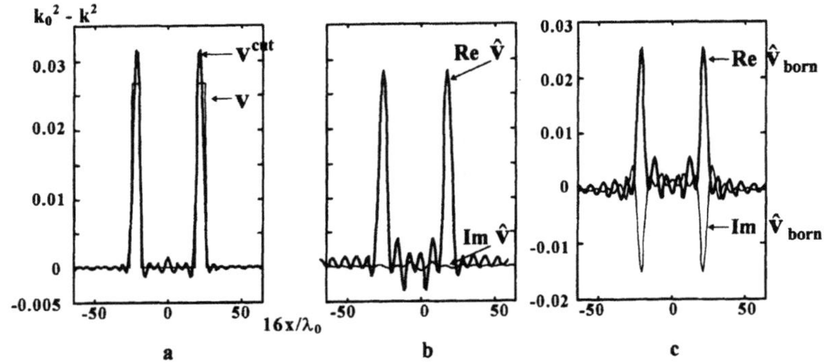

Figure 2. Central section of cylindrical scatterer creating back scattering: a) true scatterer and one with restricted space spectrum; b) the scatterer with restricted space spectrum, reconstruction without noise interference ; c) the scatterer reconstructed in Born approximation.

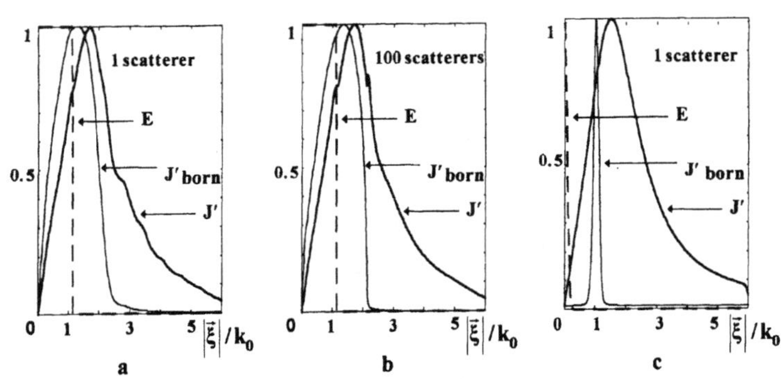

Figure 3. Normalized mean spectral density of secondary sources energy integrated over $|\bar{\xi}|$-radius ring: $J'_{born}(|\bar{\xi}|)$ - in Born approximation, $J'(|\bar{\xi}|)$ - when multiple scattering effects are taken into account. Normalized mean spectral density of scatterer energy $E(|\bar{\xi}|)$:

- Gauss shape of E, $d/\lambda_0 = 0.2, \Delta c/c_0 = -0.5$ for a single scatterer (a) and 100 ones (b);
- cylindrical shape of E, $d/\lambda_0 = 2, \Delta c/c_0 = -0.5$ for a single scatterer (c).

inverse scattering problem. For the three-dimensional one this restriction is weaker, i.e. the admissible area of the space spectrum localization is larger.

By this means, it is desirable for the practical solution of the inverse problem to preliminary estimate a degree of expansion of the secondary sources space spectrum caused by the multiple scattering effects. The estimation allows not only to determine the critical coefficient of the data redundancy and to predict the solution stability, but also to chose the optimal solution method. In this connection the statistical problem is set and solved to evaluate the maximal expansion of the secondary sources space spectrum.

Let there is an ensemble of scatterers { $v(\vec{r})$ } belonging to a certain class. The following is supposed:

- $<\tilde{v}(\vec{\xi})>_v = 0$ ($<\bullet>_v$ is the averaging over all $v(\vec{r})$);
- the mean spectral density $E(|\vec{\xi}|) \equiv <|\tilde{v}(\vec{\xi})|^2>_v$ is given a priori;
- space spectrum components $\tilde{v}(\vec{\xi})$ are correlated only inside the volume $\Delta \approx \delta^n$, where $\delta \approx 2\pi/L$, n is dimension of the problem, L is linear size of the scattering domain $\Re$:

$$<\tilde{v}(\vec{\xi}_1)\tilde{v}^*(\vec{\xi}_2)> \approx \begin{cases} E(|\vec{\xi}_1|), & if\ |\vec{\xi}_2 - \vec{\xi}_1| < \delta \\ 0, & if\ otherwise \end{cases}$$

Scatterers are insonified over all directions (index α) by plane monochromatic waves with random phases φ_α: $U_0(\vec{r},\alpha) = \exp[i\vec{k}_\alpha\vec{r} + i\varphi_\alpha]$, $|\vec{k}_\alpha| = k_0$; $<U_0(\vec{r},\alpha)>_\alpha = 0$ $\forall \vec{r} \in \Re$, $<\bullet>_\alpha$ is the averaging over directions. Space spectrum $\tilde{I}(\vec{\xi},\alpha)$ of the secondary sources $I(\vec{r},\alpha) \equiv v(\vec{r})U(\vec{r},\alpha)$ ($U(\vec{r},\alpha)$ - the full inner field) meets requirements of the Lippmann-Schwinger equation:

$$\tilde{I}(\vec{\xi},\alpha) - \frac{1}{(2\pi)^n}\int \tilde{v}(\vec{\xi}-\vec{\xi}')\tilde{G}_0(|\vec{\xi}'|,k_0)\tilde{I}(\vec{\xi}',\alpha)\,d\vec{\xi}' = \tilde{I}_{born}(\vec{\xi},\alpha), \qquad (2)$$

$\tilde{G}_0$ - the space spectrum of the classical Green function $G_0 \equiv G_0^+$ for background;

$$\tilde{I}_{born}(\vec{\xi},\alpha) = \frac{1}{(2\pi)^n}\int \tilde{v}(\vec{\xi}-\vec{\xi}')\tilde{U}_0(\vec{\xi}',\alpha)\,d\vec{\xi}'$$

- the space spectrum of the Born secondary sources. To estimate the mean spectral density of energy of the secondary sources $J(|\vec{\xi}|) \equiv <|\tilde{I}(\vec{\xi},\alpha)|^2>_{v,\alpha}$ is required, if in the Born approximation $J_{born}(|\vec{\xi}|) \equiv <|\tilde{I}_{born}(\vec{\xi},\alpha)|^2>_{v,\alpha}$. The averaging $<\bullet>_{v,\alpha}$ is double: over all scatterers and over all directions. It corresponds to statistically maximal widening of the secondary sources space spectrum.

The values $\tilde{v}$, $\tilde{I}_{born}$, $\tilde{I}$ at fixing their arguments are considered as random values with distributions near to normal and the zero-mean (in view of the propositions mentioned above). Then a multiplication the left and the right parts of the eq. (2) by the conjugated values and the averaging over v and α lead to the equation for the sought $J(|\vec{\xi}|)$:

$$J(|\vec{\xi}|) + \frac{1}{(2\pi)^{2n}}\int E_0(|\vec{\xi}-\vec{\xi}'|)\cdot|\tilde{G}_0(|\vec{\xi}'|,k_0)|^2\cdot J(|\vec{\xi}'|)\,d\vec{\xi}' = J_{born}(|\vec{\xi}|), \qquad (3)$$

$$J_{born}(|\vec{\xi}|) = \frac{1}{(2\pi)^{2n}}\int E_0(|\vec{\xi}-\vec{\xi}'|)\cdot<|\tilde{U}_0(\vec{\xi}',\alpha)|^2>_\alpha d\vec{\xi}'.$$

The physical sense of the value $E_0(|\vec{\xi}|) \equiv E(|\vec{\xi}|)\cdot\Delta$ is the mean spectral scatterer 'energy' containing in volume Δ.

The equation (3) has been numerically solved for *two models of the scatterers ensemble*. In the model I each scatterer realization $v(\vec{r})$ is represented by a single inhomogeneity with linear size l and space spectrum width $\approx 2\pi/l$. Location of the inhomogeneities fluctuates within l- size area. In the randomized model II each realization is represented by M inhomogeneities analogous to inhomogeneity in the model I. For solution of the eq. (3) distinctions of the model I and II appear as: $L^{(II)} \approx \sqrt[n]{M} \cdot L^{(I)}$, $J_{born}^{(II)} \approx M \cdot J_{born}^{(I)}$ ($\sqrt[n]{M}$ is the mean number of the inhomogeneities along a fixed direction). It is interesting to note that the value $E_0 = E \cdot \Delta$ is independent of M, because $E \sim M$, but $\Delta \sim 1/M$.

A modeling of the eq. (3) has been carried out at 2-D space for the models I and II with different parameters. Functions $J'_{born}(|\vec{\xi}|) = 2\pi |\vec{\xi}| \cdot \delta \cdot J_{born}(|\vec{\xi}|)$ and $J'(|\vec{\xi}|) = 2\pi |\vec{\xi}| \cdot \delta \cdot J(|\vec{\xi}|)$ are the secondary sources spectral energy containing in a ring of the $|\vec{\xi}|$- radius and of thickness δ. Figures 3a,b correspond to fine-scale inhomogeneities ($l \le \lambda_0$) in the model I (fig. 3a) and in the model II for M=100 (fig. 3b). Shape of the function $E(|\vec{\xi}|)$ is proposed to be a cylinder with the $(\sqrt{2}/d)$- radius, $d/\lambda_0 = 0.2$, i.e. $l \approx$ several d. Amplitude of the function $E(|\vec{\xi}|)$ is calculated in assumption of an effective contrast $\Delta c/c_0 = -0.5$. It is seen that the maximum of J' is shifted to a greater $|\vec{\xi}|$ as compared to the Born approximation. The wide space spectrum J' is caused by the wide Born space spectrum $J'_{born}(|\vec{\xi}|)$ and by a small amount of the rescattering acts. It should be pointed out that for the models I and II the sought widths of the normalized functions J' are equal, in practice.

Figure 3c corresponds to large-scale inhomogeneity ($l \approx$ several λ_0) in the model I. Proposed Gauss shape of $E(|\vec{\xi}|)$ is defined by the e^{-1} half-width equal to $(\sqrt{2}/d)$, $d/\lambda_0 = 2$, and amplitude of $E(|\vec{\xi}|)$- by the effective contrast $\Delta c/c_0 = -0.5$. Here the wide space spectrum J' is mainly caused by the multiple acts of rescatterings under the narrow Born space spectrum. The normalized function J' estimated in the model II ($M = 25$) for the same parameters does not qualitatively differ from the figure 3c. This fact that in the model II the width and shape of J' have a little dependence on the number M, if the linear size $L > (2 \div 3)\lambda_0$, has a physical explication. Indeed, the wave fields radiated by the secondary sources components with $|\vec{\xi}| > k_0$ have a strong attenuation and, therefore, an influence of these fields is local. At the same time, the wave fields radiated by the components $|\vec{\xi}| \approx k_0$ are similar to the initial plane waves with arbitrary phases. As a result, when M increases, the total energy of the secondary sources increases, but the shape of their mean spectral energy is practically invariable.

Summarized results of the researches lead to the main conclusion of the work: the method proposed and studied in [1, 2, 3] is very effective, but the area of its efficiency has the restrictions. It is important to note that the restrictions are caused not by peculiarities of the given algorithm, but by the general peculiarities of the solution of the 2-D monochromatic inverse scattering problem. The statistical estimations of the width of the secondary sources space spectrum help to control the scattering data volume required for the scatterer reconstruction. They are also instructive to solve the questions of the solution uniqueness and stability.

REFERENCES

1. P.G.Grinevich, S.V.Manakov, The inverse problem of the scattering theory for two-dimensional Schrödinger operator, $\bar{\partial}$-method and nonlinear equations, *Func. Analysis and Its Applic.*, 20:14 (1986).
2. R.G.Novikov, The inverse scattering problem on a fixed energy level for the two-dimensional Schrödinger operator, *J.Funkt.Anal.*, 103:409 (1992).
3. V.A.Burov, O.D.Rumiantseva, The functional-analytical methods for the scalar inverse scattering problem, *in*: Analytical Methods for Optical Tomography, G.G.Levin, ed., SPIE, Washington (1992).
4. L.D.Faddeev, Inverse problem of quantum scattering theory.II, *J. of Soviet Math.*, 5:334 (1976).
5. V.A.Burov, and O.D.Rumiantseva, Influence of the scattering data redundancy on uniqueness and stability in reconstruction of strong and complicated scatterers, *in*: Acoustical Imaging-22, P.Tortoli and L.Masotti, ed., Plenum Press, New York (1996).

HAUSDORFF MOMENTS METHOD OF ACOUSTICAL IMAGING

Keith S. Peat and Yaroslav V. Kurylev

Department of Mathematical Sciences
Loughborough University
Loughborough, Leicestershire, LE11 3TU
UK

INTRODUCTION

This paper is concerned with the nonlinear inverse acoustic problem of the construction of the density / speed of sound distribution within a domain, given spectral data on the boundary of the domain. The work has practical application in several fields, including ultrasonic imaging for medical diagnostics, geophysical exploration and non-destructive testing. For example, in medical diagnostics, different tissue types are characterised by differences in their density, speed of sound and attenuation of sound. Thus in principle, if one can characterise the distribution of the varying speed of sound through a body, then one can determine the nature and location of the various tissue types. In particular one could monitor for the presence and growth of malignant tissue types.

In mathematical terms, a model of this problem using non-stationary (time-dependent) data leads to an inverse boundary problem for the acoustic wave equation

$$u_{tt} - c^2 \Delta u = 0, \tag{1}$$

whereas stationary (spectral) data gives rise to an inverse problem for the acoustic operator

$$Au = -c^2 \Delta u \quad \text{or} \quad A_\rho u = -\rho^{-1} \Delta u, \tag{2}$$

with say Dirichlet boundary conditions. For a complete set of exact data one model implies the other, as they are related through the Fourier transform. In the simplified models of equations (1) and (2), attenuation has been ignored. In practice this restricts the application to acoustic and low ultrasonic frequencies. In the context of medical diagnostics the upper frequency limit would be of the order of 1-2 MHz.

In this paper, attention will focus on the spectral problem and consideration will be given to the practical issues of incomplete and error-prone data. Thus data of the form $\left(\left\{\tilde{\lambda}_k\right\}_{k=1}^{N}, \left\{\partial_n \tilde{\phi}_k\right\}_{k=1}^{N}\right)$ is used, where $\tilde{\lambda}_k, \tilde{\phi}_k$ are approximations to the eigenvalues and orthonormalised eigenvectors, respectively, and only the lowest N eigenpairs are used. Although this paper is restricted to consideration of the spectral problem of equation (2), the method can still be applied to the time-dependent problem of equation (1) where the equivalent restriction to the finite set of harmonic data is measurements of boundary pressure, say, over a finite time interval. In either case, even if the data were exact, incompleteness of the data implies that one cannot reconstruct the distribution of the soundspeed exactly. The presence of error in the data only serves to make the situation worse. The paper presents an algorithm for the approximate reconstruction of the soundspeed distribution from incomplete and erratic boundary spectral data, based upon a Hausdorff (polynomial) moments method. The paper is restricted to two-dimensional problems, although the method extends to higher dimensions. Complete details of the theoretical analysis for multi-dimensional problems, together with complete error estimates, are given in Kurylev and Starkov[1]. Further theoretical issues and preliminary numerical results for two-dimensional problems are given in Kurylev and Peat[2]. Alternative methods for this problem are the Boundary Control (BC) method[3] and the DBAR method[4], which however do not analyse the rate of convergence for the cases of incomplete or error-prone data.

The theoretical analysis considers the relative errors in reconstruction which arise from incomplete data and errors in the data. Errors in the reconstruction procedure reduce as the amount of data is increased and as errors in the data are reduced and, in the limit of complete and error-free data, precise reconstruction of the soundspeed distribution is achieved. The dependence of the error in reconstruction upon both the amount of data and the errors in it is found to be logarithmic. The error of reconstruction is also found to be non-monotonic, in that for a given level of error in the data, there is a corresponding optimal choice of the amount of data to minimise the reconstruction error.

This paper presents numerical results from two-dimensional 'model' problems. Numerical errors arise due to the fact that one cannot reconstruct the first moments exactly, and also from inversion of the Hausdorff (polynomial) moments. The latter error is dominant. The paper presents different ways in which this error can be reduced. In particular, a change from the use of Fourier transforms[2] to the use of Legendre polynomials is shown to reduce the errors of image construction. Overall errors of reconstruction of only 2% have been achieved with the use of only 50 moments, which in turn were constructed from only 180 eigenfunctions.

THEORY

Consider the inverse problem for the acoustic operator, equation (2), and denote by $\{\lambda_k\ ,\ \partial_n\phi_k|_{\partial\Omega}\}$ the eigenvalues and corresponding values on the boundary of the outward normal derivatives of the orthonormalised eigenfunctions. Each eigenpair $\{\lambda_k\ ,\ \partial_n\phi_k|_{\partial\Omega}\}$ or, more precisely, their combination $\psi_k = \{\partial_n\phi_k|_{\partial\Omega}\}/\lambda_k$ determines the Fourier coefficient α_k^h in the decomposition of a harmonic function h where

$$h = \sum_{k=1}^{\infty} \alpha_k^h \phi_k\ , \qquad \alpha_k^h = \int_\Omega \rho h \phi_k \mathrm{d}\Omega\ , \tag{3}$$

since

$$\lambda_k \phi_k = -\rho^{-1} \Delta\phi_k \tag{4}$$

from equation (2) and hence

$$\alpha_k^h = \frac{-1}{\lambda_k} \int_\Omega h \Delta\phi_k \mathrm{d}\Omega = -\int_{\partial\Omega} h \psi_k \mathrm{d}S. \tag{5}$$

The last step uses Green's theorem and it is recalled that h is a harmonic function.

Hence, since the IBSD give only the finite set of Fourier coefficients α_k^h, $k = 1,\ldots,N$, the Parseval equality enables only the approximate evaluation of the inner product of harmonic functions, as

$$\langle h, g \rangle_\rho \approx \sum_{k=1}^{N} \alpha_k^h \overline{\alpha_k^g}. \tag{6}$$

However, $\langle h, g \rangle_\rho = \int_\Omega \rho hg \,\mathrm{d}\Omega$ and thus the IBSD determine approximately area integrals of the form $\int_\Omega \rho hg \,\mathrm{d}\Omega$ for arbitrary harmonic functions h, g. Equations (5) and (6) enable one to obtain an approximation of the Hausdorff moments

$$\rho_{p,q} = \int_\Omega \rho x^p y^q \mathrm{d}x\mathrm{d}y \tag{7}$$

in terms of the IBSD. This follows since one can write $x^p y^q$ in the form

$$\left(\frac{z+\bar{z}}{2}\right)^p \left(\frac{z-\bar{z}}{2\mathrm{i}}\right)^q = (-\mathrm{i})^q 2^{-(p+q)} \sum_{a=0}^{p+q} C_{p,q}^a z^a \bar{z}^{p+q-a} \tag{8}$$

where $z^k, \bar{z}^k$ are harmonic for any integer k. One is now left with the classical problem of determining the approximate distribution of ρ given an estimate to a finite number of moments.

2.1 Fourier Transform Method

The first attempt[2] made use of the Fourier transform

$$\hat{\rho}(\xi,\eta) = \frac{1}{2\pi} \int\int_{-\infty}^{+\infty} \rho(x,y) \mathrm{e}^{i(\xi x + \eta y)} \mathrm{d}x\mathrm{d}y \tag{9}$$

to give

$$\hat{\rho}_{p,q} = \partial_\xi^p \partial_\eta^q \hat{\rho}(\xi,\eta)\Big|_{\xi,\eta=0} = \frac{\mathrm{i}^{p+q}}{2\pi} \iint_\Omega \rho(x,y) x^p y^q \mathrm{d}x\mathrm{d}y = \frac{\mathrm{i}^{p+q}}{2\pi} \rho_{p,q}\,, \tag{10}$$

where the value of ρ has been taken as zero outside of Ω. An approximation to the Fourier transform $\hat{\rho}(\xi,\eta)$ then follows from a Taylor series expansion about the origin, which has to be truncated since an approximation to only a finite number of the coefficients $\hat{\rho}_{p,q}$ is given by equation (10). The required approximation to $\rho(x,y)$ then follows from the inverse of $\hat{\rho}(\xi,\eta)$, but note that the domain of the integral of the transform is infinite and has to be truncated at some finite radius for evaluation purposes.

2.2 Legendre Polynomial Method

An improvement in the quality of the earlier results[2] is sought in this paper by the use of a Legendre polynomial technique for solution of the Hausdorff moments problem, as proposed by Talenti[5] with reference to one-dimensional problems. Let us assume that

$$\rho_{p,q} = \int_{-1}^{+1}\int \rho x^p y^q \mathrm{d}x\mathrm{d}y \tag{11}$$

which may involve a re-scaling of coordinates and the identification of $\rho(x,y)=0$ outside of the original domain. Now $\rho(x,y)$ is written in the form

$$\rho(x,y) = \sum_{i=0}^{M}\sum_{j=0}^{M} \lambda_{i,j} P_i(x) P_j(y) \tag{12}$$

where $P_i(x)$, $P_j(y)$ are Legendre polynomials and $\lambda_{i,j}$ are constants. These constants are Fourier-Legendre coefficients, which are related to the moments by

$$\lambda_{i,j} = \sum_{l=0}^{i}\sum_{m=0}^{j} C_{i,l} C_{j,m} \rho_{l,m} \tag{13}$$

where the coefficients $C_{i,j}$ are those of the Legendre polynomials.

3. NUMERICAL RESULTS

Numerical results have been obtained for a model problem with known boundary spectral data and with a simple boundary, namely a circle. A conformal transformation

$$z = (cw+1)/(w+c), \text{ c constant}$$

was used such that $\rho = \left|\frac{\mathrm{d}z}{\mathrm{d}w}\right|^2$. In this case there is no separation of variables. The transformation does not alter the eigenvalues but the eigenfunctions are changed by a factor of $\left|\frac{\mathrm{d}w}{\mathrm{d}z}\right|$.

The Fourier coefficients α_k^h in equation (5) were evaluated by numerical integration around the boundary, using 3-noded quadratic elements with a 3-point Gauss quadrature scheme for the curved boundary of the circular domain.

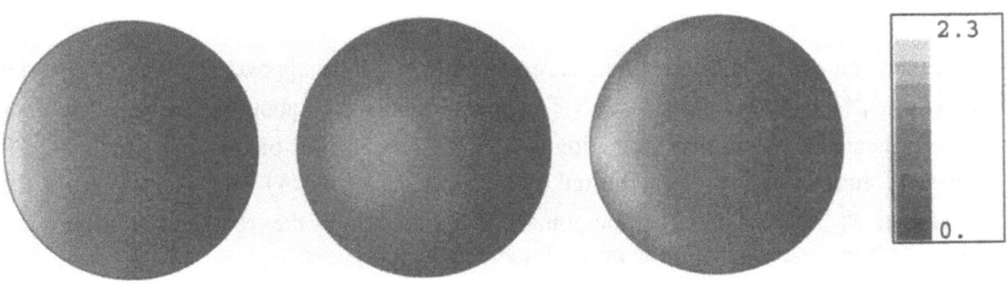

Figure 1. Reconstructions by the standard techniques. Left to right: Original; Fourier; Legendre.

In Figure 1, reconstructions of an original distribution for c=5 are shown using both the Fourier and Legendre techniques. In both cases only 180 eigenfunctions were used, with 50 moments for the Fourier method and 49 moments for the Legendre method. The error between the original and the reconstructed values of ρ averaged over the whole domain were 30% and 8.7% for the Fourier and Legendre methods respectively.

With both techniques, the domain for calculation extends beyond the physical domain and therefore incorporates the large discontinuity from the value of ρ on the boundary of the physical domain to the assumed value of zero outside. It is this discontinuity which is responsible for most of the error, especially with the Fourier technique. In real applications, it would be normal for the distribution of ρ to be known around the boundary. When this is the case, it is possible to construct a function ρ_B, say, with the known boundary distribution, evaluate its moments directly, and then solve for the difference between the original distribution and the constructed function. In this way discontinuities at the boundary are eliminated. This approach will be referred to as using a 'boundary function'. Figure 2 illustrates reconstructions of the same test case as for Figure 1 by both Fourier and Legendre techniques, using a boundary function ρ_B given by

$$\rho_B - 1 = \begin{cases} \left(\rho_{r=1,\phi} - 1\right)(2r-1), & r \geq 0.5 \\ 0, & r \leq 0.5 \end{cases}$$

The number of eigenfunctions and moments used is the same as for Figure 1. The average errors of reconstruction reduce to 4.7% and 2.5% respectively for the Fourier and Legendre techniques.

Figure 2. Reconstructions using a boundary function with a discontinuous slope. Left to right: Original; Fourier; Legendre.

The boundary function ρ_B used above incorporates a discontinuity in slope which can still be discerned in the reconstructions. Figure 3 shows similar reconstructions for a smooth boundary function of the form

$$\rho_B - 1 = \left(\rho_{r=1,\phi} - 1\right)r^2.$$

The errors of reconstruction now reduce to 2.6% and 2.1% for the Fourier and Legendre techniques, respectively, with the same number of eigenfunctions and moments as before. Thus one may conclude that the error in both techniques increases in the presence of discontinuities in the value or slope of ρ and ρ_B, but the Legendre technique is less susceptible to such errors than the Fourier technique. In addition, one has the problem with the Fourier technique of choosing a finite radius of the domain of the integral of the inverse transform. The results shown in these figures for the Fourier technique are those corresponding to the optimum value of this radius for this problem and are therefore optimum values. The optimum choice of radius is very much problem dependent but not unduly critical. However, there is no corresponding uncertainty or arbitrary choice with the Legendre technique, which is therefore to be preferred on both this ground and that of its ability to cope better with non-smooth distributions of ρ.

Figure 3. Reconstructions using a boundary function with a smooth slope. Left to right: Original; Fourier; Legendre.

The influence of measurement errors on the reconstructions could change this conclusion in some circumstances, however, as is shown next. The exact analytical IBSD which were used in Figures 1 to 3 were seeded with a random error of ±5% maximum, both in the eigenvalues and the normal derivatives of the eigenfunctions at the boundary. The reconstructions of Figure 3, that is using a smooth boundary function, were repeated for this case of error-prone boundary data and the results are shown in Figure 4.

Figure 4. Reconstructions when the boundary data has 5% error, using a smoot boundary. Left to right: Original; Fourier; Legendre.

It is clearly seen that the Legendre technique is much more susceptible to errors in the boundary data than is the Fourier technique. The average errors of reconstruction were 24.2% and 3.2% respectively. Thus smooth original distributions with smooth boundary functions are best evaluated by the Fourier technique when error-prone data is used, but the presence of discontinuities in the original distributions will make the Legendre technique more favourable at some stage, dependent upon the relative magnitudes of the discontinuities and the errors in the data. It should be noted that no regularization technique has been applied in the error-prone cases.

A second type of test case was considered in which a small region of the domain was given a constant value of 10% higher than the uniform distribution in the rest of the domain. This represents a 5% change in the speed of sound, typical of the difference found between healthy and cancerous tissue. Thus the test case is an idealised attempt at locating a small tumour, say. The analytical values of moments were calculated directly for these cases, rather than using IBSD which are difficult to evaluate in this case. The boundary function was taken to be a constant, namely the uniform background value of ρ throughout the domain. The Fourier technique gave meaningful reconstructions only when the origin of coordinates was centered on the tumour. Even if one could locate the tumour accurately enough to effect the required coordinate transformation, the technique still breaks down in the presence of two or more tumours. In contrast, the Legendre technique is effective in such cases, see Figure 5.

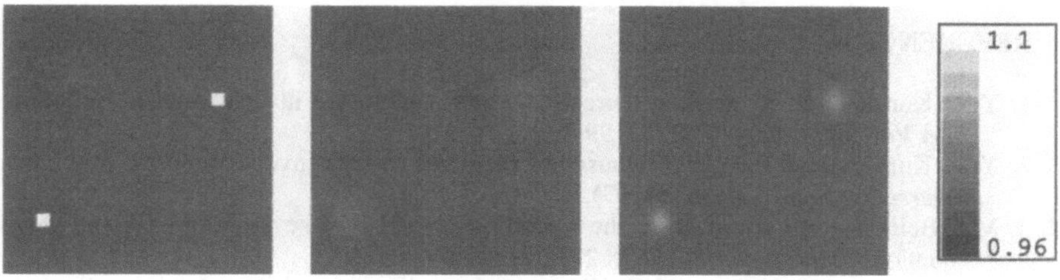

Figure 5. Reconstructions by the Legendre technique. Left to right: Original; 121 moments; 441 moments.

It is seen that the tumours are discernible with the use of only a small number of moments, but become much more distinct as the number of moments increases, as is to be expected. This effect is even more evident when the tumours are close together, see Figure 6.

Figure 6. Reconstructions by the Legendre technique. Left to right: Original; 121 moments; 441 moments.

If one relates the idealised case above to mammography, then the tumours are effectively about 5mm in diameter and there are of the order of 10^4 eigenfunctions with frequencies less than 1MHz, i.e. not unduly affected by attenuation, which is more than enough to accurately calculate several hundred moments. The main problem is that exact values of moments were used in the calculations for Figures 5 and 6 and the Legendre technique is very susceptible to error in the data.

4. CONCLUSIONS

The moments method has been shown to give accurate reconstructions in different types of model problems when exact but incomplete boundary spectral data is used. The majority of the error in reconstruction does not occur in the nonlinear step for the creation of the moments, but in the second step of the classical Hausdorff moments problem, which is linear and ill-posed. A Fourier technique for this problem is particularly prone to error when the function to be reconstructed is non-smooth. Knowledge of the value of the function on the boundary can be used to reduce these errors substantially. A Legendre technique is less susceptible to error due to non-smoothness of the function, such that it can be used to reconstruct a uniform field with isolated lumps in it, for which the Fourier technique is totally inappropriate. However the Legendre technique is more susceptible to error in the data than is the Fourier technique. The way forward would appear to be the development of some form of regularization procedure for error-prone data such that the Legendre technique remains accurate in such cases.

REFERENCES

1. Y.V. Kurylev and A. Starkov, Directional moments in the inverse acoustic problem, *IMA Vol. Math. Appl.* 90:295 (1997).
2. Y.V. Kurylev and K.S. Peat, Hausdorff moments in 2-D inverse acoustic problems, *Inverse Problems* 13:1363 (1997).
3. M.I. Belishev, An approach to the multidimensional inverse problems for the wave equation, *Dokl.Akad.Naul SSSR* 297:524 (1987).
4. A. Nachman, J. Sylvester and G. Uhlmann, An n-dimensional Borg-Levinson theorem *Comm. Math. Phys.* 115:595 (1988).
5. G. Talenti, Recovering a function from a finite number of moments *Inverse Problems* 3:501 (1987).

A GENERALIZED INVERSION OF THE HELMHOLTZ EQUATION AND ITS APPLICATION TO ACOUSTICAL IMAGING

Joie P. Jones[1] and Sidney Leeman[2]

[1] Department of Radiological Sciences
University of California Irvine
Irvine, CA 92697-5000 USA
[2] Department of Medical Engineering and Physics
King's College School of Medicine and Dentistry
London SE22-8PT, UK

INTRODUCTION

Diffraction tomography is usually formulated not as an exact inversion of the full wave equation but rather as an approximate inversion of the inhomogeneous Helmholtz equation. Under two very different, but well known, set of conditions, the inversion process leads to the Born approximation in one case and the Rytov approximation in the other. Here we develop an iteration method to obtain higher order Born approximations which we believe to be computationally efficient as well as effective. In addition, we show that the Born and Rytov inversions of the Helmholtz equation are simply extremes of a more general class of approximations which can be fashioned to fit the problem at hand and altered iteratively to match changing conditions of propagation.

PHYSICAL MODEL

For many situations, the propagation of radiation can be described by the Helmholtz wave equation[1,2],

$$\nabla^2\psi(r) + k^2\psi(r) = \rho(r)\psi(r) \tag{1}$$

Acoustical Imaging, Volume 24.
Edited by Hua Lee, Kluwer Academic/Plenum Publishers, 2000.

where $\psi(r)$ denotes the wave at spatial location r, $\rho(r)$ is the scattering interaction density, and $k = \omega/c$ with ω the circular frequency of the field and c the mean value for the wave velocity over the scattering region. The Helmholtz equation is certainly applicable to longitudinal sound waves or scalar electromagnetic waves propagating in velocity-inhomogeneous media. For simplicity, the medium will be regarded as lossless (k real).

The Helmholtz equation (Eq. 1) can also be written as the following integral equation

$$\psi(r) = \psi_0(r) + \int_V G(r, r')\rho(r')\psi(r')dr' \tag{2}$$

where ψ_0 denotes the incident field (i.e., the wave that would exist in the absence of the velocity fluctuations), V is the bounded volume in which the scattering occurs, and G is the Green's function appropriate for this scattering configuration. Here G usually takes the form of

$$G(r, r') = -\exp(ik|r - r'|) / 4\pi(|r - r'|) . \tag{3}$$

In an entirely symbolic way, this integral equation can be written as

$$\psi = \psi_0 + \psi_S \tag{4a}$$

where ψ_S is the scattered field or as

$$\psi = \psi_0 + G\rho\psi \tag{4b}$$

where G is now an integral operator representing the Green's function.

The underlying structure of the basic integral equation is now apparent and it may be solved by iteration to yield the so-called Born-Neumann expansion for the field,

$$\psi = \psi_0 + G\rho\psi_0 + \{G\rho\}^2\psi_0 + \ldots \tag{5}$$

The above equation represents a valid solution, provided that the expansion converges. Terms in this expansion involving $\{G\rho\}^n$ with $n>1$ may be interpreted as describing multiple scattering events of order n.

The full series solution for the field ψ is extremely cumbersome to evaluate in any realistic case. In practice, therefore, the series is terminated in order to give an approximate solution. In particular, the first Born approximation is given by

$$\psi \approx \psi_0 + G\rho\psi_0 . \tag{6}$$

This approximation is of some interest since the inverse problem can be solved exactly when it is valid[3]. Clearly, the validity of the first Born approximation implies that multiple scattering effects are negligible.

In general, ψ_0 may be considered known or accurately measured and the form of the integral equation suggests that the field may everywhere be considered to be the sum of the incident field and a scattered field, ψ_S, where

$$\psi_S(r) = \int_V G(r, r')\rho(r')\psi(r')dr' . \tag{7}$$

In practice, it is a set of measurements of the wave outside the scattering region that constitutes a physically meaningful data set. The first Born approximation is valid when the scattering is not too strong and the scattered field may be expressed as

$$\psi_S(r) = \int_V G(r, r')\rho(r')\psi_0(r')dr' . \tag{8}$$

If the incident wave is a plane wave directed along the unit vector, n, then $\psi_0(r) = \exp(ikn\cdot r)$ and Equation 8 reduces to

$$\psi_S(r) = (1/4\pi)\int \{\exp(ik|r - r'|) / (|r - r'|)\}\rho(r') \exp(ikn\cdot r')dr' . \tag{9}$$

Measurements of the scattered field may be considered to be performed only at very large distances from V (far-field measurements). In the far field, the scattered wave behaves as an outgoing spherical wave, modulated in amplitude and phase by the scattering amplitude, Ω, which has the form[3]

$$\Omega = (1/4\pi)\int\rho(r) \exp(-ik(m - n)\cdot r)dr \tag{10}$$

where m is a unit vector indicating the direction in which the measurement of the far field scattering is taken. It is convenient to introduce the notation $k_S = km$, $k_i = kn$, and $k = k_S - k_i$. Equation 10 can then be written as

$$\Omega(k) = (1/4\pi)\int\rho(r) \exp(-ik\cdot r)dr . \tag{11}$$

Thus, provided multiple scattering can be ignored, Ω and ρ are Fourier transform pairs and ρ can be directly inverted:

$$\rho(r) = (1/2\pi^2)\int \Omega(k)\exp(ik\cdot r)dk \tag{12}$$

Hence, $\rho(r)$ can be computed if the scattering amplitude can be measured for all values of k. Indeed, the various ways which have been devised to access all of the k-space are, in reality, different implementations of the diffraction tomography concept, and not, as often claimed, new types of scatter imaging.

Since Ω and ρ are Fourier transform pairs, the measured data (in this case Ω) can provide information about the Fourier transform of the desired density. It is in this sense that the methods of inverse scatter imaging may be linked to the central slice theorem of conventional reconstruction from projections.

COMPLEX Δ - DOMAIN

For simplicity, let us assume that the interaction density has rotational symmetry. More critically, let us further assume that the interaction density can be written as the Laplace transform of some general function or distribution, $P(\alpha)$. Then,

$$\rho(r) = \int_0^\infty P(\alpha)e^{-\alpha r}d\alpha \ . \tag{13}$$

If it is further assumed that $P(\alpha)$ is zero for all $\alpha \leq \mu$, then

$$r\rho(r) = \int_\mu^\infty \sigma(\alpha)\, e^{-\alpha r}d\alpha \tag{14}$$

where $\sigma = dP/\,d\alpha$. To ensure convergence, it will be necessary to require that $|\sigma(\alpha)\, d\alpha|$ be finite. The inversion technique to be developed will attempt to construct $\sigma(\alpha)$ from the measured data; i.e., it will aim to uncover the Laplace domain representation of the interaction from the value of the scattering amplitude.

The data acquisition configuration consists of a plane wave incident along the direction n, with the scattering into all directions being detected, the measured entity being the scattering amplitude at all scattering angles, θ, but at a fixed k (i.e., frequency) only. The assumed rotational symmetry of the scattering interaction imposes the same symmetry on the scattering amplitude, which consequently depends on one angular variable only.

It is then a relatively simple exercise in Fourier transform evaluation to show that the first Born approximation may be written as

$$f_1 = \int \sigma(\alpha)(\alpha^2 - \Delta)^{-1}\, d\alpha \ . \tag{15}$$

The behavior of f_1 in the complex Δ domain is easy to establish from this integral representation. It is analytic everywhere in the finite complex Δ - plane, except for Δ real and $\geq \mu^2$. The existence of such a branch cut along the real axis of the Δ - plane is associated with the circumstance that the denominator in the integrand of the integral in Equation 15 vanishes for these Δ values, leading to a singular integral.

The second Born term, f_2, has a more complicated structure, but it is still possible to exploit the idea that its analyticity in the complex Δ domain breaks down when the integrand in its defining integral shows a non-removable singularity. Thus, the powerful method devised by Landau[4] may be invoked to show that f_2 is analytic everywhere in the finite Δ - plane, except for a branch cut along the real axis, for $\Delta \geq (2\mu)^2$. Indeed, it may be shown quite generally[5] that the Nth Born term has a similar behavior, but with its branch cut extending over $\Delta \geq (N\mu)^2$. It turns out that the singularities of the scattering amplitude in the finite Δ - plane are the same as those of the Born expansion, a result that may be shown[5] to be *independent* of the convergence of the Born expansion. The discontinuity across the branch cut may then be defined as

$$D(\Delta) \equiv \{\, f(k;\, \Delta + i\varepsilon) - f(k;\, \Delta - i\varepsilon)\} \,/\, i\pi \ . \tag{16}$$

THE INVERSION PROCEDURE

An inversion procedure may be devised which exploits the properties of the discontinuity across the branch cut. In the range $\mu^2 \leq \Delta < (2\mu)^2$, the branch cut arises only from f_1. Thus,

$$i\pi D_{12} = \int \sigma(\alpha)(\alpha^2 - (\Delta + i\varepsilon))^{-1}\, (\alpha^2 - (\Delta - i\varepsilon))^{-1}\, d\alpha \tag{17}$$

where the notation D_{MN} denotes the discontinuity for $(M\mu)^2 \leq \Delta < (N\mu)^2$, and where the usual "iε" convention is used to imply that the limit as $\varepsilon \to 0$ is always intended. Invoking the identity

$$1 / (x \pm i\varepsilon) = P(1/x) + i\pi\delta(x) \tag{18}$$

where P denotes taking the principal value integral. It readily follows that

$$\Delta^{1/2} D(\Delta) = \sigma(\Delta^{1/2}) . \tag{19}$$

The above fixes the function σ in the range $\mu \leq \sigma < 2\mu$, thus enabling the following low resolution version of the object to be computed via

$$r\rho_{12}(r) = \int_{\mu}^{2\mu} \sigma(\alpha)\, e^{-\alpha r} d\alpha . \tag{20}$$

This finding may be extended further. We note that D_{23} arises from two contributions: (1) the first Born approximation of the density $\rho_{23}(r)$ which is, as yet, not known and (2) the second Born approximation of the density $\rho_{12}(r)$ which is known from D_{12}. The contribution (2) may be calculated from the known function ρ_{12} and this in turn may be subtracted from D_{23} to give a corrected discontinuity, D′. It then follows that

$$\Delta^{1/2} D'(\Delta) = \sigma(\Delta^{1/2}), \text{ for } (2\mu)^2 \leq \Delta < (3\mu)^2 . \tag{21}$$

In this way, the function σ may be obtained over an even wider range, and an even higher resolution version of the object may be formed, i.e., ρ_{13}. The extensions to higher orders is apparent.

SUMMARY OF INVERSION PROCESS

For clarity we now outline the method for recovering the interaction density, $\rho(r)$:

(a) Measure the scattering amplitude at all angles for a fixed frequency plane wave input.

(b) Calculate, by analytic continuation into the complex Δ domain, the experimentally determined values, $D_{EXP}(\Delta)$, of the discontinuity across the branch cut seen to lie on the real Δ-axis, for $\Delta \geq \mu^2$.

(c) Obtain a low resolution version of the object, $\rho_{12}(r)$, from $D_{EXP}(\Delta)$, in the range $\mu^2 \leq \Delta < (2\mu)^2$. Use this to calculate a (subtracted) correction to the values of $D_{EXP}(\Delta)$ in the range $(2\mu)^2 \leq \Delta < (3\mu)^2$.

(d) Obtain a higher order resolution mapping, $\rho_{13}(r)$, by utilizing the corrected values of $D_{EXP}(\Delta)$ for $(2\mu)^2 \leq \Delta < (3\mu)^2$. Note that the mapping $\rho_{13}(r)$ is obtained by adding a correction term, $\rho_{23}(r)$, to the previous estimate, $\rho_{12}(r)$.

(e) The general structure for obtaining progressively higher resolution mappings is now apparent. Note, however, that the computational effort rises rapidly as higher order corrections are incorporated.

RELATIONSHIPS BETWEEN BORN AND RYTOV APPROXIMATIONS

Under two very different conditions the inversion of the Helmholtz equation leads, in one case, to the so-called Born approximation and, in the other, to the so-called Rytov approximation. Although some connections between these two approximations have been pointed out in the literature[6-8], here we develop a further generalization which shows that the Born and Rytov inversions are simply extremes of a more general class of approximations which can be fashioned to fit the problem at hand. To demonstrate this result it will be helpful to re-write the Helmholtz equation (Eq. 1) in the following way

$$\nabla^2\psi(r) + k_0^2\psi(r) = -k_0^2 f(r)\psi(r) \tag{22}$$

where $f(r) = [k(r)/ k_0]^2 - 1$. Now let

$$\phi = \psi_0 g(\alpha), \quad \alpha = \psi/\psi_0$$

with $\psi_0 = Ae^{-i\omega t}$ and $\nabla^2\psi_0(r) + k_0^2\psi_0(r) = 0$. Then Helmholtz's equation becomes

$$\nabla^2\varphi + k_0^2\varphi = k_0 f\psi_0(dg/d\alpha)\alpha + \psi_0(d^2g/d\alpha^2)(\nabla\alpha)^2 . \tag{23}$$

Note that if $g(\alpha) = \alpha - 1$ in Equation 23, then we recover the Born approximation, i.e.,

$$\varphi = \psi - \psi_0 = \psi_S, \qquad \nabla^2\psi_S + k_0^2\psi_S = -k_1 f\psi . \tag{24}$$

If, on the other hand, $g(\alpha) = \ln\alpha$, we recover the Rytov approximation, i.e.,

$$\varphi = \psi_0 \ln(\psi / \psi_0), \qquad \nabla^2\phi + k_0^2\varphi = -k_0 f\psi_0 - \psi_0(\nabla\ln\alpha)^2 . \tag{25}$$

For the more general case, let

$$g(\alpha) = \beta(\alpha^{1/\beta} - 1). \tag{26}$$

The Helmholtz equation now becomes

$$\nabla^2\phi + k_0^2\varphi = k_0^2 f\psi_0\alpha^{1/\beta} - \psi_0(1 - 1/\beta)\alpha^{1/\beta}(\nabla\ln\alpha)^2 \tag{27}$$

with

$$\varphi = \psi_o\beta[(\psi/\psi_0)^{1/\beta} - 1] . \tag{28}$$

Note that if $\beta = 1$ we have the Born approximation. If $\beta\rightarrow\infty$ we have the Rytov approximation. Thus, the Born and Rytov are extremes of a more general approximation which can be represented by a general class of operators which might be fashioned to fit the problem at hand. A "single" inversion using such operators might well involve several different algorithms for different parts of the problem. Although this result only applies to the Helmholtz equation, a similar analysis seems to hold for the full wave equation which has important implications for acoustical imaging.

REFERENCES

1. G. Arfken. *Mathematical Methods for Physicists, 3rd edition*, Academic Press, San Diego (1985).
2. Z. H. Cho, J. P. Jones, and M. Singh. *Foundations of Medical Imaging*, John Wiley & Sons, New York (1993).
3. S. Leeman, Impediography revisited, in: *Acoustical Imaging, Vol. 9*, K. Wang, ed., Plenum Press, New York, pp. 513-520 (1980).
4. L. D. Landau, On analytic properties of vertex parts in quantum field theory, *Nucl. Phys., 13*, pp. 181-192 (1959).
5. S. Leeman, Non-local potentials and two body scattering, *Proc. Roy. Soc. Lond. A315*, pp.497-516 (1970).
6. M. Kaveh, M. Soumekh, and R. K. Mueller, A comparison of Born and Rytov approximations in acoustical tomography, in: *Acoustical Imaging, Vol.10*, J. Powers, ed., Plenum Press, New York (1981).
7. M. Kaveh, M. Soumekh, Z. Lu, R. K. Mueller, and J. Greenleaf, Further results on diffraction tomography using Rytov's approximation, in: *Acoustical Imaging, Vol. 12*, E. Ash and K. Hill, eds., Plenum Press, New York (1982).
8. Z. Lu, X. Han, Y. Zhang, and G. Chen, A comparison of intermediate, Rytov and Born transformations in acoustical tomography, in: *Acoustical Imaging, Vol. 20*, Y. Wei ed., Plenum Press, New York (1993).

REAL TIME PROCESSING OF THE RADIOFREQUENCY ECHO SIGNAL FOR ON-LINE SPECTRAL MAPS

Elena Biagi, Marco Calzolai, Massimiliano Forzieri,
Simona Granchi, Leonardo Masotti, Marco Scabia

Laboratorio di Ultrasuoni e Controlli non Distruttivi
Dipartimento di Ingegneria Elettronica
50139 Firenze, Italia

INTRODUCTION

This work is about the development of a Hardware-Software platform that completely integrates the architecture of a Personal Computer (PC) through a high speed custom interface. The complete integration of the Personal Computer brings two important advantages. The platform hooks its performances to the technological evolution in the quickly evolving PC field. Likewise it's possible to distribute the processing load between the hardware platform and the software running on the PC.

The platform is presented as a multi-parametric real time digital signal processing and multi-imaging processing support.

The platform can be used for different applications where fast signal and image processing requirement must be fulfilled. This work highlights the platform attractive characteristics for biomedical ultrasound. *F.E.M.M.I.N.A.* is the name of the platform version for this research field. This peculiar name is able to take into account the platform main features, in fact, it is the acronym of *Fast Echographic Multiparametric Multi Imaging Novel Apparatus*

In biomedical ultrasound the real time operating mode is an essential factor for clinical sperimentation of new tissue investigation algorithms. In fact, the platform programmability and modularity allow to implement and test new prototype echographic architectures offering the possibility to experiment the feasibility of different signal and image processing procedures in order to produce innovative echograph design.

At present *F.E.M.M.I.N.A.* is used to obtain spectral maps, but it can also be used for other operations such as noise reduction and image definition enhancement.

Acoustical Imaging, Volume 24.
Edited by Hua Lee, Kluwer Academic/Plenum Publishers, 2000.

In the field of biological tissue characterization, spectral analysis of backscattered ultrasonic echoes appears to be an useful tool to extract information about the internal morphological organization [1,2,3]. As an acoustic pulse propagates through soft tissues a selective frequency attenuation occurs [4,5,6]. Pathological portions of the tissue can be identified by the detection of their filtering action on the reflected or backscattered echo-signal. At the present all clinical diagnostic equipment don't imply spectral processing of the signal. The conventional B-mode images are mainly devoted to represent the topology of tissue structure, by using acoustic impedance inhomogeneities in the time domain. Nowadays it is not possible to obtain, by using commercial echographs, spectral images with frame rate suitable for medical diagnosis. In this context, a reduced configuration of the platform was employed to acquire and process echographic signals in real time, to produce spectral images.

PLATFORM DESCRIPTION

Figure 1 shows the hardware architecture of the platform. The architecture is based on 2 buses, a 32 bit fast bus, for frame data transfer, and a 16 bit control bus. The various processing modules are configured as a cascade along the fast bus; the first module in the configuration, is an acquisition module, which acquires the radiofrequency signal from an external source which in this work is an echograph.

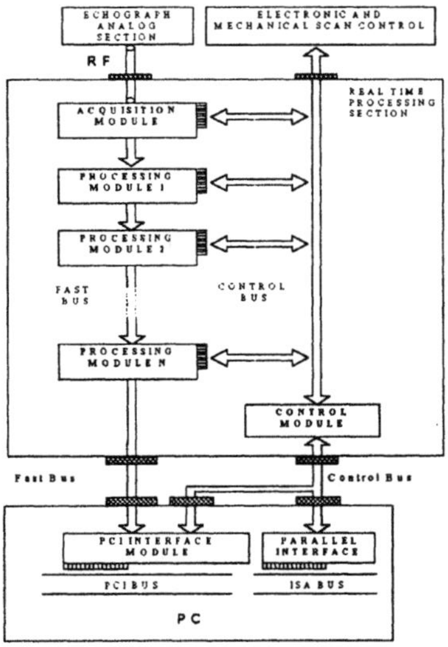

Figure 1: Platform Hardware architecture.

The PC integration into the platform was achieved by realizing a dedicated fast PCI board which is connected to both the fast bus, and the control bus, through the control module.

During the real time operation mode, the system transfers the data frames sequentially along the processing modules cascade, on the 32 bit data bus, in a pipeline mode. The control module is devoted to other operations such as programming devices in the various modules, configuring the platform, initializing each module, controlling and processing the platform "state" in real time and calibrating the processing modules.

The PC constitutes a functional block of the platform, by being completely integrated in it; this has several benefits, like taking advantage of the quick technological evolution in the PC field, simplifying the echographic architecture and reducing its cost, and making it possible to test the processing modules to obtain a higher reliability.

The PC is also used as a flexible visualization support for the echographic frames, and to perform some software signal processing, both with custom algorithms and commercial libraries.

The basic element of the modular structure is the processing module, whose structure is shown in figure 2.

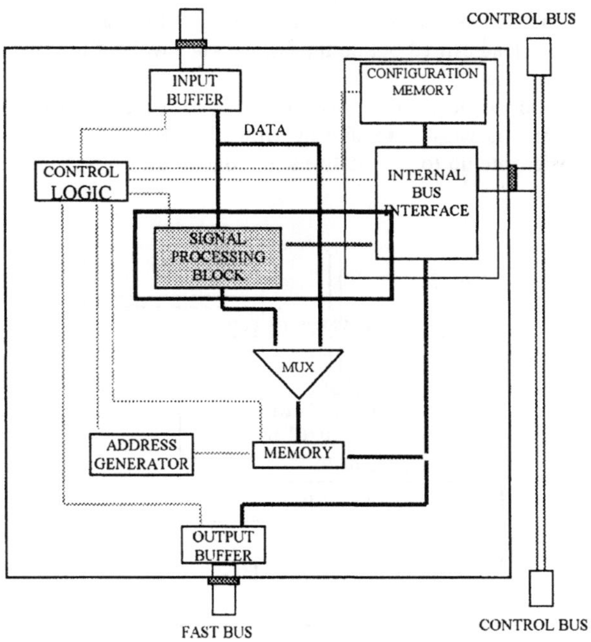

Figure 2: Processing module block diagram

All processing modules exhibit a common structure, and are differentiated only by their signal processing block, which defines the processing module functionality, and by the programming interface.

The real time and sequential operating mode is achieved by implementing in every module a dual block frame data memory reported in figure 3. A frame memory block is written with the data being calculated by the processing module itself While the other memory block is read by the following module in the processing cascade. The two block roles are inverted for each new frame.

The processing module has also a configuration memory, which is a series of registers dedicated to module identification, configuration and control. Moreover a multiplexer can render the block transparent in order to bypass it.

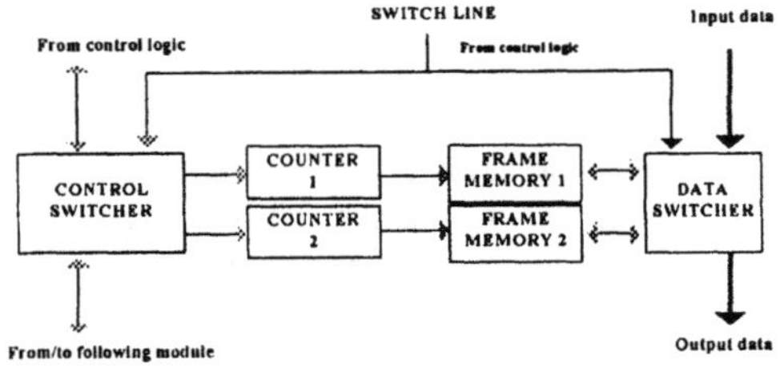

Figure 3: Dual block frame memory

The control module, programmed by the PC, during the real time processing operation, is responsible for synchronizing the various processing modules either by generating a frame synchronization signal, or by acquiring it from an external source. The frame data transfer is instead carried out on the fast data bus without handshaking, in order to maximize the transfer rate.

The PC integration in the platform was obtained by realizing a high speed custom PCI interface (Fig. 4), which is capable to directly write the data frames in the PC system memory, with a transfer rate up to 133 MByte/s, through a fast port connected to the fast bus in the platform.

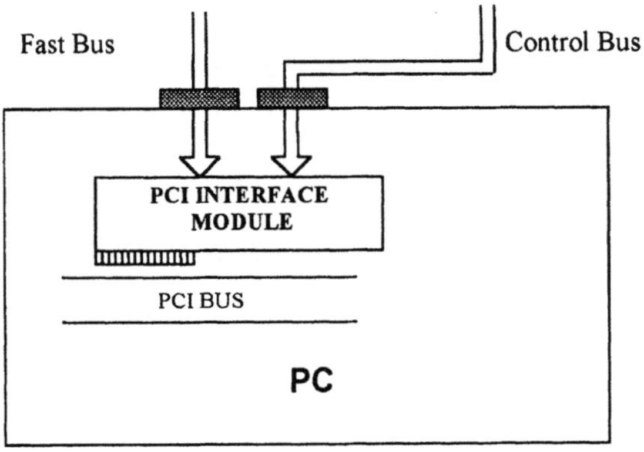

Figure 4: Dedicated PCI Board

The PCI board has also a second Control Port, connected to the control module, which was designed to be compatible with the EPP parallel port; this compatibility makes it possible to have the platform operate, in slower mode, also with PCs that cannot use the custom PCI interface, such as laptop computers. The control port, in this configuration, can reach a transfer rates up to 10 MByte/s.

The PCI board[7] was designed following the plug & play architecture criteria, and it renders the whole platform a plug & play device.

The software running on the PC (Figure 5) consists of a device driver (Windows VxD), which handles the communication with the hardware and performs the virtual to physical address conversion, and a C Library, which performs the low level calls to the driver, and encapsulates the Windows Direct Draw routines, used for fast visualization.

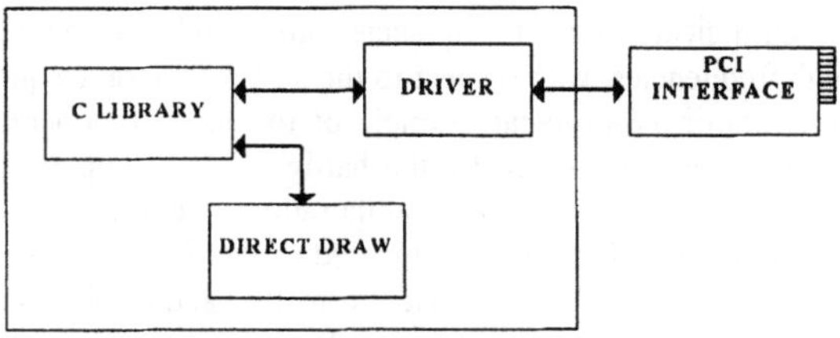

Figure 5: PC Software

The C Library is also capable of doing some software signal processing, both with custom algorithms and with commercial libraries.

EXPERIMENTAL RESULTS

Figure 6 shows a first example of configurations that was implemented to obtain spectral maps.

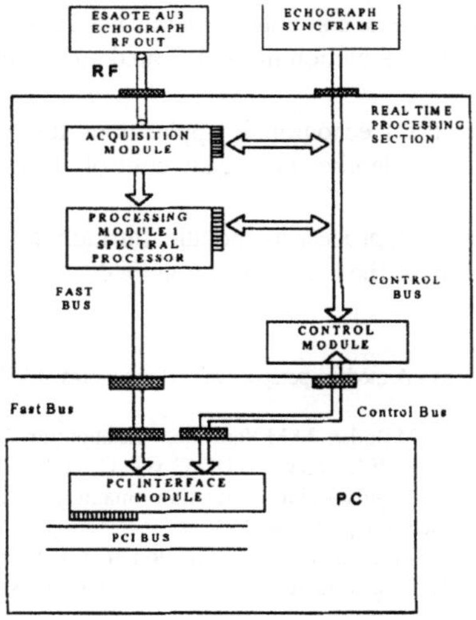

Figure 6: Platform configuration for spectral maps

This configurations consists on an acquisition module, based on a 40 MHz, 12 bit AD converter, and a spectral processing module, based on a Gray Chip GC 2011 digital filter (configuration internationally patented by Esaote S.p.A., 1997). The Digital filter is used to perform the Discrete Wavelet Transform (DWT) decomposition by successive filtering and decimation on different signal bands.

The radiofrequency signal is acquired from an Esaote AU3 echograph which uses a 5 MHz central frequency, linear array probe. The digital filter performs a spectral decomposition into 3 bands, from 5 MHz up to 20 MHz, and these frame data are then transferred to the PC, which is responsible of envelope extraction, scan converter operation, and image composition and display. The PC composes each spectral image by assigning each of the 3 bands into which the radiofrequency was decomposed, to one of the 3 basic colors, Red Green and Blue, that form up every pixel; in this way we obtain a composite image where shades of red indicate in-band frequencies, green indicates high frequencies, and blue indicates very high frequencies, with respect to the 5 MHz central frequency of the probe.

This configuration is a typical example of the mode of operation of our platform: while some operations are performed by the hardware, other operations are performed in software by the PC. In this particular configuration, the time required by the spectral processor to perform the DWT is low, if compared to the time required by the PC to execute the envelope extraction, scan converter, image formation and display, so essentially the frame rate is determined by the PC. For example, the image visualization and software scan converter of a 640x480x8 bits image takes about 25.4 ms, on an Intel Pentium II 233 MHz PC, corresponding to about 40 frames/s. Note that a 640x480 image is much larger than those typically provided by commercial echograph.

CONCLUSIONS

In this work we presented a hardware-software platform, which was designed to be expandable, modular, and adaptable to any other system which needs to produce significant and easily readable real time images. Such platform, being completely programmable, can also be reconfigured in dependence of the object being investigated.

The platform completely integrates a personal computer, and this has several advantages, like gaining benefit of the quick technological evolution in the PC field, and being able to display complex images in a flexible way.

The platform is particularly well suited for echographic applications for operations such as spectral maps calculation and display, adaptive time gain control, frequency attenuation compensation.

Moreover it is possible to reconfigure each processing module, through the control bus, in order to adjust the processing algorithm to the kind application field.

References

1. S. M. Kay, S.L. Marple, " Spectrum analysis - A modern perspective", Proceedings of IEEE, vol. 69, 1981, pp. 1380-1419.
2. F.L.Lizzi, M.Ostromogilisky, E.J.Feleppa, M.Rorke, M.M.Yaremko, "Relationship of ultrasonic spectral parameters to features of tissue microstructure", IEEE Trans. on UFFC, vol. 33, , 1986, p. 319-328.
3. S.Leeman, A.J.Healey, L.Ferrari, "A novel pulse-echo attenuation imaging techniques", Acoustical Imaging, Vol. 21, Plenum Press, New York 1995, pp.445-452.
4. T.Baldeweck, P.Laugier, G.Berger, "Ultrasonic attenuation estimation in highly attenuating medium: application to skin characterization", Acoustical Imaging, Vol.22, Plenum Press, New York 1996, pp.341-348.
5. E.J.Feleppa, W.R.Fair, T.Liu, W.Larchian, A.Kalisz, V.Reuter, A.Rosaldo, "Ultrasonic spectrum-analysis, tissue-typing image for prostate-biopsy guidance and staging", Acoustical Imaging, Vol.23, Plenum Press, New York 1997, pp.41-46.
6. F.lizzi, A.Kalisz, M.Astor, D.J.Coleman, R.H.Silverman, D.Z.Reinstein, "Very-high frequency ultrasonic imaging and spectral assays of the eye", Acoustical Imaging, Vol.23, Plenum Press, New York 1997, pp.107-112.
7. E.Solaris, G.Willse, "PCI Hardware and Software", Annabook, third edition December 1996

RECONSTRUCTION OF INNER FIELD BY MARCHENKO-NEWTON-ROSE METHOD AND SOLUTION OF MULTI-DIMENSIONAL INVERSE SCATTERING PROBLEM

V.A.Burov, S.A.Morozov and O.D.Rumiantseva

Moscow State University, Faculty of Physics, Dept. of Acoustics
Moscow, 119899, Russia

The aim of the present report is to propose a practical realization of solution method for the acoustical inverse scattering problem. The method is based on a reconstruction of sources of a scattered field and a wave field itself inside a scattering domain X with finite space sizes. This approach has its origin in the works of V.A.Marchenko and R.G.Newton[1]. In more recent time it was discussed for a far field [2] and for a near field [3,4], including the application for elastic media. Let us briefly reproduce a deduction of the main relationship in the frequency domain. The Helmholtz equations for the classical outgoing Green function G^+ and the incoming one G^- in an inhomogeneous medium (i.e. containing a scatterer) are basic. There is considered a contour Ω_z (in 2-D) or a surface (in 3-D) surrounding the scatterer - that is inhomogeneity of a sound phase velocity $c(\mathbf{z}')$ (fig.1):

$$c(\mathbf{z}') = \begin{cases} c(\mathbf{x}), & \mathbf{x} = \mathbf{z}' \in X \\ c_0, & \mathbf{z}' \notin X \end{cases}$$

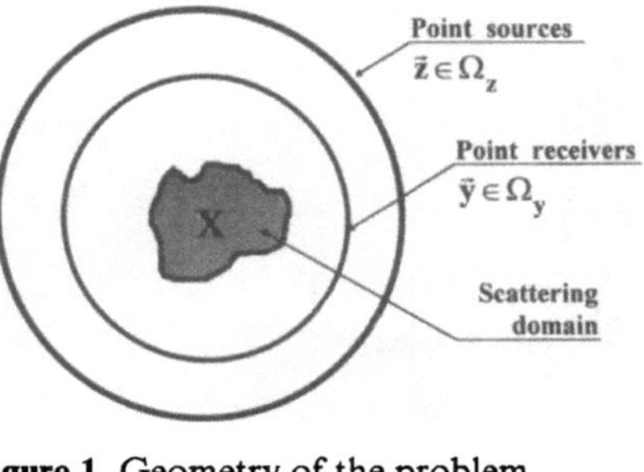

Figure 1. Geometry of the problem.

Acoustical Imaging, Volume 24.
Edited by Hua Lee, Kluwer Academic/Plenum Publishers, 2000.

Point sources with coordinates $\mathbf{z} \in \Omega_z$ lie on this contour. They radiate probing fields with an amplitude equal to 1. Point receivers with coordinates $\mathbf{y} \in \Omega_y$ lie on the contour Ω_y. It is convenient to propose that the contour Ω_y is *inside* the contour Ω_z, but outside the scattering domain X. Then wave fields - the Green functions $G^{\pm}$ - meet the requirements of the equations

$$\nabla^2 G^+(\omega,\mathbf{z}',\mathbf{y}) + \frac{\omega^2}{c^2(\mathbf{z}')} G^+(\omega,\mathbf{z}',\mathbf{y}) = \delta(\mathbf{z}'-\mathbf{y}) \tag{1a}$$

and

$$\nabla^2 G^-(\omega,\mathbf{z}',\mathbf{x}) + \frac{\omega^2}{c^2(\mathbf{z}')} G^-(\omega,\mathbf{z}',\mathbf{x}) = \delta(\mathbf{z}'-\mathbf{x}) \ . \tag{1b}$$

A multiplication of the eq. (1a) by $G^-(\omega,\mathbf{z}',\mathbf{x})$ and of the eq. (1b) by $G^+(\omega,\mathbf{z}',\mathbf{y})$ and then a subtraction of the first relationship from another one lead to the equation, in which characteristics of the medium inhomogeneity $\omega^2/c^2(\mathbf{z}')$ cancel each other. A subsequent integration over space domain $\mathbf{z}' \in S_z$ lying inside the contour Ω_z and the using of the divergence theorem give the main Marchenko-type relationship:

$$G^+(\omega,\mathbf{x},\mathbf{y}) - G^-(\omega,\mathbf{x},\mathbf{y}) = \oint_{\Omega_Z} [G^+(\omega,\mathbf{z},\mathbf{y}) \frac{\partial G^-(\omega,\mathbf{x},\mathbf{z})}{\partial n_z} - \frac{\partial G^+(\omega,\mathbf{z},\mathbf{y})}{\partial n_z} G^-(\omega,\mathbf{x},\mathbf{z})] d\mathbf{z}$$
$$\forall \mathbf{z} \in \Omega_z, \quad \forall \mathbf{y},\mathbf{x}\colon \ |\mathbf{y}| < |\mathbf{z}|,\ |\mathbf{x}| < |\mathbf{z}|. \tag{2}$$

The fields $G^{\pm}$ may be represented as a sum of undisturbed Green functions $G_0^{\pm}$ (the probing fields) for a homogeneous background medium and additional functions $G_{sc}^{\pm}$ describing the scattering effects: $G^{\pm} = G_0^{\pm} + G_{sc}^{\pm}$. An equation analogous to the eq. (2) takes place for $G_0^{\pm}$. A term-by-term subtraction of the $G_0^{\pm}$-equation from the eq. (2) gives the following relationship for the scattered fields $G_{sc}^{\pm}$:

$$\begin{aligned} & G_{sc}^+(\omega,\mathbf{x},\mathbf{y}) - G_{sc}^-(\omega,\mathbf{x},\mathbf{y}) - \\ & - \oint_{\Omega_Z} [\{G_0^+(\omega,\mathbf{z},\mathbf{y}) + G_{sc}^+(\omega,\mathbf{z},\mathbf{y})\} \frac{\partial G_{sc}^-(\omega,\mathbf{x},\mathbf{z})}{\partial n_z} - \\ & \quad - \frac{\partial \{G_0^+(\omega,\mathbf{z},\mathbf{y}) + G_{sc}^+(\omega,\mathbf{z},\mathbf{y})\}}{\partial n_z} G_{sc}^-(\omega,\mathbf{x},\mathbf{z})] d\mathbf{z} = \\ & = \oint_{\Omega_Z} [G_{sc}^+(\omega,\mathbf{z},\mathbf{y}) \frac{\partial G_0^-(\omega,\mathbf{x},\mathbf{z})}{\partial n_z} - \frac{\partial G_{sc}^+(\omega,\mathbf{z},\mathbf{y})}{\partial n_z} G_0^-(\omega,\mathbf{x},\mathbf{z})] d\mathbf{z} \end{aligned} \tag{3}$$

Consideration of all receiver coordinates $\mathbf{y} \in \Omega_y$ leads to a set of the integral equations with respect to the unknowns $G^{\pm}_{sc}(\omega,\mathbf{x},\mathbf{y})$ (or $G^{\pm}_{sc}(\omega,\mathbf{x},\mathbf{z})$, that is equivalent) for *any fixed point* $\mathbf{x} \in X$ belonging to the inner space of the inhomogeneity. All terms involving the unknowns are at the left part of the eq. (3). At the right part the known values are gathered: the undisturbed Green functions and the scattering data $G^{\pm}_{sc}(\omega,\mathbf{z},\mathbf{y})$ which can be measured directly ($G^{+}_{sc}(\omega,\mathbf{z},\mathbf{y})$) or calculated from the measurement data ($G^{-}_{sc}(\omega,\mathbf{z},\mathbf{y})$). The scattering data are also included in the left part (3), i.e. in operator acting on the unknown functions.

Furthermore, in the cited works the moving from the frequency domain to the time-domain description is carried out by the inverse Fourier transformation over the frequency variable. Additionally, the causality principle is used. Thanks to this principle the term $G^{-}_{sc}(t,\mathbf{x},\mathbf{y})$ - the incoming inner field- can be eliminated for time moments after finishing of the probing field radiation. As a result, a system of integral equations (Marchenko-Newton-Rose-, or MNR-equations) in the time domain gains a finished form and does not contain unnecessary unknowns. The system has a number of advantages being important for a practical realization. They were noted in the work [5]. Firstly, traditional in echoscopy devices pulse regime is supposed. In the initial version [4] each MNR equation is valid for delta-pulse probing signal. However, now a generalization of the equation for the case of an arbitrary probing signal is available [5]. Moreover, any experimental scheme to record the full set of the scattering data is acceptable. In particular, a circular scheme (the contours Ω_y and Ω_z are circumferences in 2-D) simplifying a numerical algorithmic realization of the MNR equation is convenient. Secondly, the MNR equation is *linear* with respect to the unknown inner field because it does not contain the characteristics of the inhomogeneity. This fact gives a hope to reconstruct the inner field without a parallel estimation of the inhomogeneity function. In the case of iterative solution methods of the inverse scattering problem (which are the most developed at present [6, 7]) such a possibility is eliminated because of the multiple scattering effects. The reconstructed inner field can be further used to immediately obtain an estimation of sought parameters of the inhomogeneity. For this purpose the wave equation or the Lippmann-Schwinger one are applied [5]. Thirdly, the method is valid for both 2-D and 3-D problems.

Unfortunately, using only the system of the time-domain equations described above (or the equivalent system of the equations (3) for multitude of frequencies simultaneously with the causality principle) does not give a possibility to reconstruct the scattered fields. The matter is the following. The solution of the eq. (2) for the inverse scattering problem in quantum mechanics reduces to the consideration of the Rieman-Hilbert problem [1] from the functional analysis in complex space. Two important properties of such a problem take place. Firstly, wave fields have some analytical properties. Secondly, an influence of the scattering processes asymptotically decreases as energy of scattered particles increases. Thanks to these two facts the inverse scattering quantum problem considered for *all* values of the energy has a unique solution. In the acoustical case considered by us, a *finite* frequency band was used in the numerical modeling. Moreover, the effect of decreasing the scattering processes as the frequency increases is absent in acoustics. The latter circumstance complicates the problem sharply, because the solution of the multi-frequency system of the type (3) becomes non-unique. Here there are some reasons. Firstly, the equation (2) is basic; however it has the trivial solution $G^{\pm}(\omega,\mathbf{x},\mathbf{y}) = 0$, which for the equation (3) turns into the solution $G_{sc}^{\pm}(\omega,\mathbf{x},\mathbf{y}) = -G_0^{\pm}(\omega,\mathbf{x},\mathbf{y})$. Secondly, there is a multitude of functions which are nullified by the operator acting on the unknowns $G_{sc}^{\pm}$ at the left part of the monochromatic equation (3). Introduction of finite-dimensional manifolds - for discrete frequencies and for discrete points of the sources and receivers - converts the operator into a matrix. The matrix has a rank half as its order. The additional using of the causality principle imposes too soft restrictions (in the form of conditions connecting the monochromatic components of the sought fields) and does not remedy the situation. In this case, a rank of matrix formed by the least square method (LSM) and acting on unknowns is essentially less than its order. As a result, the LSM-solution is not the true one. The LSM- and true solutions differ by a linear combination of vectors nullified by the matrix at the left part.

The necessity of invoking some additional relationships to provide the unique solution is pointed out in the work [8]. The *simultaneous* solution both of equations similar to (3) and of the Lippmann-Schwinger equations is proposed. The using of the latter equations brings into consideration new unknowns - scatterer characteristics

$$\eta(\omega,\mathbf{x}) = \omega^2(c_0^{-2} - c^{-2}(\mathbf{x})). \tag{4}$$

The function $\eta(\omega,\mathbf{x})$ is included into the Lippmann-Schwinger equation as the product with the unknown inner field. It makes the problem to be nonlinear with respect to the unknowns. Moreover, the necessity appears to introduce a connection between the field and the scatterer $\eta(\omega,\mathbf{x})$. For this purpose the equations similar to these for Coulomb potential at the zero frequency $\omega = 0$ (!) are used. However, a practical measurement of the scattering data of such a kind (i.e. at very low frequencies) is not convenient. Finally, the nonlinear character of the formulated problem requires iterative methods of its solution and brings up the question about a convergence domain of the solution process.

Below other way is proposed. The way consists in a transformation of the equations (3) and in addition some relationships to them, which do not distort the linear character of the problem. For this purpose additional unknowns $F^{\pm}(\omega,\mathbf{x},\mathbf{y})$ - the complex amplitude of the secondary sources - are introduced. The functions $F^{\pm}$ are represented as a product of the Green functions $G^{\pm}(\omega,\mathbf{x},\mathbf{y})$ (i.e. the outgoing or the incoming fields corresponding to the point source with the coordinates **y**) by the scatterer function $\eta(\omega, \mathbf{x})$:

$$F^{\pm}(\omega,\mathbf{x},\mathbf{y}) = \eta(\omega, \mathbf{x})G^{\pm}(\omega,\mathbf{x},\mathbf{y}) \tag{5}$$

These are the sources that form the inner fields and radiate the scattered fields which are the scattering data of the problem.

A multiplication of the left and the right parts (2) by $\eta(\omega, \mathbf{x})$ leads to the equation analogous to the eq.(2), however with respect to $F^{\pm}(\omega,\mathbf{x},\mathbf{y})$:

$$F^{+}(\omega,\mathbf{x},\mathbf{y}) - F^{-}(\omega,\mathbf{x},\mathbf{y}) = \oint_{\Omega_z} [G^{+}(\omega,\mathbf{z},\mathbf{y})\frac{\partial F^{-}(\omega,\mathbf{x},\mathbf{z})}{\partial n_z} - \frac{\partial G^{+}(\omega,\mathbf{z},\mathbf{y})}{\partial n_z}F^{-}(\omega,\mathbf{x},\mathbf{z})]d\mathbf{z}$$

$$\forall \mathbf{z} \in \Omega_z, \quad \forall \mathbf{y},\mathbf{x}: \ |\mathbf{y}| < |\mathbf{z}|,\ |\mathbf{x}| < |\mathbf{z}|. \tag{6}$$

Additionally, linear relationships of the connection are also used:

$$\int_{x' \in X} G_0^{\pm}(\omega, |\mathbf{x}-\mathbf{x}'|)\ F^{\pm}(\omega, \mathbf{x}', \mathbf{y})\, d\mathbf{x}' - G_{sc}^{\pm}(\omega, \mathbf{x}, \mathbf{y}) = 0\ ; \quad \mathbf{y} \in \Omega_y\ , \ \mathbf{x} \in X \qquad (7a)$$

$$\int_{x' \in X} G_0^{\pm}(\omega, |\mathbf{z}-\mathbf{x}'|)\ F^{\pm}(\omega, \mathbf{x}', \mathbf{y})\, d\mathbf{x}' = G_{sc}^{\pm}(\omega, \mathbf{z}, \mathbf{y})\ , \qquad \mathbf{y} \in \Omega_y\ ; \ \mathbf{z} \in \Omega_z \qquad (7b)$$

The equations (7a) considered for the inner points of the scattering domain $\mathbf{x} \in X$ connect the unknown inner fields and the unknown secondary sources amplitudes. The equations (7b) considered for the sources points $\mathbf{z} \in \Omega_z$ give the relation with the observed fields. As a result, an extended system (3), (6), (7a, b) takes place allowing to estimate the inner field. It is important to emphasize once more that all equations of this system are *linear*. A complication during the solution of the system is its *non-locality* because of the connection equations (7a,b). Then it is necessary to find the inner field for the total set of inner points **x** *simultaneously*. Therefore, algebraic dimension of the extended system essentially increases. However, the solution becomes possible even for one frequency, i.e. the causality principle is not be used.

The first results have been obtained for numerical realization of the problem concerning the estimation of the inner field on the base of the scheme described (3), (6), (7a,b). Cylindrical scatterer is considered. Its refractive index is $c_0 / c = 0.8$ and its diameter is one wave length λ_0. The scattered field is observed in 64 points **y** on the circumference Ω_y with the $7.5\lambda_0$- radius, for each of 64 point sources lying on the circumference Ω_z with the $8\lambda_0$- radius. On figures 2a,b the real parts of the true and reconstructed scattered fields are shown as functions of radial distance r from the origin - the scatterer center. The reconstructed values are obtained by the LSM - solution of the monochromatic algebraic system. The system includes the equations (3), (6) for the inner fields and the secondary sources amplitudes, and the connection equations (7a,b) considered for 35 intermediate values of the radial coordinate $r = |\mathbf{x}'|$ within the interval from 0 to $0.875\lambda_0$. Some discrepancy of the results might be caused by a rather rough scheme for discreting the problem. In future it would be possible to rule out the LSM-estimations and passing to the solution of a variation problem with the connections. Since the connections have linear character, an introduction of them will not lead to necessity of iterations.

The most essential moments of the present reports seem as follows. The possibility appears for the rescattering effects to be taken into account without help of iterative schemes and with keeping the linearity of the solved system. Such an approach can be applied to 2-D or 3-D inverse scattering problem in monochromatic or pulse regime.

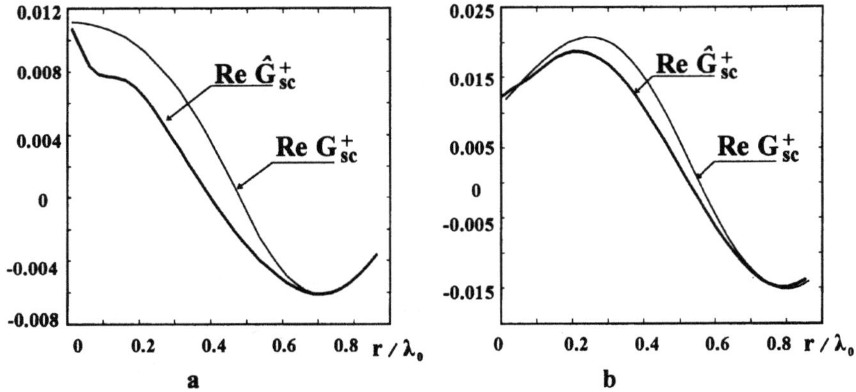

Figure 2. True (G^{+}_{sc}) and reconstructed ($\hat{G}^{+}_{sc}$) fields scattered by cylindrical inhomogeneity: a) scattering in straightforward direction; b) scattering at $\pi/2$ angle.

REFERENCES

1. R.G.Newton, Inverse scattering II. Three dimensions, *J. Math. Phys.*, 21:1698 (1980).
2. J.H.Rose, M.Cheney, and B. De Facio, Determination of the wave field from scattering data, *Phys. Rev. Letters*, 57:783 (1986).
3. M.Cheney, G.Beylkin, E.Somersalo, and R.Burridge, Three-dimensional inverse scattering for the wave equation with variable speed: near field formulae using point sources, *Inverse Problems*, 5:1 (1989).
4. D.Budreck, and J.H.Rose, Three-dimensional inverse scattering in anisotropic elastic media, *Inverse Problems*, 6:331 (1990).
5. V.A.Burov, E.E.Kasatkina, and O.D.Rumiantseva, Statistical estimations in inverse scattering problems, *in*: Acoustical Imaging-22. P.Tortoli and L.Masotti, ed., Plenum Press, New York (1996).
6. S.A.Johnson, Y.Zhou, M.L.Tracy, M.J.Berggren, and F.Stanger, Inverse scattering solutions by a sinc basis, multiple source, moment method. Pt.III: Fast algorithms, *Ultrason. Imaging*, 6:103 (1984).
7. V.A.Burov, M.N.Rychagov, and A.V.Saskovets, Account of multiple scattering in acoustic inverse problems of tomographic type, *in*: Acoustical Imaging-19, H.Ermert and H.-P.Harjes, ed., Plenum Press, New York, (1992).
8. J.H.Rose, M.Cheney, Self-consistent equations for variable-velocity three-dimensional inverse scattering, *Phys. Rev. Letters*, 59:954 (1987).

RF ULTRASOUND ECHO DECOMPOSITION USING SINGULAR-SPECTRUM ANALYSIS

Carlos Dias Maciel[1] and Wagner Coelho de Albuquerque Pereira[1]

[1]Programa de Engenharia Biomédica - COPPE/UFRJ
Cx. Postal 68510 cep 21945-970 Rio de Janeiro - RJ, Brazil
maciel@peb.ufrj.br

INTRODUCTION

During the last decades investigators turned their attention to the problem of quantitative ultrasound characterization. More recently statistical signal processing is being used to provide parameters from RF ultrasound echoes opening a new branch of studies. Several works (Simon et al.(1997)) have been done in biological tissue characterization using spectral analysis to quantify parameters like mean particle diameters and mean scatterer spacing. The main approach in these cases is to provide a separation between the periodic and non-periodic structures represented by the discrete and continuos part of the signal spectrum.

Vautard and Ghil (1989, 1992) showed that the expansion in Principal Component Analysis (PCA) yields other powerful tool for time-series analysis different from the common approaches. In particular, near-equality of a pair of eigenvalues is associated with signal periodic activity. The singular-spectrum analysis (SSA) can easily and automatically localize intermittent oscillations. The shape of these oscillations is determined adaptively from the data, which makes SSA more flexible and better suited for analysis of nonlinear, anharmonic oscillations.

The main purpose of this work is to provide a qualitative decomposition of the signal into significant and noise components of ultrasound biomedical echoes. In addition, a

Acoustical Imaging, Volume 24.
Edited by Hua Lee, Kluwer Academic/Plenum Publishers, 2000.

temporal basis analysis is made to distinguish what are the significant components associated with quasi-periodic behavior. This process is to be made along the whole RF ultrasound echo signal.

THEORY

The SSA as a data-analysis method has been used for years in nonlinear dynamics (Broomhead & King,(1986)) applied into several signal processing areas (Wax & Kailath, 1985; Therrien (1992)). SSA is based on classical Principal Component Analysis (PCA) and gives the principal axes of a sequence vector x by expanding it with respect to an orthonormal basis (E^k, $1 \leq k \leq M$):

$$x_{i+j} = \sum_{k=1}^{M} a^k{}_i E^k{}_j \ , \ 1 \leq j \leq M \qquad \textbf{eq. 1}$$

The projection coefficients a_i^k are called the Principal Components (PCs), and the basis vector E^k the empirical orthonormal functions (EOFs). The vectors E^k are the eigenvectors of the covariance matrix (T_x) of the sequence x. This formula is an application of the general Karhunen-Loève orthogonal expansion and is most used for information compression and signal to noise ratio enhancement (Broomhead & King (1986), Johnson & Wichern (1992)).

We can assume T_x as a toeplitz matrix, all eigenvalues λ_k are non-negative and organized in descending order, $\lambda_1 \geq \lambda_2 \geq \ldots \geq \lambda_M$, and the eigenvectors are either simetric or anti-simetric. Truncating the sum, eq. 1, at a order $p < M$ reduces the information to the first p Principal Components instead of the M original components. This truncation is made following an optimal criterium: the first p components describe the largest fraction of the total variance that one can obtain using a projection into p orthogonal vectors (Broomhead & King (1986), Johnson & Wichern (1992)).

As cited in Vautard and Ghil (1989, 1992), the eigenvalues for a purely-periodic oscillation of angular frequency ω are:

$$\lambda^S = \frac{\sigma_x{}^2}{2}\left(1 + \frac{\sin(2\omega\tau)}{2\omega\tau}\right) \text{ and } \lambda^A = \frac{\sigma_x{}^2}{2}\left(1 - \frac{\sin(2\omega\tau)}{2\omega\tau}\right) \qquad \textbf{eq. 2}$$

where λ^S (λ^A) indicates eigenvalue associated with a(n) (anti)simetric eigenvector and τ is the window length. Other eigenvalues are zero. Depending on the value of τ, the values of λ^S and λ^A vary periodically. When τ is multiple of $\pi/2\omega$ there is, in term of PC analysis, a degeneracy of the problem. Our interest is to analyze the behavior of the system when τ is big enough to neglect the second term of the sum (eq. 2). In this situation, the eigenvalues are nearly equal and represent a periodic behavior.

The spectral properties of the principal components also play an important role. The power spectrum of the kth PC ($Px_k(f)$), is (Vautard and Ghil (1989, 1992)):

$$P_{x_k}(f) = |\rho^k(f)|^2 \, P_x(f) \qquad \textbf{eq. 3}$$

$$\text{where } \rho^k(p) = \frac{1}{M}\sum_{m=1}^{M} E_m{}^k e^{2i\pi mp/N} \qquad \textbf{eq. 4}$$

$P_x(f)$ is the power spectrum of x.

ρ^k is the transfer function of the kth filter associated with the kth eigenvector. SSA concentrates the transfer functions in regions where sharp peaks in the power spectrum occur, facilitating the detection of such oscillations.

For signal detection and reconstruction, we wish to identify the EOF's dominated by signal and that are associated with a periodic behavior. If a component of noise or signal that is not associated with a periodic behavior has been misidentified as an oscillatory EOF pair, a broadband signal will appear.

To realize the correct pair separation we begin our analysis using the same criterion as Vautard et al., 1992. This criterion is: if oscillating pairs of eigenvalues (k, k+1) represent the same oscillatory behavior the associated eigenvectors must be spectrally located around the same frequency, or

$$\delta f_k = |f_k - f_{k+1}|, \qquad \textbf{eq. 5}$$

has to be small. Although necessary, this criterion is not sufficient, since the amplitude of the peaks must also be high. If the presumed oscillatory pair completely resolves a frequency f^* between f_k e f_{k+1}, the response function of a reconstruction filter based on components k and k+1 must be close to 1. At least, the pairs indicate a periodic behavior at frequency f^* if the signal energy at that frequency is greater than two thirds of the variance of x.

SIMULATION

The simulation adopted was to obtain a one dimension RF echo from tissue based on the convolution of the tissue amplitude impulse response and the RF incident pulse, as described in Simon et al. (1997)). Figure **1** shows two signals as an example, (a) one can see the backscattered echo in which the amplitude due to the regular part of the tissue is 8 dB higher than the echoes from the diffused part. Item (a) also shows a zoom view of the echo and its amplitude spectrum from which one can visualize the spectral lines related to the periodic part of the original echo. Figure **1** (b) is similar, but now the relation of regular signal to diffuse one is -3 dB. The spectral lines are still visible, but the loss of quality is evident.

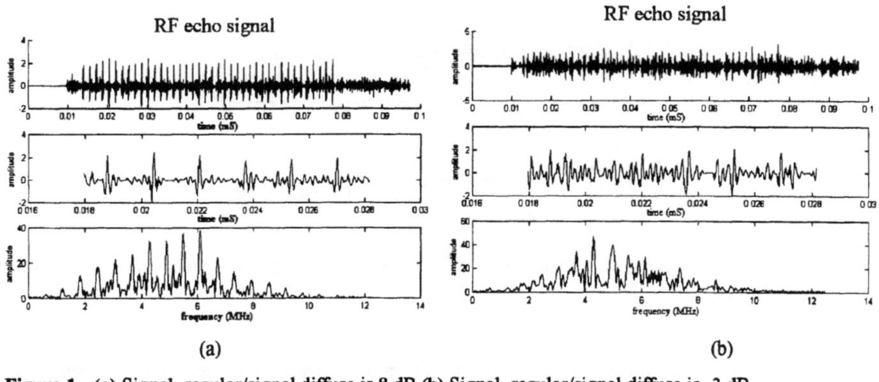

Figure 1 - (a) Signal regular/signal diffuse is 8 dB (b) Signal regular/signal diffuse is -3 dB.

The next step is to apply SSA to the RF signals shown in the 2nd row of Figure 1 and obtain the eigenvalues. In this case, a covariance matrix of dimension 80x80 was built resulting in 80 eigenvalues. In Figure 2 it is possible to observe, in both cases, (a) and (b), that there are a lot of eigenvalue pairs that could represent a periodic behavior. As cited in Vautard and Ghil (1989) even noise could generate pairs of eigenvalues. In a first approach it is necessary to separate signal from the noise. In systems where the noise is white, the tail of the singular spectrum is flat, meaning that all eigenvalues are identical and associated with noise subspace.

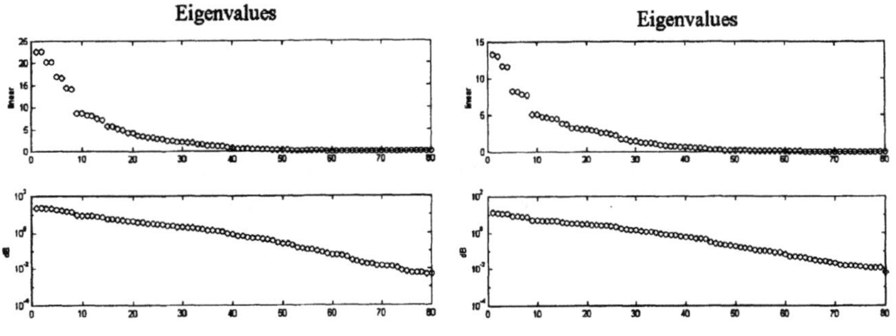

(a) (b)

Figure 2 - eigenvalues using M=80 (a) SNR = 8 dB (b) SNR = -3 dB, both in linear and logarithmic scales.

In our analysis, the diffuse component is treated as colored noise. In this situation we simply made a Whitening Transformation (Figure **3**) to whiten the diffuse medium backscattering signal (Therrien (1992)). The objective of this transformation is obtain a flat tail at the end of singular spectrum as shown in Figure 3

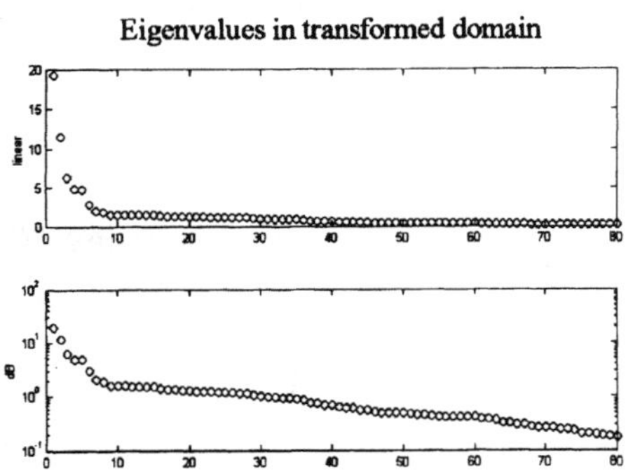

Figure 3 - Eigenvalues using M = 80 and SNR = -3 dB after Whitening Transformation, both in linear and logarithmic scales.

For the periodic signal reconstruction, we may consider that the periodic signal is generated by the regular structure of the tissue. The diffuse signal has a more complex composition. It is associated directly with diffuse structure of tissue, but other characteristics influence this signal. The first and more obvious one is the noise level of the instrument. The second one is the tolerance in the position distribution of the regular particles in the tissue.

This tolerance influences in a dramatic way most of the spectrum analysis, even for a high amplitude backscattering of the regular particles. Thus the mere visual inspection of the spectrum is not sufficient to analyse the signal.

To obtain the harmonics identification we may consider the signal to noise separation of the regular part of the signal using the eigenvalues. The algorithm adopted was the Information Theoretic Criteria (ITC) proposed by Wax & Kailath, 1985. It is based on hypothesis tests and maximum likelihood when we deal with white noise. When applying ITC to our problem in the transformed domain (after the Whitening Transformation), we get the signal subspace and in this subspace we use the eigenvalue pair criterium to separate the quasi-periodic behavior.

Figure 4 shows two analysis cases based on signals present at Figure 1(a) and (b). In both cases we consider the first 8 eigenvalues (4 pairs), as shown in Figure 2 and Figure 3, to the reconstruction the original periodic series. The first one, Figure 4(a), shows the reconstruction of the regular signal (first row and first column) and the diffused signal (second row and first column). In both cases we have the power spectrum using FFT in the second column of Figure 4(a).

The second one, Figure 4(b), shows the reconstruction of regular signal (first row and third column) and the diffused signal (second row and third column). In both cases we have the power spectrum via FFT is in the fourth column of Figure 4(a).

From Figure 4 we can also observe in both cases (SNR (a) 8 dB and (b) -3 dB) the performance of the periodic signal reconstruction. In the worst case, Figure 4(b), harmonics levels are lower than in the 8 dB case. Even in this situation, the method performed well the separation.

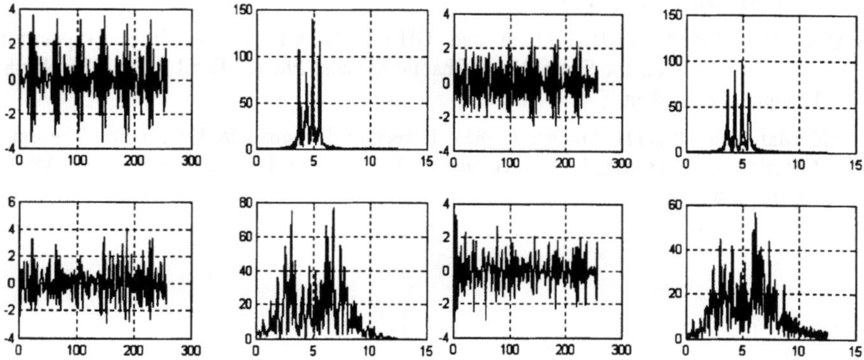

Figure 4 - Signal and noise reconstruction (a) SNR = 8 dB (b) SNR = -3 dB, with respectives amplitude spectra.

CONCLUSION

This application of SSA shows a simple way to study the decomposition of RF ultrasound signals. This approach is based on signal subspace models. We are working in more robust criteria to eigenvalues separation using surrogate data tests and Monte-Carlo analysis.

In a second step of this work, some index could be proposed to identifying, for example, liver diseases by studying how periodic the normal tissue is expected to be. This approach could be done by the study of chaotic dynamics as cited in Pueyo, 1997, for exemple. In a common approach the periodic signal reconstruction maybe used to quantify tissue using a parameter known as mean scatterer spacing (MSS) as cited in Simon (1997). Diffused signal does not describe the irregular particles in the medium but it may be used to quantify how disordered is the medium.

Preliminary results show good agreement among theory, simulation and initial phantom RF ultrasonic echo experiments.

REFERENCE

BROOMHEAD, D.S. and KING, Gregory (1986) "Extracting Qualitative Dynamics from Experimental Data" Physica 20D p. 217-236 North-Holland, Amsterdam.

JOHNSON, Richard and WICHERN, Dean (1992) "Applied Multivariate Statistical Analysis" Prentice Hall, Englewood Cliffs, New Jersey.

PUEYO, Salvador (1997) "The study of chaotic dynamics by means of very short time series" Physica D 106, p 57-65, North-Holland, Amsterdam.

SIMON, Claudio, SHEN, Jian, SEIP, Ralf, EBBINI, Emad S. (1997) "A Robust and Computationally Efficient Algorithm for Mean Scatterer Spacing Estimation" IEEE Trans. on UFFC vol. 44 n. 4 .

THERRIEN, Charles (1992) "Discrete Random Signals and Statistical Signal Processing" Prentice Hall Processing Series, Englewood Cliffs, New Jersey.

VAUTARD, Robert and GHIL, Michael (1989) "Singular-Spectrum Analysis in Nonlinear Dynamics With Applications to Paleoclimatic Time Series" Physica D 35 p. 395-424 North-Holland, Amsterdam.

VAUTARD, Robert, YOIU, Pascal and GHIL, Michael (1992) "Singular-Spectrum Analysis: a Toolkit for Short, Noisy Chaotic Signals" Physica D 58 p. 95-126 North-Holland, Amsterdam.

WAX, Mati & KAILATH, Thomas (1985) "Detection of Signals by Information Theoretic Criteria" IEEE Transactions on Acoustics, Speech, and Signal Processing, vol ASSP-33, no. 2, april .

HIGH-PERFORMANCE COMPUTING IN REAL-TIME ULTRASONIC IMAGING

D.F. García Nocetti,[1] J. Solano González, [1] M.F. Valdivieso Casique, [1] R. Ortiz Ramírez, [1] E. Moreno Hernández [2]

[1] DISCA, IIMAS, Universidad Nacional Autónoma de México, P. O. Box 20-726, C.P. 01000, México D.F., México
[2] Centro de Ultrasónica, ICIMAF, Calle 15 No. 551, Vedado, 10400, Habana, Cuba

INTRODUCTION

Ultrasonic imaging using digital computers has been a very active research field in recent years (Wells, 1996). Requirements for fast and accurate image construction from ultrasonic signals cover a wide field of applications ranging from signal acquisition through pre-processing, formation and displaying of the ultrasonic image to its enhancement and analysis (Fish, 1990). Custom built systems where algorithms are directly implemented in special purpose hardware tend to be expensive in design time, have a limited market and flexibility with improvements in technology and algorithms (Cavaye, et al., 1993). The availability of parallel architectures offer new opportunities for the realisation of low-cost, flexible, faster and more reliable systems.

In the field of ultrasonic imaging, the image quality achieved depends on many factors (Powis et al., 1984). Particularly important are the transducer (e.g. its sensitivity and focusing) and the "front end" electronics (i.e. electronic noise level and accuracy of the digitized signal). Preprocessing (signal dynamic range and interpolation) and postprocessing (grey scale maps and image smoothing) are adjusted to produce the best possible image quality (Russ, 1995).

This paper presents an ultrasonic imaging system based on a parallel processing transputer architecture. The system exploits the different forms of intrinsic parallelism often found within the process of ultrasound imaging (Webber, 1992). Processes such as beam focusing and scanning, formation, displaying and postprocessing are designed to be executed in parallel in order to achieve the performance required in real-time ultrasonic imaging applications. Beam focusing and scanning techniques are developed and implemented for adjusting the depth at which the transducer beam is focused and for increasing the spatial resolution of the ultrasonic image. With respect to image formation, parallel interpolation techniques are applied to improve the original image through the use of mathematical approximation functions which generate intermediate samples that together with the original ones will produce more detailed images to the observer.

Acoustical Imaging, Volume 24.
Edited by Hua Lee, Kluwer Academic/Plenum Publishers, 2000.

Furthermore, the flexibility and scalability of the system allows to incorporate new developments in technology and opens a greater scope for developing and utilising better and more complex algorithms that can improve the image visual appearence and achieve a real-time response.

ULTRASONIC IMAGING

The "pulse-echo" principle is the basis of ultrasonic imaging. In practice, a series of acoustic pulses are transmitted along the ultrasound beam, with the transducer "listening" for echoes after each pulse. The time interval between successive transmit pulses must be such that all the echoes from one pulse have died away before the next one is transmitted.

In the process of formation of an ultrasonic image, pulses are transmitted sequentially along adjacent lines of sight, and dots are placed at the appropiate points on the display wherever echoes are detected, see figure 1.

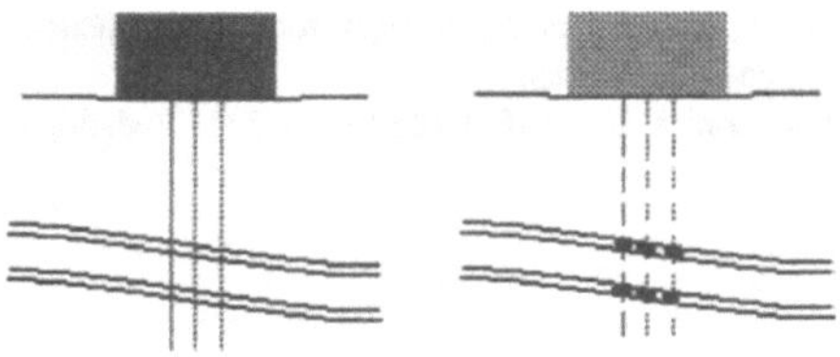

Figure 1. Formation of an ultrasonic image.

Parallel lines of sight (as produced by linear array transducers) give rise to a rectangular scan area. Lines of sight with common origin point (as produced by electronic phased array transducers) generate a sector scan area. In a linear array, as depicted in figure 2a, the beam is formed using a group of tranducer elements, it can therefore be transfered laterally by dropping one element from one end of the group and picking up an extra element at the other end. In a phased array, figure 2b, electronic time delays are used to scan as well as to focus the beam.

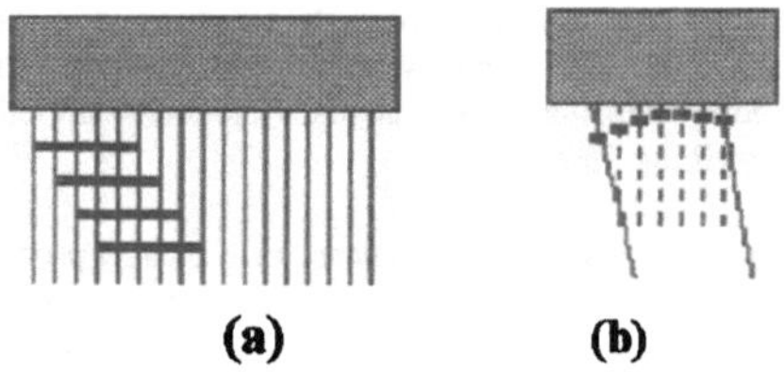

Figure 2. Methods for scanning an ultrasound beam.

Figure 3 shows an ultrasonic imaging machine in block diagram form. Functions under user control include: transmission power, transmission focal depth (for electronically focused machines), receiver gain, time gain compensation and depth gain compensation settings, preprocessing (signal dynamic range), postprocessing (grey scale maps and image smoothing), image size, zoom, etc.

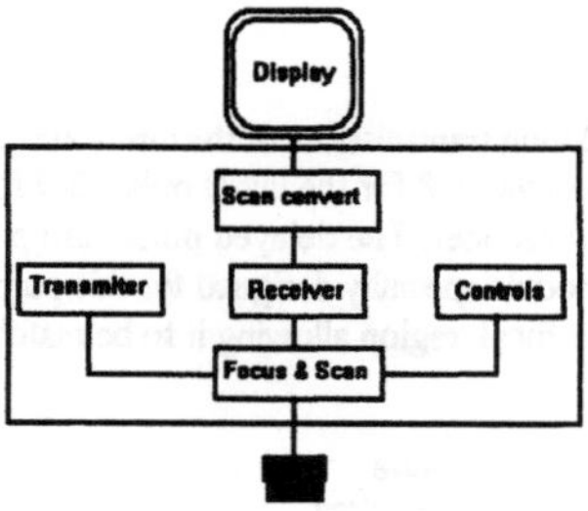

Figure 3. Block diagram of an ultrasonic imaging system.

One important restriction in ultrasound imaging applications is that the transmission power of the pulses should be kept as low as possible for safety reasons, and still obtain the necessary information. Therefore, the receiver gain, preprocessing and postprocessing should be adjusted adjusted to produce the best image quality for a given examination.

SYSTEM DESCRIPTION

The work described here concerns with de development of a parallel processing ultrasonic imaging system, aiming to scan, construct and display high quality ultrasonic images in real-time with a minimum rate of 25 frames/second. Figure 4 shows a block diagram of the system that is based on a parallel processing transputer architecture.

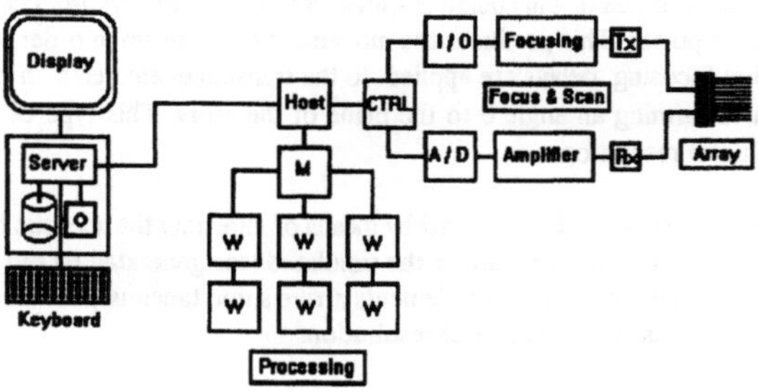

Figure 4. Parallel processing ultrasonic imaging system.

Three main blocks are processed in a pipelined fashion: focusing and scanning, processing and displaying. A transputer controls the focusing and scanning stage for acquiring the input ultrasonic data. It also provides the required processing power to construct the image to be diplayed. The INMOS T805-30 transputer architecture, programmed in the OCCAM language, is connected to a host computer which provides input, output and file system services. The host computer communicates with the first transputer (host node) by running a server program.

Focusing and Scanning

The system achieves focusing during transmission of the ultrasonic pulse by means of introducing delays to each group of elements (e.g. elements 1-8 for the initial pulse, 2-9 for the second and so on) out of the 64 elements-3.5 Mhz linear array transducer. The delayed pulses are produced by a programmable digital input/output transputer-controlled module specially designed for this purpose. These delays are programmed electronically to vary the depth of the focal region allowing it to be matched to the depth of the region under examination, see figure 5.

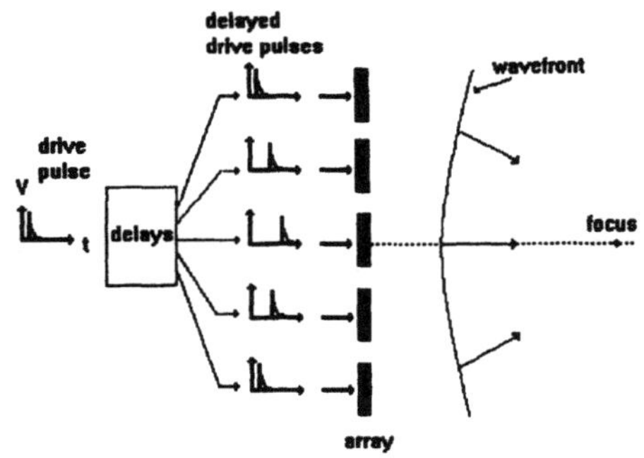

Figure 5. Electronic beam focusing during transmission

Three types of beam focusing techniques have been implemented: linear, dihedral and cylindrical. For linear focusing, a group of neighbouring transducer elements are used together to form a larger transducer which projects a rectangular beam. This beam is shifted laterally by one transducer element at the time for successive transmission pulses. This produces the movement of the beam in order to generate the scanning process. For dihedral focusing, delays are applied to the transducer elements in order to divide the array in two semi-apertures forming an angle θ to the plane of the array. This type of focusing is intended to increase the image lateral resolution.

Finally, cylindrical focusing can be achieved by means of directing the acoustic beams to a central point. This effect can be achieved by the curvature of the simulated lens generated by the time delays. The beams generated by the pulses from the transducer elements arrive sumultaneously to the central point. This type of focusing is used to increase the image axial resoltution.

Independently of the beam focusing technique utilised, pulse echoes are received by the transducer array as a result of each one of the shots (56 in total). The received echoes are summed up, amplified and the output signal is digitized by means of an A/D converter transputer-controlled module.This module takes 256 samples of the signal and encodes it into a gray scale according to the amplitude of each of the samples. At this stage, a new set of transducer elements can be fired, and this generates a new set of echoes and its corresponding encoded array. An image frame is then constructed with the 56 vectors produced by this process. The resulting matrix (256x56) is transmitted to the host transputer for further processing and the system is now ready to acquire a new frame.

Processing

The matrix generated by the focusing and scanning stage is received by the host transputer, buffered and sent to a parallel processing subsystem which transforms the original matrix into an augmented matrix (256x168) to produce a more detailed image to be diplayed. In order to generate the augmented matrix, different interpolation algorithms have been implemented such as, linear splines, quadratic splines and cubic convolution (Russ, 1995).

Each one of these algorithms have been implemented using a processor farm model of parallelism (Galletly, 1990). Under this scheme a **master** process controls and assigns tasks to its subordinate **worker** processes by means of a so called **farmer** process. The workers execute identical tasks and upon completion of a task, further work is assigned until the whole set of tasks has been exhausted.

Figure 6 shows the farm topology used for the implementation described in this work. Note that the master also includes a worker process which allows it to interpolate its own subframes whilst it is waiting for results from the rest of the workers. When all the subframes are processed, the master integrates the augmented data frame consisting of 256x168 elements and sends it to the host transputer which in turn sends it to the host computer for displaying purposes.

Displaying

The displaying process is conducted by the host computer which runs a server program for this purpose. This program interfaces the computer to the host transputer, providing the parallel system with input-output facilities and file system services. Interpolated data frames, consisting of 256x168 elements are encoded into a gray scale level and transferred to the host video RAM. An user program displays the images on a VGA type display .

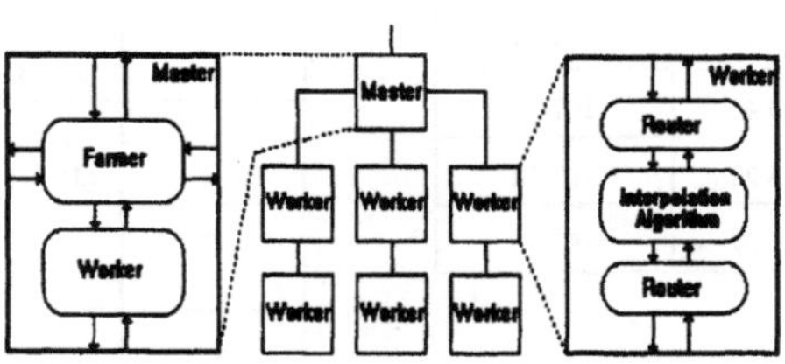

Figure 6. System processing farm model.

RESULTS AND ANALYSIS

As described earlier, the ultrasonic imaging system aims to acquire, construct, process and display images with a minimum rate of 25 frames/second. In order to achieve this, different parallel processing schemes have been applied to the different stages that integrate the system.

For the acquisition stage depicted in figure 4, pulses are fired from the transducer array constituted by 64 elements and 3.5 MHz central frequency. Pulses are produced and controlled by a transputer which adds time-delays according to the focusing scheme utilised. Echoes received by the transducer array as a result of each one of the shots (56 in total) are summed up, amplified and passed through a 1 MHz A/D converter. The 256-bit digitized signal is then passed to the transputer which encodes it into a gray scale according to the amplitude of each of the samples of the received signal. As it can be observed in the block diagram depicted in figure 4, a transputer network has been used to process the 256x56 original image. In order to produce a more detailed image to the observer (i.e. a 256x168 image), three of the most common interpolation algorithms have been implented (i.e. linear splines, quadratic splines and cubic convolution) using for this purporse a Processor Farm parallel scheme.

However, it is well known (Webber, 1992) that treatment of images can become cumbersome due the great amount of data that needs to be passed through the different processors in a parallel network. A granularity and communications study has been conducted in order to determine the grain size which minimizes the communication and synchorisation time according to the interpolation algorithms considered. For this study, the original image is buffered by the master transputer and only the interpolated data (i.e. 256x112) is communicated back by the worker processors after the intepolation algorithm has been applied. Table 1 shows the results where N is the number of 56-element arrays transmitted, and the columns show the communication+sequential processing times for the total image.

Table 1. Sequential Processing + Communication times for interpolation algorithms.

N	linear splines	quadratic splines	cubic convolution
1	130.91	173.30	211.27
2	120.19	165.29	206.05
3	117.98	164.93	199.16
4	116.84	164.90	199.15
5	114.70	163.41	197.72
8	114.50	163.23	197.70
16	124.20	163.07	197.62

Table 2. Performance results. Execution times (ms) & frames/s.

number of workers	linear splines	quadratic splines	cubic convolution
1	100.80 (9.92)	147.76 (6.76)	183.78 (5.44)
2	50.56 (19.77)	74.30 (13.45)	92.41 (10.82)
3	34.08 (29.34)	49.72 (20.11)	61.83 (16.17)
4	28.63 (34.92)	37.92 (26.36)	46.72 (21.4)
5		31.92 (31.32)	38.04 (26.28)
6			33.4 (29.94)

It can be observed that for the quadratic splines and cubic convolution the optimum grain size is 16, whereas for the linear splines algorithm the optimum grain size is 8. It is clear that the processing time, in the case of linear splines, is the smallest compared with the communication of data involved and therefore requires a smaller grain size. Considering these results, a Processor Farm implementation has been realised using a star topology (to make the most of the 3 tansputer links available) and involving the farmer in the process of computing part of the tasks (i.e. an active farm).

The results shown in Table 2 indicate the execution times and the equivalent number of frames/second obtained for each of the implemented algorithms. It can be observed that the goal rate of at least 25 frames/second has been achieved in all cases by only varying the number of processors required for each implementation. It is also important to point out that the scalability obtained from the implementations has been very high, achieving efficiencies up to 98%, 97% and 96% for the linear,quadratic and cubic convolution algorithms respectively.

Figures 7 and 8 show the resultant images before and after appliyng cubic convolution interpolation using a cylindrical focusing techique. A 35 mm diameter pipe has been used for this purposed, placing it at different depths ranging from 1 to 13 cm.

In the ultrasonic image observed in figure 7, the pipe is seen as an elipse slice. Due to the cylindrical focusing applied, the center of the object appears clearer than the edges. Figure 8 shows the ulrasonic image after applying cubic convolution interpolation. The quality of the image is excellent allowing the observer to distinguish more details which is an important feature in many systems, such as in medical tomography.

Figure 7. Ultrasonic original image.

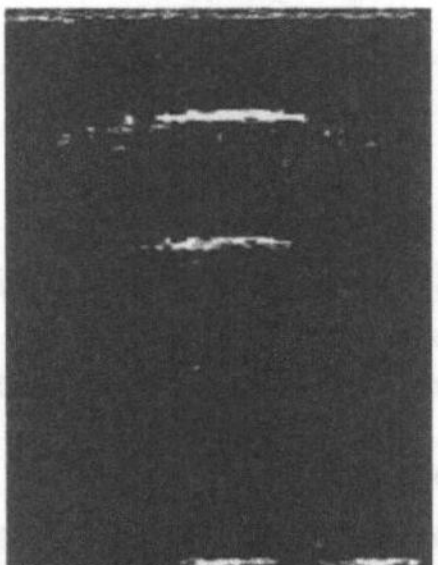

Figure 8. Ultrasonic image applying cubic convolution interpolation.

The system described is also capable of using the multiprocessing block for later analysis of a particular image chosen by the user. It is possible then to apply postprocessing techniques off-line whilst continuing acquiring new information.

CONCLUDING REMARKS

The work presented here describes an ultrasonic imaging system based on a parallel processing architecture which acquires, constructs, processes and displays images with a minimum rate of 25 frames/second. The Intrinsic parallelism often found in ultrasonic imaging has been succesfully exploited by means of developing and implementing a parallel processing subsystem for controlling the focussing and scanning stage in the acquisition of ultrasonic data, and for providing the required capability to the system for processing and constructing high quality images in real-time.

Beam focusing and scanning techniques have been also developed and implemented in order to improve the image's spatial resolution. The system can be programmed electronically to vary the depth of the focal region to match the depth of the region under consideration.

Considering the poor resolution of the original acquired image, parallel interpolation algorithms have been considered to produce a more detailed image to the observer. Algorithms such as linear splines, quadratic splines and cubic convolution have been implemented using for this purpose a farm model of paralelism. A study of granularity, considering execution and communication times for the interpolation algorithms, has been fundamental for the effectiveness of the approach. The scalabitity obtained has been very high achieving efficiencies up to 98%, 97% and 96% for the linear splines, quadratic splines, and cubic convolution algorithms. The flexibility and scalability presented by the system developed, allows us to incorporate emergent technologies and opens a greater scope for developing and utilising more complex algorithms in order to improve the image visual appearence and still achieve a real-time response.

ACKNOWLEDGMENTS

The authors acknowledge M. Fuentes, A. Jiménez for their participation on this work. Also to DGAPA-UNAM (PAPIIT-IN106796), CONACYT(Proy.2146P-A9507) and CONACYT (Prog. México-Cuba No.8.17) for their financial support.

REFERENCES

Cavaye, D.M. and R.A.White (1993). *Arterial Imaging-Modern and Developing Technology.* Chapman&Hall Medical Series. London. U.K.

Fish, P. (1990). *Physics and Instrumentation of Diagnostic Medical Ultrasound.* John Wiley & Sons, Chichester. U.K.

Galletly, J. (1990). *Occam 2.* Pitman Publishing. U.K.

Powis, R.L. (1984). *A Thinker's Guide to Ultrasonic Imaging.* Urban&Schwarzenberg. Baltimore. USA.

Russ, J.C. (1995). *Image Processing Handbook.* 2nd. Edition, CRC Press.

Webber. H.C. (1992). *Image Processing and Transputers.* IOS Press.

Wells, P.N.T. (1996), State-of the-Art of Ultrasound Imaging in Medicine and Biology. *Acoustical Imaging* Vol. 22, Plenum Press. New York.

CAUSALITY REVISITED

Sidney Leeman[1], Joie P. Jones[2], and Andrew J. Healey[3]

[1]Department of Medical Engineering and Physics
King's College School of Medicine and Dentistry
London SE22 8PT, U.K.
[2]Department of Radiological Sciences
University of California Irvine
Irvine, CA 92697-5000, U.S.A.
[3]EEIE
City University
London ECIV 0HB, U.K.

INTRODUCTION

Causality is a well established principle in physics, and, like many other such principles that are thought to be rigorously obeyed, it should be mirrored in ultrasound physics - and more so, in medical ultrasound. This paper examines that hypothesis, and seeks to review some of the constraints that the causality principle imposes on the physics of pulse propagation, in particular,

Unlike most other physics laws, causality does not have a universally exact mathematical prescription. It is often stated as "the effect cannot precede the cause", but this is occasionally quite difficult to write in a mathematically pragmatic way. We shall refer to this somewhat non-analytical statement as 'exact causality', in order to distinguish it from some of the other, perhaps less restrictive formulations, that follow below. We shall confine ourselves to the consequences for medical ultrasound pulse propagation.

ANALYTICITY AND CAUSALITY

Consider the situation that a pressure disturbance $p(0,t)$, is created at (one-dimensional) spatial location $x=0$, starting at time $t=0$. Formally, this disturbance may be written as a Fourier transform

$$p(0,t) \equiv f(t) = \int F(\omega) \exp(i\omega t) d\omega \qquad (1)$$

Acoustical Imaging, Volume 24.
Edited by Hua Lee, Kluwer Academic/Plenum Publishers, 2000.

where ω denotes the circular frequency. Formally, the integral extends over the range $(-\infty, \infty)$ and may be extended to complex ω values. Close the integral with large radius contour in the upper half ω-plane, and it can be seen that $f(t)$ will be 0 for negative t, provided F is well behaved in the upper half-plane (no poles; finite at ∞). This is, of course, as it should be: the disturbance cannot exist before it is created.

More interesting conclusions can be made if the forward-travelling pulse from this disturbance is examined. The emitted pulse is given by

$$p(x,t) = \int F(\omega)\, \exp(i\{\omega t - kx\} - \alpha x) d\omega \tag{2}$$

with $k = k(\omega)$ and $\alpha = \alpha(\omega)$. α denotes the frequency dependent attenuation (actually: absorption) in the medium, and the dispersive acoustic speed is embodied in k. The same manipulations in the complex ω-plane as above yield that $p(x,t) = 0$ provided that $x \geq C_0 t$, where C_0 -- the non-dispersive velocity of the pulse front -- is determined by $|F(\infty)|$. Also, convergence of the integral in Equ.(2) requires that $k(\omega)$ and $\alpha(\omega)$ are related (dispersion relations).

It thus turns out that exact causality implies that a pulse propagates with a sharp *front*, travelling at a non-dispersive speed, even in lossy, dispersive media. We shall refer to this result as 'propagative causality'. Another result, which is not explicitly proven here because it is, in fact, well-known, is that the frequency dependencies of attenuation and velocity are related via dispersion relations.

Some observations are relevant to pulse simulations: Gaussian pulses, much used in computer simulations, are non-causal because the pulse front is not sharp, but extends to ∞! Also, arbitrary choices of dispersive attenuation and sound speed are unlikely to obey causality. Moreover, there is a fundamental sonic speed in addition to the phase and group velocities: the non-dispersive pulse-front speed will be termed the 'signal velocity'. Many of these observations may be proven by considering the moderately simple case, where the propagation medium is modelled as a simple Maxwell model, as discussed below.

THE SIMPLE MAXWELL MODEL

In this case, the acoustic density perturbations relate to the acoustic pressure via

$$\rho_A = \frac{1}{C^2} + a \int^t p(x,t')dt' \tag{3}$$

where C and a are constants. When utilised in the linear approximation, this leads to the wave equation

$$\frac{\partial^2 p}{\partial x^2} - \frac{1}{C^2}\frac{\partial^2 p}{\partial t^2} - ap = 0 \tag{4}$$

For this medium, the following holds (proven in the usual way)

$$\alpha(\omega) = \frac{1}{\sqrt{2}}\frac{\omega}{C}\left[\sqrt{1 + \frac{a^2C^4}{\omega^2}} - 1\right]^{\frac{1}{2}} \tag{5}$$

$$V(\omega) \equiv \frac{\omega}{k} = \frac{2a}{\alpha} \quad \text{(phase velocity)} \tag{6}$$

In actual calculations, it turns out that the pulse-front travels with a non-dispersive velocity C -- the signal velocity [Leeman, 1980].

While the simple Maxwell model does not provide a good fit to all soft tissues, it does, at least, manage to fit some absorption data, *viz* that measured by Brady *et al* [1976] for the mouse testis, so it cannot be regarded as entirely unrealistic. The model predicts three important sound speeds: the phase velocity $\frac{\omega}{k}$, the group velocity $\frac{\partial\omega}{\partial k}$, and the signal velocity, C. While it is not suggested that such a model is suitable for human soft tissues in general, there is the suggestion that extreme care should be taken when measuring sound speed in *dispersive* media in general, in order to establish which sound speed is actually being accessed by the experimental setup.

It is interesting to consider a modified wave equation

$$\frac{\partial^2 p}{\partial x^2} - \frac{1}{C^2}\frac{\partial^2 p}{\partial t^2} - ap + b\frac{\partial^2 p}{\partial x \partial t} = 0 \tag{7}$$

This equation exhibits essentially the same dispersive absorption as the simple Maxwell model, but the (phase) velocity dispersion is altered [Leeman, 1980]. Moreover, its pulse-wave solutions satisfy propagative causality. This example shows that propagative causality does not *uniquely* fix the velocity dispersion from the absorption!

SUPER-LUMINALITY

It has been shown that causality predicts that the canonical wave equation

$$\frac{\partial^2 p}{\partial x^2} - \frac{1}{C^2}\frac{\partial^2 p}{\partial t^2} = 0 \tag{8}$$

has transient solutions with pulse fronts that travel at sonic speed C. On the other hand, so-called 'X-wave' solutions have been found with super-luminal (sound speed > C) components, and it may be argued that, for some reason, causality does not hold in this case.

However, we show here that, depending on the measurement technique, superluminal components may be observed with 'ordinary' medical ultrasound pulses in a water tank, in the laboratory. Consider a uniformly excited planar disc radiating into an ideal, lossless, dispersion-free medium obeying the canonical wave equation (water, say, for low megahertz pulses). Few would gainsay the observation that, in these circumstances, causality is observed, and that the sound speed (only a single value is needed because the medium is non-dispersive) is equal to C. It is well known that the emitted pulse may be regarded as consisting of two components: the edge-wave (emanating as an toroidal wave from the transducer edge, and which spreads out as it propagates) and the direct-wave (a replica of the transmitting surface, which propagates without change of shape). When a point hydrophone is placed on the beam axis near to the transducer face, it first measures the arrival of the direct-wave, and then, a short while later, the edge-wave component is detected, provided that the excitation is short enough for the two components to remain separated from one another. A typical measurement, albeit with a weakly focused transducer in a water tank, is shown in Figure 1. It is clear that the two components are distinguished from one another.

On the other hand, if the measurement is repeated at some distance from the transducer face, it is found that the two components have coalesced, as is also shown in Figure 1. Since it is known that the direct wave travels at the speed C, appropriate to water, the only construction that can be put on these measurements is that, as measured on the beam axis,

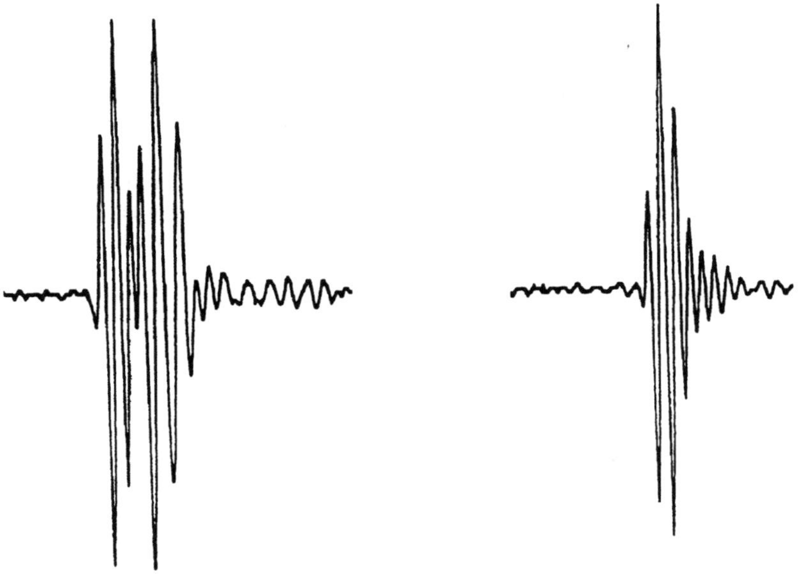

Figure 1. On-axis time-domain pulse from typical transducer, as measured with a point hydrophone. *Left*: Close to the transducer face: the direct- and edge-wave components are clearly separated. *Right*: Relatively far from the transducer face: the edge-wave component has caught up with the direct-wave component, and the two have coalesced.

the edge-wave component 'catches up' with the direct-wave component (on axis) *i.e.* it moves faster than the direct-wave component (on axis). With this measurement configuration, it appears that edge-wave component's speed is distance dependent, super-luminal, and tends to C at far distances. However, a careful examination of the geometry of the situation in three dimensions shows that the effect is a *geometrical* artifact, dependent on measurement with a point hydrophone, and not an indication of true sonic super-luminality. Similar considerations hold for statements about super-luminality of X-wave pulses: the true 3D wave travels at a sonic speed appropriate to the ideal medium, and causality is not infringed.

SOME CONCLUSIONS

Causality is strictly obeyed by real physical systems. Propagative causality, as defined here, is appropriate for sonic pulses, and imposes a finite upper limit on the signal velocity, which is non-dispersive even in dispersive media. Causality implies dispersion relations, but these do not enforce uniqueness on the relationship between absorption and velocity dispersion. Beware when constructing simulations!

ACKNOWLEDGEMENTS

The Wellcome Trust is thanked for some financial support.

REFERENCES

Brady, J.K., Goss, S.A., Johnston, R.L., O'Brien, W.D., and Dunn, F., 1976, *J. Acoust. Soc. Am.*, 60:1407.
Leeman, S., 1980, Ultrasound pulse propagation in dispersive media, *Phys. Med. Biol.*, 25:481.

RESOLUTION ANALYSIS OF ACOUSTIC TOMOGRAPHIC IMAGING WITH FINITE-SIZE APERTURES BASED ON SPATIAL-FREQUENCY COVERAGE

Stephanie Lockwood and Hua Lee

Department of Electrical and Computer Engineering
University of California at Santa Barbara
Santa Barbara, CA 93106

ABSTRACT

The objective of this paper is to present a comprehensive analysis of resolving capability of acoustic diffraction tomography using finite-size receiving apertures based on the span of spatial spectral coverage. The analysis is conducted on both the transmission and reflection modes, and can be generalized into various data acquisition configurations. In addition, the analysis provides an important link to other tomographic imaging modalities such as the classical X-ray tomography and synthetic aperture radar systems.

INTRODUCTION

Tomography is probably the most exciting topic in computed imaging is recent years. The applications of tomographic imaging can be found in many areas, ranging from X-ray and magnetic resonance imaging (MRI) to synthetic-aperture radar (SAR) and nondestructive evaluation (NDE) [1, 2, 3, 4]. Among them, diffraction tomography, which is directly related to acoustical imaging, is widely regarded as the most interesting development for its elegance associated with wave propagation as well as the complexity due to the variety of imaging configurations.

System analysis of diffraction tomography in acoustical imaging is commonly performed based on ideal cases in a superficial manner, under the assumption of infinite-size aperture. There is a lack of fundamental quantitative analysis of system performance taking into account the effect of aperture size when it becomes finite. In this paper, we provide a comprehensive treatment of diffraction tomography based on the approach of *backward propagation image formation.* We first briefly review the spectral distribution of coherent acoustical imaging systems, and subsequently illustrate the formation of spectral coverage of diffraction tomography of both transmission and reflection modes. Then, from the perspective of backward propagation image formation, we incorporate the effect of finite-size apertures into the analysis in a simple and direct manner. By doing so, it allows us to fully visualize the image degradation due to finite aperture size, through the reduction of spatial-frequency bandwidth coverage.

This analysis of acoustic tomography also provides us with important links to other tomographic imaging systems. As shown in this paper, classical X-ray tomography can be described as a limiting case of transmission-mode diffraction tomography, with an interesting generalization of the *Central Projection Theorem*. At the other end, spotlight mode synthetic aperture radar (SAR) can be regarded as a limiting case of the reflection-mode diffraction tomography as the receiver aperture reduces to a single position.

SPATIAL FREQUENCY DISTRIBUTION

Spatial frequency content is one of the governing parameters of resolving capability of an imaging system. Therefore, we start this paper with a brief overview of the spatial frequency distribution associated with the propagation and scattering of coherent wavefields. It is widely known that the formulation of coherent wavefields is governed by the Helmholtz equation. The spectral content of the coherent wavefield is concentrated over a sphere of radius $1/\lambda$, where λ is the wavelength. From the perspective of a linear system, the transfer function associated with the propagation can be described as

$$H(\mathbf{f}) = \delta(|\mathbf{f}| - 1/\lambda) \tag{1}$$

where $\mathbf{f} = [f_x, f_y, f_z]$ is the spatial-frequency vector and f_x,f_y, and f_z, are spatial frequencies in the x, y, and z directions, respectively [5].

In this paper, presentations are given in two dimensions for simplicity. The spectral distribution of the resultant coherent wavefield can be illustrated as Fig. (1). In the ideal case of data acquisition with an infinite linear receiving aperture, the detected acoustic wavefield is a lowpass signal with the cutoff spatial frequency reaching the maximum span of $1/\lambda$. By backward propagating the received wavefield to the source region, we reconstruct the wavefield distribution corresponding to a spectral content of the semicircle toward the direction of the aperture span, as shown in Fig. (2) [6, 7]. When detected by using an aperture array of finite size, only part of the circularly-distributed wavefield spectrum can be obtained for image formation. More specifically, with a finite aperture, the spectral content of the received wavefield will be limited to an arc of even smaller angular span due to the reduction of effective spatial-frequency bandwidth.

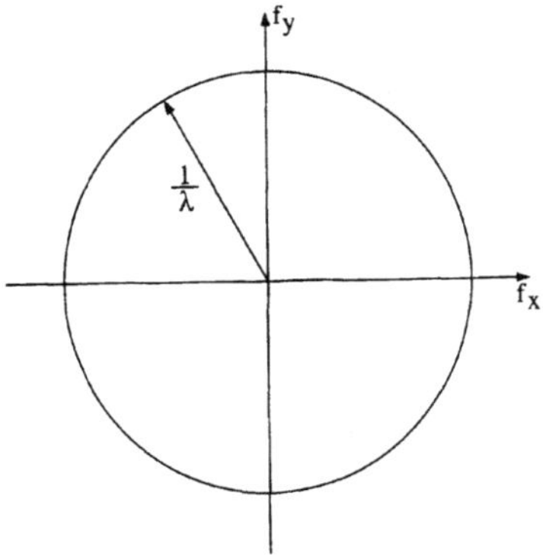

Figure 1. Spectral distribution of the resultant coherent wavefield.

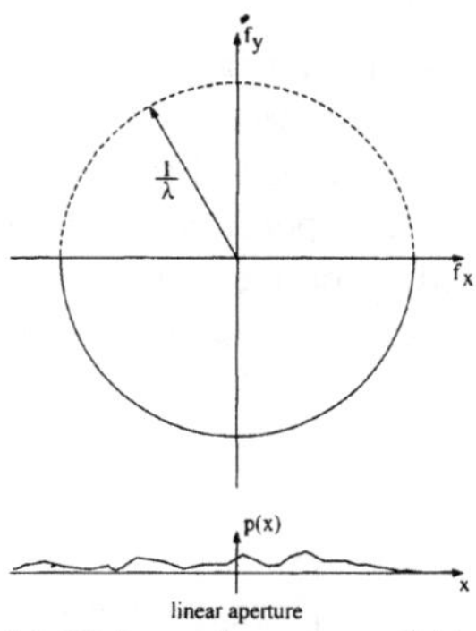

Figure 2. Spectral distribution of the resultant coherent wavefield after backward propagating the received wavefield to the source region.

ACTIVE ILLUMINATION

In diffraction tomography, the configuration of the imaging systems normally requires active illumination. Mathematically, we describe the received wavefield as the result of the secondary reactive radiation caused by the acoustic illumination waveform.

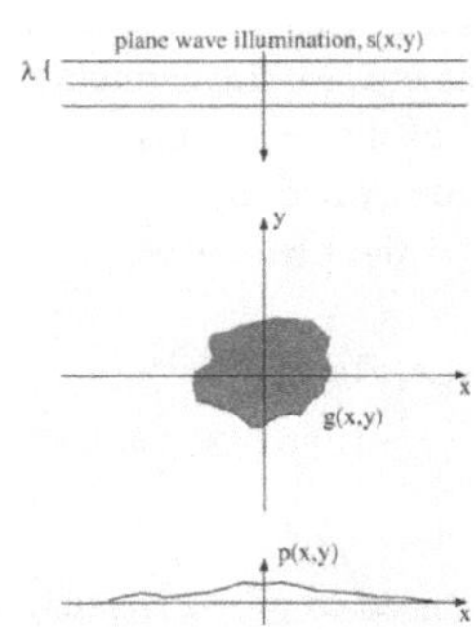

Figure 3. Transmission-mode imaging scheme using coherent plane wave illumination.

Consider the following imaging scheme shown as Fig. (3) where the object distribution $g(x, y)$ is illuminated by a coherent plane wave of wavelength λ, and the spatial frequency of the plane wave is $(f_x, f_y) = (0, 1/\lambda)$ corresponding to the direction of the active illumination. Mathematically, the wavefield generated by the illumination plane wave can be written as

$$s(x, y) = \exp(j2\pi y/\lambda). \tag{2}$$

The object is then modulated by the illumination wavefield and becomes a secondary source, and the resultant wavefield is then detected at a linear aperture, assumed to be of infinite size at this point, with the aperture span along $y = y_0$. The detected wavefield along the linear aperture is termed a *projection*, and denoted as $p(x)$. The configuration illustrated in Fig. (3) is typically referred to as the transmission imaging mode where the receiver array is on the opposite side of the transmitter.

As indicated in the previous discussion, the projection $p(x)$ is a lowpass signal with cutoff frequency of $1/\lambda$. By backward propagating the projection $p(x)$ to the object region, an image subcomponent is then formed, and the spectral distribution of this image component is a semicircle toward the direction of the aperture, as shown in Fig. (2). The other half of the circular spectrum corresponds to the reflection-mode wavefield, which can be obtained by placing an aperture on the same side of the transmitter. We should also note that the transmission and reflection modes are responsible for different spectral distributions of the overall resultant wavefield.

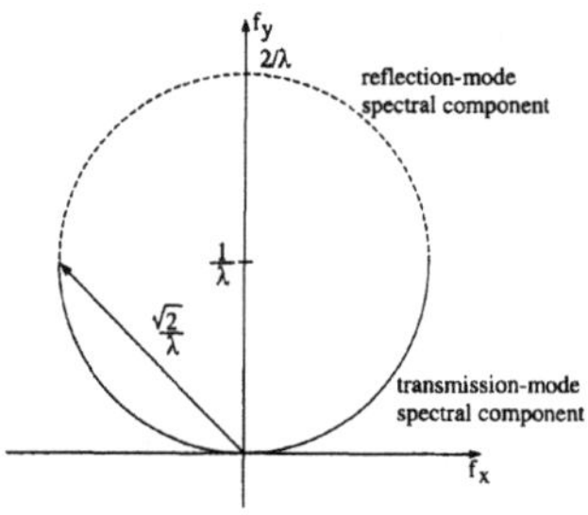

Figure 4. Spectrum of the demodulated and backward propagated image component for the system of Fig. (3).

Because the resultant wavefield is caused by the scattering due to the illumination instead of direct radiation from the object $g(x, y)$, the semicircular spectrum is associated with a *secondary source function* instead of the object distribution itself. Mathematically, the secondary source function can be described by $g'(x, y)$ of Eq. (3) as the object distribution modulated by the wavefield of the plane-wave illumination.

$$\begin{aligned} g'(x, y) &= g(x, y)s(x, y) \\ &= g(x, y)\exp(j2\pi y/\lambda) \end{aligned} \tag{3}$$

To perform image formation, we first need to properly relate the backward propagated image component to the object distribution, by removing the contribution from the illumination wavefield. Because of the use of plane wave for illumination, the demodulation procedure is relatively simple. It can be performed in the space domain, or even more effectively in the spatial-frequency domain by shifting the spectral distribution backward by $1/\lambda$ toward the transmitter. As a result, the spectrum of the demodulated backward propagated image component is an offset semicircle, shown in Fig. (4).

The semicircle spectral distribution corresponding to the transmission-mode image component always passes through the origin, independent of the operating wavelength. As the operating frequency of illumination signal increases, the operating wavelength becomes shorter, the span of the semicircle becomes larger, and the curvature of the circle decreases accordingly. When the operating frequency of an acoustic tomography system approaches infinity, the spectral content of a backward propagated image component from one single projection, especially in the low spatial-frequency region, approaches a linear segment passing through the origin. This means, in terms of the spectral content as well as image formation, the transmission-mode diffraction tomography is approaching the classical case of X-ray tomography as the operating frequency approaches infinity. In addition, from a different perspective, we can describe the conventional X-ray tomography as a limiting case of the generalized transmission-mode diffraction tomography, and the spectrum analysis presented here in this paper can be viewed as the generalized version of the *Central Projection Theorem.*

TOMOGRAPHIC RECONSTRUCTION

As it is illustrated in the previous section, the spectral coverage of the image component reconstructed from one single projection is quite limited, and so is the resolving capability, consequently. In order to improve the resolution of the system, one approach is to perform a rotational scan to generate additional coverage of the spectral content, which can be achieved by rotating either the data acquisition system or object specimen.

Because Fourier transformation is rotation invariant, as the object rotates, the spectral distribution of the object function also rotates accordingly. Therefore, the rotational scan enables us to collect a sequence of semicircular spectral components from various angles and form a circular spectral patch through tomographic superposition. For transmission-mode systems, the resultant spectral coverage due to the rotational tomographic scan is a circular disk and the radius is $\sqrt{2}/\lambda$. For the reflection mode, the final spectral coverage is a circular band, and the upper and lower bounds of the band are $2/\lambda$ and $\sqrt{2}/\lambda$, respectively. Fig. (5) shows the spectral coverages of tomographic reconstruction by transmission and reflection modes.

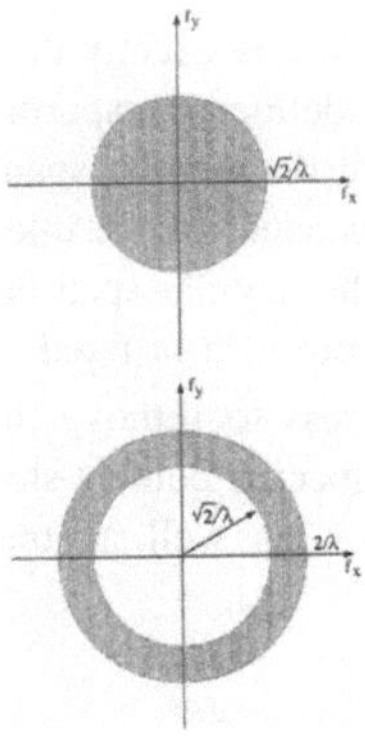

Figure 5. Spectral coverages of tomographic reconstruction by transmission and reflection modes.

It is important to point out that the transmission-mode imaging produces a spectral coverage completely different from that generated by reflection-mode operations. These two spectral coverages are mutually orthogonal without any overlapping. The spatial frequency band of the transmission-mode tomographic images lies in the lowpass region, while the reflection-mode covers the bandpass domain. Respectively, the point spread functions of the transmission and reflection mode systems are

$$\begin{aligned} h_{tr}(r) &= \frac{1}{r} J_1\left(2\sqrt{2}\pi r/\lambda\right) \\ h_{ref}(r) &= \frac{1}{r} J_1(4\pi r/\lambda) - \frac{1}{r} J_1\left(2\sqrt{2}\pi r/\lambda\right) \end{aligned} \tag{4}$$

It is also interesting to note that both the transmission and reflection modes cover exactly the same amount bandwidth area in the spatial frequency domain.

$$Area = \frac{2\pi}{\lambda^2} \tag{5}$$

This means, even though the point spread functions of these two operating modes are different in nature, the levels of information content are identical, and therefore, these two modes have the same level of resolving capability.

FINITE-SIZE APERTURE

The analysis so far has been limited to the case of infinite receiving aperture. However, the size of the receiving aperture is normally finite in practice. When the size of the receiving aperture becomes limited, the detected wavefield does not extend to the full span of the spectral coverage as in the cases of infinite-size aperture. The cutoff frequency of the lowpass wavefield is consequently reduced from the upper limit of $1/\lambda$ to

$$f_c = \frac{1}{\lambda} \sin\left(\frac{\theta}{2}\right) \tag{6}$$

where the term $\sin(\theta/2)$ is a factor given by the Rayleigh criteria as a function of aperture size and range distance [8, 9]. When it is projected into the two-dimensional space in the spatial-frequency domain, the spectrum covers only an arc segment instead of a semicircle. The span of the arc is directly related to the size of the receiving aperture, as given in Eq. (6). According to the formulation of the lowpass cutoff frequency in the resolution analysis, the angular span of the arc is exactly θ. This means the angular span of the arc is the same as angular coverage defined by aperture with respect to the source region. It is also important to point out that the angular span of the wavefield spectrum is governed by the aperture size and is independent of the operating mode, transmission or reflection, of the imaging system. Hence, the angular span is identical for both the transmission and receiving modes when the aperture size is fixed.

After the demodulation process to remove the contribution due to the illumination wavefield, the spatial-frequency spectral content shifts. Fig. (6) shows the resultant spectral distributions, of the transmission as well as the reflection-mode wavefield, after the demodulation process.

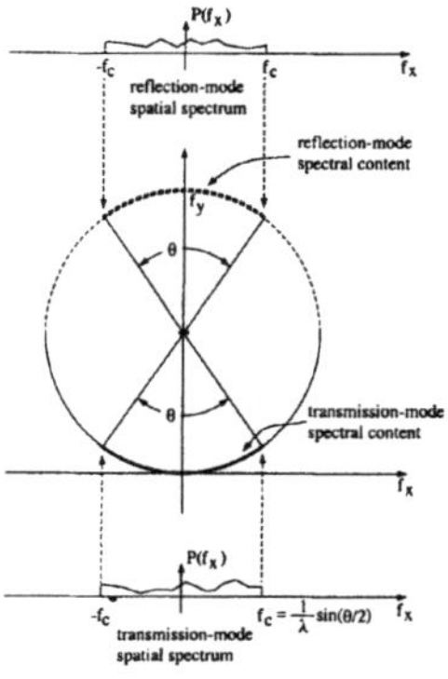

Figure 6. Spectral distributions of the transmission and reflection-mode wavefields after demodulation.

A tomographic angular scan is then conducted in a similar manner to expand the spectral coverage. As a result, the complete spectral distribution of the transmission-mode is now a disk of reduced area, and the reflection-mode coverage becomes an annular band with reduced width. Quantitatively, the radius of the circular patch corresponding to the transmission-mode operation is

$$R = \frac{2}{\lambda} \sin\left(\frac{\theta}{4}\right) \tag{7}$$

and the upper and lower bounds of the annular band of the reflection-mode spectrum are

$$\begin{aligned} R_{max} &= \frac{2}{\lambda} \\ R_{min} &= \frac{2}{\lambda} \cos\left(\frac{\theta}{4}\right), \end{aligned} \tag{8}$$

respectively. Once again, the transmission and reflection modes, with limited aperture size, produce different spectral coverages without any overlapping area. The transmission mode covers the low-frequency domain while the reflection mode covers the bandpass region. It is also important to note these two spectral distributions have identical total spatial-frequency coverage area of

$$Area = \frac{2\pi}{\lambda^2}\left(1 - \cos\frac{\theta}{2}\right) \tag{9}$$

and thus have the equivalent level of resolving capability. With a finite-size receiving aperture, the point spread functions of the transmission and reflection-mode tomographic reconstruction become

$$\begin{aligned} h_{tr}(r) &= \frac{R}{r} J_1(2\pi R r) \\ h_{ref}(r) &= \frac{R_{max}}{r} J_1(2\pi R_{max} r) - \frac{R_{min}}{r} J_1(2\pi R_{min} r) \end{aligned} \tag{10}$$

It can be easily seen that, if the angular coverage approaches the full 180-degree span, the spectrum reaches the maximum content corresponding to the case of infinite-size aperture, as we expect.

The illustration of spectral coverage has been concentrated mainly on the transmission or reflection modes in this paper. With slight modifications, we can generalize the analysis into various data acquisition configurations, monostatic or bistatic. One can easily visualize that the spectral content of an arbitrary data acquisition scheme will be an annulus. The radial upper and lower bounds of the annulus are $\sqrt{2}/\lambda$ and zero, respectively. The cross-sectional width of the annulus, equivalently the variation of the upper and lower bounds, is governed by the angular coverage defined by the aperture as well as the angular offset between the direction of illumination and the location of the receiving aperture.

Yet, the most intriguing development from this analysis is when the imaging system configuration approaches the other extreme where the data acquisition is performed with a single transmitter-receiver in monostatic operating mode. In this case, the angular span of the spectral content, θ, becomes zero, which becomes a limiting case of the reflection mode. As a result, the spectral coverage is reduced from a circular band to a ring of radius $2/\lambda$. This result is in complete agreement with that derived from the coherent synthetic-aperture radar (SAR) operating in spotlight mode [10].

In synthetic-aperture radar imaging, the resolving capability is improved by expanding the spectral coverage from a ring to an annulus with wideband illumination such as chirp waveforms. And, as illustrated in this paper, a similar expansion of spectral coverage can be achieved with a designated aperture size. This gives the equivalence and a direct trade-off relationship between the temporal bandwidth of illumination signals and spatial span of receiving apertures. This relationship allows us to further the optimization of system resolution by combining the consideration of aperture configurations with signaling schemes.

CONCLUSION

The main objective of this paper is to establish a basic framework for the resolving capability of acoustic tomography with finite-size receiving aperture. For consistency, the resolution analysis is parameterized in terms of spectral coverage in the spatial frequency domain. To provide the formulation in a comprehensive manner, we first illustrated the spectral coverage of diffraction tomography of the ideal case of infinite-size receiving aperture for both transmission and reflection modes. Subsequently, we extended the analysis by including the degradation factor corresponding to finite aperture size. Based on the formulation through the measure of spectral coverage, we provided the direct relationship of resolving capability with the aperture size due to the change of effective bandwidth area. The analysis provided in this paper concentrates mainly on diffraction tomography in either transmission or reflection modes. Nevertheless, with slight modifications, the analysis can be extended well into various data acquisition configurations, monostatic or bistatic.

The usefulness of this resolution analysis of diffraction tomography is not confined within the field of acoustical imaging. Because of the simplicity and generality, the analysis also provided theoretical linkage with, on one hand, the classical X-ray tomography as a limiting case of the transmission-mode diffraction tomography and synthetic-aperture radar (SAR) imaging on the other hand as a limiting case of reflection-mode tomography as we reduced the aperture size to a single receiving element.

ACKNOWLEDGEMENTS

This research is supported by the National Science Foundation under Grant No. CMS-9309775.

REFERENCES

1. H. Lee and G. Wade. *Imaging Technology*. IEEE Press, New York, 1986.
2. H. Lee and G. Wade. *Modern Acoustical Imaging*. IEEE Press, New York, 1986.
3. J. E. Mast, H. Lee, and J. P. Murtha. Application of microwave pulse-echo radar imaging to the nondestructive evaluation of buildings. *International Journal of Imaging Systems and Technology*, 4(3):164–169, 1992.
4. R. Y. Chiao and H. Lee. Scanning tomographic acoustic microscopy. *IEEE Transactions on Image Processing*, 4(3):358–369, Mar. 1995.
5. H. Lee and G. Wade. Alternative viewpoint on wave propagation. *IEEE Transactions on Sonics and Ultrasonics*, SU-30(5):331–332, Sept. 1983.
6. A. J. Devaney. Inverse source and scattering problems in ultrasonics. *IEEE Transactions on Sonics and Ultrasonics*, SU-30(6):355–364, Nov. 1983.
7. H. Lee. Formulation of the generalized backward projection method for acoustical imaging. *IEEE Transactions on Sonics and Ultrasonics*, SU-31(3):157–161, May 1984.
8. J. W. Goodman. *Introduction to Fourier Optics*, pages 129–131. McGraw-Hill, New York, 1968.
9. H. Lee. Formulation for quantitative performance evaluation of holographic imaging. *Journal of the Acoustical Society of America*, 84(6):2103–2108, Dec. 1988.
10. D. L. Mensa. *High resolution radar imaging*, pages 104–116. Artech House, Dedham, Massachusetts, 1981.

RADIATION FORCE DOPPLER EFFECTS ON CONTRAST AGENTS

Piero Tortoli, Francesco Guidi, Emiliano Maione, Francesca Curradi[1]
Vittorio Michelassi[2]

[1] Electronic Engineering Department
[2] Energetics Department "Sergio Stecco"
University of Florence
via S.Marta 3, 50139 Firenze, Italy

INTRODUCTION

Encapsulated microbubbles are of widespread use as ultrasound contrast agents, because of their inherent efficiency in backscattering ultrasound energy. They have been shown capable of providing two-dimensional echo-images of better contrast as well as color-Doppler images where blood flow is detected even in difficult conditions.

When contrast agents are injected in human blood to enhance scattering, it is implicitly assumed that they flow at the same velocity as blood erythrocytes. In some cases, however, it was observed that the spectrum of Doppler frequencies turns out to be notably widened in presence of additional microbubbles. This effect was attributed to the improved signal-to-noise ratio, allowing a wider portion of spectrum to emerge above noise, or to transit time effects resulting from the destruction of insonated bubbles[1].

We recently observed that a different phenomenon seems to happen when the acoustic temporal average intensity impinging on particles is relatively high[2]. In fact, the Doppler spectrum originated in such cases presents a shape which is completely different to that obtained at low intensity levels. In particular, some portions of the original spectrum are strongly enhanced, while others are not. By considering the correspondence between Doppler frequencies and particle velocities, this behavior suggests that the movement of injected contrast agents may be not the same as that of the original fluid.

A series of *in vitro* tests were performed to explain the physical origins of this phenomenon and to evaluate the influence of factors such as frequency of operation, pulse repetition frequency, pressure amplitude and length of the transmitted burst.

In this paper, we first describe these experiments, performed *in vitro* by using Levovist (©Schering, Berlin, Germany) particles suspended in distilled water flowing at a steady rate. The results we obtained from Doppler analysis look consistent with the recent finding that radiation force has the ability to move microbubbles in a sound field[3]. This movement is shown to be not coincident with the movement of fluid in which they are injected, although dependent on its direction.

EXPERIMENTAL SET-UP

Our main task was to investigate whether contrast agents mixed in a moving fluid always move at the same velocity as the fluid itself or, in particular conditions, they don't. To achieve this goal, we used a flow phantom consisting of a 10 mm diameter plexiglas tube, together with an ultrasound Pulsed Wave Transmitter-Receiver (TX-RX) connected to a Personal Computer (PC). A suspension of Levovist particles at 300 mg/ml concentration in distilled water, was forced to flow in the tube at a steady rate.

For this application, it was important to be able to control the way a wideband transducer was excited. The TX unit could produce sinusoidal bursts of programmable duration. By changing this duration, it was correspondingly changed the length of the resolution cell, i.e. the spatial region contributing to the backscattered signal at the time instant it is sampled in the receiver. The frequency of these bursts could be either 4 MHz or 8 MHz, which are relatively close to- and far from- the Levovist resonant frequency, respectively. The Pulse Repetition Frequency (PRF) could be selected between 8 kHz and 16 kHz.

In particular, it was possible to change the amplitude of the excitation signal over a wide range, such as to produce, in the transducer focus, pressure amplitudes from 100 kPa to over 500 kPa. The values indicated in the following figures represent the *in-situ* amplitude pressures, estimated on the basis of measurements performed *in water* with calibrated needle hydrophones.

For each transmitted pulse, the received echo signals were demodulated on two quadrature channels. The demodulation frequency could be made coincident either with the transmission frequency or with its second harmonic. The two cases will be referred to as *fundamental* and *harmonic* operating modes, respectively[4]. The baseband outputs from the demodulators were directly sampled at the PRF rate. Since in this application no significant clutter was expected, we didn't use any high-pass filter. In this way, Doppler spectral contributions down to zero frequency could be detected.

In phase and quadrature Doppler samples were digitized through two 16-bit Analog-to-Digital converters and stored in the PC memory. The data stream was finally post-processed through 512-point FFTs with 50% overlap. Doppler spectra were finally ensemble averaged to produce a dramatic reduction in spectral variance.

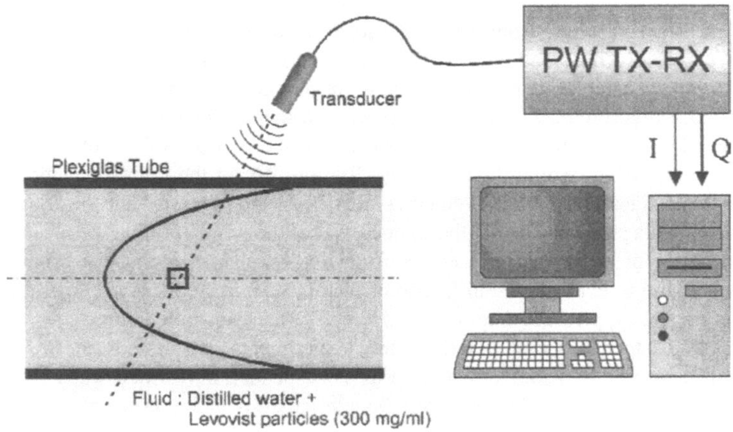

Figurel. Experimental set-up: steady flow in a plastic tube was analyzed through a custom transmitter-receiver unit connected to a personal computer.

EXPERIMENTAL RESULTS

The typical "harmonic" spectra which were obtained by insonifying the tube with a 45° incidence angle, at different pressure amplitudes, is shown in Figure 2. The resolution cell was 2 mm long and located in the center of a laminar flow directed toward the transducer. The spectrum obtained at the lowest pressure, 200 kPa, presents the typical shape expected in these experimental conditions. It is clearly symmetrical, with a width depending on the combination of the so-called geometrical broadening with the velocity gradient within the range cell[5]. Its mean frequency, $f_D \approx 750$ Hz, corresponds to a mean velocity component, v, of about 10 cm/s within the resolution cell, according to the well known[6] Doppler equation:

$$f_D = \frac{2}{\lambda} v \cos\theta$$

where λ is the wavelength of the transmitted beam and θ is the angle between the beam axis and the flow direction (45° in our experiments).

When pressure increases, the Doppler spectra produced by microbubbles tend to asymmetrically enlarge: while upper frequency remains unchanged, the lower frequency progressively decreases. In particular, when the pressure amplitude, P_A, is higher than 320 kPa, negative Doppler spectral components are also originated. From Doppler equation, this may happen if velocity components directed at an angle $\theta > 90°$ are intercepted within the resolution cell. It should be observed that these components suggest the presence of a particle movement away from the transducer, even though a *forward* flow was observed.

Since these spectra were obtained in harmonic mode, in principle they could be due to some uncontrolled nonlinear process originating spurious harmonics in the received signal. To evaluate this possibility, we repeated the experiment by demodulating the Doppler echoes with the same frequency (4 MHz) used in transmission. Results obtained in this case, and shown in Fig. 3, are clearly equivalent to those obtained in harmonic mode, the only difference lying in the Signal-to-Noise ratio, which, of course, is better in fundamental mode.

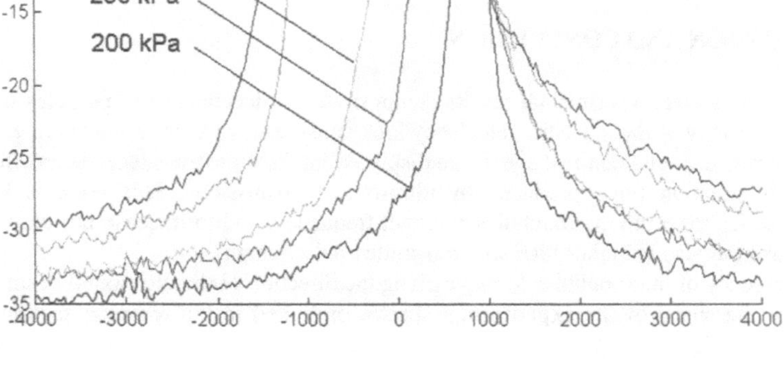

Figure 2. Typical ensemble averaged chirp spectra obtained at increasing pressures (*harmonic* mode)

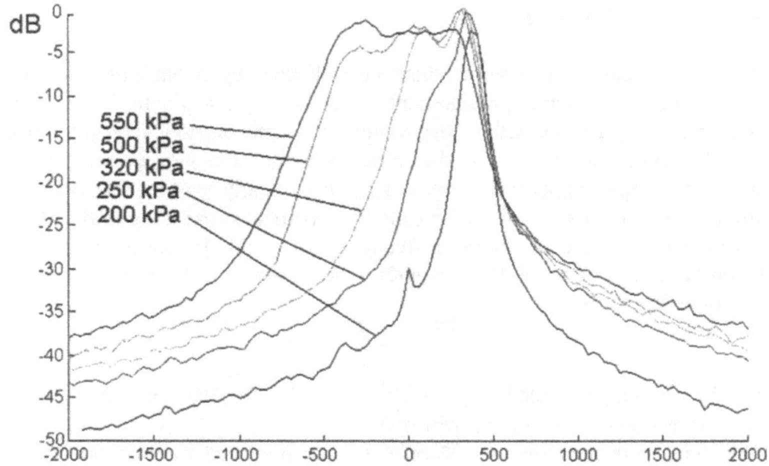

Figure 3. Typical ensemble averaged chirp spectra obtained at increasing pressures (*fundamental* mode). Note that the horizontal axis covers one half the range used in Figure 2 because double Doppler frequency shifts are involved in harmonic mode.

When the transmission frequency was increased up to 8 MHz, the bandwidth enlargement almost disappeared. This demonstrates that the phenomenon under investigation strongly depends on the frequency of the transmitted burst.

We also verified the dependence of the detected Doppler bandwidth on other parameters contributing to the intensity of the ultrasound field. By changing the PRF, it was observed that the bandwidth enlargement is less visible at 8 kHz than at 16 kHz. This suggests the influence of the acoustic intensity, since a higher PRF involves that each scatterer is subjected to the pulsed field a higher number of times for each second.

Similar results were achieved by changing the length, T_x, of the sinusoidal burst applied to the transducer. Significant examples are shown in Figure 4a and Figure 4b, where the spectra obtained at T_x =1.3 μs and T_x =2.6 μs, respectively, can be compared. The longer the transmitted pulse, the longer the time during which each particle is under the acoustic field, and the more evident is the bandwidth enlargement.

DISCUSSION AND CONCLUSION

The above experimental results demonstrate the dependence of particles movement on the intensity of the acoustic field. They look consistent with the recent finding, based on optical methods, that contrast agents are displaced far from the transducer as an effect of the primary radiation force produced by ultrasonic transmission[3]. This force is known to decrease far from the microbubbles resonant frequency, and to increase not only at higher pressures but also at higher PRF and transmitted pulse lengths.

The tendency of microbubbles to move along the direction of the ultrasound beam axis may explain the shape of the experimental spectra presented in the previous section. In fact, when this effect is negligible (i.e., at low acoustic pressures), the particles move along

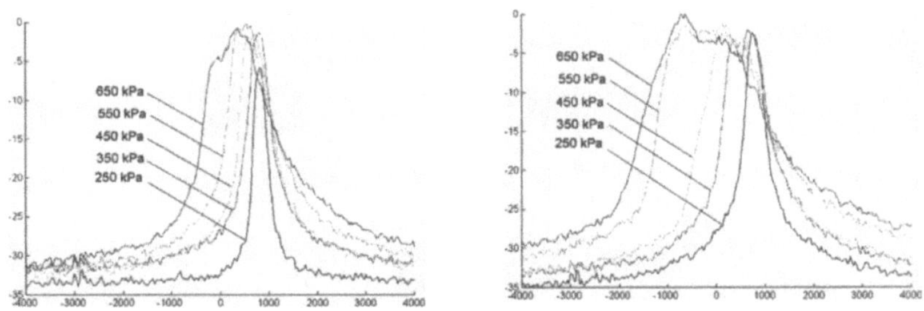

Figure 4. Doppler spectra produced in fundamental mode by transmitting a 4 MHz sinusoidal burst of length 1.3μs (left) and 2.6 μs (right)

directions parallel to the tube, in a paraxial flow fashion, as confirmed by the narrow spectra detected at 200 kPa. At higher pressures, the effect of radiation force is expected to be such that the particle trajectory tends to be bent. In a *forward flow*, this means that microbubbles entering under the ultrasound beam with velocity v and Doppler angle θ, are progressively decelerated (i.e., v decreases) and deflected (i.e., θ increases). Both these effects involve a gradual decrease in the Doppler shift frequency produced by particles moving along the bent trajectory. This behavior finally explains the *chirp* spectra which may be observed in Figures 2-4: they are due to the frequency modulation yielded in the received Doppler signal by gradual changes in both amplitude and angle of the local Doppler vectors produced during the particle transit under the ultrasound beam.

This is further confirmed by another experiment where a *reverse flow*, i.e., a flow receding from the ultrasound transducer, was investigated. In this case it is expected that the microbubbles trajectory is still bent by radiation force, with the effects, opposite to that discussed for *forward flow*, of increasing their velocity while decreasing the Doppler angle. The negative Doppler shift initially produced by contrast agents, is therefore progressively increased while they move across the ultrasound beam. This behavior is confirmed by the results shown in Figure 5, where *chirp* spectra are still observed, although they are now suggestive of increased (negative) velocities, i.e. of particles acceleration.

These results demonstrate that microbubbles used as contrast agents may move differently from the fluid where they are injected. This happens especially when the *in situ* pressure amplitude is higher than 200 kPa, a value above which the phenomenon of bubble collapse is also predicted[7]. This observation suggests the possibility that the effects of radiation force are even higher on collapsing bubbles than in normal (non-collapsing) bubble conditions.

It is important to underline that the temporal average acoustic intensities we used to produce visible spectral broadenings are within the ranges recommended by FDA. The maximum intensity we used was only 420 mW/cm^2, corresponding to P_A= 550 kPa, Tx=2.6 μs and PRF=16 kHz.

If similar conditions are reproduced *in vivo*, the movement of microbubbles may be not the same as that of blood erythrocytes. Deformations of spectra due to the radiation force would be such that in a *forward flow* the mean Doppler frequency is decreased although the positive upper frequency is almost unchanged. On the other hand, when a *reverse flow* is interrogated, both the mean Doppler frequency and the negative upper frequency would be increased.

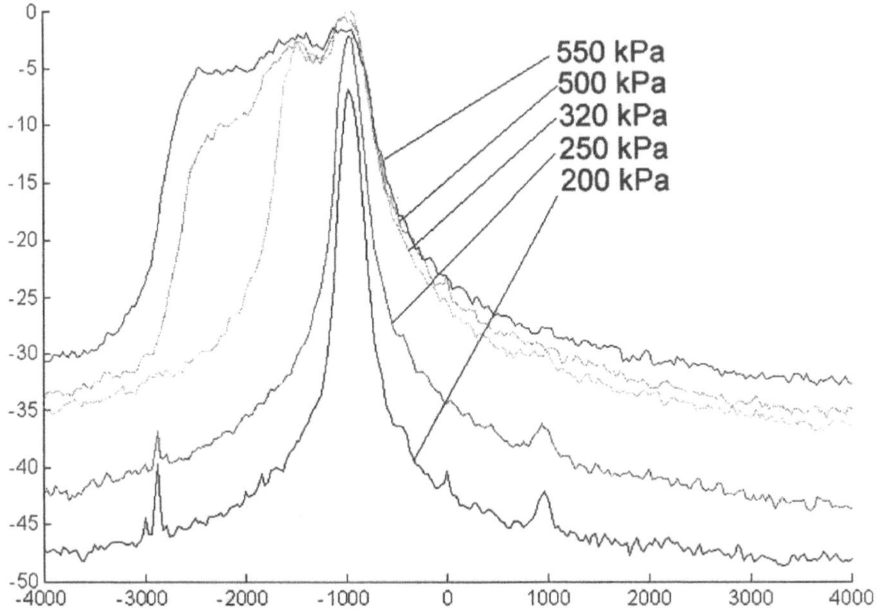

Figure 5. Average Doppler spectra obtained by interrogating a *reverse* flow with a transducer producing different pressures.

Acknowledgements

The authors wish to acknowledge valuable help by Daniele Bagnai in obtaining the experimental results.

REFERENCES

1. W.T.Shi, F.Forsberg, H.Oung, Spectral broadening in conventional and harmonic Doppler measurements with gaseous contrast agents, *1997 IEEE Ultrasonics Symposium Proceed.* p.1575-1578, (1997).
2. P. Tortoli, , D. Bagnai, D. Righi, Quantitative analysis of Doppler spectrum modifications yielded by contrast agents insonified at high pressure, accepted for publication in *IEEE Trans. on Ultrasonics, Ferroelectrics & Freq. Control*, (1999).
3. P.A.Dayton,K.E.Morgan, A.L.Klibanov,G.Brandenburger, K.R.Nightingale and K.W.Ferrara, "A preliminary evaluation of the effects of primary and secondary radiation forces on acoustic contrast agents", *IEEE Trans. on Ultrasonics, Ferroelectrics & Freq. Control*, Vol.44, N.6, p.1264-1277, (1997).
4. P.N.Burns, J.E.Powers et al., "Harmonic power mode Doppler using microbubble contrast agents: an improved method for small vessel flow imaging", *1994 IEEE Ultrasonic Symposium Proceed.*, pp. 1547-1552, (1994).
5. G.Guidi, V.L.Newhouse, P.Tortoli, "Effects of the simultaneous presence of geometrical and velocity broadening on the Pulsed Doppler spectrum", *Acoustical Imaging vol.22*, P.Tortoli & L.Masotti Eds., Plenum Publishing Company, New York, pp.413-418, (1996).
6. D.H.Evans, W.N.McDicken, R.Skidmore and J.P.Woodcook. *Doppler Ultrasound : Physics, Instrumentation And Clinical Applications.* Chichester. Wiley, (1989).
7 . N. de Jong, P. Frinking, F. Ten Cate, P. van der Wouw, "Characteristics of contrast agents and 2D imaging", *1996 IEEE Ultrasonic Symposium Proceed.*, pp.1449-1458, (1996).

B-MODE SPECKLE TEXTURE: THE EFFECT OF SPATIAL COHERENCE

Jeffrey C. Bamber[1], Ronald A. Mucci, Donald P. Orofino[2] and Karl Thiele

Hewlett Packard Company, Medical Products Group, Imaging Systems
Andover, MA, 01810-1099, USA
[1]present address: Physics Department, Institute of Cancer Research and Royal Marsden NHS Trust, Downs Road, Sutton, Surrey, SM2 5PT, U.K.
[2]present address: The Mathworks, Natick, MA 01760-9889, USA

INTRODUCTION

Most published theoretical work on B-mode speckle is limited to predicting the statistical properties of speckle at the focal depth or in the far field of plane sources (Wagner et al, 1983; Wagner et al, 1988). As a result, several practical observations associated with the systematic depth dependence of speckle brightness, probability density function and correlation functions (Flax et al, 1981; Smith et al, 1982; Oosterveld et al, 1985) have yet to be fully discussed and explained in a quantitative manner.

In this preliminary report it is shown that incoherent scattering conditions (which are also required for fully developed speckle) result in a systematic variation in the spatial coherence of the received acoustic field as a function of transducer-target distance. These systematic variations in spatial coherence could account for the observed speckle properties in the near field and beyond the focus. Eventual outcomes of further work may include:

a) a better understanding of the relationship between alternative metrics for the distortions due to aberration,
b) the development of new approaches to the measurement of imaging system performance
c) development of schemes for improving the visual uniformity of ultrasound images and
d) developing corrections for systematic errors in tissue characterization methods such as texture analysis, blood flow velocimetry and tissue elasticity imaging.

THEORETICAL BACKGROUND AND METHODS

Further understanding and explanation of the systematic depth dependence of speckle properties is being sought through the study of the spatial coherence of the backscattered field. The first step has been, therefore, to gain an appreciation for the systematic component of the depth dependence of the spatial coherence, as described by the spatial correlation function of the received backscattered wave.

General approach

In the first instance, and as a first approximation, this is being done by a method that makes use of the van Cittert-Zernike (vCZ) theorem. This is a classical theorem of statistical optics that describes how the degree of spatial coherence of a field radiated from an incoherent source varies with both the geometry of the source and the distance between it and the point at which the field is observed (Goodman, 1985). When applied to backscatter ultrasonic imaging (Mallart and Fink, 1991) the theorem predicts that, so long as the receiver is in the far field of an incoherent scattering volume, the spatial correlation function of the field at the receiver is proportional to the two dimensional (2D) spatial Fourier transform of the transmit intensity distribution over the scattering plane at the depth of interest. This is expected to give a reasonably accurate prediction when the scattered wave originates from the geometrical transmit focal depth but should increasingly overestimate the degree of spatial coherence as the scattering region moves away from this position.

As an illustration of this approach, consider incoherent scattering from an axial distance, z, equal to the radius of curvature, r, of a wave emerging from a rectangular aperture of length, a_x, and elevation height, a_y. The intensity distribution in the focal plane is given by

$$\frac{a_x a_y}{r^2}\,\mathrm{sinc}^2(a_x u)\,\mathrm{sinc}^2(a_y v) \qquad (1)$$

where $u = x/\lambda r$ and $v = y/\lambda r$, λ is the acoustic wavelength and x and y are the position variables for the focal plane. Taking the spatial Fourier transform of this expression, according to the vCZ theorem, the spatial correlation function $R(l_x, l_y, f)$ of the backscattered wave (at the plane of the aperture now acting as a receiver) is given by

$$R(l_x, l_y, f) = \frac{\chi_0(f)}{r^4}\Lambda\left(\frac{x}{a_x}\right)\Lambda\left(\frac{y}{a_y}\right) \qquad (2)$$

where $\Lambda(g) = 1 - |g|$, when $|g| < 1$ and is zero otherwise, l_x and l_y are the spatial lags in x and y in the plane of the receiver, and $\chi_0(f)$ is a measure of the average strength of scattering. The spatial coherence function of the wave backscattered from the focal plane of a linear array is therefore a triangle whose base is equal to twice the array length. This is in agreement with the result obtained by Mallart and Fink (1991), that the spatial correlation function of the field incoherently backscattered from the focal region is proportional to the autocorrelation function of the transmit aperture. Equation 2 indicates that this is also true for the elevation direction and that the elevation aperture size should not affect the lateral spatial coherence, and vice versa.

Specific technique

To study the spatial coherence for backscattering from other depths the 2D acoustic intensity distributions expected from a realistic phased array transmit aperture were calculated using a continuous wave numerical simulation. The 2D spatial Fourier transform of these distributions was then computed using a fast Fourier transform algorithm. To study the depth dependence of the results each spatial coherence function was then reduced to a single measure of coherence length. Two measures of coherence length were calculated, the spatial lag at which the correlation drops to one half ($l_{0.5}$) and the spatial lag at which it drops to zero (l_0).

RESULTS

The results to be reported are for a simulated phased array with the characteristics listed in Table 1. Figure 1 provides a reference result obtained for backscattering from the focal plane (z=100 mm) at a frequency of 3.5 MHz, verifying that the simulation provides the same result as that predicted analytically (Eq. 2). Further confirmation was provided in that this result was found to be independent of frequency or focusing f-number.

Table 1. Characteristics of the transducer simulated

Lateral aperture size	20.32 mm
Elevation aperture size	12.0 mm
Number of elements	128
Transmit focal distance	100.0 mm
Elevation focal dist	100.0 mm
Apodization function	rectangular or Hamming

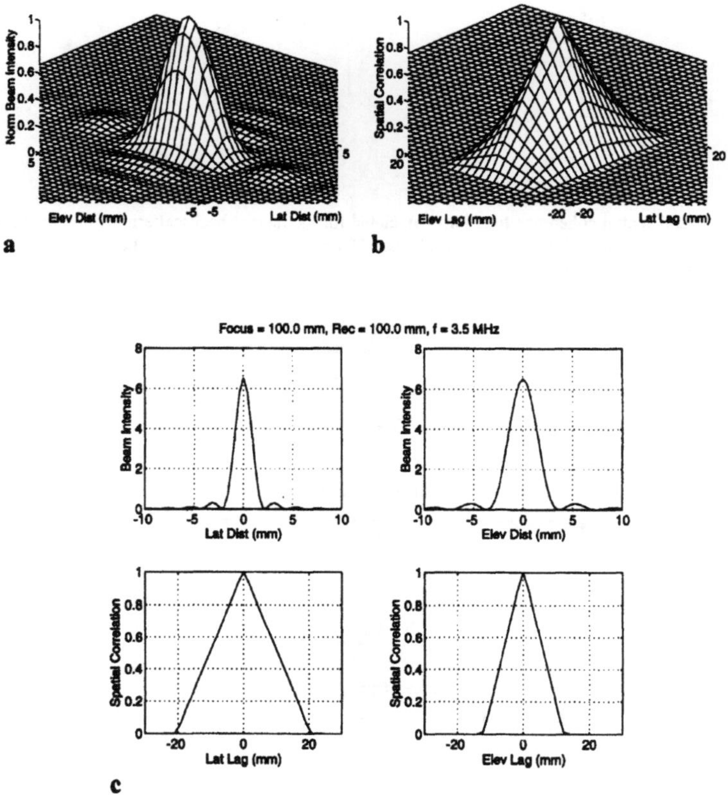

Figure 1. (a) Calculated 2D intensity distribution at the focal plane of the simulated transmitting aperture described in Table 1. (b) Predicted spatial correlation function for the wave incoherently backscattered from this region of the beam. (c) Cross-sectional profiles, in the lateral and elevation directions, for both transmitted intensity and reflected spatial correlation functions.

Such intensity distributions and spatial coherence functions were also generated at 10 mm depth intervals for depths from 50 mm to 200 mm. Selected examples of the 1D spatial coherence profiles are shown in Fig. 2, illustrating the observation that the shape of the spatial correlation function varies considerably with the transducer to scatterer distance, being triangular at the focus but becoming increasingly concave both anterior and posterior to the focus. The length of the base of the function, l_0, however, remains approximately constant at twice the length of the aperture over a large range of depths.

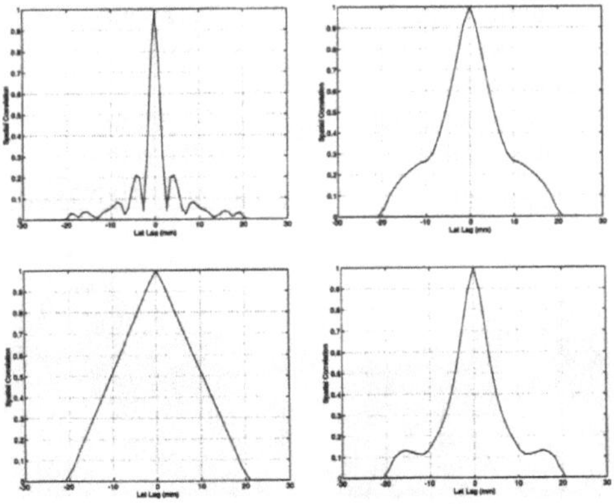

Figure 2. Lateral spatial coherence functions calculated for incoherent backscattering from depths of 50 mm (top left), 80 mm (to right), 100 mm (bottom left) and 150 mm (bottom right), for the source aperture defined in Table 1 transmitting at 3.5 MHz.

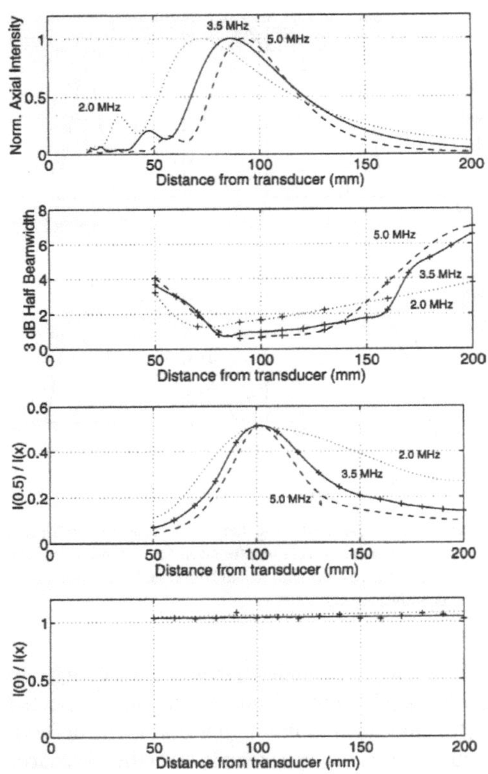

Figure 3. On-axis intensity (top), -3 dB half beamwidth (upper middle), normalized lateral coherence length $l_{0.5}$ for half correlation (lower middle) and normalized lateral coherence length l_0 for zero correlation (bottom), as a function of depth of the incoherent scattering volume.

As illustrated more clearly in Fig. 3, it is the half coherence length, $l_{0.5}$ (shown normalized to the lateral aperture size) that better demonstrates the loss of spatial coherence for incoherent targets positioned away from the transmit focal zone. For comparison, the on-axis intensity and the lateral -3 dB half width of the transmit beam have also been plotted as a function of depth. Finally, also in Fig. 3 are the same set of beam and spatial correlation parameters measured when the frequency was varied both above and below 3.5 MHz, by an amount that is typical of being at the extremes of the useful bandwidth of a reasonable pulse-echo transducer.

Other results (not shown) indicated that even outside of the focal plane the spatial correlation in the lateral direction was not influenced by changes in the focused elevation aperture size. Apodization of the transmit aperture was found to result in a lower maximum coherence length but one that varies more smoothly with depth of the scattering volume.

DISCUSSION

Limitations

This approach to computing the system related depth dependence of spatial coherence has limitations, which must be considered when comparisons are made with experimentally measured backscatter spatial coherence functions. Some of the main ones are:

(a) The vCZ theorem breaks down whenever the far field approximation is violated.

(b) The vCZ theorem requires a condition of quasimonochromatic waves.

(c) The affects of finite sized receiving elements are not accounted for.

A full discussion of these and other potential difficulties is beyond the scope of this communication but preliminary studies indicate that the method should be adequate for investigating the general trends to be expected in medical imaging. Liu and Waag (1994), for example, have found that the quasimonochromatic assumption is not substantially violated by typical ultrasound imaging systems. Indeed, preliminary experimental observations of spatial coherence of waves backscattered from a diffuse foam phantom show many of the trends reported here (see companion paper, Bamber et al, 1998).

B-mode speckle

The results demonstrate that unlike the transmit beam, for which the minimum width and peak on-axis intensity are at frequency and aperture dependent positions near the focus, the position of maximum spatial coherence of echoes from an incoherent backscattering medium is independent of frequency and occurs at the geometrical transmit focal depth. It is shown in a companion paper (Bamber et al, 1998) that the performance of a conventional beamformer, and therefore of a single large area coherent sensor, suffers when the coherence of the received wave is reduced. Thus although in a non-attenuating medium, the total area under the transmit beam intensity distribution (i.e. acoustic power) is independent of depth, application of Eq. 3 of Bamber et al (1998) to these spatial coherence functions predicts that average speckle brightness will be low in the near field and increase towards a maxium at the focal depth, decreasing thereafter. This is not a manifestation of the increased on-axis transmit beam intensity at the focus and indeed Oosterveld et al (1985), who observed the phenomenon, were not able to explain it.

The process of beamforming an incoherent wave is statistically equivalent to the coherent summation of signals of random magnitude and phase that produces the backscattered interference pattern that is known as speckle. A reduction in backscatter spatial coherence can therefore be expected to have no effect on the probability density function (PDF) of speckle. As a result, measures of the PDF such as the signal to noise ratio (SNR) should be expected to be independent of depth in a focused field unless the scattering medium is not completely incoherent. In such a case, characteristics closer to those of a

Rayleigh distribution should be observed in the near field and beyond the focus but at the geometrical transmit focus the SNR should pass through a minimum. This too was observed by Oosterveld et al (1985).

Finally, the lateral autocorrelation function (ACF) of B-mode speckle at the focal depth is usually explained on the basis that the interference pattern sampled by the receiving aperture will be correlated to the pattern obtained for an adjacent scan line by the degree to which the scattering volumes (i.e. beam widths) for the two lines overlap. Thus the correlation between echo magnitudes sampled from adjacent scan lines varies as the ACF of the round-trip beam profile. On the other hand, the results presented here lead us to suggest that outside of the focal zone the rapidly varying magnitude and phase in the low coherence backscattered waves will tend to reduce the correlation between adjacent beamformed signal lines, compared to that expected from the beamwidth at a given depth. In other words we hypothesize that, although the beam may be broad in these regions, a fine speckle pattern should be expected. The fine grain of near field speckle is indeed a common observation (Flax et al, 1981; Smith et al, 1982; Oosterveld et al, 1985). It is less commonly remarked upon for the region beyond the focus but Oosterveld et al were at a loss to explain their observation that the lateral width of the speckle ACF as a function of depth passed through a maximum just beyond the imaging focus. It is, however, close to the result to be expected from examining Fig.3. The effects of many other factors, such as frequency dependent attenuation and the relationship between dynamic versus fixed receive focusing, have yet to be studied.

Aberration metrics and other applications

Some scalar properties, such as $l_{0.5}$ or the area under $R(l)$, would appear to be better metrics of spatial coherence (and hence perhaps of aberration) than others, such as l_0. In general, however, a multivariate characterization of the shape of $R(l)$ would be desirable.

Simple apodization of the transmit aperture reduces the dependence of spatial coherence on incoherent target distance but at the expense of the maximum achievable coherence. A dynamic receive aperture controlled from measured or predicted $R(l)$ may be of value.

REFERENCES

J.C. Bamber, R.A. Mucci, D.P. Orofino, Spatial coherence and beamformer gain, *companion paper in this volume* (1998).

S.W. Flax, G.H. Glover, N.J. Pelc, Textural variation in B-mode ultrasonography: a stochastic model, *Ultrasonic Imaging*. **3**:235-257 (1981)

J.W. Goodman, *Statistical Optics*, Wiley, New York (1985).

D.L. Liu, R.C. Waag, About the application of the van Cittert-Zernike theorem in ultrasonic imaging, *IEEE Trans Ultrason Ferroelec and Frequ Control*. **42**:590-601 (1995).

R. Mallart, M. Fink M, The van Cittert-Zernike theorem in pulse echo measurements, *J Acoust Soc Am*. **90**:2718-2727 (1991).

B.J. Oosterveld, J.M. Thijssen, W.A. Verhoef, Texture of B-mode echograms: 3-D simulations and experiments of the effects of diffraction and scatterer density, *Ultrasonic Imaging*. **7**:142-160 (1985).

S.W. Smith, J.M. Sandrik, R.F. Wagner, O.T. von Ramm, Measurements and analysis of speckle in ultrasound B-scans, pp. 195-211 in: P. Alais, A.F. Metherel (eds.) *Acoustic Imaging*. **Vol.10**, Plenum Press, New York (1982).

R.F. Wagner, S.W. Smith, J.M. Sandrik, H. Lopez, Statistics of speckle in ultrasound B-scans, *IEEE Trans Sonics Ultrasonics*. **US-30**:156-163 (1983).

R.F. Wagner, M.F. Insana, S.W. Smith, Fundamental correlation lengths of coherent speckle in medical ultrasonic images, *IEEE Trans Ultrason Ferroelec and Frequ Control*. **35**:34-44 (1988).

EXTENDING THE BANDWIDTH OF THE PYRAMIDAL DETECTOR

L.R. Sahagun[1], S. Isakson[2], F. Mendoza-Santoyo[1], G. Wade[1,2]

[1] Centro de Investigaciones en Optica, Leon, GTO, Mexico
[2] ECE Dept., University Of California, Santa Barbara. CA 93106

INTRODUCTION

Figure 1 diagrams a typical Scanning Laser Acoustic Microscope (SLAM). The acoustic energy passes through the object and travels to a coverslip, producing perturbations. The signal is transferred to a scanning laser beam on the coverslip and the detector converts the signal on the beam to an electrical signal. It is the purpose of this paper to illustrate how the spatial frequency transfer function from the object to the electrical signal output can be maximized in a stable detector system.

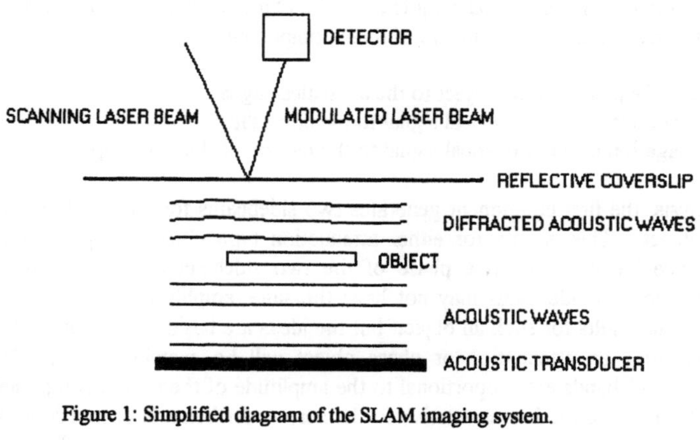

Figure 1: Simplified diagram of the SLAM imaging system.

While the Knife-Edge Detector (KED) has proven to be a useful detector for many applications, its limitations become apparent in tomographic and wide spatial bandwidth applications.[1] The source of these limitations is its limited and anisotropic bandwidth and also the creation of multiple images due to oblique insonification. Attempts to overcome these limitations have focused primarily in two different directions, interferometric techniques[1,2,3] and the pyramidal detector (PD).[4]

Interferometric methods are promising as they provide isotropic detection of surface waves and a spatial frequency bandwidth limited only by the achievable spot size and system limitations common to all detectors. This characteristic allows the detectors to be used in tomographical applications with higher image resolution then can be achieved with the KED. Unfortunately, they are also very sensitive to system vibrations and path length stability, making them difficult use in practice.

Recently the idea of a pyramidal mirror detector has been proposed[4] and provides a nearly isotropic detector bandwidth and possesses the stability of the KED. These properties allow this detector to be used in a system, such as SLaTHAM[3,5], to image an object quickly with rigid alignment between the microscope parts. Its one limitation is the inability to detect a zero spatial frequency surface wave thereby producing a hole in the spatial frequency transfer function of the detector. Additionally the size of this hole also increases when the detector is adjusted to detect higher spatial frequencies.

This paper analyzes this detector further. The first part of this paper examines the fundamental resolution limitations caused by the SLAM system. Next the effects of the pyramidal detector will be examined in comparison to those of the KED. Finally methods to extend the capabilities of the detector are presented.

SPATIAL FREQUENCY LIMITATIONS TO THE COVERSLIP

Before the spatial frequency limitations of the detector are examined it is important to know the limitations of the remainder of the SLAM system. For this purpose we will first examine the transfer function from a given spatial frequency in the object to the vertical motion in the coverslip. This transfer function is made up of three components:

1. The spatial frequency in the object to the acoustical signal.
2. The propagation of the acoustical signal to the coverslip.
3. The propagation of the acoustical signal to the motion of the coverslip.

In small signal analysis, the first component generates two sidebeams for any single spatial frequency in the object. This is true for either attenuation type objects or phase only objects; the difference being the relative phase of the two sidebeams. If you have a combination object, the two sidebeams may not have the same amplitude and it is more difficult to illustrate the results for such an object, but the ideas are basically the same. For simplicity, however, only an attenuation or phase object will be examined here. The amplitudes of the two sidebands are proportional to the amplitude of the spatial component in the object and the angle of the beams relative to the zero order beam is proportional to the spatial frequency.

If the propagating medium is not lossy then the second component has no effect on the amplitude of the transfer function for non-evanescent sideband components. If a component is evanescent, then there is a phase reversal in the detected signal and the detected signal will be highly dependent on the distance from the object to the coverslip. Since we will see there is an unavoidable zero in the transfer function before these evanescent frequencies are reached, for the purposes of this paper, it will be assumed that all evanescent components have died out.

Since the scanning laser beam is only sensitive to the vertical motion of the coverslip, then the third component of the overall transfer function is simply the cosine of the angle between the wave and the vertical axis. If we assume that the plane of the object is parallel to the plane of the coverslip, then this portion of the transfer function is not effected by the orientation of the object's spatial frequency in the plane of the object. This assumes there is no impedance mismatch between the propagating medium and the coverslip. When this last criteria is not totally true, it usually only causes slight changes in the overall transfer function.

When all of these components are put together, the amplitude of the transfer function of figure 2a is obtained if orthogonal insonification is assumed. This displays the relative amplitude of the coverslip motion to any particular spatial frequency in the plane of the object. No laser beam detector system will be able to improve upon this "ideal" transfer function. Figure 2b gives a cross-section of this transfer function and compares it to the transfer function if the object is insonified at an angle of 10°, which is common with the KED. While there is a slight shift in the waveform, the general character is about the same.

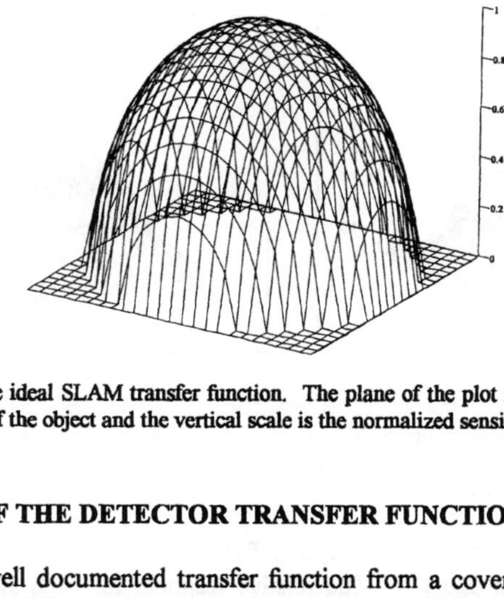

Figure 2. The ideal SLAM transfer function. The plane of the plot represents the spatial frequencies of the object and the vertical scale is the normalized sensitivity of the system.

THE EFFECT OF THE DETECTOR TRANSFER FUNCTIONS

The KED has a well documented transfer function from a coverslip surface wave to the detector output.[1,2] When this transfer function is combined with the system transfer function, given above for 10° insonification, the overall system sensitivity is seen in figure 3. It is apparent from this plot that the maximum spatial frequency reasonably detectable is significantly reduced and the frequency of maximum sensitivity is reduced as well.

The PD with orthogonal insonification produces a broader transfer function with higher sensitivity.[4] When combined with the remainder of the system transfer function, it produces the overall transfer function displayed in figure 4a. While generally an improvement over the KED, it does have the disadvantage of little response to the zero and low spatial frequency components of the object. While orthogonal insonification has distinct advantages, oblique insonification will overcome this problem when both sidebands are present, which is usually the case. Fortunately the optimal angle of this detector is less; only 5° from orthogonal for this detector. This results in better response and fewer artifacts created in the imaging process. The results of this oblique insonification are shown in figure 4b.

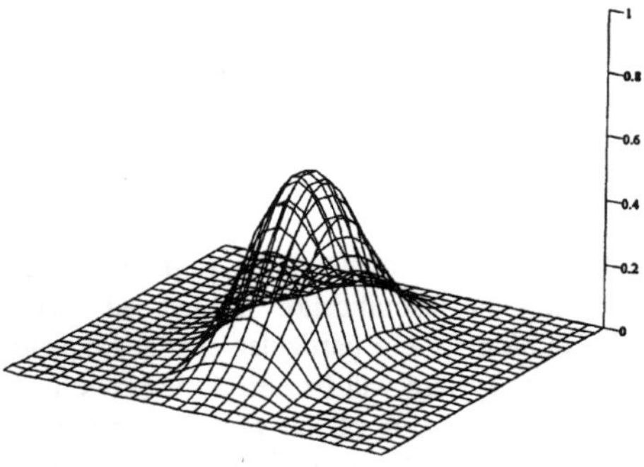

Figure 3. The response of the SLAM system with a KED. Note the narrow bandwidth and reduced gain of the system.

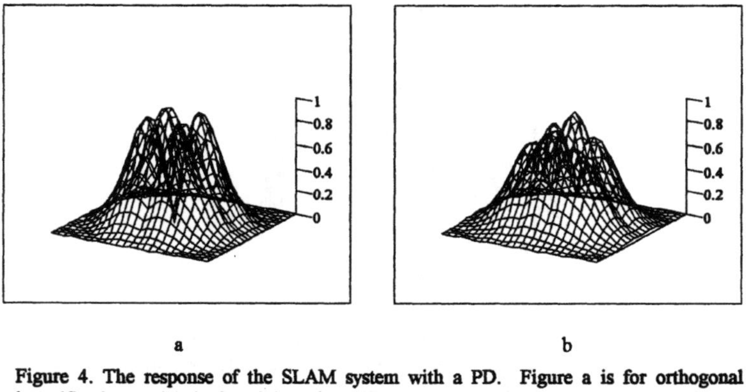

Figure 4. The response of the SLAM system with a PD. Figure a is for orthogonal insonification and b is for oblique insonification. Note the increased bandwidth and higher gain of the system.

A cross-sectional comparison the KED and PD detectors at their optimal angles of insonification is shown in figure 5. The relative response for each system is shown for spatial frequencies along the KED preferred axis. It is seen that the PD has a much broader response though at a slight cost in response at low spatial frequencies.

FURTHER PD RESPONSE IMPROVEMENTS

While the PD response is improved over the KED response, more improvements can still be achieved to get closer to the ideal response. One method is to scan the surface with a second spot of larger diameter. This can be done by using a second scan or by using a second optical frequency and separating the frequencies at the detectors. Figure 6 gives the response of the second spot with both orthogonal and oblique insonification. It is seen that the response for the larger spot tends to improve the low spatial frequency response, but does not contribute to the high frequency data.

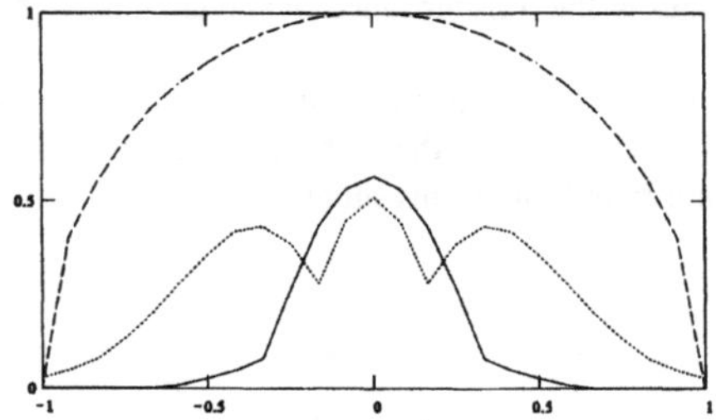

Figure 5. Cross-sectional comparison of the KED (solid) and PD (short dashed) systems. The response of the ideal system (long dash) is given for reference. The ordinate represents spatial frequencies in the x direction (the direction favored by the KED) and the abscissa is the relative response.

Obviously this could be pushed further by using more and more scanning spots. Smaller ones improving the high frequency response and the larger spots improving the low frequency results. Particularly for orthogonal insonification, this can make the impact of the low response at low spatial frequencies an arbitrarily small region. And it can push the high frequency response ever closer to the ideal curve and even emphasis the evanescent wave detection ability (not shown in figure 6). However this technique will not completely remove the zeros at zero spatial frequency and the limit between normal and evanescent wave generation. While oblique insonification does remove them partially, the zero spatial frequency defect is still seen to effect the transfer function with reduced gain and the evanescent wave defect is at most shifted, but never eliminated.

Furthermore, not all objects produce two equal sidebands for a given spatial frequency. The oblique insonification method counted on there being two sidebands in order to eliminate the zero at zero spatial frequencies. If the object modulates both intensity and phase, then a single sideband may be formed. Under those circumstances the zero might be moved, but will always appear at some spatial frequency. Additionally the overall response is reduced with oblique insonification.

Fortunately there are a few other techniques that can frequently improve the situation. If a small amount of information about the object is known, then some superresolution techniques can be applied. Classically, if it is known that the object is of limited extent, then analytic continuation can be used to reconstruct the missing spatial frequencies. While the objects in SLAM are always of limited extent, this technique works best for very small objects due to the noise sensitivity of the algorithm. However, the more that is known of the object apriori, the more this information can be applied to the reconstruction of the object and spatial spectrum.

Another technique is available if multi-angle tomography is the application. Since the zero is a point in the plane and oblique insonification is used, the location of the zero shifts in each view. The hole in one view will be filled in from another view in the final image.

If additional hardware is reasonable, focused insonification can be scanned with the laser beam. It was previously assumed that insonification of the object was by plane waves. By focusing the insonifying beam, the transfer function is convolved with a second function that blurs the sharp discontinuities in the transfer function and eliminates the zeros.

Finally, it is reasonable to ask if the information is even needed to reconstruct the object. If computer processing is to be used, one of the first steps is to find the edges. Information at low spatial frequencies contribute little to this situation.

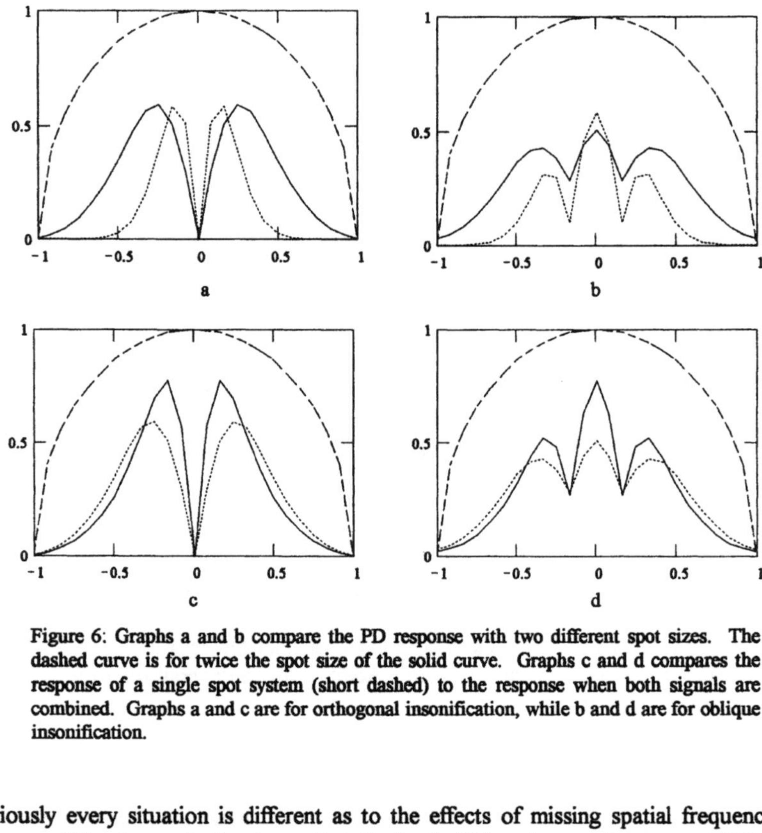

Figure 6: Graphs a and b compare the PD response with two different spot sizes. The dashed curve is for twice the spot size of the solid curve. Graphs c and d compares the response of a single spot system (short dashed) to the response when both signals are combined. Graphs a and c are for orthogonal insonification, while b and d are for oblique insonification.

Obviously every situation is different as to the effects of missing spatial frequency data. However, it has been clearly shown that the bandwidth and general response of the PD is much greater than the KED and the PD can more easily use the preferred orthogonal insonification. Those defects that still remain can be further mitigated by the techniques shown here.

REFERENCES

[1] Rylander, R.L., "A Laser-Scanned Ultrasonic Microscope Incorporating a Time Delay Interferometric Detector," Ph.D. dissertation, University of Minnesota (1982).

[2] Whitman, R.L., and Korpel, A., "Probing of Acoustic Surface Perturbations by Coherent Light," *Applied Optics*, vol. 8, pp. 1567-1576 (1969).

[3] A. Meyyappan, S. Isakson, F. Mendoza-Santoyo, R.A. Duarte, L.R. Sahagun, G. Wade, *Detection Systems for Scanning Laser Tomo-Holographic Acoustic Microscopy*, International Journal of Imaging Systems and Technology, Vol 101. No 7 pp 131-135 (1996).

[4] L.R. Sahagun, F. Mendoza-Santoyo, G. Wade, S. Isakson, *The Pyramidal-Mirror Detector for Scanning Laser Acoustic Microscopy*, International Journal of Imaging Systems and Technology, accepted for publication.

[5] Meyyappan, A. and Wade, G., "Scanning Tomographic Acoustic Microscopy with Multiple Transducers and Frequencies," *Coloque de Physique*, Colloque No. 2, pp. C2-307 - C2-310 (1990).

THE USE OF A REFERENCE-BEAM DETECTOR APPLIED TO THE SCANNING LASER ACOUSTIC MICROSCOPE

Moisés Cywiak[1], Cristina Solano[1], Glen Wade[1] and S. Isakson[2]

[1]Centro de Investigaciones en Optica, A.C.
Apdo. Postal 1-948, León, Gto, México 37000
[2]Department of Electrical and Computer Engineering
University of California, Santa Barbara, CA

Scanning laser acoustic microscopy (SLAM) has become a reliable option for obtaining high-quality images on a microscopic scale with good resolution and contrast. Is the only acoustic microscope that produces it's images in real time. The conventional technique for image data readout in SLAM is to deflection-modulate a laser beam by scanning it over a solid surface containing the acoustic field scattered from the object and then demodulate the beam by means of a knife-edge detector (KED). The probing beam is reflected form the surface of a mirrored coverslip onto which scattered acoustic waves, transmitted through the object, impinge. The image information is encoded as deflection modulation on the reflected beam and is detected by the KED. The system consists basically of a knife-edge placed in the focal field of a lens with a photodiode positioned behind the knife-edge.

The KED is sensitive not only to the spatial frequencies of the sinusoidal components in the object but also to the direction in which the sinusoidal variations take place. Unfortunately, the KED transfer function is anti symmetric with zero response at zero spatial frequency and negative response at negative frequencies [1]. Because of the nature of the transfer function, flexibility in choosing the optimum laser spot size is limited. In addition an obliquely incident insonification at an optimum angle of incidence is ordinarily employed in SLAM, with a resultant Doppler shift in the detected frequency of the transmitted zero-order acoustic beam, along with single side band detection. In spite of the use of these measures, the bandwidth of the spatial frequency spectrum used in reconstructing the image must be restricted so that it does not extend beyond the positive part of the transfer function and avoids the null response at zero frequency. Therefore, the resolution of the reconstructed image is correspondingly limited. Thus, while the KED provides acceptable results for imaging when the spatial variations in the object are in the preferred direction and at the preferred frequencies, it is quite unsatisfactory for variations at other spatial frequencies and in other directions and has a severely limited resolution.

The need for higher resolution and more sensitive methods of acoustic data appropriation and image reconstruction of microscopic objects has grown. In this sense we describe and analyze the new reference-beam detector (RBD).

Reference beam detection is inherently superior to the knife-edge detection currently in use in scanning laser acoustic microscopy. This new detector makes use of a reference beam, retarded 90 degrees, which is mixed coherently in a photodiode with the acoustically-obtained image-modulated beam. This new detector has an isotropic transfer function with no negative response anywhere, which is circularly symmetrical around its highest value, namely the zero-frequency point in the spatial spectrum. This property makes it possible to detect spatial frequencies in all directions simultaneously and with equal sensitivity and simplifies the associated electronics. It also makes possible the employment of acoustic evanescent-wave detection so that ultrasound of low temporal frequency can be used and at the same time high spatial frequency can be detected for obtaining high resolution. Oblique insonification, required for best operation in the knife-edge detector, is thus not preferred in the reference-beam detector and the resultant Doppler shift in the detected frequency of the transmitted zero-order acoustic waves is avoided.

Evanescent-wave detection is possible with the RBD. The maximum spatial frequency for the detected evanescent waves is limited only by the cross-sectioned radius of the scanning laser beam. This frequency can be very high even though the temporal frequency of the sound is low. Thus the propagating acoustic wavelengths can be large, the acoustic temporal frequency (and hence the attenuation in the object) low, and the same time the fine structure within the object can be highly resolved due to the detection of evanescent waves. For example a helium-neon laser beam can be focused to a radius of about half a micron. If a 100 MHz acoustic frequency is used in water, the acoustic wavelength is 15 microns. However, if the object is mounted in the very-near field, the RBD could detect evanescent acoustic waves 15 times smaller than the acoustic wavelength. This provides a much higher resolution of the object that can be produced by conventional means at this acoustic wavelength.

Reference

1. R. Mueller and R. Rylander, *New demodulation scheme for laser-scanned acoustical-imaging systems*, J. Opt. Soc. Am., **69**, 407 (1979).

ACOUSTIC MICROSCOPY EVALUATION OF ENDOTHELIAL CELLS MODULATED BY FLUID SHEAR STRESS

Yoshifumi Saijo[1], Hidehiko Sasaki[1], Hiroaki Okawai[1], Noriyuki Kataoka[2], Masaaki Sato[2], Shin-ichi Nitta[1], and Motonao Tanaka[3]

[1]Department of Medical Engineering and Cardiology, Institute of Development, Aging and Cancer, Tohoku University
4-1 Seiryomachi, Aoba-ku, Sendai 980-8575, JAPAN
[2]Biomechanics Laboratory, Graduate School of Mechanical Engineering, Tohoku University
Aobayama 01, Aoba-ku, Sendai 980-8579, JAPAN
[3]Tohoku Welfare Pension Hospital
1-12-1 Fukumuro, Miyagino-ku, Sendai 983-0005, JAPAN

INTRODUCTION

The shape and elasticity of the cultured endothelium are modulated by shear stress. The inverted microscopy demonstrates morphological change of the cell, from spherical to elongated shape, after exposition to shear stress. The change of cell structure has been observed by electron microscopy, and the cytoskeletal structure is closely related to the morphological change. The elasticity of the cell surface can be measured by atomic force microscopy (AFM) or micropipette aspiration of a single cell. By these methods, it has not yet been fully understood whether the change of the elasticity occurs in only the surface of the cell or whole part of the cell.

We have developed a scanning acoustic microscope (SAM) system for medicine and biology. SAM has some advantages for assessing morphological and physical properties of the cultured cell compared with the conventional methodology. One is that SAM can obtain the morphological information of the living cell, because it does not require pathological staining techniques. The other is that sound speed of the high frequency ultrasound through the cell, contains the information of elasticity of the entire cell. The objective of the present study is to visualize the shape of the endothelial cells exposed to the fluid shear stress by SAM. The study also includes the detection of acoustical change of the endothelium modulated by shear stress.

Acoustical Imaging, Volume 24.
Edited by Hua Lee, Kluwer Academic/Plenum Publishers, 2000.

MATERIALS AND METHODS

Endothelial Cells

Endothelial cells from bovine aorta were cultured on 80 mm x 80 mm glass plates in the Dulbecco's modified Eagle's medium with 10% heat-incubated bovine serum. The incubator was maintained at 37 °C and filled with 95% air and 5% CO_2. After 4days of culture, cells were found to be in the confluent state by inverted microscopy. The cells have well grown and contacted each other in the confluent state.

Then the cells were exposed to the shear stress of 2 Pa using a parallel-plate flow chamber. Figure 1 is the illustration of the flow chamber. A 0.5 mm thickness of silicone gasket was placed between flow chamber and the glass plate on which the endothelial cells were cultured. The height of the flow chamber was 0.5 mm, and the width was 60 mm.

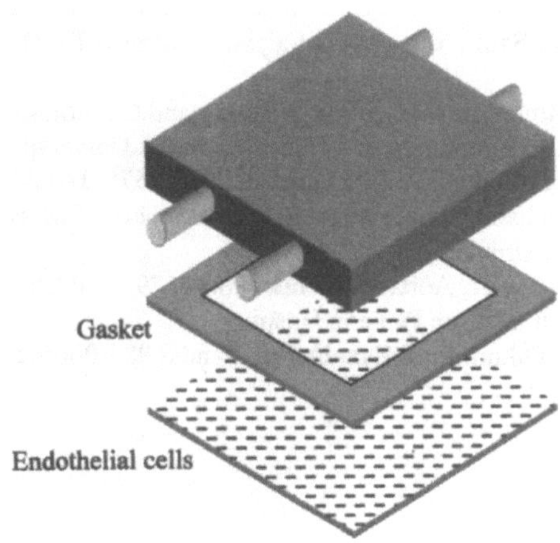

Figure 1. A schematic illustration of the parallel plate flow chamber. The height of the flow chamber is 0.5 mm, and the width is 60 mm. The thickness of the gasket is 0.5 mm.

The fluid shear stress was calculated as equation (1).

$$\tau_w = \frac{6\mu Q}{h^2 b} \quad \cdots \quad (1)$$

(τ_w: fluid shear stress, μ: viscosity of the fluid, Q: flow volume, h: height of the flow chamber, b: width of the flow chamber)

The cells were observed by both inverted and acoustical microscopy before and after exposure to the shear stress.

Scanning Acoustic Microscope System

Figure 2 is the block diagram of our SAM system[1-2,4], which comprises of five parts. The acoustic focusing element comprises a ZnO piezoelectric transducer with a sapphire lens. The ultrasonic frequency is variable over the range of 100 to 210 MHz and the beam width at the focal volume ranges from 5 μm (at 210 MHz) to 10 μm (at 100 MHz). The focusing element is mechanically scanned at 60 Hz, in the lateral direction (x) above the specimen, which remains stationary on the specimen holder, while the holder is scanned in other lateral direction (y) in 8 sec, thus providing two-dimensional scanning. The mechanical scanner is so arranged that the ultrasonic beam be transmitted for every 1 μm interval over a 0.5 mm width. The number of sampling points is 480 in one scanning line and 480 x 480 points make one frame. Both amplitude and phase images can be obtained in a field of view 0.5mm x 0.5mm. Either reflection or transmission mode is selectable in the system. In the present study, the reflection mode was equipped. Phosphate buffer saline, which was maintained at 37°C, was used for the coupling medium between the transducer and the specimen.

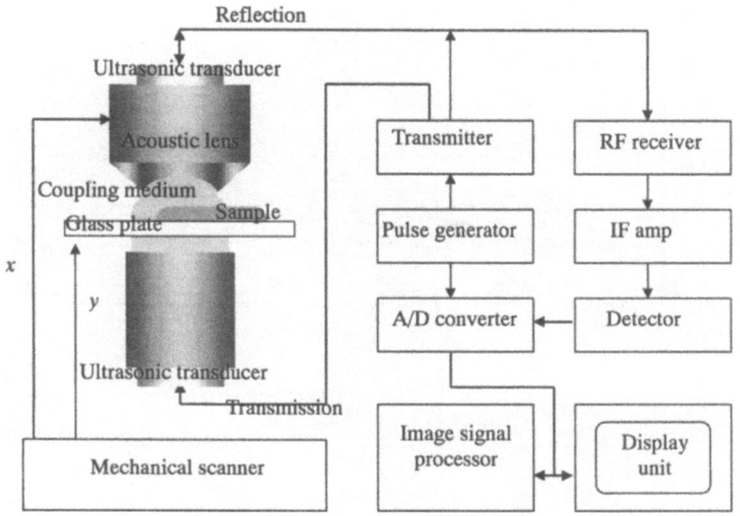

Figure 2. A block diagram of the scanning acoustic microscope system developed at Tohoku University.

The region of interest (ROI) was determined by observing the two-dimensional amplitude image. Both the amplitude and phase in the ROI were measured at every 1 MHz interval between 100 to 210 MHz to obtain the quantitative values of acoustic properties. Thus, the frequency varying change of the amplitude and phase were obtained.

Equation (2) shows the relationship between the frequency, amplitude and phase due to the interference between the reflection of the surface and bottom of the specimen. By fitting the obtained frequency varying curve with the calculated values, the thickness of the cell in the ROI were obtained.[3, 5]

$$
\begin{aligned}
y_r &= y_s + y_b \\
&= e^{-(\alpha + j\frac{2\pi f}{c})d} \\
&= e^{-j\frac{2\pi f}{c}d} \cdot e^{-A_0 f^n d} \quad \cdots \quad (2)
\end{aligned}
$$

(y_r: received signal, y_s: reflection from the surface of the specimen, y_b: reflection from the bottom of the specimen, f: frequency, c: sound speed of the specimen, d: thickness of the specimen)

Figure 3 shows the graphs showing the relationship between the frequency and the amplitude or phase in 6 μm thickness specimen.

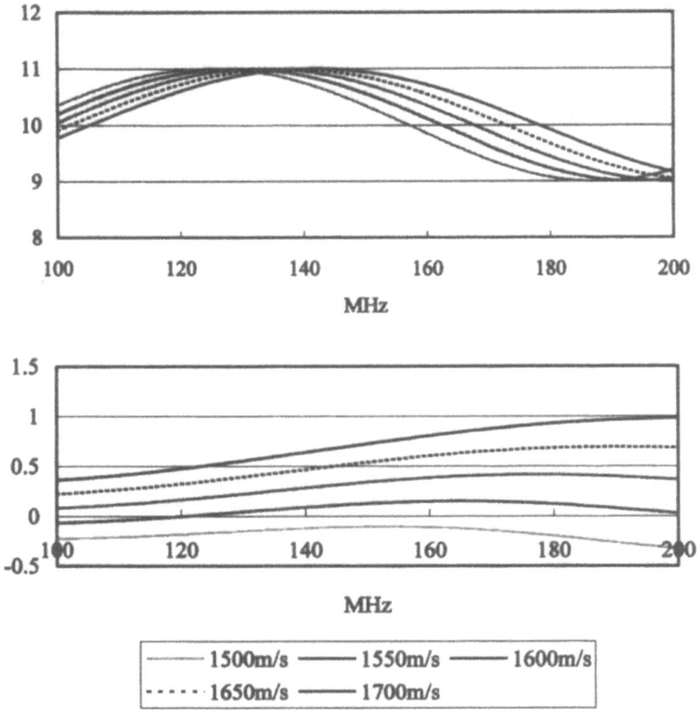

Figure 3. The graphs showing the relationship between frequency and the amplitude (upper) or phase (lower) in 6μm thickness specimen.

Image Analysis

For the quantitative analysis of the endothelial morphology before and after exposure to fluid shear stress, the shape index and the angle of the orientation were analyzed. Acoustic images were input into the personal computer system (Power Macintosh 9600/233; Apple Computer) and the off-line analysis was performed. In the present study, the following two parameters were calculated by using NIH Image 1.61 (Free software for image analysis developed at NIH).

The shape index is defined by the equation (3).

$$Shape\ Index = \frac{4\pi A}{P^2} \quad \cdots \quad (3)$$

(A: area of the cell, P: perimeter of the cell)

Figure 4 represents the angle of the orientation of the cell corresponding with the direction of the flow.

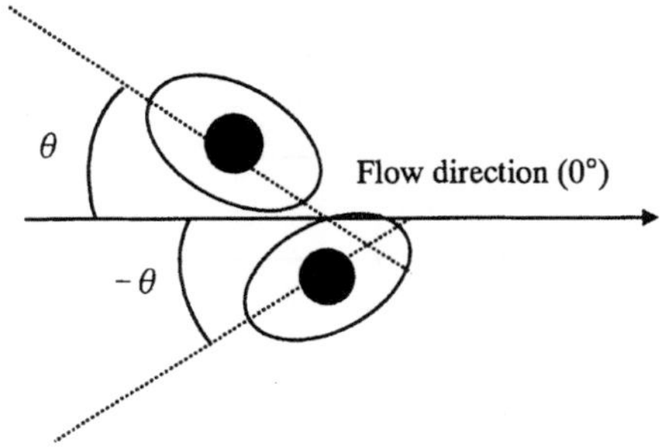

Figure 4. A schematic illustration of the cell orientation and the flow direction.

RESULTS

Figure 5 shows the phase images of the endothelium before (left) and after (right) exposure to fluid shear stress. Cells exposed to shear stress for 24 hrs became elongated, and the direction of the long axis of the cell aligned parallel to the flow.

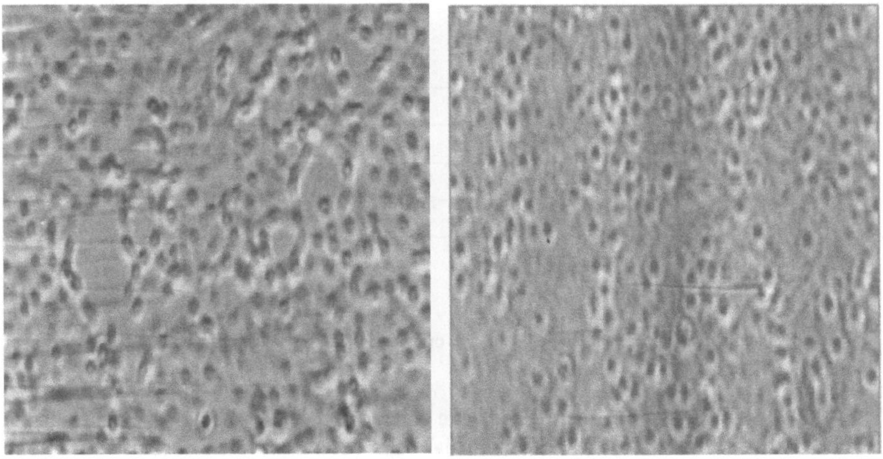

Figure 5. Acoustic microscopic images of the endothelium before (left) and after (right) exposure to fluid shear stress.

Figure 6 shows the shape index (upper) and the cell orientation (lower) before and after fluid shear stress. If the shape of the cell were completely round, the shape index would be 1.0. In the analysis, the mean value of the shape index of the cells after fluid shear stress was decreased to 0.79 ± 0.08 whereas that was 0.87 ± 0.07 before the stress. The result quantitatively represents the elongation of the endothelial cells by fluid shear stress. Also, the mean angle of the cell orientation was decreased from 48.7 ± 33.1° to 4.6 ± 3.4°, which means that the orientation of the endothelium was aligned with the direction of the flow.

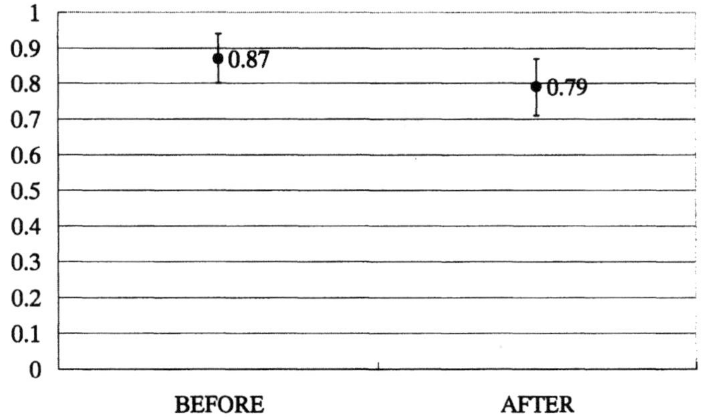

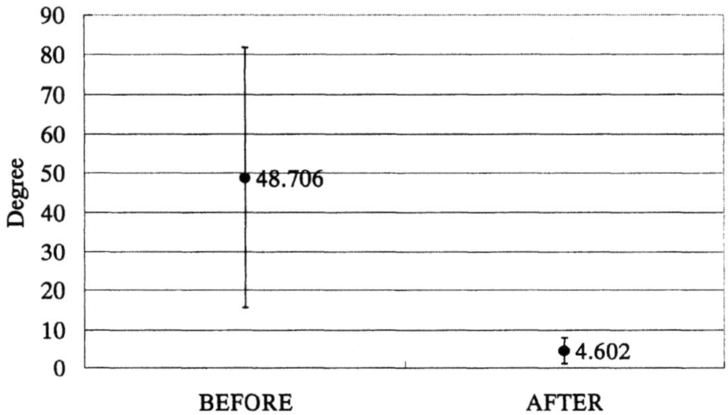

Figure 6. Shape index (upper) and angle of the cell orientation (lower) of the endothelium before and after fluid shear stress.

The ROI for the quantitative acoustic parameters were placed at the center of the nucleus and the cytoplasm in the endothelial cells before and after shear stress. The frequency varying amplitude and phase curves at 4.0, 4.5, 5.0, 5.5, 6.0, 6.5, 7.0 μm were drawn, and the obtained data were fitted to those curves. The thickness of the nucleus was determined as 6.0 μm and that of the cytoplasm was 5.5 μm before shear stress. These values did not show significant change after the stress.

The intensity was higher in the nucleus than in the cytoplasm in phase image. The three dimensional shape of the cell was eclipse so that the center of the cell was thickened. However, considering the thickness, the sound speed of the nucleus was 1610 ± 10 m/s and that of the cytoplasm was 1570 ± 20 m/s before shear stress. After the exposure to the flow, the sound speed in the cytoplasm increased to 1590 ± 20 m/s whereas that was remained 1610 m/s in the nucleus.

DISCUSSION

The SAM system equipped in the present study requires 8 sec to make one frame, since the *y*-direction scan needs 8 sec. In order to test the stability of the cells during SAM measurement, two consecutive scans with 8 seconds' interval were done. No significant difference was seen in the acoustic microscopic images between two consecutive scans. However, further evaluation of the effect of tightly focused ultrasound to the structure and function of the living cells should be needed, for applying the SAM technology to the cellular biology.

In the previous studies, the attenuation and sound speed image were obtained since the thickness of the specimen could be assumed to uniform in the field of view. In the present study, the sound speed at only one point could be obtained because the surface of the endothelium was not flat. However, the point analysis of the sound speed has provided the important information of the physical properties of the cells. The average sound speed of the cytoplasm before flow exposure was 1570 m/s, and the value after exposure increased to 1590 m/s. Micropipette aspiration has revealed that the viscoelasticity of the cell surface becomes stiffer after applying a shear stress. The increase of the F-actin filaments is considered to be one of the reasons for this viscoelastic change. In the SAM investigation, the sound speed of the cytoplasmic region increased after the shear stress. The result suggests that the increase of the elasticity is not limited in the cell surface, but also in a whole cell.

CONCLUSIONS

Scanning acoustic microscopy (SAM) was applied to assess the change of shape and acoustical parameter of the endothelium after exposure to the shear stress. SAM is useful for assessing the properties of the living cells since it can detect morphological and acoustical changes without pathological staining or fixation techniques.

REFERENCES

1. Y. Saijo, M. Tanaka, H. Okawai, F. Dunn, The ultrasonic properties of gastric cancer tissues obtained with a scanning acoustic microscope system, *Ultrasound Med Biol*. 17:709(1991).
2. Y. Saijo, H. Okawai, H. Sasaki, T. Naganuma, M. Tanaka, Intravascular ultrasound and scanning acoustic microscopy evaluation of aortic wall, *Acoustical Imaging*. 21:423(1995).
3. Y. Saijo, H. Sasaki, H. Okawai, M. Tanaka, Development of ultrasonic spectroscopy for biomedical use, *Acoustical Imaging*. 22:335(1995).
4. Y. Saijo, M. Tanaka, H. Okawai, H. Sasaki, S. Nitta, F. Dunn, Ultrasonic tissue characterization of infarcted myocardium by scanning acoustic microscopy, *Ultrasound Med Biol*. 23:77(1997).
5. Y. Saijo, H. Sasaki, H. Okawai, S. Nitta, M. Tanaka, Visualization of living cells by acoustic microscopy, *Acoustical Imaging*. 23:7(1997).

ULTRASOUND IMAGING OF HUMAN TEETH USING A DESKTOP SCANNING ACOUSTIC MICROSCOPE

Y.P. Zheng, E.Yu. Maeva, A.A. Denisov, and R.G. Maev

Center for Imaging Research and Advanced Materials Characterization,
School of Physical Science, University of Windsor, Ontario, Canada, N9B 3P4

ABSTRACT

The diagnosis of teeth caries is currently a difficult and experience-dependent practice in clinical dentistry. Meanwhile, ultrasonic imaging technique is believed to be a prospective method for inspection of layered structure of teeth. In this paper, we present results of our investigation on ultrasound imaging of human teeth with different diseases using a desktop acoustic microscope with an intermediate frequency of 25—100 MHz. Acoustic images and optical images of sections along different directions of teeth are compared. The difference of acoustic impedance of enamel, dentine, and other areas can be easily observed in acoustic imaging. Some details in teeth can be resolved by acoustic imaging, though they cannot be observed in the optical image. Imaging of teeth with caries, fillings, and implants as well as results of quantitative studies on acoustic impedance of different parts of teeth will be reported.

INTRODUCTION

Dental caries is one of the most important reasons for the loss of human teeth[1]. Due to the complex anatomic conditions of teeth, the diagnosis of caries is still a difficult and experience-dependent practice in clinical dentistry. Two clinical methods of caries detection in use at present are radiography and visual examination[2]. Since radiography cannot reliably distinguish caries, which is confined to the enamel, the visual inspection, with a mirror and an explorer, is the most popular diagnostic method of the detection of dental caries. However, studies have shown that only a low rate of caries can be correctly detected by dentists using visual inspection[1]. Moreover, such a subjective diagnosis cannot provide high reproducibility even for the same dentist. This makes teaching-learning extremely difficult and imprecise. A more reliable noninvasive objective diagnostic method for caries is very much desired in the dental practice.

Optical microscope and spectroscope have been used for the in vitro characterization of human teeth with and without caries for many years[3,4]. However, it is impossible to introduce those optical characterization methods to the *in vivo* dental diagnosis. On the other

Acoustical Imaging, Volume 24.
Edited by Hua Lee, Kluwer Academic/Plenum Publishers, 2000.

hand, ultrasonic imaging technique has a high prospect of visualizing the layered structure of teeth, and has been used for the measurement of teeth since the early 1960's[5]. Early investigations demonstrated that ultrasonic pulse-echo system could be used to detect the enamel-dentine junction and the dentine-pulp interface[6]. It has been shown that a high frequency acoustic microscope can be used to resolve the region with caries lesion from health parts in teeth[7]. Recently, more efforts have been made on the investigation of ultrasonic dental inspection method aiming to provide a noninvasive clinical tool for dentists[2,8]. Despite all these investigations, ultrasonic studies on teeth are still lacking in the literature compared with its high potential for the dental caries detection. A systematic ultrasound study on human teeth with various conditions will be greatly helpful for the further development of dental diagnosis tools using ultrasonic technique.

The overall objective of the current investigation is to study the feasibility for mechanical characterization of human teeth using ultrasonic imaging technique. In this paper, we present results of the ultrasound imaging of human teeth with different conditions using a desktop acoustic microscope with an intermediate frequency of 25—100 MHz. Acoustic images and optical images of sections along different directions of teeth are compared. Imaging of teeth with caries, fillings, and implants as well as results of quantitative studies on acoustic impedance of different parts of teeth will be reported.

METHODS AND MATERIALS

A pulse acoustic microscope with an intermediate frequency range has been earlier developed for the characterization of various materials[9,10]. It includes the ultrasonic equipment, precision scanner, and personal computer plus the control software for scanning movement, ultrasonic unit adjustment and display processing. All the parameters about the acoustic microscope are adjusted via the software in the controlling personal computer, and can be saved and recalled later. The precise scanner has a drive system with a step width of 0.05 mm across the fast scanning axis and 0.01 mm across the slow axis, and with a scanning area of 80 × 120 mm. Rather than relying on the immersion of the investigated sample in a water-filled tank[7], a continuous stream of liquid is placed alongside the ultrasonic probe serving as the coupling medium. For a small scanning area, liquid drops can also be used. The lightweight probe holder enables precise manual vertical adjustment of the ultrasonic probe. A set of focusing ultrasonic probes with frequency of 25, 50, and 100 MHz have been carefully designed for different materials and different penetration depths. Lithium niobate transducers are used and tightly attached to the quartz lenses. Probes have different focal lengths ranging from 2 to 15 mm and different aperture angles of 20°, 40°, and 60°, selected according to the ultrasonic velocity in tested samples. With all probes the focal diameter of the order of a single wavelength can be achieved.

The human teeth used in this investigation were extracted due to orthodontic needs and were provided by a local dentist. Five teeth with different conditions were collected. Flat surfaces of vertical or transversal section were obtained by removing some parts of the teeth. The surfaces were casually polished by papers. Two samples were mounted using a thermosetting material with mineral-filled Diallyl Phthalate (Buehler No. 2-3330-080), and prepared with a temperature of 90 °C. Other samples were prepared at room temperature without mold. The optical image of the sample was obtained directly using a color scanning device (HP Scan Jet 4C) which was connected to a personal computer.

RESULTS OF IMAGING

The imaging of two samples prepared with high temperature is shown in Figure 1 and Figure 2. Figure 1a and 1b show the optical image and acoustic image, respectively, of the sample with a cross section along the transaction of crown and root of a molar. The acoustic image of the sample obtained using a 25 MHz lens was well correlated with its optical image. Interfaces between different areas within the teeth could be more easily observed in the acoustic image. This effect is more obvious in Figure 2, which shows the imaging of the sample with a vertical section of a pre-molar. The part in the up-left corner has been broken and lost during the preparation of the sample, and the pulp cavity has been exposed. The area of enamel and dentine can be easily identified in the acoustic image with different gray levels representing different acoustic impedance, and their interface is clearly defined. In addition, the impedance of the dentine in the outer area of the root is quite different from the other part of the dentine. Since we did not observe similar phenomenon in other samples prepared under room temperature and we confirmed that it was not caused by the surface wave during defocusing, this effect associated with the enhanced interface between enamel and dentine area might be caused by the high temperature treatment during the sample preparation.

Figure 3 shows the images of a vertical section of a canine with artificial plastic crown and two metal screws for the attachment. A 100 MHz lens was used to obtain the acoustic image. Part of the tooth was removed, and the surface was casually polished. One of the screws reached the surface, and the other was located underneath the surface. Since the plastic crown was transparent, two screws could both be observed in the optical image. On the contract, only one screw, which reached the surface, could be identified in the acoustic image. In addition, the crack within the dentine area along the screw, and the interface between the artificial crown and the remained dentine have been defined more clearly in the acoustic image than in the optical image. However, the part of the tooth without any preparation presented a quite irregular pattern in the acoustic image as shown in the lower portion of Figure 3b. This demonstrates that flat surface is important for the acoustic scanning. It was also found that the acoustic impedance of the plastic crown was just similar with that of the dentine.

Images of a vertical section of another canine with serious caries are shown in Figure 4. A 50 MHz lens was used to obtain the acoustic image. Since the sample was broken after the acoustic scanning, a crack in its optical image is shown, which was obtained afterwards. As in Figure 2, the enamel and dentine could be differentiated clearly. Some specific optical patterns could be observed in the central portion of the sample, but most of them could not be identified in its acoustic image. This demonstrates that optical and acoustic images represent different parameters of the tooth tissue. In addition, the acoustic impedance (representing by the gray level) of the dentine along the caries interface was a little different from other dentine area around it. This provided the evidence that ultrasound imaging could be used to resolve the sound dentine from that with the early caries, which condition is similar to that of the dentine along the serious caries as shown in Figure 4b.

Figure 5a to 5d show images of a pre-molar with fissure caries. From the outer surface of the enamel, the fissure caries represented only as dark lines in optical image (Fig 5a). Optical and acoustic images of a vertical section along the dotted line in Figure 5a are given in Figure 5b and 5c, respectively.

Acoustic images were obtained using 100 MHz lenses. Similar with results introduced above, the enamel and dentine area can be identified clearly from the gray level in the acoustic image. The fissure caries has been developed along two directions within the enamel in this cross section of the tooth. In addition, a crack from one root of caries in

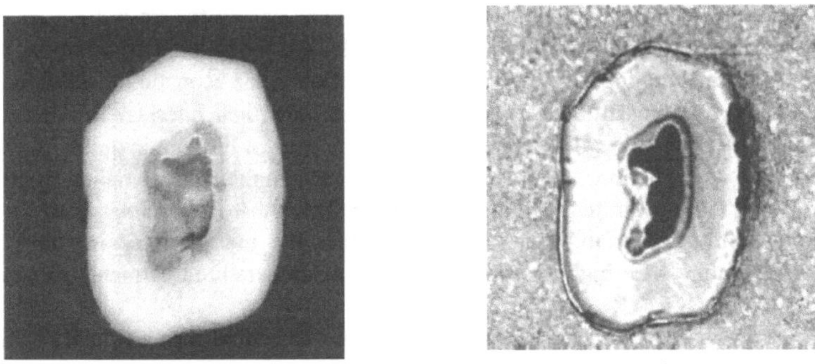

(a) (b)

Figure 1. A cross section of a human molar mounted in a mold. (a) optical image, and (b) acoustic image with 25 MHz lens.

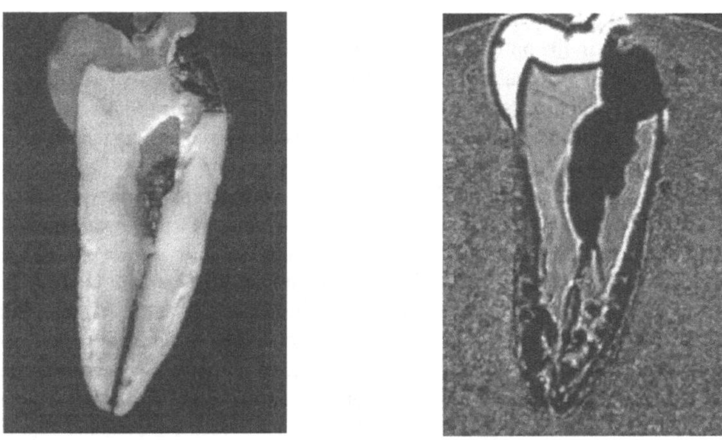

(a) (b)

Figure 2. Vertical section of a human pre-molar mounted in a mold. (a) optical image, and (b) acoustic image with 25 MHz lens.

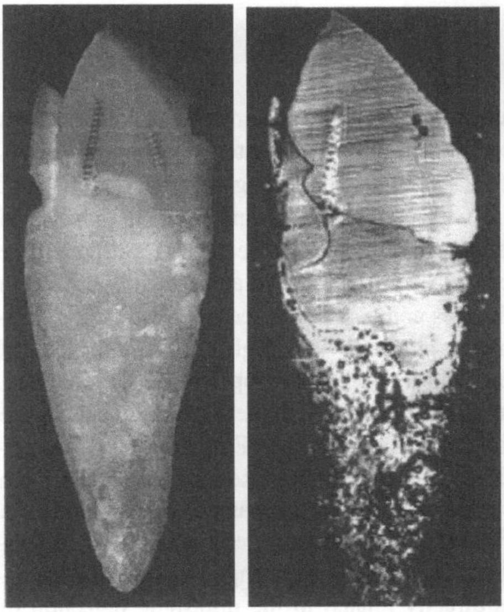

(a) (b)

Figure 3. Vertical section of a human canine with artificial plastic crown and metal screws. (a) optical image, and (b) acoustic image with 100 MHz lens.

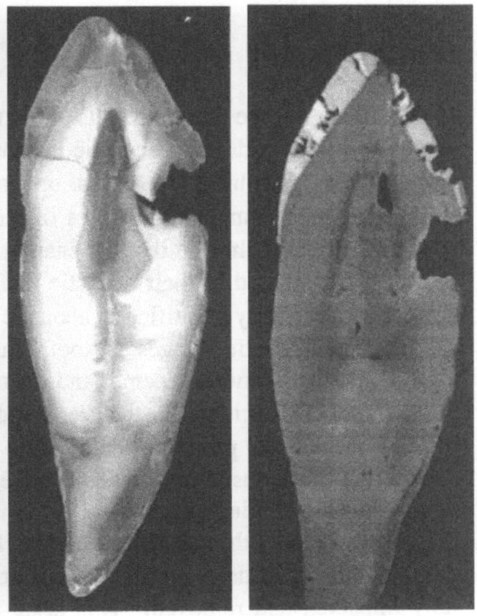

(a) (b)

Figure 4. Vertical section of a human canine with serious caries. (a) optical image, and (b) acoustic image with 50 MHz lens.

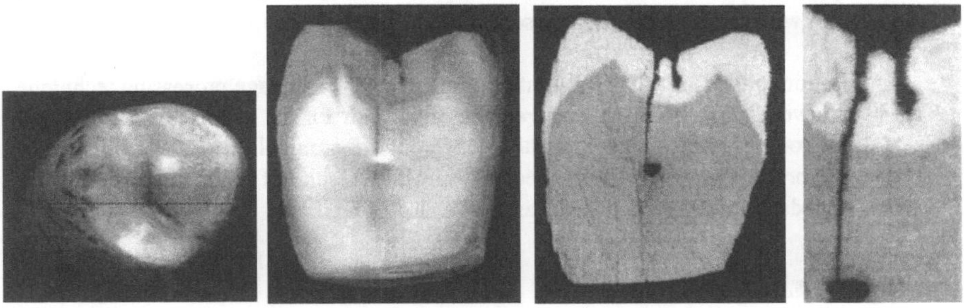

(a) (b) (c) (d)

Figure 5. Human pre-molar with fissure caries. (a) optical image from top of the crow, (b) optical image of a vertical section along the line in (a), (c) corresponding acoustic image with 100 MHz lens, and (d) magnification of part of (c).

enamel to the pulp cavity has induced during the preparation of the sample. This crack as well as the caries could be more clearly observed in the acoustic image as shown in Figure 5c and 5d. However, no obvious change in the acoustic impedance of the enamel along the caries was observed.

CONCLUSION

In this paper we present results of the ultrasound imaging of human teeth with different conditions using a desktop acoustic microscope with an intermediate frequency of 25, 50, and 100 MHz. Acoustic images and optical images of sections along different directions of teeth have been compared. Images of teeth with caries, filling, and implants have been analyzed. Results showed that the enamel and dentine area in the section of teeth could be clearly differentiated in their acoustic images. In addition, cracks and caries within teeth could be more easily identified in acoustic images compared with their corresponding optical images. Acoustic images of tooth samples prepared with high temperature showed clearer interface between enamel and dentine, and non-uniform properties in dentine area. Since it is a critical issue for the preparation of tooth samples, the temperature dependence of tooth mechanical properties deserves to be further investigated. Furthermore, it was observed that the acoustic impedance changed in the dentine along a serious caries in a tooth comparing the dentine area around it. However, a similar result has not been obtained in the enamel of a tooth with fissure caries. Further studies are required on this issue to confirm how the acoustic impedance of enamel and dentine can be changed before they are totally impacted and removed. This issue is directly related to the ultrasonic detection of early caries, which is becoming more and more important in clinical dentistry[2].

Despite its high perspective, the ultrasound imaging technique may not mature enough in current stage for the *in vivo* diagnosis of dental caries since the complicated geometry of teeth. However, it should be possible to characterize teeth with different diseases *in vitro* by using ultrasound technique. Comparing with optical methods, ultrasonic methods can provide teeth properties more related to their mechanical parameters. Results obtained by ultrasonic characterization of teeth are important foundations for the further investigation of ultrasonic diagnostic tools for dentistry. For example, how the acoustic impedance can be affected by the early caries in enamel and dentine is crucial to the ultrasound *B*-Scan or *3D*-Scan of teeth with caries. The study presented in this paper is our first step towards this direction.

In addition to the *C*-scan imaging introduced, an advanced image processing technique have been developed based on *volume rending algorithms* for the reconstruction of a real shape of the tooth surface and its internal holes through *B*-scan collections. The results of the imaging processing will be presented. Meanwhile, results of quantitative studies on acoustic impedance of different parts of teeth will also be reported.

ACKNOWLEDGMENT

Authors would like to acknowledge Dr. Raffaele A. Giannotti from the Windsor Dental Centre for supplying the tooth samples used in this study.

REFERENCES

1. A. Lussi. Validity of diagnostic and treatment decisions of fissure caries, *Caries Res.* 25: 296–303 (1991).
2. S.Y. Ng, P.A. Payne, and M.W.J. Ferguson. Ultrasonic imaging of experimentally induced tooth decay, in: *Acoustic Sensing and Imaging*, Conference Publication No. 369, IEEE: 82–86 (1993).
3. A.I. Darling. Studies of the early lesion of enamel caries with transmitted light, polarised light and radiography, *British Dental Journal* 101: 289–297 (1956)
4. R.R. Alfano, W. Lam, H.J. Zarrabi, M.A. Alfano, J. Cordero, D.B. Tata, and C.E. Swenberg. Human teeth with and without caries studied by laser scattering, fluorescence, and absorption spectroscopy, *IEEE J. Quantum Electronics* 20: 1512–5 (1984).
5. G. Kossoff and C.J. Sharpe. Examination of the contents of the pulp cavity in teeth, *Ultrasonics*, 4: 77–83 (1966).
6. F.E. Barber, S. Lees, and R.R. Lobene. Ultrasonic pulse-echo measurements in teeth, *Archs Oral Biol.* 14: 745–760 (1969).
7. A. Briggs. *Acoustic Imaging*, Science Publications, Oxford (1992).
8. R.Wichard, J. Schlegel, R. Haak, J.F. Roulet, and R.M. Schmitt. Dental diagnosis by high frequency ultrasound, in: *Acoustical Imaging* Vol.22, Plenum Press, New York: 329–334 (1996).
9. K.I. Maslov. Acoustic scanning microscope for investigation of subsurface defects, in: *Acoustical Imaging* Vol.19, Plenum Press, New York: 645–649 (1992).
10. K.I. Maslov, R.G. Maev, and V.M. Levin. New methods and technical principles of low frequency scanning acoustic microscopy, in: *Acoustical Imaging* Vol.20, Plenum Press, New York (1993)

THE ACOUSTIC PARAMETERS MEASUREMENT BY THE DOPPLER SCANNING ACOUSTIC MICROSCOPE

R. G. Maev and S. A. Titov

Centre for Imaging Research and Advanced Material Characterization,
School of Physical Sciences, University of Windsor, Windsor,
N9B 3P4, Canada

INTRODUCTION

The scanning acoustic microscopy is a well–known nondestructive method for high–resolution analysis of mechanical properties. Usually, the results of the analysis are represented as an image of the sample. The acoustic image characterizes the spatial distribution of the mechanical discontinuities and generally it is difficult to estimate acoustical parameters from the images [1].

It is possible to measure such parameters by V(z) method. In this method the output signal of the microscope V is recorded as a function of the distance Z between the lens focus and the sample surface. The shape of V(z) curve depends on the elastic properties of the surface area under investigation. The velocity and attenuation of leaky surface acoustic waves (SAW) can be estimated from the analysis of the specific oscillations observed in V(z) [2].

Developed wave theory states the relationship between the measurable response V(z) and reflectance function $R(\theta)$, where θ is the angle between the direction of an incidence plane wave and the normal, to the interface surface [3]. According to this approach reflectance function can be restored by the spectral analysis of the complex V(z). Moreover, the phase measurement of the tone burst signal is not trivial from the technical point of view.

Another approach to obtain both amplitude and phase of the V(z) function, is based on the use of a continuous wave reflection scanning acoustic microscope [4,5]. In this device, the continuous wave V(z) curve is defined as an electrical reflection coefficient of the transducer, and is measured as a function of the sample position. The resulting curve is a mixture of standing waves in the water between the lens and the specimen. Therefore, some difficulties in the estimation of the quantitative parameters have arisen.

It was proposed to use the Doppler effect for the formation of an output signal in the continuous wave microscope [6]. In this work the reflectance functions for materials with known acoustical parameters were restored with help of this device. The experimental data were used for the direct leaky Rayleigh wave velocity determination. Also, as it will be shown in this paper, the density and the velocities of the longitudinal and transverse waves can be estimated by the fitting to the measured phase of the reflectance functions.

MEASUREMENT SYSTEM

The operation principle of the Doppler acoustic microscope is illustrated by fig. 1. The transducer of the device is exited by the continuous wave with the frequency of w_0 . The lens element is moved at a constant velocity toward the sample surface. Due to the Doppler effect, the reflected ultrasonic wave has a frequency shift in comparison to the probing signal. Therefore, it can be separated from the powerful electrical breakthrough and reverberation signals in the frequency domain.

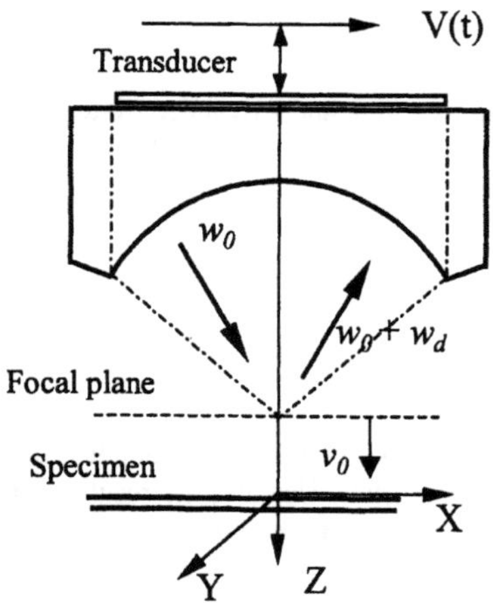

Figure 1. Experimental setup.

Each angular spectrum component of the reflected wave undergoes the Doppler frequency shift which depends on the incidence angle θ:

$$w_d = 2k_z v_0 \tag{1}$$

Here, $k_z = k_0 \cos(\theta)$ is z component of the wave number $k_0 = w_0/c$, c – sound velocity in the immersion liquid, v_0. – velocity of the lens motion.

The unitary reflected signal occupies a band of frequencies, which are determined by the maximum aperture angle θ_m of the acoustic lens: $(2k_0 v_0 \cos(\theta_m), 2k_0 v_0)$. The signal, reflected from the moving surface twice, undergoes a double frequency shift accordingly, and occupies a frequency band $(4k_0 v_0 \cos(\theta_m), 4k_0 v_0)$, and so on. If $\cos(\theta_m) > 0.5$, which is almost always fair to say, then the spectra of these signals are not overlapped, and by filtering a unitary reflected signal, the problem of standing waves in the liquid can be solved.

On the other hand, the amplitudes and the phases of these angular spectrum components are changed according to the reflection function of the surface immersion liquid – sample. It was shown [6] that the output signal of the Doppler continuous wave microscope can be expressed for isotropic specimens as:

$$V(t) = \frac{1}{2v_0} \int_{-\infty}^{\infty} H_0\left(\frac{w_d}{2v_0}, w_0\right) R\left(\frac{w_d}{2v_0}, w_0\right) \exp\{-i(w_d + w_0)t\} dw_d \tag{2}$$

Here $H_0(w_d/2v_0,w_0)$ is measurable transfer function of the system. The spectrum of this output signal is equal, neglecting the constant factor:

$$S(w - w_0) = H_0\left(\frac{w_d}{2v_0}, w_0\right) R\left(\frac{w_d}{2v_0}, w_0\right) \tag{3}$$

Thus, the spectrum of the Doppler signal is determined by the product of reflectance function and transfer function of acoustic system.

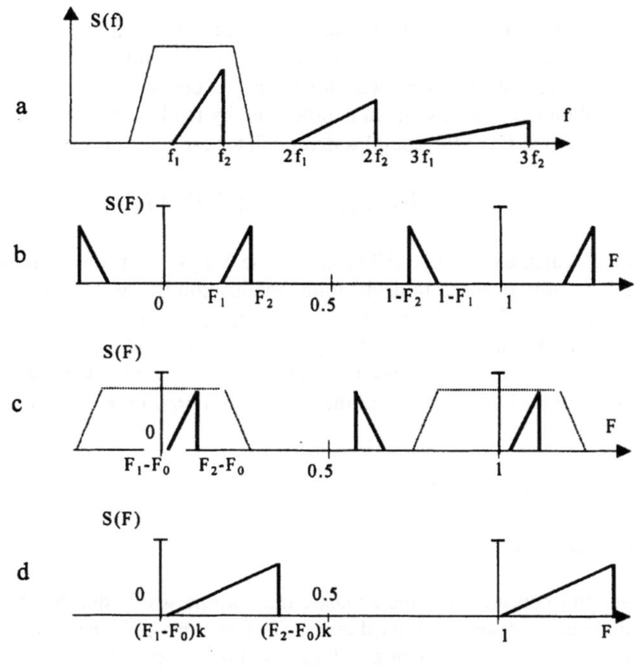

Figure 2. Spectra of the received Doppler signal at the analog (a), and digital (b,c,d) processing stages.

The considered theoretical principles were applied to the design of the point–focus continuous wave Doppler microscope with the operating frequency of 300 MHz. The excitation of an acoustic lens is made by the continuous wave low–noise generator. The sum of the signal, reflected from the sample surface, and the powerful electrical breakthrough are processed by the mixer.

The low-frequency output signal of the mixer contains Doppler contributions that correspond to several reflections. According to the relationship (1), the reflected signals occupy the band of frequencies (nf_1, nf_2), where $f_1=k_0v_0\cos(\theta_m)/\pi$, $f_2=k_0v_0/\pi$ and n is the number of the reflection. The structure of the Doppler signal spectrum is represented on fig.2(a). For the given parameters of the system, the unitary reflected signal has a frequency band from 0.9 to 1.2 kHz .

This Doppler signal is selected by the analog bandpass filter, and then converted into digital form for computer processing. To obtain a sufficient accuracy, a laser interferometer was used for scanning coordinate measurements and for the generating of the ADC triggering signal. Thus, in this system, the Doppler signal is sampled with spatial frequency $2/\lambda_d$, where λ_d=0.633μm is the wavelength of the laser beam.

The digital Doppler signal is the real part of a sampled complex signal $V(z)$:

$$V_j = Re\{V(\lambda_d j/2)\}; \quad j = 0, \pm1, \pm2,.. \pm[D/\lambda_d]. \tag{4}$$

Spectrum of this signal is symmetrical and it looks like the sample of digital representation shown on Fig.2(b). The band of the normalized digital frequencies of the signal is determined by the ratio of the light and acoustic wavelengths:

$$F_1 = \lambda_d k_0 cos(\theta_m)/2\pi = 0.096; \qquad F_2 = \lambda_d k_0/2\pi = 0.129. \tag{5}$$

At the stage of digital processing, the signal is exposed to the complex demodulation and its spectrum is analyzed. The signal V_j is multiplied on a complex continuous signal $exp(i2\pi F_0 j)$ thus, its spectrum is displaced along the frequency axis on the value F_0 (fig.2(c)). Then, the obtained complex signal is processed by the low–pass filter with a response h_j. The output signal V'_j of the complex demodulator can be expressed as:

$$V'_j = [\{V_j cos(2\pi F_0 j)\} * h_j] + i\,[\{V_j sin(2\pi F_0 j)\} * h_j] \,. \tag{6}$$

It is easy then to calculate module |V(z)| and phase arg{V(z)} using complex parts. Finally, the signal is usually exposed to decimation, apodization by the Hamming window function and FFT (fig. 2(d)). Decimation, i.e. reducing the sample rate, can be thought of as only using every Nth digital filter output for spectrum analysis, where N is the decimation factor. The decimation causes the spectrum spreading in N times therefore the relationship between the digital frequency F and the value $m = cos(\theta) = k_z / k_0$ can be stated as:

$$m = (F/N + F_0)/F_2 \,. \tag{7}$$

EXPERIMENTAL RESULTS

Developed Doppler continuous wave acoustic microscope was tested by the investigation of aluminum, hardened steel and fused quartz. The example of Doppler signal that is recorded as a function of the displacement z and the result of the spectral analysis is shown on fig. 3.

The envelopes of the signals |V(z)| have typical oscillations, the period of which is related to leaky SAW velocity C_R and used usually for its determination [2]. The phase arg{V(z)} also has the oscillations with the same period. According to (3) the spectrum S(m) of the signal is an estimation of the reflectance function that is multiplied by the transfer function of the system.

Transfer function was obtained by the analysis of the material which reflectance function is basically constant within the angular aperture of the lenses. Investigation showed that the phase of the transfer function is equal approximately to zero for used lens system [6]. Therefore, the phase of a measured spectrum arg{S(m)} is determined by the phase of the reflectance function.

The phase of the reflectance function changes by almost 2π radians around SAW critical angle θ_R ; R=$cos(\theta_R)$. The critical angel locations were determined from the maximum of the phase derivative and the Rayleigh wave velocities c_R are then deduced by

Snell's law. The experimental relative error was estimated as $3\delta/<c_R>$, where $<c_R>$ is the mean value of measured value and δ is a standard deviation (Table 1).

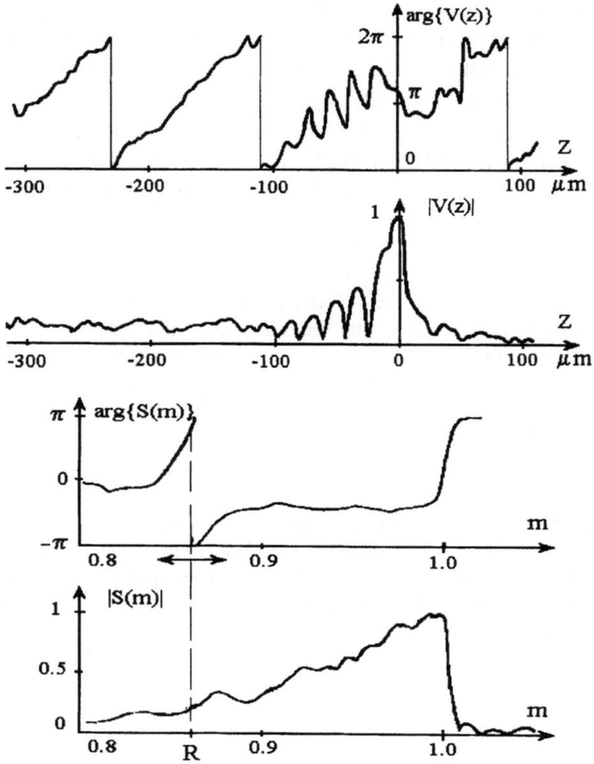

Figure 3. Experimental magnitude |V(z)| and phase arg{V(z)} of the signal recorded for aluminum; magnitude |S(m)| and phase arg{S(m)} of the calculated spectrum; $m=k_z/k_0$.

Table 1. Results of the acoustic parameters measurement

Material	Acoustic parameter	Known value[1]	Measured value $<c, \rho> \pm 3\delta$	Errors $3\delta/<c_R, \rho>$
Aluminum	C_R, m/s	2844	2855 ± 34	0.012
	C_L, m/s	6420	6240 ± 420	0.067
	C_S, m/s	3040	3095 ± 70	0.023
	ρ, g/cm^3	2.7	2.9 ± 0.24	0.08
Steel hardened	C_R, m/s	2945	2931 ± 28	0.0096
	C_L, m/s	5874	5950 ± 272	0.046
	C_S, m/s	3179	3220 ± 80	0.025
	ρ, g/cm^3	7.8	7.5 ± 0.35	0.045
Fused quartz	C_R, m/s	3410	3425 ± 47	0.014
	C_L, m/s	5970	5900 ± 395	0.067
	C_S, m/s	3765	3780 ± 91	0.025
	ρ, g/cm^3	2.2	2.21 ± 0.2	0.09

Besides SAW velocity of C_R, the uniform and isotropic solid sample is characterized by the density ρ, the velocity of the longitudinal wave C_L and the velocity of the transverse wave C_S. It is possible also to determined these parameters (ρ, C_L, C_S) from the measured Doppler signals.

The parameters determination was done by the nonlinear fitting of the theoretical reflection function to the experimental data. Only the phase of the Doppler spectrum arg{S(m)} was used in the estimation algorithm. The set of the experimental points was selected in the area of the critical Rayleigh angle θ_R, where their values depend strongly on the acoustical properties. The theoretical function to be fitted was defined as a phase of the well–known expression for the reflection coefficient [7]. It was supposed for the initial estimates of parameters: ($\rho = 3$ g/cm^3, C_L=2 C_R^*, $C_S = C_R^*$), where C_R^* – is the Rayleigh wave velocity measured previously.

The results of the calculations are submitted in Table 1. The estimated parameters are in reasonable agreement with the known values. Also, it is noticed that the relative error of the estimation for the longitudinal wave velocity is larger than the error for the transverse one. This can be explained by the fact [1] that the transverse wave velocity is in a strong relation with the Rayleigh velocity ($0.87< C_R / C_S <0.95$), which error is comparatively small due to the rapid phase change at the critical angle.

CONCLUSION

Doppler continuous wave scanning acoustic microscope was developed for the investigation of the reflectance function at a frequency of 300 MHz. By testing aluminum, steel and fused quartz samples it was shown that the errors of the acoustic parameters estimation are: 9% for density, 7%, 2.5%, 1.4% – for longitudinal, shear and leaky Rayleigh waves velocities correspondingly.

REFERENCES

1. A. Briggs. *Acoustic microscopy*, Clarendon Press, Oxford (1992).
2. J. Kushibiki and N. Chubachi, Material characterization by line-focus–beam acoustic microscope, *IEEE Trans. Sonics Ultrason., vol. SU–32, pp.189–212*, (1985).
3. K.K. Liang, G.S.Kino and B.T. Khuri–Yakub, Material characterization by the inversion of V(z), *IEEE Trans. Sonics Ultrason., vol. SU–32, no.2, pp.213–224*, (1985).
4. S. Sathish, G. Gremaud and A. Kulik, Theory of continuous wave scanning acoustic microscope, *Acoustical Imaging, vol. 20, edited by Y. Wei, and B. Gu, Plenum Press, New York, pp 259–264*, (1993).
5. S. Sathish, G. Gremaud, A. Kulik and P. Richard, V(z) of continuous wave reflection scanning acoustic microscope, *J. Acoust. Soc. Am., vol. 96, no.5, pp.2769–2775*, (1994).
6. S.A.Titov, R.G.Maev, Doppler continuous wave acoustic microscope, *Proceedings 1997 IEEE Ultrasonic Symposium, ed. by S.C.Shneider, M.Levy, B.R.McAvoy, pp. 713–718.*
7. L.M. Brehovskikh. *Waves in Layered Media*, Academic, New York (1980).

QUANTITATIVE CONTACT SPECTROSCOPY BY ATOMIC-FORCE ACOUSTIC MICROSCOPY

U. Rabe, E. Kester, V. Scherer, and W. Arnold

Fraunhofer-Institute for Nondestructive Testing (IZFP)
Bldg. 37, University,
D-66123 Saarbrücken, Germany

INTRODUCTION

In Atomic Force Microscopy (AFM) the deflection of a microfabricated elastic beam with a sensor tip at its end is used to generate high-resolution images of surfaces. Dynamic modes, where the cantilever is vibrated while the sample surface is scanned, are standard in commercial instruments. Various techniques, such as Force modulation microscopy[1,2]. Ultrasonic Force Microscopy[3,4], Atomic Force Acoustic Microscopy[5], Scanning Local Acceleration Microscopy[6], Microdeformation Microscopy[7] or Pulsed Force Microscopy[8] enables one to obtain images which depend on the elasticity of the sample surface. However, quantitative determination of Young's modulus of a sample surface with AFM is still a challenge, especially for stiff materials such as hard metals or ceramics. In this contribution the evaluation of the cantilever vibration spectra at ultrasonic frequencies is presented in order to discern local elastic data quantitatively.

AFM cantilevers are elastic beams of several 100 µm length, a width of several tens of µm and a few µm thickness. One end of the beam is fixed on a substrate and the free end holds the sensor tip. Having a homogeneous rectangular cross section, the dominant acoustical vibrations of the beam are flexural, extensional and torsional modes[9]. When the sensor tip approaches a sample surface, the tip-sample interaction forces, such as Van-der-Waals forces, electrostatic forces, repulsive contact forces or damping forces, represent a nonlinear spring-dashpot system. The forces change the boundary conditions of the cantilever and, consequently, its resonance frequencies. Acoustic vibrations can be coupled from the vibrating sample surface into the cantilever via the spring and also from the vibrating cantilever into the sample. Resonance frequencies are shifted relatively to the free resonances when they are in force interaction with a surface. This fact can be exploited to determine the contact stiffness quantitatively[9]. The spatial derivative of the interaction forces relative to the stiffness of the cantilever determines which resonance is most sensitive to small variations in the interaction forces. The terms „Ultrasonic Force Microscopy" or „Atomic Force Acoustic Microscopy" (AFAM) comprise different techniques, all of which

Acoustical Imaging, Volume 24.
Edited by Hua Lee, Kluwer Academic/Plenum Publishers, 2000.

have in common that the sample and the cantilever are submitted to mechanical oscillations or waves in the frequency range of several tens of kHz up to the MHz range. These methods can be differentiated according to the frequency range employed, the type of excitation signal, such as sinusoidal, amplitude modulated, or pulsed excitation, and the type of waves which are generated, i.e. flexural waves in the cantilever, longitudinal or shear waves in the sample and surface waves. Furthermore, either linear or nonlinear signals[3,4,10] may be used for detection. At a previous Acoustical Imaging conference, reviews have been presented describing these techniques[11,12].

MEASUREMENT TECHNIQUE

Commercial AFMs (AFM/LFM and Dimension 3000, Digital Instruments, Santa Barbara, Ca) and additional ultrasonic equipment were used to excite flexural vibrations in the cantilever (Fig. 1). A frequency generator HP 33120A provides a sinusoidal signal which is applied to a conventional ultrasonic transducer (e.g. Panametrics V110 5 MHz center frequency) attached to the sample. The ensuing sample surface vibration couples into the cantilever when the sensor tip is in force interaction. Alternatively the excitation signal can be applied to the piezoelectric transducer integrated in the cantilever holder and the tip-sample forces change the boundary conditions of the cantilever and hence its frequency response. The cantilever vibrations are measured either with an external detector[5] or with the internal detector of the instrument, provided its bandwidth is in the MHz regime. The vibration signals of the cantilever are fed to a heterodyne down-converter which in addition receives a reference signal at the excitation frequency from the frequency generator. It operates at frequencies from 75 kHz to 10 MHz and shifts the desired signal to a fixed 20 kHz intermediate frequency. The down-converted signal is evaluated by a lock-in amplifier. The set-up is controlled using a Labview program enabling one to change the excitation frequency, to read the lock-in output, and to display the spectra.

First, the resonances of the cantilever are measured in free vibrations without contact to a sample surface. Depending on the stiffness of the cantilever and the spacing of its resonances, several free resonance frequencies of the cantilever between 30 kHz and 3 MHz can be found corresponding to the calculated values[9]. Second, the cantilever is adjusted so

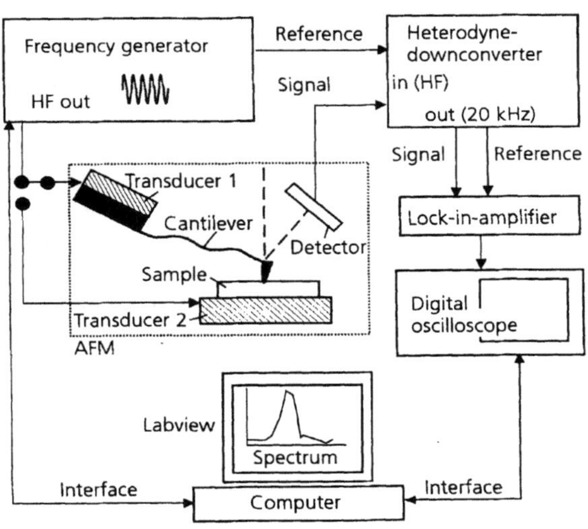

Fig. 1: Principle of Atomic Force Acoustic Microscopy

that it rests on the surface with an applied load that is given by the product of its actual deflection and the cantilever spring constant. The software program for the acquisition of the contact spectra is then started. The excitation amplitude must be kept small in order to stay in the linear range of the tip-sample interaction forces. To be able to apply static loads higher than adhesion forces, we used stiff cantilevers with spring constants of about 40 - 80 N/m. We made a series of measurements with different static deflections of the cantilever between 10 nm and 40 nm. For a cantilever with a spring constant of about 50 N/m, the force F_N is varied between 500 nN and 2000 nN. The shift of the contact resonances with the applied load helps to distinguish the cantilever resonances from parasitic resonances of the system.

MATERIALS

Silicon wafers oriented in <100> direction were examined as a reference material. Oxide layers were removed by cleaning the sample in ultrasonic baths of acetone and methanol, followed by a 10 min etching in 50% HF. The elastic modulus is $E_s = 1.3 \times 10^{11}$ N/m^2 and the Poisson number is $\nu_s = 0.181$. As the contact stiffness depends very much on the geometry of the interacting bodies - the tip and the surface - we only used samples with a mean surface roughness $R_a \approx 1$ nm or less.

Thin films of nanoscaled ferrites with spinel structure can be used in magneto-optic and magnetic recording systems. Their properties depend largely on the control of their preparation and their nanostructure. It has been shown, for example, that the coercivity is not only correlated with the chemical composition, but also with the grain size[13], the quenching temperature, and the cation distribution[14]. To understand the coercivity as a function of the nanostructure, it has been demonstrated by XPS spectroscopy and X-ray diffraction[15] that the coercivity is related to the chemical gradients induced by the thermal evolution and by mechanical gradients leading to a variation of the elastic moduli on a nanoscale. In order to measure the elastic properties locally, the AFAM technique described here is used. On thin films of magnetite Fe_3O_4 and maghemite γFe_2O_3 with spinel structure, the influence of the deviation from the stoichiometry on the elasticity for a given grain size is determined.

Among other applications, PZT ceramics are used for ultrasonic transducers. Their effective piezoelectric constant is determined by the homogeneity of the polarization and the piezoconstant. We use the AFAM-technique to image the local displacements which depict the local piezoelectric constant and hence the polarization through the inverse piezoelectric effect.

EXPERIMENTAL RESULTS

Fig. 2 shows the spectra taken on a Si single crystal oriented in <100> direction. The Si-cantilever had dimensions of 135×32×5.2 µm^3 and a stiffness of 77 N/m. The two first free flexural resonance frequencies were at 322 kHz and 1.968 MHz, respectively. When the tip contacts the surface, the resonance frequency shifts by Δf to a higher value (b), increasing with increasing load applied by the cantilever leading to a higher contact radius between tip and surface (c-e). The Q factor of the resonance drops from $\approx$ 500 for the free resonance to less than 50 in contact. We found no difference in Δf between cleaned surfaces and the surface of the wafers as obtained. Fig. 3 shows the spectra obtained on the magnetic nanocrystalline films. With the same experimental conditions (cantilever radius r [nm]: $30 < r < 50$, spring constant $k_c = 38$ N/m, second free resonance $f_2 = 908$ kHz), the shift of the second contact resonance on the magnetite film is larger than on the maghemite film.

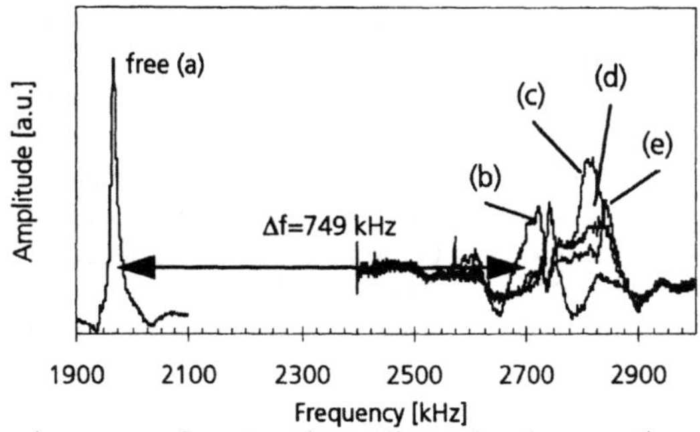

Figure 2: Second flexural resonance of a rectangular cantilever when the sensor tip contacts a <100> silicon surface. The resonant frequency is of a free cantilever is lowest (a). The contact resonance frequency increases with increasing normal laod F_N: 770 nN (b), 1540 nN (c), 2310 nN (d) and 3080 nN (e).

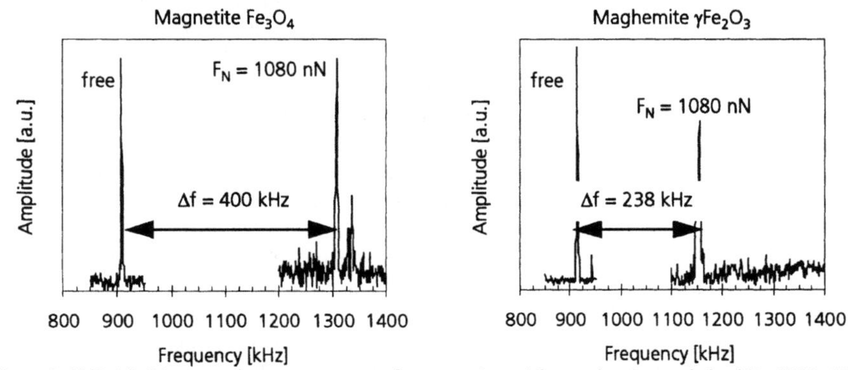

Figure 3: Shift Δf of the second contact resonance for magnetite and for maghemite; static load F_N=1080 nN.

EVALUATION OF DATA AND DISCUSSION

When the cantilever is in contact with the sample, vertical and lateral forces act on the sensor tip which are highly nonlinear with distance. For small vibration amplitudes of the cantilever close to its equilibrium position, the tip-sample interaction can be approximated by linear vertical and lateral[16] spring-dashpot systems as shown in Fig. 4 (a). k^*, γ and k^*_{lat}, γ_{lat}, are the vertical and the lateral contact stiffness and damping constants, respectively. L is the length of the cantilever, k_c is the spring constant, and h is the tip length. The damping constant η_{air} of the free beam is given by dissipation of energy, mainly due to air friction and by radiation of ultrasound into the surrounding air. The characteristic equation from which the resonance frequencies of the system are obtained as a function of contact stiffness can be found analytically[9,17]. In principle, lateral and vertical contact stiffness could be determined from the measured resonance frequencies, if torsional cantilever vibrations were measured additionally. For simplicity, we assume that $k^* = k^*_{lat}$ and $\gamma = \gamma_{lat}$. In this case, the characteristic equation for the flexural resonance frequencies can be inverted to determine k^* and γ, provided the free resonance frequencies, their Q-factors, and the stiffness of the cantilever are known. Additionally, the cantilever is assumed to be parallel with the sample, i.e. $\alpha_0 = 0$ in Fig. 4 (a). However, the $\alpha_0 \neq 0$ situation can also be treated[18].

The theory predicts that during contact, the resonance frequency of each mode is shifted to higher frequencies relative to the free resonance. Increasing the normal force, the

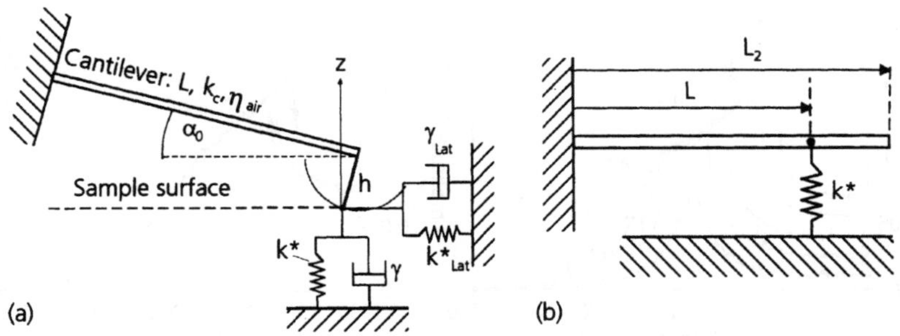

Fig. 4: Linear model used for determination of k*. a: including lateral and vertical tip-sample interaction forces. b: taking into account that the tip is located at $L < L_2$.

contact area between the tip and the sample increases. The contact force as predicted by a Hertzian contact[19] for a spherical tip and a flat surface sample is:

$$F_N = \frac{4E^*}{3}\sqrt{(-z_e)^3 R} \tag{1}$$

The contact stiffness is given by:

$$-\left.\frac{\partial F_C}{\partial z}\right|_{z=z_e} = k^* = \sqrt[3]{6 \cdot E^{*2} \cdot R \cdot F_C} \tag{2}$$

where E* with $1/E^* = (1-\nu_t^2)/E_t + (1-\nu_S^2)/E_S$, is the reduced Young's modulus, E_t, E_s, ν_t, and ν_s are the elastic moduli and the Poisson ratios of the tip and the surface, respectively. R is the radius of the sensor tip (measured with a calibration standard), and z_e is the vertical tip deflection in force equilibrium before insonification (note that z_e is negative). Eq. (2) shows that k* increases as a function of normal load which is in fact observed (Figs. 2 and 3). Though the Q factor of ≈ 500 in air can decrease by more than one order of magnitude in contact because of tip-sample damping, the influence of the damping on Δf was low in our experiments. Hence $\gamma = \gamma_{lat} = 0$ was set, for the calculating of the Young's modulus of the surface being examined.

In the case of Si-crystal oriented in <100> direction, the Young's modulus calculated from the shift of the two first resonances and taking into account the normal force between the tip and the sample is $1.2 \pm 0.6 \times 10^{11}$ Pa. It is in good agreement with the known value of 1.3×10^{11} Pa[20].

The Young's moduli of magnetite and maghemite measured with AFAM are $1.2 \pm 0.2 \times 10^{11}$ and $0.6 \pm 0.2 \times 10^{11}$ Pa, respectively. Having the same crystalline structure and the same grain size, magnetite and maghemite differ only in stoichiometry. Indeed, maghemite results from the complete oxidation of the Fe^{2+} cations initially present. The oxidation proceeds by the fixation of new oxygen ions which is associated with the simultaneous creation of new unit cells in which the oxygen sites are occupied and the cation sites are empty. Thus maghemite contains cation vacancies decreasing its stiffness.

There is a systematic decrease of the contact stiffness with mode number[21], which is probably due to the fact that the model employed does not reflect the geometry of the cantilever exactly. Assumptions, such as constant cross section for the total length of the cantilever, infinitely stiff clamping at one end, and the sensor tip being exactly at the end of the cantilever, are not realized for currently avalaible cantilevers. The influence of the sensor tip position along the cantilever on the frequency shift of the modes has been examined earlier[17]. As described above, we applied normal loads on the sample. In this case, adhesion can be treated like an additional external force, or it can even be neglected if the normal load is high enough. The forces can cause pressure, on a sharp tip of small radius, that exceeds

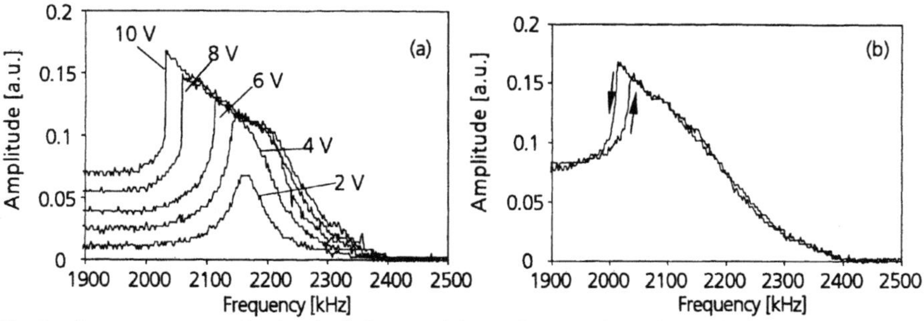

Fig. 5: Contact resonance showing the influence of the nonlinearity of the tip-sample force. a: variation of the excitation amplitude applied to the transducer below the sample. b: hysteresis of the frequency scan

the yield strength of the material of the tip. This can be easily calculated when a maximal yield strength of 1/100 of the Young's modulus is assumed and using the Hertzian model to calculate the contact area. In fact, when we start experiments with a new tip, we observe that the contact stiffness gradually increases until a saturation is reached. If one assumes a limiting tip radius, a value R = 211 ± 100 nm is obtained whereas the tip radius of new cantilevers is specified to 5 - 10 nm by the manufacturer. As a result there is a limit in resolution in measuring the contact stiffness by our technique, which can only be overcome if reliable data of the adhesion value is obtained by additional measurements. A known adhesion value would allow to take its equivalent spring constant into account in the total contact stiffness, so that one can work with low static forces. We think that this limitation in resolution *and* quantitative determination of the contact stiffness we encountered equally apply to the other techniques discussed[3,6,8].

There are additional effects. In Fig. 5a the result of an experiment is shown, where spectra were taken at different excitation amplitudes of the transducer of 2, 4, 6, 8 and 10 V, resulting in a corresponding increase of the surface displacement of the sample. The normal load F_N was 1020 nN, the sample was a 200 nm thick film of nanocrystalline Fe_3O_4 on a silicon substrate (same whose contact resonance is shown in Fig. 3a). At an excitation amplitude of 2V, a symmetric resonance peak is obtained showing that the linear approximation is still valid. When the amplitude is increased, the resonance curves become asymmetric, the maxima shift to lower frequencies and a jump in amplitude develops. This bending of the resonances to the left hand side is caused by a softening nonlinearity[21]. The resonances now show hysteresis (Fig. 5b). The frequency at which the amplitude jump occurs depends on the direction of the frequency scan. An equation of motion for the flexural vibrations in the cantilever beam with a nonlinear tip-sample interaction force has been already solved numerically[10], allowing to extend the solutions developed to explain the observation made here by taking into account the behavior of nonlinear oscillators[22,23] possibly allowing one to separate adhesion from contact stiffness.

Conventional AFM is well suited to image the domain and grain structure of ferroelectric crystals by an indirect method. One can image the domain structure by polishing the sample using some polishing liquids that etch the domains of different orientation anisotropically. This results in a topography reflecting the domain structure[24]. Without previous etching, domains walls can be imaged if one applies an ac voltage between tip and sample while scanning the tip across the sample[24,25]. The locally induced ac field causes a periodical extension of each domain by the inverse piezoelectrical effect and this out-of-plane oscillation excites the cantilever equally. As the extension depends on the orientation of the domains relative to the direction of the applied field, the contrast in this so-called piezomode AFM also changes in each domain. Until now this method was successful

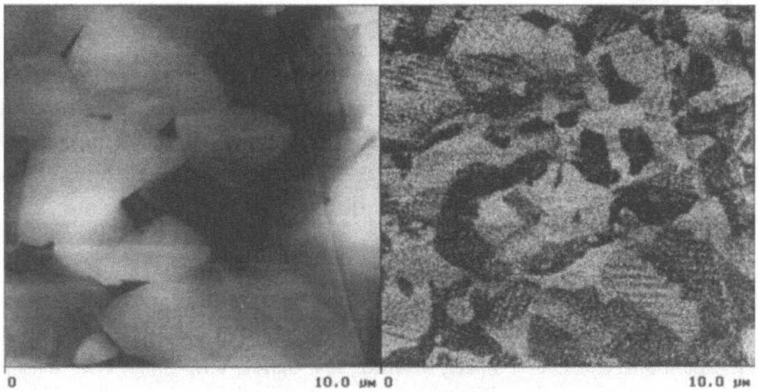

Fig. 6a Topography of a polished surface of PZ29, gray scale 70 nm

Fig. 6b Phase image of PZ29 in the piezo-mode; voltage tip sample $4V_{pp}$, f = 160 kHz

in $BaTiO_3$, Guanidine Aluminum Sulphate Hexahydrate (GASH), Triglycerine Sulfate (TGS) or in thin films of lead-zirconate-titanite (PZT) but not in bulk PZT. Fig. 6 shows an image of domains of polished commercial PZT-ceramics. Fig. 6a, of 10 x 10 μm^2 scan size, shows the topography. The gray scale comprises height differences of up to 70 nm. The topographic scan shows that the polishing causes height differences of up to 50 nm between individual grains, whereas the topography on the single grains is nearly unperturbed. Measured roughness values on single grains are on the order of Ra $\approx$ 1 nm. Fig. 6b represents the same scanned area of the PZT material. Now, the different shades of gray describe the phase difference between the sinusoidal electrical excitation and the mechanical ac response of the cantilever. One clearly recognizes the subdivisions of the crystallites into domains. The examined ceramics are of morphotropic, i.e. tetragonal and rhombohedric composition with a dominantly tetragonal phase. Within these ferroelectrics one typically observes both lamellar and random domain forms of which the domains of the rhomboedrical phase seem to provide the better contrast. The lamellar domains have a width of a few hundred nm, and they are detectable with the lateral resolution of AFM.

CONCLUSION

The AFAM technique is a sensitive method for the quantitative measurement of local elasticity, here on a scale of 200 nm. On a silicon single-crystal oriented in <100> direction, the results for the measured Young's modulus are in good agreement with the known value. In nanoscaled thin films of ferrites with a grain size of about 70 nm, we were able to detect the difference in the elasticity caused by the deviation from the stoichiometry. With the AFAM technique, surfaces with a large enough difference in elastic modulus can be distinguished reliably. However, for reproducible quantitative measurements in the linear as well as in the nonlinear force range, the problem of sensor tip stability is not yet solved. This could be made possible either by using more stable diamond tips with larger radii or by working in a controlled environment to reduce adhesion by meniscus forces. Modeling the nonlinear behavior of the contact resonances, as for example the hysteresis, would allow one to gain further information about the interaction forces. Finally, there are two techniques to image ferroelectric domains in PZT material. They can be made visible by anisotropic etching. In that case one can only image a frozen-in state of the surface. A life image, however, can be obtained by the inverse piezoelectric effect allowing one to examine the domain dynamics.

ACKNOWLEDGEMENTS

One of us (E.K.) was supported by the EU within the HCM program. We thank the Volkswagen Foundation for support and the German Science Foundation for grants within the SFB „Grenzflächenbestimmte Materialien“ at the University of Saarbrücken and the „Schwerpunkt Multifunktionswerkstoffe“. It is a pleasure to thank K. Dransfeld, University of Konstanz and J.D. Niepce, University of Dijon for discussions.

REFERENCES

[1] P. Maivald, H.T. Butt, S.A. Gould, C.B. Prater, B. Drake, J.A. Gurley, V.B. Elings, P.K. Hansma, Using force modulation to image surface elasticities with atomic force microscopy, *Nanotech.* 2:103 (1991).

[2] M. Radmacher, R.W. Tillmann, and H.E. Gaub, Imaging viscoelasticity by force modulation with the atomic force microscope, *Biophys. J.* 64:735 (1993).

[3] K. Yamanaka, H. Ogiso, and O. Kolosov, Ultrasonic force microscopy for nanometer resolution subsurface imaging, *Appl. Phys. Lett.* 64:178 (1994).

[4] W. Rohrbeck and E. Chilla, Detection of surface acoustic waves by scanning force microscopy, *Phys.Stat.Sol(a)* 131:69 (1992)

[5] U. Rabe, W. Arnold, Acoustic microscopy by atomic force microscopy, *Appl. Phys. Lett.* 64:1493 (1994)

[6] N.A. Burnham, G. Gremaud, A.J. Kulik, P.-J. Gallo, and F. Oulevy, Materials properties measurement: choosing the optimal scanning probe microscope configuration, *J. Vac. Sci. Tech.* B14:1308 (1996).

[7] B. Cretin and F. Sthal, Scanning microdeformation microscopy, *Appl. Phys. Lett.* 62:829 (1993).

[8] A. Rosa, E. Weilandt, S. Hild, O. Marti, The simultaneous measurement of elastic, electrostatic andadhesive properties by scanning force microscopy:pulsed-force mode operation, *Meas. Sci. Tech.* 8:1 (1997).

[9] U. Rabe, K. Janser, and W. Arnold, Vibrations of free and surface-coupled atomic force microscope cantilevers: theory and experiment, *Rev. Sci. Instrum.* 67:3281 (1996).

[10] S. Hirsekorn, Transfer of mechanical vibrations from a sample to an AFM-cantilever - a theoretical description, *Appl. Phys.* A 66:S249 (1998).

[11] U. Rabe, K. Janser, and W. Arnold, Acoustic microscopy with resolution in the nm range in *Acoustical Imaging*, P. Tortoli and L. Masotti, eds., Plenum Press, New York, 22:669 (1996).

[12] O. Kolosov, A. Briggs, K. Yamanaka, and W. Arnold, Nanoscale imaging of mechanical properties by ultrasonic force microscopy, P. Tortoli and L. Masotti, eds., Plenum Press, New York, *Acoustical Imaging* 22:665 (1996).

[13] A.J. Koch and J.J Becker, Permanent magnets and fine particles. Influence of crystallite size on the magnetic properties of acicular γFe_2O_3 particles, *J. Appl. Phys.* 39:1261 (1968).

[14] E. Kester, B Gillot, and Ph. Tailhades, Analysis of the oxidation process and mechanical evolution in nanosized copper spinel ferrites. Role of stresses on the coercivity, *Mat. Chem. Physics* 51:258 (1997).

[15] E. Kester and B. Gillot, Cation distribution, thermodynamic and kinetics considerations in nanoscaled copper ferrite spinels. New experimental approach by XPS and new results both in the bulk and on the grain boundary, *J. Phys. Chem. Solids*, (1998), to be published

[16] P.E. Mazeran, and J.L. Loubet, Force modulation with a scanning force microscope: an analysis, *Trib. Lett.* 3:125 (1997).

[17] U. Rabe, J. Turner, and W. Arnold, Analysis of the high-frequency reponse of atomic force microscope cantilevers, *Appl. Phys. A*66:S277 (1998)

[18] K. Yamanaka, A. Noguchi, T. Tsuji, T. Koike, and T. Goto, Quantitative material characterization by ultrasonic AFM, *Surf. Interface Analysis,* (1999) to be published.

[19] K.L. Johnson, in: *Micro/Nanotribology and its Applications*, Ed. B. Bushan, Kluver Academic Publishers, Dordrecht (1997), pp. 151 ff.

[20] M. Kulatov, Morphology of an as-grown surface of SiC-determination of step-flow direction, *J. Cryst. Growth* 158:261 (1996).

[21] U. Rabe, E. Kester, and W. Arnold, Probing nonlinear tip-sample interaction forces by atomic-force acoustic microscopy, *Surf. Interface Analysis,* (1999) to be published.

[22] A.H. Nayfeh and D.T. Mook, *Nonlinear Oscillations*, John Wiley, New York, 1995.

[23] P.R. Nayak, Contact vibrations, *J. Sound and Vibration* **22**, 297 (1972).

[24] K. Franke, J. Besold, W. Haessler, and C. Seegebarth, Modification and detection of domains on ferroelectric PZT films by scanning force microscopy, Surf. Sci. Lett. 302:L283

[25] M. Abplanalp, L.M. Eng, and P. Günter, Mapping the domain distribution at ferroelectric surfaces by scanning force microscopy, *Appl. Phys.* A66:S231 (1998)

DOUBLE FOCUS TECHNIQUE FOR SIMULTANEOUS MEASUREMENT OF SOUND VELOCITY AND THICKNESS OF THIN SAMPLES USING TIME-RESOLVED ACOUSTIC MICROSCOPY

V. Hänel and B. Kleffner

Fraunhofer Institute for Biomedical Engineering
Ultrasound Department
Ensheimer Str. 48
D - 66386 St. Ingbert
Germany

INTRODUCTION

Simultaneous measurement of sound velocity and thickness of thin coatings is a well-known technique in time-resolved Acoustic Microscopy.[1,2] The principle of this technique is shown in Fig. 1. However, the applicability is restricted, because a substrate in contact with the back of the sample is required. Furthermore, the substrate must have an area without coating, because a reference signal from the substrate's surface is needed.

In order to overcome these performance limitations, a double focus technique has been developed for thin specimens not attached to a substrate (Fig. 2). Focusing the lens on the front and afterwards on the back of the sample, only signals from the sample itself are required. Based on a simple geometrical model the relevant acoustical equations for sound velocity and thickness were deduced analytically and tested experimentally.

THEORY

Figure 1 schematically shows a spherical, focusing lens of an acoustic microscope and a sample attached to a substrate. The ultrasound pulses are displayed as arrows. On the right hand side the reflected echoes are shown in the time domain. The time intervals Δt_1 and Δt_2 determine both the sample sound velocity and the sample thickness, following from eqs. (1) and (2). Using the classic approach the lens does not have to be focused on the sample's surface and on the substrate's surface, respectively. However, this procedure has inherent disadvantages. The applicability is restricted, because a substrate in contact with the back of the sample is required. Furthermore, the substrate must have an uncoated area,

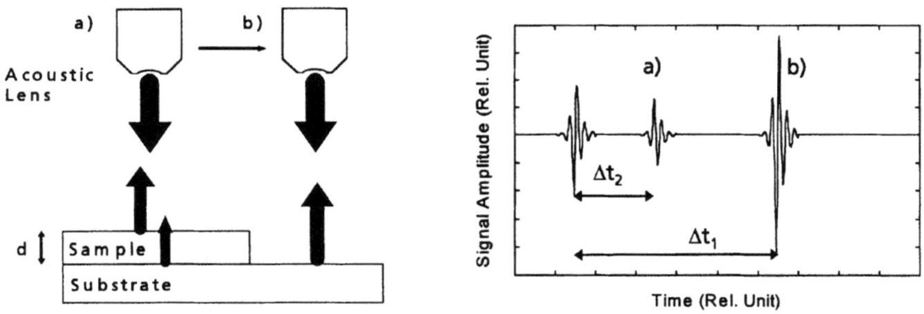

Figure 1. Simultaneous measurement of sound velocity and thickness (classic approach); left: principle, right: signals in the time domain (model)

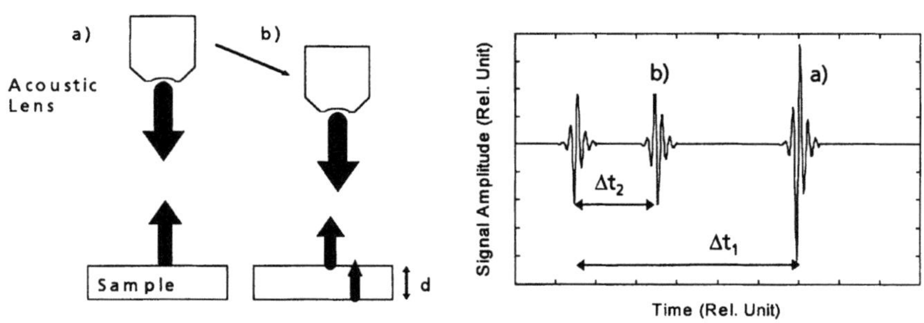

Figure 2. Simultaneous measurement of sound velocity and thickness (double focus approach); left: principle, right: signals in the time domain (model)

because the reference signal from the substrate's surface is needed in order to calibrate the measurement.

$$d = \frac{1}{2} c_{Sample} \Delta t_2 = \frac{1}{2} c_1 \Delta t_2 \tag{1}$$

$$d = \frac{1}{2} c_{H_2O} \Delta t_1 = \frac{1}{2} c_0 \Delta t_1 \tag{2}$$

As a result the double focus technique has been developed only based on signals received from the sample itself. As distinct from Fig. 1 now the lens is focused on the sample's front (a) and on the rear of the sample (b) (see Fig. 2).

Taking into consideration Fig. 3, eqs. (1) and (3) are identical. With distance $\overline{AC}$ being the necessary defocus in order to focus the lens on the back of the sample, eq. (4) follows. Equations (3) and (4) yield eq. (5).

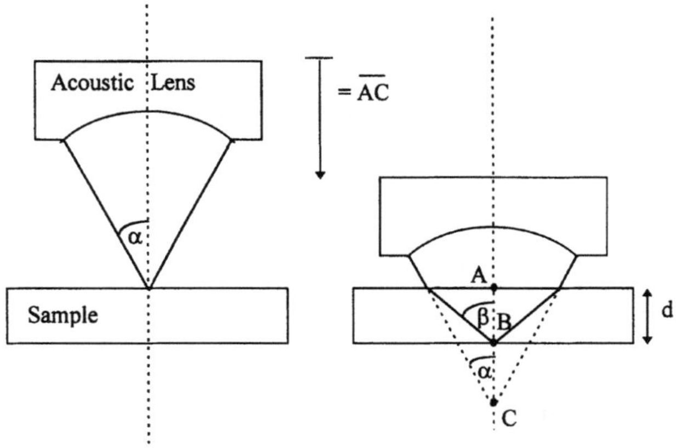

Figure 3. Model of double focus theory

$$d = \overline{AB} = \frac{1}{2}\, c_1\, \Delta t_2 \tag{3}$$

$$\overline{AC} = \frac{1}{2}\, c_0\, \Delta t_1 \tag{4}$$

$$\frac{\overline{AB}}{\overline{AC}} = \frac{c_1\ \Delta t_2}{c_0\ \Delta t_1} \tag{5}$$

There is a second possibility to calculate the quotient $\overline{AB}/\overline{AC}$ using the geometrical relations of Fig. 3,

$$\frac{\tan\alpha}{\tan\beta} = \frac{\overline{AB}}{\overline{AC}} \tag{6}$$

and Snell's law

$$\frac{\sin\alpha}{\sin\beta} = \frac{c_0}{c_1}. \tag{7}$$

This results in

$$\frac{\overline{AB}}{\overline{AC}} = \frac{1}{\cos\alpha}\sqrt{\frac{c_0^{\,2}}{c_1^{\,2}} - \sin^2\alpha} \tag{8}$$

Angle α lies between 0 and α_{max} producing a spherical aberration that can not be neglected. Figure 4 schematically shows the principle of spherical aberration (left). On the right hand side the point of intersection with the lens axis is shown in dependence on the

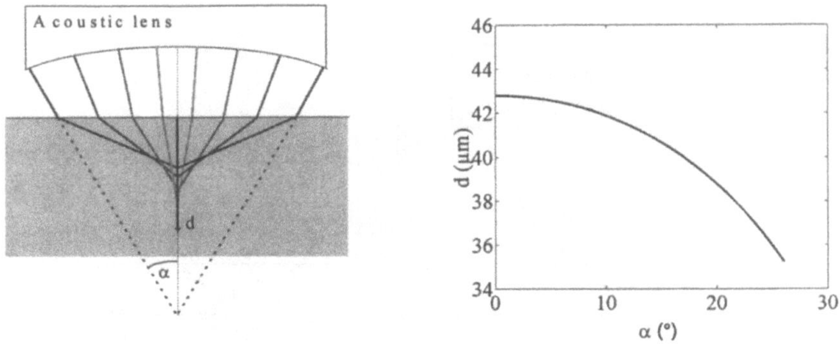

Figure 4. Spherical aberration

angle (for a typical defocus). A lens with 30° semi-aperture-angle was used yielding a maximum semi-angle in water of 26°. The resulting aberration was 8 μm. This is a relatively large value compared to the maximum focus depth of 43 μm.

As a consequence, the mean has to be determined in order to calculate the sample sound velocity. The equation that determines the sample sound velocity is shown in eq. (9).

$$c_1 = \frac{\Delta t_1}{\Delta t_2} \frac{c_0}{\alpha_0} \lim_{\alpha_0 \to 0} \int_{\alpha_0}^{\alpha_{max}} \frac{1}{\cos\alpha} \sqrt{\frac{c_0^2}{c_1^2} - \sin^2\alpha}\; d\alpha \qquad (9)$$

EXPERIMENTS

In order to experimentally confirm the applicability of the model, it was applied to thin PVDF film. The experimental setup is shown in Fig. 5. The short time pulses of a scanning acoustic microscope were used to drive an external acoustic lens with a frequency

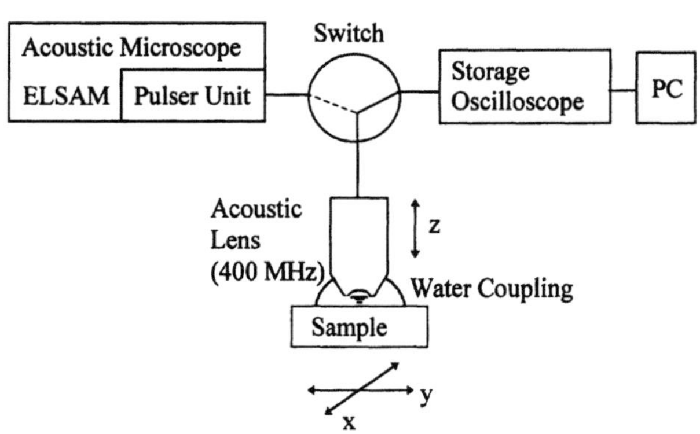

Figure 5. Experimental setup

of 400 Megahertz. The lens was installed in an external scanning facility with a three-dimensional stepping motor unit (step size one micrometer). The echoes were analyzed using an appropriate storage oscilloscope and a Personal Computer.

Figure 6 shows the measured signals. Above, the front echo and the back echo of the sample with the lens focused on the back are illustrated. Below, the front reflection with the lens focused on the sample's front is shown. The time-interval Δt_2 can be determined very precisely, because it does not depend on the defocus.

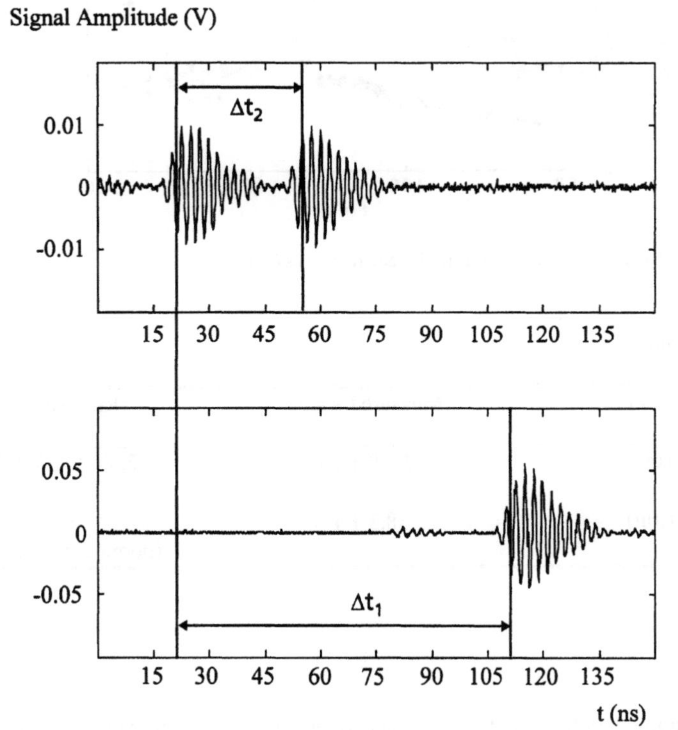

Figure 6. Measured signals in the time domain

However, for the measurement of Δt_1 those points of time have to be determined, where the lens is focused on the sample's front and on the back of the sample, respectively. These points of time are characterized by a signal amplitude reaching the maximum values for both the front and the back reflection (Fig. 7).

So defocusing the lens, the maximum values of both the front and the back signal were measured in dependence of the arrival time. The rear peak is wider than the front peak, because of the spherical aberration being higher in a solid than in water. The time-difference between the maxima of both peaks is $\Delta t_1 - \Delta t_2$.

The sample sound velocity can be calculated using eq. (9), the sample thickness is then given by eq. (3). The results are shown in table 1.

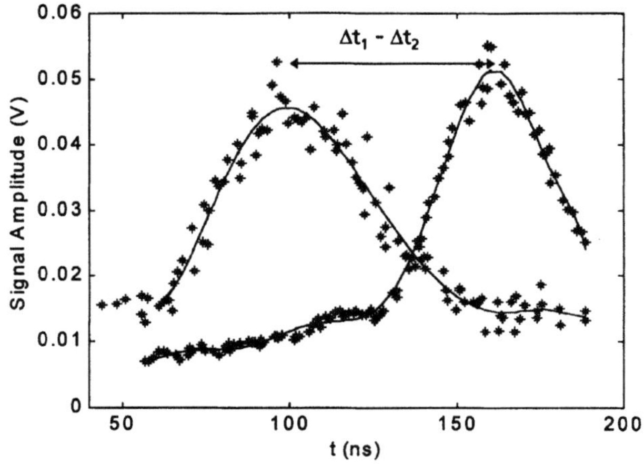

Figure 7. Signal amplitudes at front focus (right) and rear focus (left)

Table 1. Results

PVDF Film	Measured Values	References
c_l (m/s)	2400 ± 100	2200,[3,4] 2300 [5]
d (μm)	38.3 ± 1.5	40 (nominal thickness)

SUMMARY

Using the double focus technique sound velocity and thickness of thin samples can be determined directly and simultaneously. As a result no sample preparation is required, it is not necessary to attach the samples to a substrate in order to receive a reference signal. Currently, the model is being refined aiming at technical applications.

REFERENCES

1. A. M. Sinton, G. A. D. Briggs and Y. Tsukahara, in *Acoustical Imaging*, Vol. 17, Plenum, New York, 1989, pp. 87-95
2. G. A. D. Briggs, *Acoustic Microscopy*, Clarendon Oxford, 1992, p. 155
3. M. Platte, Ph.D. thesis, RWTH Aachen (Germany), 1984
4. Piezoelectric polymer transducers, technical manual, AMP Inc., Piezo Film Sensors, Valley Forge
5. A. Selfridge, Ultrasonic Devices Inc., Los Gatos, CA 95033, 1996

A NEW METHOD FOR 3-D VELOCITY VECTOR MEASUREMENT USING 2-D PHASED ARRAY PROBE

Tsuyoshi Shiina and Naotaka Nitta

Institute of Information Sciences and Electronics
University of Tsukuba
Tsukuba 305-8573, JAPAN

INTRODUCTION

The velocity information provided by ultrasonic Doppler techniques is just the velocity vector component along the beam direction. In order to estimate the true flow velocity, some a priori information such as the transducer measurement angle must be known. However, because flow in many arteries is of a complex three-dimensional (3-D) nature, conventional color flow Doppler image can be difficult to interpret correctly.

To break through the problem several approaches have been proposed[1-8]. Some methods are based on the tracking of speckle patterns in B-mode image or rf signals[4]. However, one problem with correlation search methods such speckle tracking is that considerable processing time is spent in correlating and searching.

The 3-D computed velocimetry by proposed by Ogura et al[5]. has a merit of separating composite velocity of each scatterer since it uses only linear signal processing. However, a disadvantage of this method is that much computation time is required to execute 2-D and 3-D Fourier transforms for each measurement point.

More practical systems to measure transverse flow velocity components are based on multiple transducers gathering data at multiple positions[6,7]. However, a problem with multi-transducer systems is that positioning of the desired measurement location may be difficult due to the multiple transducer geometry. Another alternative is to use a single array transducer and to split the array into multiple synthetic receive apertures. We previously proposed a method for measuring the 3-D velocity vector using four rf echo sequences received by a symmetrically positioned local aperture onto a 2-D array probe[8]. The method has the merit of rapidly obtaining three components when measurement points are scanned due to the symmetrical layout of the aperture. However a problem with the method is that aperture size is still large for measuring blood flow under a restricted aperture size such as the cardiovascular region. Moreover, measurement sensitivity of the transverse velocity component depends on the position of the four receiver apertures.

For the purpose of overcoming these problems, we propose a new method for measuring a 3-D blood flow velocity vector. The method employs a 2-D phased array

Acoustical Imaging, Volume 24.
Edited by Hua Lee, Kluwer Academic/Plenum Publishers, 2000.

probe and effectively extracts the information on velocity vector from the Doppler phase shift distribution on the receiver aperture so that the large measurement space with a single and smaller aperture can be realized by the sector scan. Another merit of the method is that the measurement sensitivity and precision of the velocity vector are independent of the direction of transverse component. In this paper, we describe the principle of the proposed method and evaluate the basic performance by computer simulation.

METHOD

Principle of 3-D velocity vector measurement

Considering the use of a 2-D phased array probe, the *x-y* plane is set on the 2-D array aperture and the z-axis is set to the depth direction as shown in Fig.1. The transmitting and receiving apertures on the array are represented by concentric disks whose center is located at the origin O on the *x-y* plane. The beam direction is determined by the angle α between the beam direction and the z-axis, and by the angle β around z-axis, by means of a spherical coordinate system. Each symbol is denoted as follows:

r_T : radius of transmitting aperture
r_R : radius of receiving aperture
$Q(x,y)$: element of array probe located at (x,y) on the receiving aperture
$P(x_p,y_p,z_p)$: measuring point located at (x_p,y_p,z_p) on which the transmitted beam is focused
T : pulse repetition time

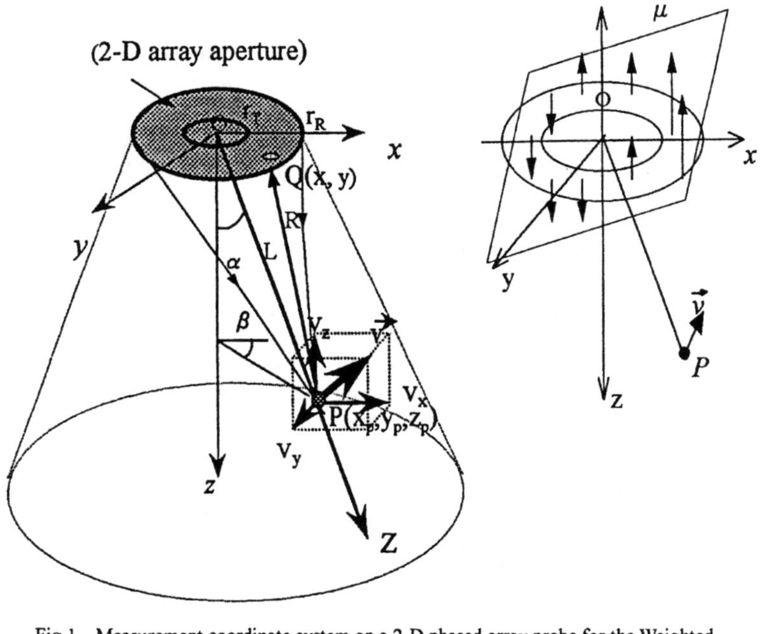

Fig.1 Measurement coordinate system on a 2-D phased array probe for the Weighted Phase Gradient Method (left) and the plane μ in 3-D space (x,y, μ) (right).

L : distance between the center of the transmitter and measuring point P

R : distance from P to Q

$\tau\,(x,y)$: time delay defined by the propagation time of scattered waves from point P to the receiver element Q(x,y)

$u(t;x,y)$: echo signal scattered at P and received by element Q, which is shifted in the time domain by $\tau\,(x,y)$ in order to focus on P

$\boldsymbol{v}$: velocity vector with x,y,z componets (v_x, v_y, v_z)

Each scatteter at point P scatters the initial pulse, then reaches point P_1 with velocity vector $\boldsymbol{v}$ after time T and scatters the second pulse. At intervals of T, the scatterer similarly moves to the next point P_n with velocity v and scatters the pulse. Each echo are scattered at P_n and received by element Q is similarly compensated by $\tau\,(x,y) + nT$ and is represented by $u_n(t;x,y)$. The difference of the Doppler phase shift between $u_n(t;x,y)$ and $u_{n+1}(t;x,y)$ is represeted as follows,

$$\phi(x,y,L) = kT\,\{(\boldsymbol{l}_T, \boldsymbol{v}) + (\boldsymbol{l}_R, \boldsymbol{v})\}$$

$$= kT\left\{\left(1+\frac{L}{R}\right)\frac{v_x x_p + v_y y_p + v_z z_p}{L} - \frac{v_x x + v_y y}{R}\right\} \qquad (1)$$

where k is the wave number, and $\boldsymbol{l}_T$ and $\boldsymbol{l}_R$ are unit vectors in the direction of OP and PQ, respectively. As a special case, the value of ϕ at the origin O is expressed by the velocity component along the beam direction, V_Z, as

$$\phi(0,0,L) = 2kT\,\frac{v_x x_p + v_y y_p + v_z z_p}{L} = 2kT \cdot V_Z \qquad (2)$$

Equation (1) can be rewritten as eq.(3) by using eq.(2)

$$\frac{1}{2kT}\{2R\phi(x,y,L) - \phi(0,0,L)(L+R)\} = -v_x x - v_y y \qquad (3)$$

Here, if the left-hand side of eq.(3) is express by $\mu\,(x,y,L)$ as eq.(4), it indicates that $\mu\,(x,y,L)$ constitutes a plane with respect to two variables x and y in 3-D space (x,y,μ) as shown in the right of Fig.1.

$$\mu\,(x,y,L) = -v_x x - v_y y \qquad (4)$$

Then, we can obtain the x and y component of scatterer velocity by a gradient of μ as

$$(v_x,\ v_y) = -\mathrm{grad}\,\mu\,(x,y,\mathrm{L}) \qquad (5)$$

In practice, the gradient can be easily calculated by performing a linear regression on the μ at N receiving elements $Q(x_k, y_k, 0)$ over 2-D array aperture. Therefore, three components of a scatterer velocity can be derived as follows:

$$v_x = \frac{S_{YY}\sum_{k}^{N} x_k\mu_k - S_{XY}\sum_{k}^{N} y_k\mu_k}{S_{XX}S_{YY} - (S_{XY})^2} \qquad (6)$$

$$v_y = \frac{-S_{XY}\sum_k^N x_k\mu_k + S_{XX}\sum_k^N y_k\mu_k}{S_{XX}S_{YY} - (S_{XY})^2} \tag{7}$$

$$v_z = -v_x \tan\alpha\cos\beta - v_y \tan\alpha\sin\beta - \frac{V_Z}{\cos\alpha} \tag{8}$$

where $S_{XX} = \sum_k^N x_k^2,\ S_{YY} = \sum_k^N y_k^2,\ S_{XY} = \sum_k^N x_k y_k$ and $\mu_k = \sum_k^N \mu(x_k, y_k, L)$

The above processing implies that a 3-D velocity vector is obtained by the gradient of $\mu\ (x,y,L)$ or the phase shift with weights determined by geometrical relationship between the receiving elements and the measurement point. Therefore, we refer to the technique as the 'Weighted Phase Gradient Method'. The proposed method enables us to employ a smaller-sized single aperture as compared with the conventional techniques, which is investigated in the next section. The sensitivity and precision of the velocity measurement are independent of the direction of transverse flow. Furthermore, real-time measurement can be realized since the autocorrelation technique is utilized for the phase shift extraction as well as the present Doppler method[9].

Evaluation of performance

The accuracy and precision of the velocity measurement are expected to depend on the size of the aperture. Regarding to this point, the multi-transducer system is compared with the proposed approach. The transverse component of velocity, for example, v_x is obtained with the echoes received at two positions and described as

$$v_x = \frac{1}{2kT}\frac{R_a}{a}(\phi_1 - \phi_2) \tag{9}$$

where $2a$ corresponds to the distance between two receiving aperture and R_a is the distance between the measurement point and receiving apertures. When the measurement error of phase in each element is uncorrelated and its power is expressed as δ^2, the mean square error of v_x by the multi-transducer system is derived as follows:

$$\varepsilon_0^2 = \frac{1}{4k^2T^2}\left(\frac{R_a}{a}\right)^2 \delta^2 \tag{10}$$

To simplify the expression of the mean square error of v_x by the proposed method, let us consider the case where elements of receiving aperture are located to be symmetrical about the x and y-axis, and the measurement point is placed on the z-axis. Then the mean square error is given by

$$\varepsilon^2 = \frac{1}{k^2T^2}\frac{\sum_k \{R_k{}^2 + (\frac{L+R_k}{2})^2\}x_k{}^2}{(\sum_k x_k{}^2)^2}\delta^2 \tag{11}$$

Letting the radius of receiving aperture $r_R = a$, then $|x| < a$ and $R(x,y) < R_a$ for the receiving point $Q(x,y)$. Considering these relation, from eq.(10) and (11), the following inequality is derived:

$$\frac{\varepsilon^2}{\varepsilon_0^2} \leq \frac{4}{N} \frac{a^2}{a_M^2} \tag{12}$$

where N is the number of receiving elements, and a_M represents the minimum radius. This indicates ε^2 can become much smaller than ε_0^2 if N is large enough, in other words, a single and smaller-sized receiving aperture can be applied. Although the position of receiving elements is chosen to be symmetrical for a simplification in above case, we can put them more flexibly since it is not restricted in theory. In practice, therefore, it is reasonable that elements within a circular region with small radius are used for calculating axial components and elements at the surrounding annular area on the receiver are used for transverse components of velocity vector.

Simulation analysis

The basic performance of the method was evaluated by computer simulation. The transmitting aperture on a 2-D array probe was a circle of 3mm diameter. As to the receiving aperture, its diameter was 30mm and its element pitch was 3mm. Transverse velocity components, v_x and v_y, were obtained using signals acquired at the elements at the annular part of 20mm inner diameter. The total number of the elements on the receiving aperture was 60. The transmitted pulse was a burst wave modulated by Gaussian shape envelope with a center frequency of 5MHz and with the duration of 10 wavelength.

The flow field was assumed to be uniform and parabolic distribution of scatterers in the tube of 2mm inner diameter. The surrounding medium was modeled as random scatterers. The sample volume width was set to 2.0mm. The measurement point was located as L=61mm, α=10.5°, β=26.5°. The velocity vector was given as (v_x,v_y,v_z)= $(0.61|v|, 0.61|v|, 0.5|v|)$ for $|v|$=10, 30, 50 cm/s. The axial component v_z and transverse one v_x of the measured value of velocity vector is compared with the predicted value as shown in Fig.2. In each case, the mean and standard deviation were calculated based on ten times measurements. Although S.D. has a tendency to increase as $|v|$ increases, this indicates that the measured velocity components coincide with the predicted lines.

Figure 3 shows the mean and standard deviation obtained as varying the diameter of receiving aperture for the case of $|v|$=50 cm/s. It can be seen that S.D. increase as the

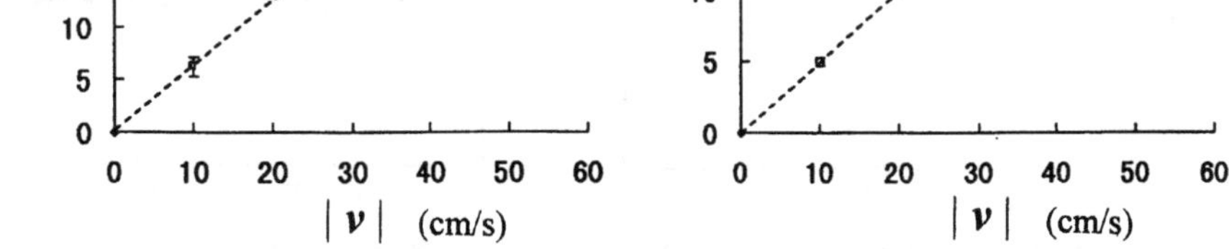

Fig.2 The measured velocity vector for given values. Horizontal axis indicates the set velocity vector magnitude and vertical axis indicates measured velocity component v_x(left) and v_z(right). The broken line indicates predicted values.

diameter of the aperture decreases, while the bias was almost zero except for the smallest diameter. The results indicate that practical accuracy can be attained by the proposed method with an aperture diameter of 30 mm although a more detailed investigation is required to estimate the optimal conditions for practical use.

Next, the flow imaging simulation is performed. Two types of flow field were simulated. One is the uniform flow of $|v|$=30 cm/s in the horizontal straight tube of 2 mm diameter. The other is the vortex flow at the center of which the magnitude of velocity becomes maximum, i.e., $|v|$=70 cm/s. The axial scan angle α was changed from -45° to 45° , and the lateral scan angle was β =45° . A frame was composed of 271 scan lines, and pulse repetition number was 30. Other parameters such as the size of aperture are the same as the above.

Figure 4(a) shows the image of the uniform flow obtained by the proposed method. Although three images for (v_x, v_y, v_z) were obtained, the magnitude of velocity is alone shown on account of limited space. Similarly, the image of the vortex flow by the proposed method is shown in Fig.4 (b). Figure 4 (c) and (d) are obtained by conventional Doppler color flow map, where a grayscale is used in the stead of a blue-red color bar. These indicate that the proposed method can uniformly display the flow pattern in the different directions, while as is often observed in clinical cases, the conventional method fails to reconstruct a flow image when the angle between the beam and flow direction is around 90° . As to the vortex flow, Fig.4 (d) seems to be the two flows which are parallel and in opposite directions. These results validate that the proposed method can measure the 3-D flow pattern more precisely than the conventional method.

CONCLUSION

To overcome the limitation of the conventional Doppler flowmetry, a new method of 3-D velocity vector measurement was proposed. By detecting the gradient of the weighted phase shift distribution on the receiving aperture, the proposed method realizes a single and smaller-sized aperture, and the sensitivity independent of the direction of the transverse flow as compared with the multi-beam techniques. Moreover the real-time measurement can be expected since the conventional autocorrelation technique is utilized for the phase shift extraction. The basic performance of the method was evaluated and

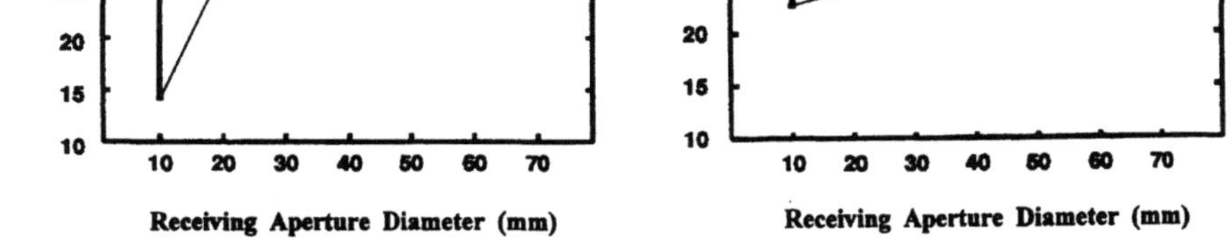

Fig. 3 Relationship of accuracy and precision on the velocity measurements with the aperture diameter in the case of the uniform flow with a velocity of 50 cm/s for the transverse component v_x (left) and axial component v_z (right). The broken line represents the predicted value.

the sector scan imaging simulation was executed. The results showed the high feasibility of the proposed method for measuring the 3-D velocity vector. As the future work, a more detailed investigation of the method must be undertaken for a developing practical system.

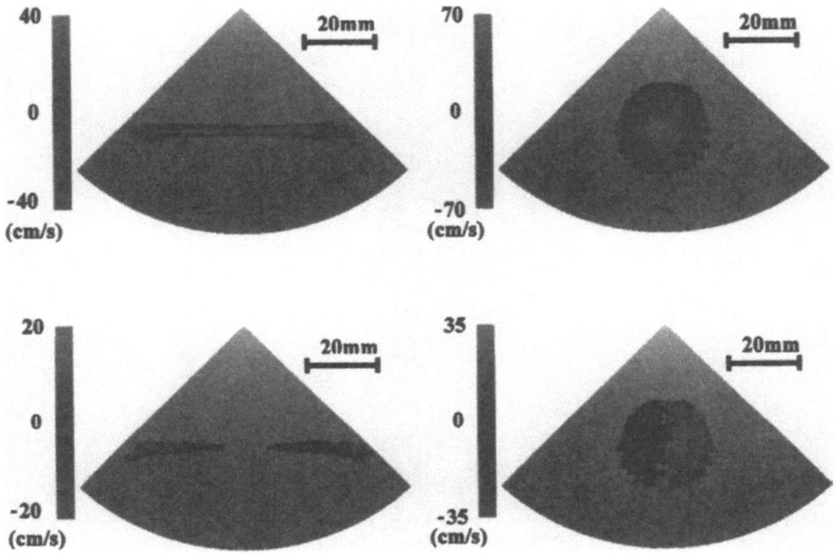

Fig.4 Comparison of sector scan image between the proposed method and the conventional color Doppler method for uniform flow (left) and vortex flow (right). The top is the magnitude of velocity obtained by the proposed method. The bottom is the conventional Doppler flow image, where a grayscale is used in the stead of a blue-red color bar.

REFERENCES

1. I.A.Hein, W.D.O'Brien, Current time –domain methods for assessing tissue motion by analysis from reflected ultrasound echoes—A review, *IEEE Trans. Ultrason. Ferroelectr. & Freq. Cont.* 42: 889 (1995).
2. O.Bonnefous, Measurement of complete (3D) velocity vector of blood flows, *Proc. 1988 IEEE Ultrason. Symp.* 795 (1988).
3. V.L.Newhouse, D.Cathignol and J.Y.Chapelon, Study of vector flow estimation with transverse Doppler, *Proc. 1990 IEEE Ultrason. Symp.* 1259 (1990).
4. B.S.Ramamurthy and G.E.Trahey, Potential and limitations of angle-independent flow detection algorithms using radio-frequency and detected echo signals, *Ultrasonic Imaging* 13:252 (1991).
5. Y.Ogura, K.Katakura and M.Okujima, A method of ultrasound 3-D computed velocimetry, *IEEE Trans. Biomed. Eng.* 44:823 (1997).
6. H.F.Routh, T.L.Pusateri and D.D.Waters, Preliminary studies into high velocity transverse blood flow measurement, *Proc. 1990 IEEE Ultrason. Symp.* 1523 (1990).
7. M.D.Fox and W.M.Gardiner, Three-dimensional Doppler velocimetry of flow jets, *IEEE Trans. Biomed. Eng.* 35: 834 (1988).
8. N.Nitta, K.Hagihara and T.Shiina, Experimental Investigation of 3-D blood flow veleocity measurement, *Jpn. J. Appl. Phys.* 35:3126 (1996).
9. C.Kasai, K.Namekawa, A.Koyano and R.Omoto, Real-time two-dimensional blood flow imaging using an autocorrelation technique, *IEEE Trans. Son. & Ultrason* 32:458 (1985).

ACOUSTIC VELOCITY PROFILING OF A SCATTERING MEDIUM: SIMULATED RESULTS

Marissa A. Rivera Cardona[1]; Wagner C. A. Pereira[1] and João Carlos Machado[1]

[1]Biomedical Engineering Program
Federal University of Rio de Janeiro - COPPE/UFRJ
P.O.Box 68510 - Rio de Janeiro -RJ - Brazil - 21945-970
marissa@peb.ufrj.br, wagner@peb.ufrj.br, jcm@peb.ufrj.br

INTRODUCTION

During the last three decades many investigations were conducted to provide quantitative ultrasonic characterization, such as wave propagation velocity (WPV), attenuation and backscattering coefficient, of the irradiated medium. Knowledge of these parameters, besides providing a better description of the propagating medium, would indeed improve the image quality. Recent advances on microcomputer hardware and signal processing techniques made it possible to computer simulate the propagating medium (phantom), as well as the transducer response, and to obtain simulated results of the wave scattered by the medium. This work presents a method to obtain simulated results of the WPV profile, along the penetration depth of a transmitted pulsed wave, for a weakly scattering medium. The method may be divided in three steps: initially, signals scattered by spherical particles, placed inside a given medium showing a specific WPV, and received at two different observation points are simulated (see Figure 1).

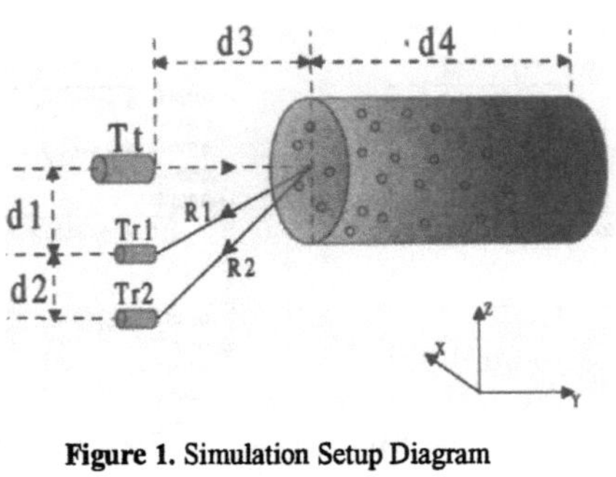

Figure 1. Simulation Setup Diagram

Acoustical Imaging, Volume 24.
Edited by Hua Lee, Kluwer Academic/Plenum Publishers, 2000.

Following, delay profiles (local delay profiles) for every pair of scattered signals are estimated by means of applying the Sliding Window Cross Correlation Technique. Finally, to determine the WPV profile, a Geometrical Acoustic Model for multilayer medium is applied to the delay profile.

SCATTERED SIGNAL SIMULATION

Let's consider a medium, containing randomly distributed spherical particles, with a propagating wave velocity varying as a function of the wave penetration depth. Figure 1 depicts a typical situation, of the simulation setup, containing the transmitting (Tt) and receiving (Tr) transducers and the simulated phantom. A coupling layer (water) of known thickness (d3) and acoustic wave velocity is interposed between them.

The waves scattered from each particle and collected by Tr, at two different locations, are calculated, using formulations given by Faran, (1951) and Hickling (1962). Those formulations were computer implemented by Pereira (1997). According to Hickling's equations, at observation point r_o, the pressure scattered by a spherical particle in response to an incident pressure (Equation 1), is given in expression (2).

$$P_{inc} = P_o \exp[ikr\cos\theta] \tag{1}$$

$$P_{esp} = \left(\frac{P_o a}{2r_o^2}\right) f(x, x_1, x_2) \exp[ix(2R - (ct/a))] \tag{2}$$

where:

$x = \frac{kr_o}{R}$; and $R = \frac{r_o}{a}$;

$x_1 = k_1 a$; $x_2 = k_2 a$

a is the spherical particles radius.

k is the medium wavenumber, k_1 and k_2 are particle's longitudinal and transversal wavenumbers;

c is the medium wave propagation velocity;

r_o is the distance from the particle to the observation point (in this case, to the source);

$f(x, x_1, x_2)$ function that provides the particle scattering information

The total scattered signal coming from a group of particles is found by considering each particle contribution. Examples are shown on Figure (2).

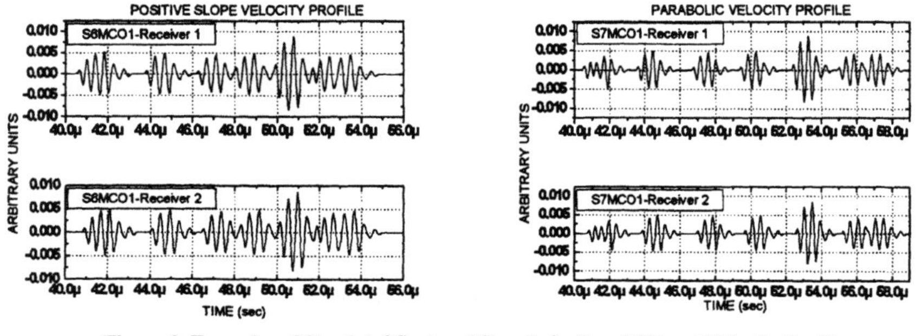

Figure 2. Examples of Simulated Scattered Signals for Four Different Velocity Profiles

The phantom is simulated by randomly locating spherical particles all over the volume of a cylindrical container filled with a medium presenting a one dimension WPV profile. Considering first order scattering, the whole volume formed by the total number of particles must be less than 1% than the total phantom volume. Particle's positions follow an uniform distribution. For the simulation, a plane wave of 2.2MHz , particle's with mean radius of 260µm (of a material of known parameters) and sample frequency of 100MHz are assumed. Several pairs (position 1 and 2) of the scattered signals are acquired by scanning the medium with respect to the transducer setup (XZ plane).

TIME DELAY PROFILE ESTIMATION

Once several pairs of scattered signals are simulated, for every pair, the time delay profile (local delay profile), between the signals received by Tr1 and Tr2 is determined, using the Sliding Window Cross-Correlation Technique (Equation 3). To increase time resolution in delay profile determination, interpolation techniques are applied (Hassab and Boucher, 1981). Finally a set of local delay profiles is obtained, one for every pair of the analyzed scattered signal.

$$\hat{R}_{xy}(m) = \frac{\sum_{t=0}^{T-1} x(t)y(t+m)}{T\sqrt{\sum_{t=0}^{T-1} \mathrm{var}(x(t))\sum_{t=0}^{T-1} \mathrm{var}(y(t+m))}} \tag{3}$$

where **x(t)** and **y(t)** are the studying signals (or in this case, the scattered signals); **m** is the time shift imposed by the cross correlation definition; **var** is the square root of the signal standard deviation.

From the transducers' point of view, the phantom is always presenting the same velocity profile. The differences that appear between scattered signals (and respective local delay profiles) during scanning, come from the fact that a different set of particles (on different positions) is being irradiated. Each local delay profile is related to a different region on the phantom. The delay curve (Figure 3) that characterizes the phantom comes from averaging all local delay profiles. Theoretical delay profiles are found by considering scattered signals coming from ponctual scatterers located along the acustical field main axis (y axis), and calculating its delay profile.

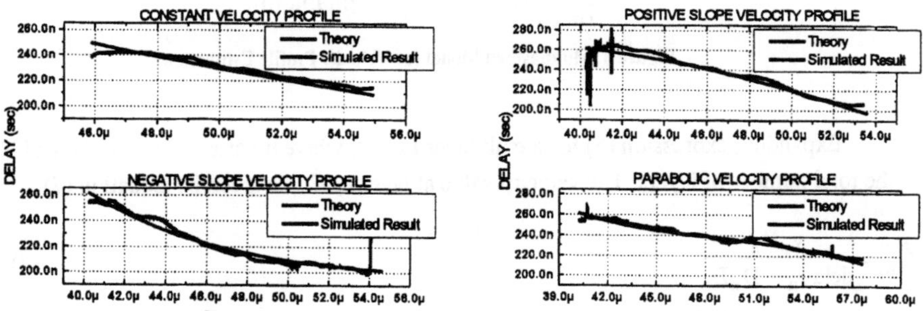

Figure 3. Averaged Time Delay Curves for the Four Different Velocity Profiles.

WAVE PROPAGATION VELOCITY PROFILE ESTIMATION

To determine the WPV profile, in this case, the simulated phantom is considered as a multilayer medium (with N layers of equal thickness) and the Geometrical Acoustics Approach (Pereira, 1996) is applied (Figure 4). The wave velocity is calculated recursively for each layer (7). Therefore, the result for the velocity of a layer N depends on the results for the previous layers. Equations (4) to (6) are easily found by the geometry shown on Figure 4. The Geometrical Acoustic Model uses the average delay profile obtained from the population of time delay profiles. In Figure 4, the coupling layer parameters, known thickness (Z_o) and known wave propagation velocity (C_o), are considered as reference values. However, C_o can also be estimated using $delay_o$, as the first value of the averaged delay profile.

$$T_1 = \frac{y_o}{C_o}\left(1+\frac{1}{\cos(\gamma_1)}\right); \text{ and } T_2 = \frac{y_o}{C_o}\left(1+\frac{1}{\cos(\gamma_2)}\right) \tag{4}$$

where: $\cos(\gamma_1) = \dfrac{y_o}{\sqrt{y_o^2+d_1^2}}$ and $\cos(\gamma_2) = \dfrac{y_o}{\sqrt{y_o^2+(d_1+d_2)^2}}$

$$C_o = \frac{\sqrt{y_o^2+(d_1+d_2)^2}-\sqrt{y_o^2+(d_1)^2}}{delay_o} \tag{5}$$

$$\text{where: } delay_o = T_2 - T_1 = \frac{y_o}{C_o}\left(\frac{1}{\cos(\gamma_2)}-\frac{1}{\cos(\gamma_1)}\right) \tag{6}$$

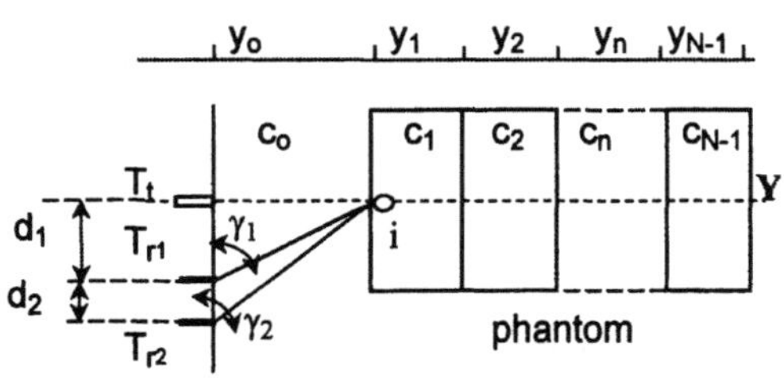

Figure 4. Multilayered Model for Velocity Profile Estimation

Expanding expression (5) for a multilayer model, where n varies from 1 to N, and N is the total number of layers. The velocity value at layer n, $\hat{C}_n$, is given by Equation (7)

$$\hat{C}_n = \frac{y_n B_n}{delay_n - D_n B_n} \tag{7}$$

where y_n is the sampling space (layer thickness) and $delay_n$ is taken from the average delay profile, and:

$$D_n = \sum_{i=0}^{n} \left(\frac{y_{i-1}}{\hat{C}_{i-1}} \right) \tag{8}$$

$$B_n = \frac{\sqrt{\left(\sum_{i=0}^{n} y_i \right)^2 + (d_1 + d_2)^2} - \sqrt{\left(\sum_{i=0}^{n} y_i \right)^2 + (d_1)^2}}{\sum_{i=0}^{n} y_i} \tag{9}$$

PRELIMINARY RESULTS

Preliminary results, as well as the theoretical values for the wave velocity profile, are shown in Figure 5, for the four different phantoms (constant, linear - negative and positive slopes - and parabolic wave velocity profiles). During the simulations the phantom thickness was 10 mm and the coupling layer was considered with a wave velocity of 1500 m/s and thickness of 30 mm, except for the situation with a constant velocity phantom where the coupling layer wave velocity was 2200 m/s and the thickness was 50 mm.

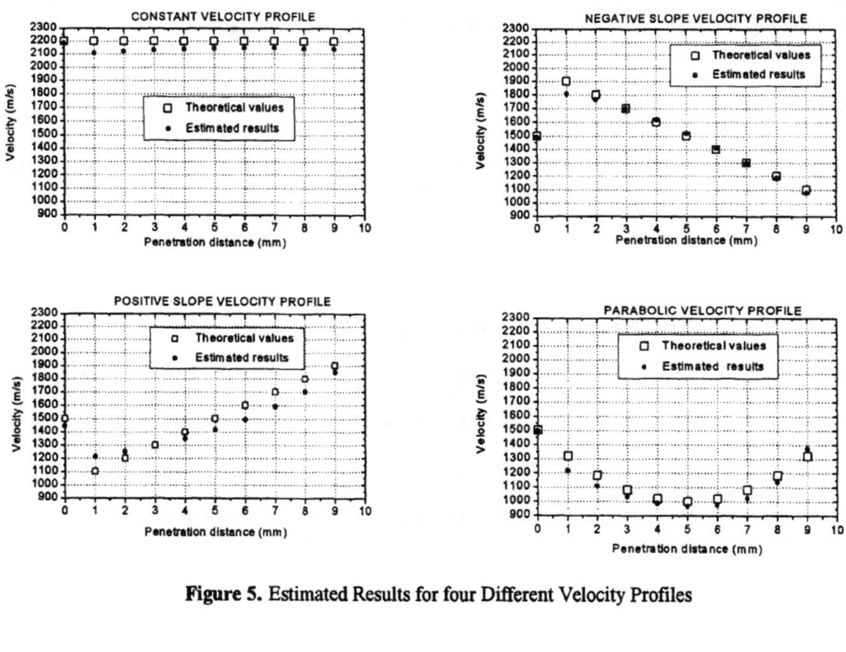

Figure 5. Estimated Results for four Different Velocity Profiles

To test the sensibility of the method, the whole signal processing was repeated considering the scattered signals contaminated with random white gaussian noise. For SNR's equal or greater than 20 dB, the results were close to those obtained with a noise free signal. Results were analyzed using the mean RMS error (Equation (10)). According to this, errors of the order of 2.88%, 1.99%,5.31% e 9.82%, were found for constant, positive and negative slope, and parabolic velocity profiles, respectively.

$$\left[\overline{\text{Error RMS}(\%)}\right] = 100\sqrt{\frac{1}{N}\sum_{i=1}^{N}\frac{\left(\text{Vexp}_i - \text{Vest}_i\right)^2}{\left(\text{Vexp}_i\right)^2}} \tag{10}$$

where: $Vexp_i$ is the velocity expected value for layer i
$Vest_i$ is the estimated velocity value for layer i
N is the total number of layers

DISCUSSION

Because velocity values are recursively calculated, their accuracy depends on earlier velocity values and these from a well estimated averaged delay profile (smooth and noise free). So a critical point is found during delay profile estimation. High temporal resolution and a great number of local delay profiles are necessary to obtain a good averaged delay profile estimation with high resolution. For averaged delay profiles overestimated we found velocity profiles underestimated, the opposite is also true. Regarding the multilayer model, for each layer, a constant velocity value is considered. Continuous velocity profile is obtain when layer thickness approaches zero. Other analysis of error sources as well as experimental implementation of the phantoms and setup are in progress.

Acknowledgements

The authors would like to thank Prof. David Martin Simpson, Ph. D., for his helpful comments and discussions about this work, and also CNPq, for the financial support.

REFERENCES

Faran, J.J., 1951, Sound scattering by solid cylinders and spheres, *Journal Acoust. Soc. Am.* vol 23 No4, pp. 405- 418.

Hassab, J. C., Boucher, R E., 1981, Analysis of discrete implementation of generalized cross correlator, *IEEE Trans. Acoust., Speech and Sig. Proc.*, vol. 29, 609-611.

Hickling, R, 1962, Analysis of echoes from a solid elastic sphere in water. *Journal Acoust Soc. Am.* vol. 34 No 10, pp 1582-1592.

Pereira, W.C.A., Greco, A.V. D., and Machado, J. C., 1996, Ultrasonic velocity mapping of multilayered media, *Acoustical Imaging*, vol. 22, pp. 63-68.

Pereira, F.R., 1997, Determinação da Velocidade de Propagação do Som em Meios Contínuous *Exame de Qualificação ao Doutorado do Programa de Engenharia Biomédica /COPPE/UFRJ.*

ULTRASONIC VELOCITY MEASUREMENT IN VISCOELASTIC MATERIAL USING THE WAVELET TRANSFORM

Eduardo Moreno[1,2], Fabian García[1], Martha Castillo[2], Alejandro Sotomayor[2], Victor Castro[2] and Martín Fuentes[1]

[1] DISCA-IIMAS-UNAM
Apdo Postal 20-726 Admon. No. 20 01000 México D.F México
[2]Centro de Ultrasónica, ICIMAF, CITMA,
Calle 15 No 551 Vedado, La Habana 10 400, Cuba

Abstract

The measurement of the time propagation of ultrasonic pulses using techniques based on digital signal processing are now used like the Pulse-overlapping method. In this case there are the methods that use the Cross-Correlation Function (CCF) and Hilbert Transform (HT). However, for the case of viscoelastic material, it is not possible to apply this new methods, as a consequent of the viscoelastic dispersion. In this paper, we present a method based on the Continuous Wavelet Transform (CWT). Experiments were made in steel and acrylic. The results show no differences in steel and significant differences in acrylic as a viscoelastic material . In this last case the CWT shows better properties over the CCF and HT for the evaluation of time propagation.

INTRODUCTION

The development of fast A/D acquisition cards made possible the use of methods based on digital signal processing like the Cross-Correlation Function (CCF) and the Hilbert Transform (HT)[1], for the measurement of mechanical wave velocity with the pulse method, in contrast with the classic Pulse-Echo Overlap Method (PEO)[2]. The digital methods are also used in the estimation of time delay in phase aberration correction for ultrasound imaging[3] .

For elastic materials the CCF and HT are equivalent for the measurement of time propagation between two consecutive pulse echoes. In both cases, it is assumed that the echoes have the same shape. For viscoelastic materials there are dispersion curves, then the two consecutive echoes will have different shapes according to this relation. The methods of CCF and HT are no longer valid, because they suppose the same shape for the pulses.

In this paper, we propose a new method with the Continuous Wavelet Transform (CWT). This Transform has been successfully used in different applications as an alternative to the Fourier Transform (FT)[4]. It is the most recent technique for signal processing with time-varying spectra , in which a signal is represented as a linear combination of wavelets in contrast with the sinusoidal bases of the FT[5].

THEORY

A. The Wavelet Transform

The Continuous Wavelet Transform (CWT) of a signal *x(t)* is defined as[6]:

$$CWT\{x(t):a,b\} = C_{a,b} = \int x(t)\Psi^*_{a,b}(t)dt \tag{1}$$

where the complex conjugate wavelets Ψ^*, are obtained from:

$$\Psi_{a,b}(t) = (\sqrt{a})^{-1}\Psi[(t-b)/a] \tag{2}$$

The continuos variables $a \in R+$ and $b \in R$ are the dilation (scale) and translation parameters, respectively. The general function $\Psi(t)$ is referred as mother wavelet and can be any function which satisfied the condition[4]:

$$\int_0^\infty \frac{|\psi(\omega)|^2}{\omega} d\omega = C < +\infty \tag{3}$$

The Wavelet Transform (WT) was born as an alternative to the Short Time Fourier Transform (STFT), to describe transient signals[4] . The WT uses the concept of time-scale plane instead of the time-frequency plane. This representation obtains detail of the signal on time (also frequency) domain. If we associate a detail of an ultrasonic pulse in correspondence to its head, then it is possible to make a time measurement at this point of the pulse. The wavelet transform, like other transforms, consists of computing the coefficients that are the inner products of a given signal *x(t)* and a family of wavelets. The equation (1) can be viewed as a correlation between the signal and the various scaled wavelets .

In this paper, we choose for the mother wavelet, the so called Morlet Wavelet defined as[4] :

$$\Psi = e^{-t^2/2} e^{i\omega t} \tag{4}$$

The application of the CWT in the time measurement between two successive echoes is based on the application of the expression (1) to the signal *x(t)* that can be expressed as the sum of the two components:

$$x(t)=x_1(t)+x_2(t) \qquad (5)$$

where x_1 and x_2 are the acoustic signals of the first and the second echoes respectively. If the sample under test is an elastic material, then x_1 and x_2 have the same shape. For viscoelastic material the correspondence dispersion law make that x_2 have some difference in the shape to x_1 as shown in Fig. 1(b).

The modules of the coefficients of the CWT, as a function of the scale parameter **a** and the time parameter **b** will show different details of the signal $x(t)$ in a 3-D plot. From this graph, it is possible to obtain a projection at constant scales. If we consider that x_1 and x_2 represent pulses, then the values of the coefficients of the CWT will be different from zero in its time intervals over the b axes. The value of the CWT for a given time will depend on the correspondence of the scale with the frequency content of the signal. As demonstrated in the next section, these properties will be used for time measurements.

B. Cross-Correlation Function

For $x(t)$ and $y(t)$ the cross correlation function is defined by [1]:

$$C_{xy}(\tau) = Lim_{T\to\infty}\frac{1}{T}\int_0^T x(t)\,y(t+\tau)dt \qquad (6)$$

If we consider the signal $y(t)$ as the second echoes relative to $x(t)$ with a time delay δt then:

$$y(t)=x(t-\delta t) \qquad (7)$$

For (6) the expression (5) correspond to the autocorrelation of $x(t)$. As a consequent of the even properties of the autocorrelation function with a maximum at the origin for a signal $x(t)$, then for a signal $y(t)$ given by (6), this properties will be similar with a maximum at $t=\delta t$. Thus the autocorrelation function of two consecutive echoes can be used for time propagation measurement according to the maximum position.

C. Hilbert transform

The relation between the first and the second echo can be written as follows:

$$y(t)=h(t)*x(t) \qquad (8)$$

where $h(t)$ is the impulse response of the system and * denotes the convolution. The Hilbert Transform (HT) can be described by[1] :

$$HT[h(t)]=g(t)*1/\pi(t-\tau) \qquad (9)$$

where $g(t)$ is a function with the attenuation and dispersion information of the media. Thus, the HT has a maximum at $t=\tau$ which corresponds to the time delay between the two echoes. The impulse response $h(t)$ can be obtained from the Inverse Fourier Transform of the echoes Fourier Transform ratio.

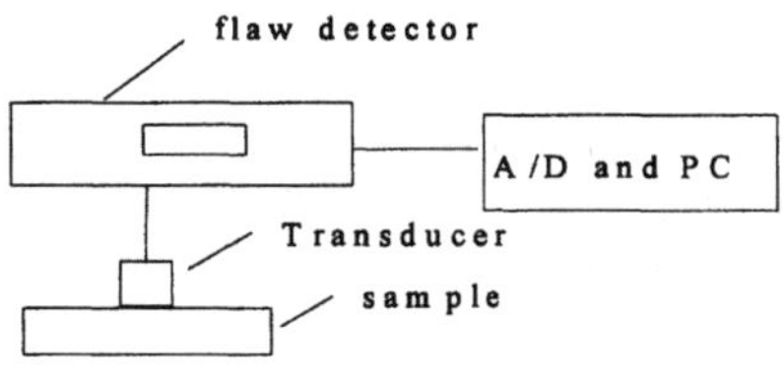

Fig 1. Schematic diagram used.

MATERIALS AND METHODS

The experimental configuration is shown in Fig. 1, where two RF echoes are sampled at a frequency of 32 MHz. A 2 MHz transducer driven by a flaw detector was used for this purpose. The digital signals are stored on the hard disk of a personal computer (PC AT 586-133 MHz). The experiment was performed on steel and acrylic with 21.3 mm and 20.0 mm thickness respectively, to compare the elastic and viscoelastic properties. The simulation and data evaluation was made in the Wavelet and Signal Processing Toolboxes of MATLAB [7,8] .

SIMULATION RESULTS

Figs. 2(a) and 2(b) show two echoes simulated for elastic and viscoelastic materials similar as used in reference[1] with the same time delay given in sampled unit . In our paper, we introduce a difference in the second echo for the simulation of viscoelastic influence.

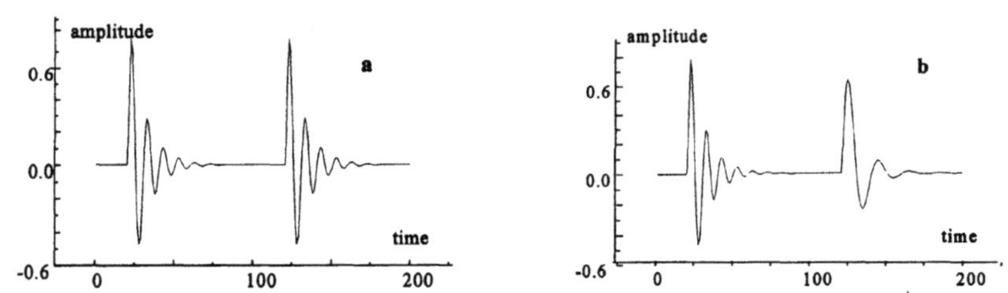

Fig 2. Simulated echoes for elastic (a) and viscoelastic (b) conditions.

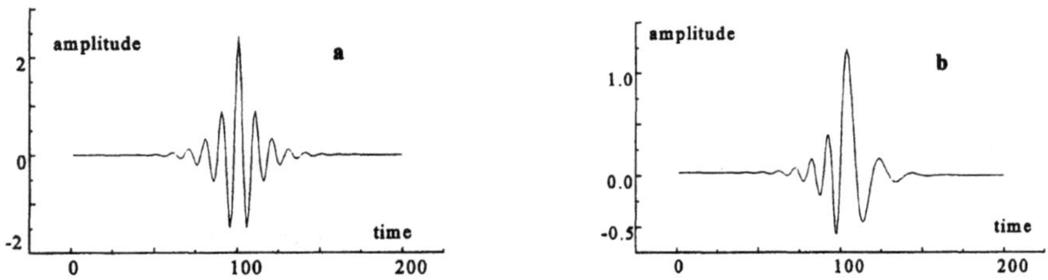

Fig 3. Simulation for CCF for elastic (a) and viscoelastic condition (b)

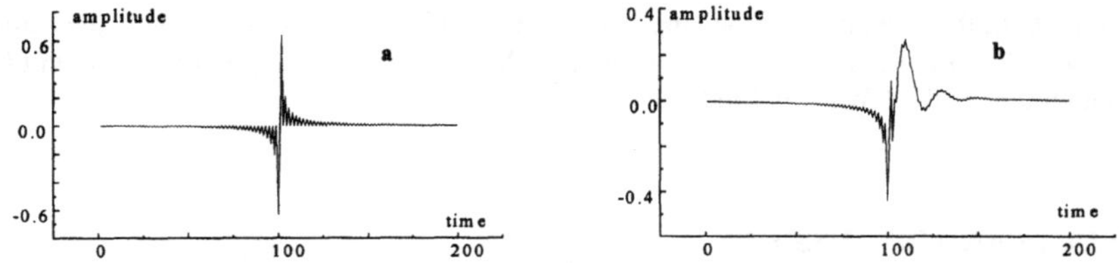

Fig 4. Simulation for HT for elastic (a) and viscoelastic condition (b).

Figs. 3 and 4 show the CCF and HT simulation obtained from Figs. 1(a) and 1(b) respectively. Both figures show that the CCF and HT has no difference in the elastic case for time measurement (time=100). However, in the viscoelastic case, there are differences and it is difficult to determine this parameter.

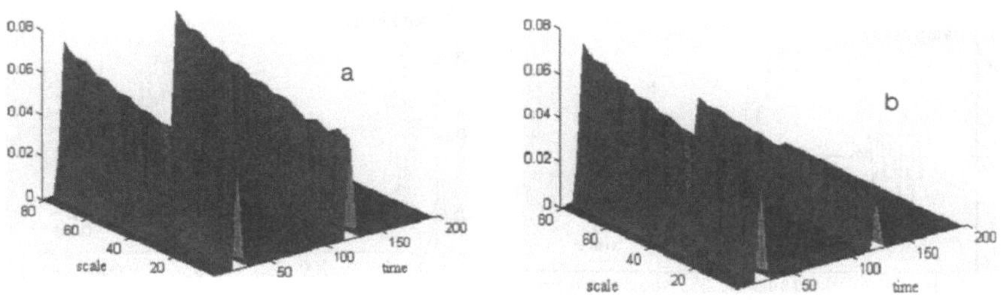

Fig 5. Modules of CWT for elastic (a) and viscoelastic condition (b) in scale-time plane

Fig. 5 shows the result for the modules of the CWT in the scale-time plane for the same cases. Fig. 6 shows the same result by the superposition of the projection at constant scale simultaneously. From this figure, it is clear that the change of the second echo to the first one, influence only in the high of the second maximum. The time delay between these two maximum correspond to the time propagation.

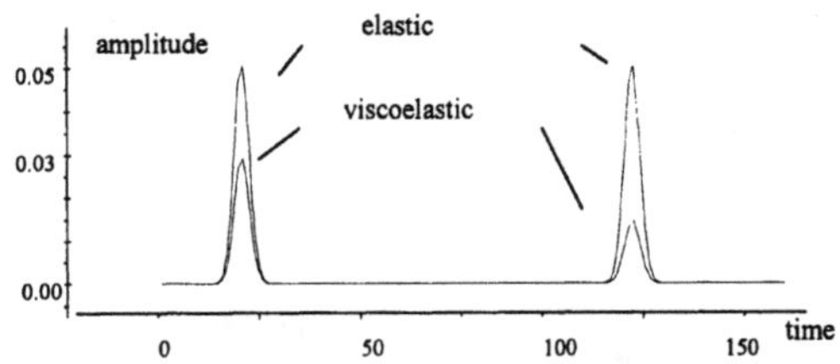

Fig 6. Superposition of Modules of CWT for elastic and viscoelastic at constant scale

This results can be explained by the viscoelastic effect on the high frequency components of the pulses. In this material exists a high attenuation for the high frequency. Thus, the pulse in the propagation through the material, lost these components. The beginning of the pulse has not the same high slope as it is in the elastic case. Then the CCF and HT are not as good as a consequent that the shape of the two pulses are no longer the same. However, for CWT the loss of high frequency components in the pulse has only a consequent in the maximum of the amplitude of it modules. The time situation is the same. The CWT detect always the beginning of pulses. According to the slope of this one will be the maximum of the CWT amplitude.

RESULTS AND DISCUSSIONS

Fig. 7(a) and 7(b) shows the consecutive echoes obtained in the steel and acrylic sample. The results for the CCF and HT for steel are shown in Fig. 8 respectively and in Fig. 9, it is shown the CWT methods for two scale values for this material. In this Figs. time corresponds to sampled number of the signal. The real time is obtained with the sampling frequency of 32 MHz. This situation will be the same for the results obtained in the acrylic sample.

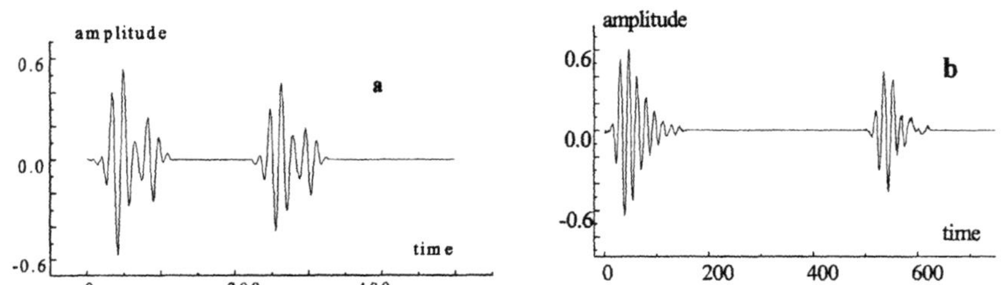

Fig 7. Consecutive pulses in steel (a) and acrylic (b)

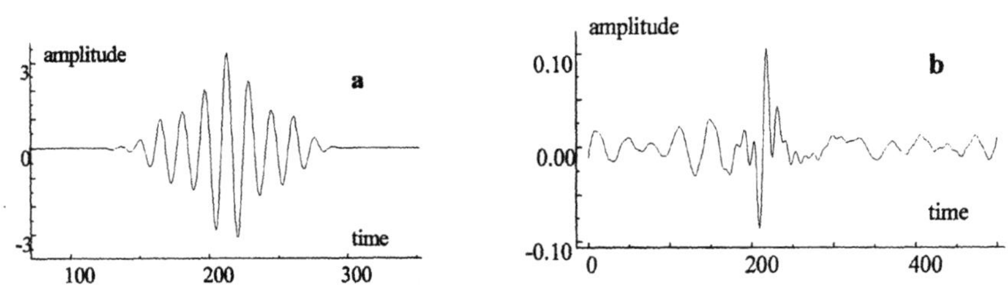

Fig 8. CCF (a) and HT (a) result for steel.

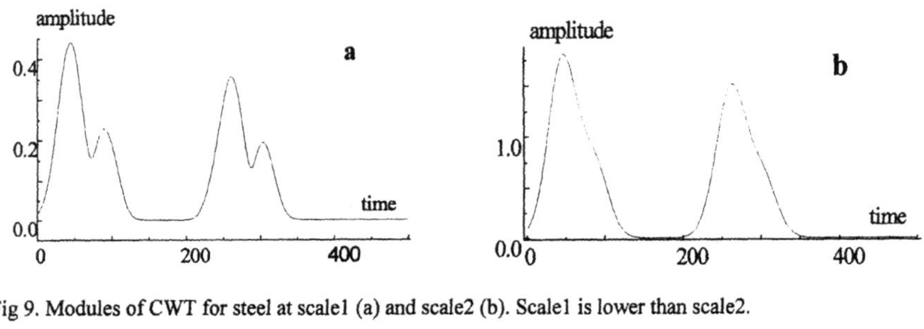

Fig 9. Modules of CWT for steel at scale1 (a) and scale2 (b). Scale1 is lower than scale2.

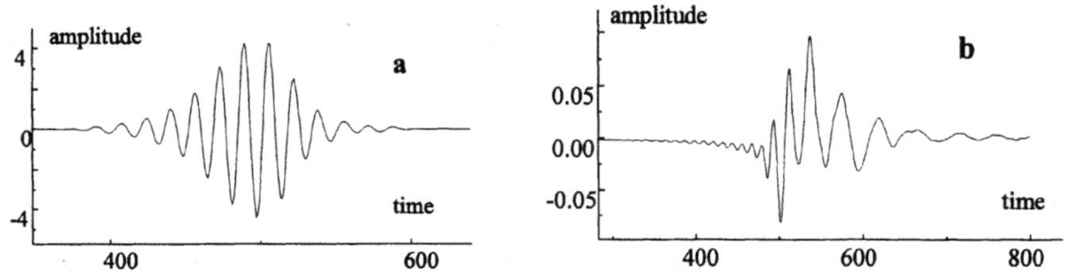

Fig 10. CCF (a) and HT (b) for acrylic

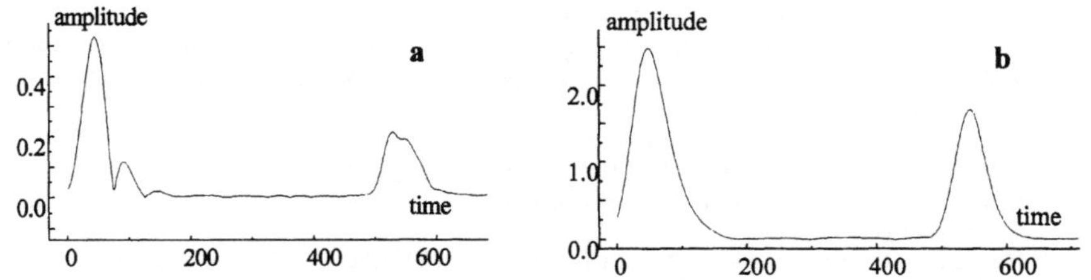

Fig 11. Modules of CWT for acrylic at scale1 (a) and scale2 (b). Scale1 is lower than scale2

In Figs. 10, 11, it is shown the same results for acrylic. Table I show the resume of the time and the velocity measurements obtained from the above Figs. For steel, the three methods are the same, according to the precision of the sample frequency of 32 MHz (1/32 MHz =0.031μseg). For acrylic the CWT show a significant difference relative to the CCF and HT. The use of low scale is better because the detail correspond to the high frequency component that are always present in the beginning of a pulse. Then the method of the CWT show a velocity measurement according to high frequency, and this result agrees with the viscoelastic dispersion curve, with a high velocity at high frequency[9]. This value is higher than reported by other methods[10].

Table I. Velocity and Time measurements. m/s (μseg)

Sample	CCF	HT	CWT scale1	CWT Scale2
Steel	5982 (6.687)	5953 (6.719)	5982(6.687)	5926(6.750)
Acrylic	2782 (15.312)	2538 (16.781)	2811(15.156)	2748(15.500)

CONCLUSIONS

In this paper, a new velocity measurement is reported. The Continuous Wavelet Transform, with the Morlet wavelet, is particularly suitable for viscoelastic material, where other methods like the Hilbert Transform and Cross-Correlation Function have problems determining these parameters. The possibilities to use different scale make attractive this method for a possible evaluation of viscoelastic characteristic. It is recommended to experiment with another material and bases in the Wavelet Transform

REFERENCES

1. R. Guerjoma et al., Non destructive evaluation of graphite by ultrasonic velocity measurement using cross-correlation and Hilbert transform methods, *Proc. IEEE Ultrasonics Symposium., 829-832,* (1992).
2. E. Papadakis, The measurements of ultrasonic velocity, in: *Physical Acoustics*, chapter II vol. XIX Thurston R and Pierce A. Academic Press Inc. (1990).
3. L. Wang, K. Kirk and O. Camps, Two bit correlation. An adaptive time delay estimation, *IEEE Trans. Ultrason., Ferroelec., Freq. Contr.*, 43: 473-481 (1996).
4. C. Chui, *An Introduction to Wavelets*. Academic Press Inc. (1992).
5. K. Ramchandran et al., *Wavelets, Subband Coding, and Best Bases,* Proceedings of the IEEE, 84(4):(1996).
6. O. Rioul and P. Duhamel, Fast Algorithm for Discrete and Continuous Wavelet Transform, *IEEE Trans. in Inform. Theory*, 38(2) (1992).
7. M. Misiti et al., *Wavelet Toolbox,* The MathWorks, Inc. (1996).
8. T. Krauss et al., *Signal Processing Toolbox,* The MathWorks, Inc. (1996).
9. E. Moreno, *Propagation of Mechanical Waves in Composite Plate Elements,*. Ph.D. Thesis. ICIMAF. La Habana, Cuba. 1994.
10. B. Hung and A. Goldstein, Acoustic Parameters of Commercial Plastics, *IEEE Trans. Sonics and Ultrasonics*, SU-30:249-254 (1983).

AN ULTRASONIC CIRCULAR APERTURE TECHNIQUE TO MEASURE ELASTIC CONSTANTS OF FIBER REINFORCED COMPOSITE

S. Arnfred Nielsen, H. Toftegaard and P. Brøndsted

Materials Research Department, Risø National Laboratory
DK-4000 Roskilde, Denmark

INTRODUCTION

Ultrasound may be used as a non-destructive evaluation (NDE) tool to classify discontinuities in fiber reinforced composite. The discontinuities may be classified *qualitatively* according to size, type and location using a suitable ultrasonic image technique (e.g. B-, C-scanning or tomography). The mentioned techniques can be based on transmitted or reflected pulses, corresponding to a desired image information of attenuation or reflectivity, respectively. However, ultrasound may also be used as a *quantitative* tool to map the elastic stiffness constants of the composite. The ultrasonic time-of-flight and corresponding phase velocity are directly related to the elastic constants of the composite. The number of elastic constants is related to the degree of anisotropy. For example, an orthotropic composite has nine independent elastic constants and a transverse isotropic composite has five independent elastic constants.

In a previous paper a synthetic circular aperture array was used to obtain qualitative tomographical images of isotropic materials (NIELSEN *et al.*, 1997*a*). The circular aperture array was realized by object rotation and by moving two transducer elements along a circular aperture. In this paper a continuation of this work is given by presenting a quantitative evaluation procedure for anisotropic materials. The current work shows how the elastic stiffness constants may be measured for transverse isotropic composites using the circular aperture array. The two transducers are used to measure the delay time for a refracted ultrasonic pulse in different planes of the composite. From these measurements quasi-longitudinal and quasi-transverse wave velocities are calculated. Finally, nonlinear least-square-fits are applied to each plane and the independent elastic constants are calculated.

In this work the five independent elastic constants for a unidirectional glass/PET reinforced composite were determined and compared with results from destructive mechanical experiments. It is discussed what parameters influence the correlation of the ultrasonic experiment.

ELASTIC CONSTANTS OF UNIDIRECTIONAL COMPOSITES

A unidirectional composite consist of aligned fibers in a matrix (MALLICK, 1993; KEDWARD, 1996). Such a material is assumed to be transverse isotropic with the plane of isotropy (x_2-x_3) perpendicular to the direction (x_1) of the fibers as indicated in Fig. 1.

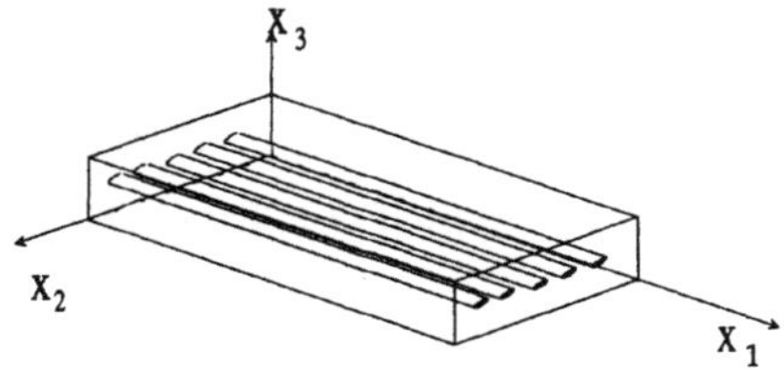

Figure 1. Transverse isotropic composite with fibers aligned in the x_1-direction perpendicular to the (x_2-x_3) plane of the composite.

The stiffness matrix is given by

$$\begin{bmatrix}\sigma_1\\\sigma_2\\\sigma_3\\\sigma_4\\\sigma_5\\\sigma_6\end{bmatrix}=\begin{bmatrix}C_{11}&C_{12}&C_{13}&0&0&0\\C_{12}&C_{22}&C_{23}&0&0&0\\C_{13}&C_{23}&C_{33}&0&0&0\\0&0&0&C_{44}&0&0\\0&0&0&0&C_{55}&0\\0&0&0&0&0&C_{66}\end{bmatrix}\begin{bmatrix}\varepsilon_1\\\varepsilon_2\\\varepsilon_3\\\varepsilon_4\\\varepsilon_5\\\varepsilon_6\end{bmatrix}\tag{1}$$

where σ_i and ε_j (i,j =1,2,3,4,6) are the volume average stresses and strains, respectively. Transverse symmetry imposes the following constraints on the elastic stiffness constants

$$C_{11},\ C_{22} = C_{33},\ C_{55} = C_{66},\ C_{12} = C_{13}, \text{ and } C_{23} \qquad (C_{44} = (C_{22} - C_{23})/2) \tag{2}$$

ELASTIC CONSTANTS MEASURED BY ULTRASOUND

Two mobile ultrasonic transducers, one acting as a transmitter and the other as a receiver, were used to synthesize the circular aperture array. Both the transmitter (T) and receiver (R) could rotate around the specimen at a well-defined distance as shown in Fig. 2. The rotation fixture was positioned in a temperature controlled water tank. By using two transducers only, a flexible setup was obtained. It was easier to vary the transducer configuration (center frequency, focal zone etc.). This was important because attenuation in the composite is frequency dependent and the ultrasonic beam determines the effective area of the composite to be examined. Hence, a focused narrow beam was important in order to examine a relative small part of the composite. In this experiment two Panametric® transducers were used with center-frequency 2.5 MHz (transmitter) and 5 MHz (receiver), respectively. The bandwidth of the receiving transducer was relative large in order to take frequency shift into account. Both transducers had an aperture of 10 mm in diameter.

The experimental procedure was the following: First, the specimen was positioned perpendicular to the transmitter and receiver. A broadband pulse was emitted from the transmitter and recorded by the receiver. The receiver was then rotated by a step motor with 1° step resolution, and the time-of-flight of the pulse was recorded as a function of the refracted angle. Secondly, the specimen was rotated using a second step motor in order to change the incident angle and the first procedure repeated.

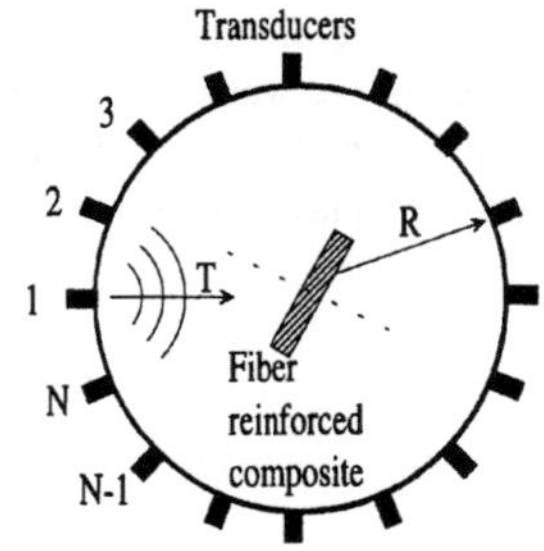

Figure 2. Schematic representation of a circular aperture array with N uniformly distributed transducer elements; one transducer (T) transmits a pulse into a fiber reinforced composite and a second receiver (R) records the refracted pulse.

Finally, from these time-of-flight measurements quasi-longitudinal and quasi-transverse phase velocities, v_p, were calculated for different angles of incidence. An expression for the phase velocity in terms of the elastic constants was then obtained by solving the Christofel equation (NIELSEN *et al.*, 1997*b*). The solution is given by

$$\begin{aligned}&\left(\Gamma_{11}-\rho v_p^2\right)\left(\Gamma_{22}-\rho v_p^2\right)\left(\Gamma_{33}-\rho v_p^2\right)+2\Gamma_{12}\Gamma_{23}\Gamma_{13}\\&-\Gamma_{13}{}^2\left(\Gamma_{22}-\rho v_p^2\right)-\Gamma_{23}{}^2\left(\Gamma_{11}-\rho v_p^2\right)-\Gamma_{12}^2\left(\Gamma_{33}-\rho v_p^2\right)=0\end{aligned} \tag{3}$$

where ρ is the density of the composite and the Christoffel stiffnesses are given for an orthotropic material by

$$\begin{aligned}\Gamma_{11}&=C_{11}n_1^2+C_{66}n_2^2+C_{55}n_3^2 &\qquad \Gamma_{23}&=(C_{23}+C_{44})n_2n_3\\ \Gamma_{22}&=C_{66}n_1^2+C_{22}n_2^2+C_{44}n_3^2 &\qquad \Gamma_{13}&=(C_{13}+C_{55})n_1n_3\\ \Gamma_{33}&=C_{55}n_1^2+C_{44}n_2^2+C_{33}n_3^2 &\qquad \Gamma_{12}&=(C_{12}+C_{66})n_1n_2\end{aligned} \tag{4}$$

where $n_1=\cos\theta\cos\phi$, $n_2=\sin\theta\cos\phi$, $n_3=\sin\phi$. θ and ϕ are the angles from the fiber direction in-plane and out-of-plane, respectively.

Finally, the phase velocities were fitted to the expression (3) in the sense of least squares fit to give the elastic constants: C_{11}, C_{33}, C_{13}, C_{55} in the (1-3)-plane; C_{11}, C_{22}, C_{12}, C_{44} in the (1-2)-plane and C_{22}, C_{33}, C_{23}, C_{66} in the (2-3)-plane.

ELASTIC CONSTANTS MEASURED BY COMPRESSION AND VIBRATION TESTS

The elastic constants C_{11}, C_{22}, C_{12} and C_{23} were measured by (destructive) compression tests and the constants C_{55} was measured by (nondestructive) vibration tests. The tests were performed on specimens cut from the same composite as the specimens used for the ultrasonic measurements.

The compression tests (TOFTEGAARD, 1996 and 1997) were performed with prism shaped specimens with the length direction either in the x_1-, x_2- or x_3-direction. The specimens were compressed in the length direction between parallel steel platens. During compression the strains in the specimen were measured with 0°/90° strain gauge couples on each of the four side faces. From the measured load and strains average curves of stress versus longitudinal strain (i.e. stress-strain curves) and of transverse strain versus longitudinal strain (i.e. strain-strain curves) were obtained. These curves were differentiated and the middle part of each differentiated curve was extrapolated to zero strain to estimate the engineering constants (Young's moduli and Poisson's ratios) of the material. The elastic constants C_{11}, C_{22}, C_{12} and C_{23} were finally calculated from the estimated engineering constants, e.g.

$$C_{11} = \left(1 - \nu_{23}^2\, E_{33}/E_{22}\right)\frac{E_{11}}{V} \tag{5}$$

where

$$V = \left[1 - \nu_{12}\left(\nu_{12}\, E_{22}/E_{11} + 2\nu_{23}\nu_{13}\, E_{33}/E_{11}\right) - \nu_{13}^2\, E_{33}/E_{11} - \nu_{23}^3\, E_{33}/E_{22}\right]$$

and where E_{11}, E_{22}, E_{33} are the Young's moduli and ν_{12}, ν_{13}, ν_{23} the Poisson's ratios.

The averaging of strains and the extrapolation of the differentiated stress-strain and strain-strain curves were performed to minimize the errors from misalignment and from friction between the specimen and the steel platens.

With the vibration technique the resonance frequencies of a freely vibrating plate are determined. The resonance frequencies are measured by a numerical model on the basis of the dimensions of the plate and assumed elastic constants. An optimization scheme is used to obtain the set of elastic constants giving the best fit to the measured resonance frequencies. This technique was used to estimate the constants C_{55} (FREDERIKSEN, 1997 and 1998).

MATERIAL AND TEST SPECIMENS

Specimens to be measured by ultrasound, compression and vibration were all cut from a 32 mm thick plate made from a fiber-reinforced. The composite consists of aligned E-glass fibers (16 μm in diameter) in a polymer matrix of polyethylene teraphtalate (PET). The composite was manufactured by stacking layers of aligned fibers and PET film sheets. Each fiber layer is actually a weave, where a small part of the fibers (1.5 % by weight) serves as weft fibers to ensure that the fiber layer can be handled. The composite was consolidated in an autoclave and the resulting fiber fraction and the composite density was 66 % by volume and 2298 kg/m^3, respectively.

The coordinate system of Fig. 3 gives the principal directions of the composite plate, with x_1 along the main fiber direction, x_2 along the direction of the weft fibers and x_3 in the thickness direction. Since the fraction of weft fibers is very low the composite is almost unidirectional and expected to be (nearly) transverse isotropic with the x_2-x_3 plane as the special plane of isotropy.

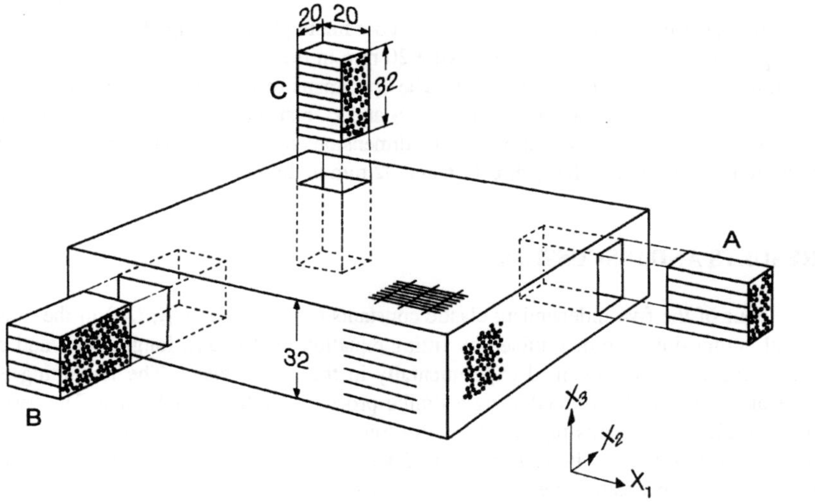

Figure 3. The investigated composite plate consists of woven glass fiber layers in a PET matrix. All test specimens were cut parallel to the coordinate planes. This was also the case for the specimens of type A, B and C used for compression testing.

Fig. 4 shows a microscope (ESEM) picture of a cross-section parallel to the x_2-x_3 plane. It is seen that the glass fibers have a circular cross-section. Moreover, the fiber array appears to be quite random, although, some areas tend towards a hexagonal packing. Three weft fibers are seen in the middle of the Figure.

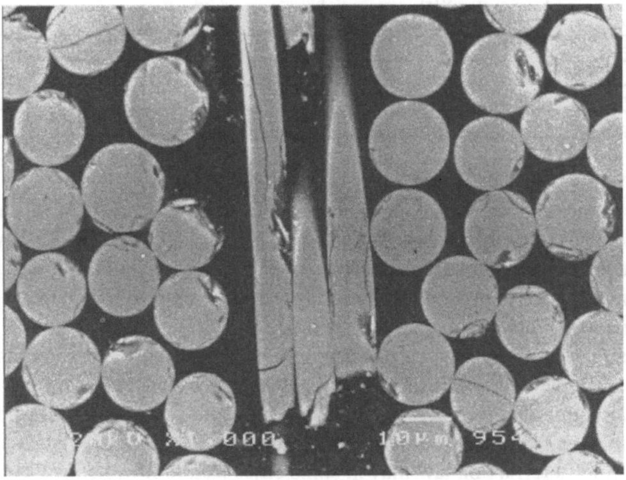

Figure 4. Microscope (ESEM) picture of a unidirectional glass/PET composite with three weft fibers. The glass fibers are white and the PET matrix dark.

All specimens were cut parallel to the coordinate planes of Fig. 3. The dimensions of the specimen for the ultrasonic tests were 20 mm in the x_1 direction, 220 mm in the x_2 direction and 32 mm in the x_3 direction. The specimens of type A, B and C in Fig. 3 were used for compression tests in direction x_1, x_2 and x_3, respectively. Three specimens of each type were tested. For the vibration test the dimensions of the specimen were 180 mm in the x_1 direction, 144 mm in the x_2 direction and 32 mm in the x_3 direction.

RESULTS AND DISCUSSION

Four of the five independent elastic constants (i.e. C_{11}, C_{33}, C_{13}, C_{55}) in the unidirectional composite were determined by fitting equation (3) to quasi-longitudinal and quasi-transverse velocities obtained experimentally in the (x_1-x_3)-plane. The remaining elastic constant, C_{23}, was determined using a similar procedure in the (x_2-x_3)-plane. The five independent elastic constants measured by ultrasound and mechanical experiments agree well as seen from Table 1. The full matrix is determined by the assumption that the unidirectional composite was transverse isotropic in symmetry, so that: $C_{33}=C_{22}$, $C_{55}=C_{66}$, $C_{13}=C_{12}$ and $C_{44}=(C_{22}-C_{23})/2$. Thus, rather than using three planes of symmetry, which in general must be evaluated to determine nine elastic constants in an orthotropic composite, only two planes of symmetry were necessary to evaluate the unidirectional composite used in this work. Theoretical values given in Table 1 were calculated in NIELSEN *et al.* (1997*b*) using analytical expressions by HASHIN *et al.* (1964).

Table 1. Experimental and theoretical determined elastic constants

Elastic constants	Ultrasonic experiment [GPa]	Mechanical experiment [GPa]	Theory [GPa]
C_{11}	61.1	59.3	53.5
$C_{33}=C_{22}$	30.3	27.6	23.2
$C_{55}=C_{66}$	8.7	9.4	6.3
C_{23}	10.2	12.1	8.0
$C_{13}=C_{12}$	11.4	10.6	8.4
$C_{44}=(C_{22}-C_{23})/2$	10.0	7.7	7.6

For the homogeneous composite specimen, several factors influenced the determination of elastic constants e.g. the assumption of unidirectionality and the quality of the composite. However, the accuracy in phase velocity is the most important in the ultrasonic experiment. The accuracy in phase velocity is a function of: receiver position, pulse distortion due to dispersion and attenuation, and temperature.

The accuracy of the phase velocity was influenced by being measured as the time-of-flight of the group velocity vector. Thus, in order to make an accurate time-of-flight measurement it was important to place the receiving ultrasonic transducer in a position where the energy emerges from the specimen, not in an anticipated position based on Snell's refraction law. Snell's refraction law may in some cases lead to lower elastic constants due to a shorter propagation path. For the (x_1-x_3)-plane of the composite, the group velocity vector may be found as the normal to the slowness surface (or inverse velocity surface), as indicated in Fig. 5 (NIELSEN *et al.*, 1998).

The time-of-flight measurement was in some cases also deteriorated by pulse distortion. Thus, the received pulse shape was distorted and the resultant shift in crossover position gave rise to an apparent reduction of wave velocity. The pulse distortion was seen as a shift to lower frequencies for especially wave incidence normal to the fiber direction. This may be explained by an enhanced dispersion or attenuation of the ultrasonic pulse compared to wave incidence along the fiber direction.

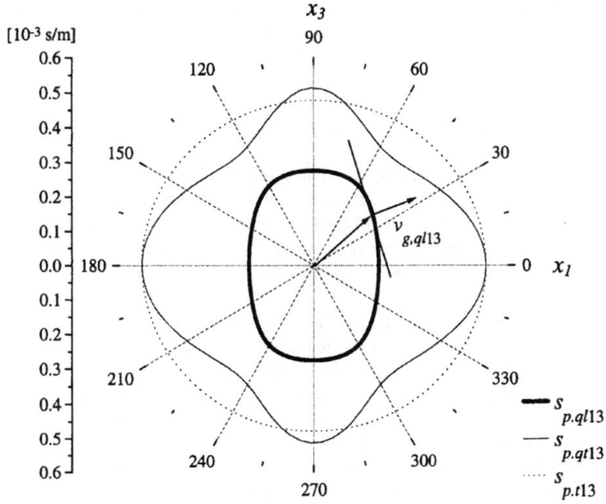

Figure 5. Slowness surface of the quasi-longitudinal ($s_{p,ql13}$), quasi-transverse ($s_{p,qt13}$) and transverse wave ($s_{p,t13}$) in the (1-3)-plane of a unidirectional (transverse-isotropic) composite. The group velocity vector is normal to the slowness surface [10^{-3} s/m].

The unidirectional composites were described as unidirectional (transverse isotropic), with an isotropy plane. However, the arrangement of plies and fibers induces a small anisotropy in this plane. Thus, depending on the measurement accuracy and the precision sought in terms of the elastic constants, unidirectional composites should be considered as orthotropic materials.

CONCLUSIONS

In this study an experimental quantitative NDE procedure is presented to measure the elastic stiffness constants using a circular aperture technique. The results for a unidirectional glass/PET reinforced composite are compared with results from destructive mechanical experiments and good agreement is found.

The results indicate that material characterization may be substantially improved through the use of the proposed method and that this method may be used to determine the elastic constants of orthotropic fiber reinforced composites in general.

REFERENCES

FREDERIKSEN, P.S. (1996). Advanced techniques for the identification of material parameters of composite plates, In *Proceedings of the First International Conference on Composite Science and Technology*, ed. S. Adali & V.E. Verijenko, Durban, 143-148.

FREDERIKSEN, P.S. (1997). Numerical studies for the identification of orthotropic elastic constants of thick plates. *Eur. Jour. Mech. A/Solids*, Vol. 16, No. 1, 117-140.

HASHIN, Z. and ROSEN, B.W. (1964). The elastic moduli of fiber-reinforced materials. *Journal of Applied Mechanics*, Vol. 31, 223-232.

KEDWARD, K.T. (1996). Lecture notes to *Design of Composite Structures*, UCSB.

MALLICK, P. K. (1993). *Fiber-Reinforced Composites: Materials, Manufacturing, and Design.* Marcel-Dekker.

NIELSEN, S.A. and BJØRNØ, L. (1997*a*). Bistatic circular array imaging with gated ultrasonic signals. *Acoustical Imaging*, Vol. 23, Edited by Lees and Farrari, Plenum Press, 441-446.

NIELSEN, S.A. and ANDERSEN, S.I. (1997*b*). Ultrasonic determination of elastic constants of fiber reinforced composites using a circular aperture array of transducers. *Proceedings of the 18th International Symposium on Materials Science: Polymeric Composites - Expanding the Limits*, Ed. S.I. Andersen *et al.*, Roskilde, Denmark, 455-464.

NIELSEN, S.A.; ANDERSEN, S.I. and BRØNSTED, P. (1998). Immersion ultrasonic method to measure elastic constants in fiber reinforced composite material. *4th European Conference on Composites, ECCM CTS-4 Testing and Standarisation*, Portugal, 71-79.

TOFTEGAARD, H. (1997) Elastic constants from simple compression tests. *Proceedings of the 18th International Symposium on Materials Science: Polymeric Composites - Expanding the Limits*, Ed. S.I. Andersen *et al.*, Roskilde, Denmark, 497-502.

TOFTEGAARD, H. (1998). Elastic constants from simple compression tests evaluated by numerical simulation. *4th European Conference on Composites, ECCM CTS-4 Testing and Standardisation*, Portugal, 310-319.

APPLICATION OF HEAT SOURCE MODEL AND GREEN'S FUNCTION APPROACH TO NDE OF SURFACE DEFECTS

Tsutomu Hoshimiya

Tohoku Gakuin University
Department of Applied Physics
13-1, Chuo-1, Tagajyo, Miyagi 985, JAPAN

INTRODUCTION

Photoacoustic (PA) imaging has been applied to nondestructive evaluation (NDE) of surface and undersurface defects of structured materials in these decades[1-5] . In the present paper, we proposed and extended a simplified heat source model and an approach of Green's function and applied them to a problem of NDE of surface defects by the use of gas-microphone PA imaging. We adopted wedge-shape as a model of a vertical surface defect, the depth of which is increasing linearly.

ANALYSIS BASED ON A HEAT SOURCE MODEL

The principle of the gas-microphone PA imaging is based on the fact that the PA effect arises when a periodically modulated optical beam generates a periodic heat flow from the solid sample to a surrounding ambient gaseous environment and then a modulated pressure wave is detected by a condenser-microphone.

In this section, analysis of a PA signal of a vertical wedge-shaped surface defect was done based on a simplified heat source model. This simplified model is based on an assumption that a PA signal amplitude is proportional to the surface area of the defect irradiated by a laser beam. We assumed a vertical defects aligned along the y-axis starting at position y=0 and increasing its depth linearly up to $y=L_1$ and then decreasing depth linearly and ending at $y=L_1+L_2$. The heat source is distributed like a belt with width 2a in a U-shape along both side walls ($x=-d/2,+d/2$; $0<z<D(y)$) and the bottom of the defect ($-d/2<x<d/2$; $z=D(y)$). If the depth function of the vertical defect along the y-axis is taken to be D(y) and the diameter 2a of the divergence-free laser beam is larger than the defect width d, then the effective irradiated area of the defect S will be given by

$$S=\pi a^2+2D(y)\cdot 2[a^2-(d/2)^2]^{1/2} \qquad (1)$$

instead of πa^2 for a normal surface. can easily be understood because PA signal proportional to Eq. (1) varies linearly with depth D(y) which in turn varies linearly with the additional first term πa^2. A graphical solution of Eq. (1) is schematically shown in Fig. 1(a). The PA signal proportional to the irradiated area given by the

Acoustical Imaging, Volume 24,
Edited by Hua Lee, Kluwer Academic/Plenum Publishers, 2000.

geometrical calculation described above was compared with the experimental data[5] shown in Fig. 1 (b). The simplified model explains experimental data qualitatively well.

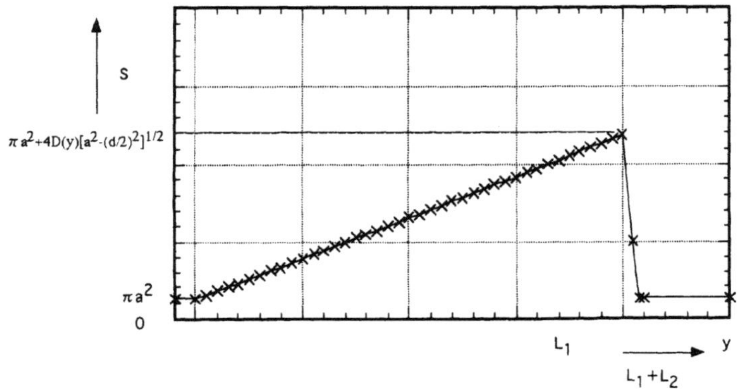

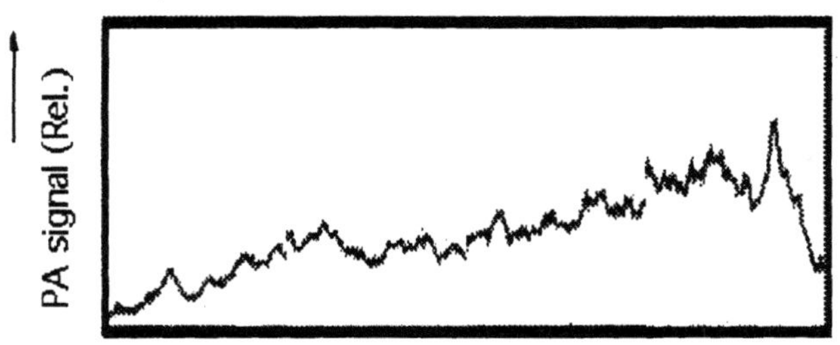

Fig.1 (a)Upper: Typical plot of surface area distribution calculated with Eq. (1).
(b) Lower: Typical PA signal distribution along a vertical wedge defect(ref.5).

APPROACH BASED ON THE GREEN'S FUNCTION

In this section, analysis of the PA signal was done by the use of the Green's function approach.[6] Generally speaking, when a sample is heated by the point heat source located at $\mathbf{r}$=(x,y,z), the sample temperature observed at observation point $\mathbf{r}'$=(x',y',z') is given by the Green's function G($\mathbf{r}$;$\mathbf{r}'$), which satisfies Helmholz-type equation

$$\nabla^2 G(\mathbf{r};\mathbf{r}') + \sigma^2 G(\mathbf{r};\mathbf{r}') = -\delta(\mathbf{r}-\mathbf{r}') \qquad (2)$$

According to the theory developed by Thomas et. al.,[1] a parameter σ is a complex wave number given by

$$\sigma = \sqrt{\frac{i2\pi f\rho c}{\lambda}} \equiv \frac{1+i}{\mu_s} \qquad (3).$$

where λ, ρ, c and f denote thermal conductivity, density, heat capacity of the solid specimen, and modulation frequency, respectively.

The solution of the Eq. (2) in three-dimensional free space (free space Green's function) is given by

$$G(\mathbf{r};\mathbf{r}') = \frac{1}{4\pi}\frac{\exp[-\sigma|\mathbf{r}-\mathbf{r}'|]}{|\mathbf{r}-\mathbf{r}'|} \qquad (4).$$

A PA signal obtained when a solid sample is excited with a periodically modulated laser beam at a plane surface z=0 can be calculated by a half plane Green's function ($G_{HP}(x,y,z=0;x',y',z')$) of the heat diffusion equation (2) which is given as

$$G_{HP}(x,y,0;x',y',z') = \frac{1}{2\pi}\frac{\exp[-\sigma\sqrt{(x-x')^2+(y-y')^2+z'^2}]}{\sqrt{(x-x')^2+(y-y')^2+z'^2}} \qquad (5)$$

by the image method. (The similar expressions will be easily obtained for the half-plane Green's functions with other irradiated surfaces.)

We considered combinations of planes in which heat source is generated at location and planes on which exists. Integration over observation point coordinate and summed up over possible combinations, generated PA signal can be derived by the general form:

$$V(y) = \iiint \sum_{path} d\mathbf{r}' \iint G(x,y,z;\mathbf{r}')dxdz \qquad (6)$$

where summation is taken over all possible path combination from the heat source located at $\mathbf{r}$=(x,y,z) to the observation point $\mathbf{r}'$.

We extended an approach based on a Green's function , or propagator, considering combinations of source surfaces and observation point surfaces. To idealize the problem, we assume that a laser beam diameter 2a just coincides with a defect width d. In this case, combination of heat source planes and observation planes are divided into 12 cases, which is shown in Fig. 2 and tabulated in Table I. Furthermore, we made an idealization that a laser beam distribution along defect direction (y-axis) is squeezed to be a delta function. This assumption reduces a procedure of convolution treatment of laser intensity distribution over a beam on the defect area.

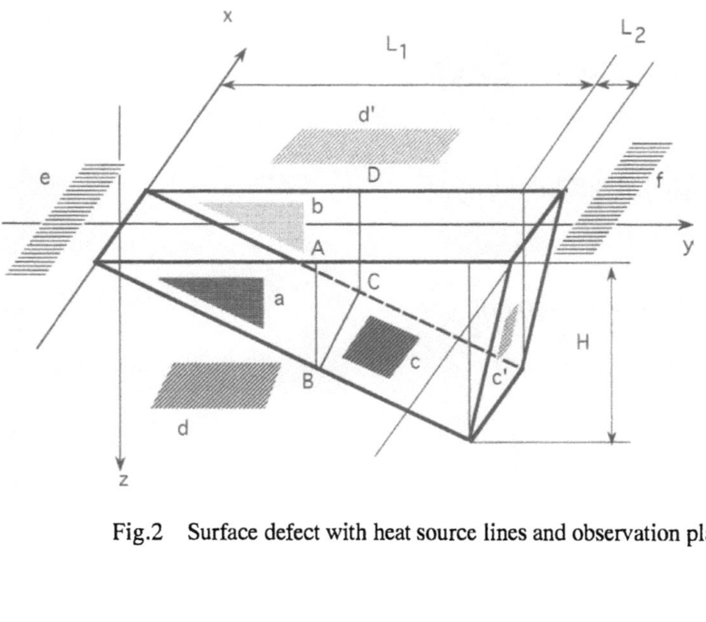

Fig.2 Surface defect with heat source lines and observation planes.

The rigorous expression of PA signal distribution function V(y) for the present case is shown as follows;

$$V(y)=\left(\int_0^{L_1}dy'\int_0^{D_1(y')}dz'+\int_{L_1}^{L_1+L_2}dy'\int_0^{D(y')}dz'\right)\left(\int_{-D_1(y)}^{D_1(y)}dz+\int_{-D(y)}^{D(y)}dz\right)G_{HP}\left(-\frac{w}{2},y',z';-\frac{w}{2},y,z\right)$$

$$+\left(\int_0^{L_1}dy'\int_0^{D_1(y')}dz'+\int_{L_1}^{L_1+L}dy'\int_0^{D(y')}dz'\right)\left(\int_{-D_1(y)}^{D_1(y)}dz+\int_{-D(y)}^{D(y)}dz\right)G_{HP}\left(\frac{w}{2},y',z';\frac{w}{2},y,z\right)$$

$$+\left(\int_{-\infty}^{\frac{w}{2}}dx'\int_0^{L_1+L}dy'\right)\left(\int_{-D_1(y)}^{D_1(y)}dzG_{HP}(x',y',0;-\frac{w}{2},y,z)+\int_{-D_2(y)}^{D(y)}dzG_{HP}(x',y',0;-\frac{w}{2},y,z)\right)$$

$$+\left(\int_{\frac{w}{2}}^{\infty}dx'\int_0^{L_1+L}dy'\right)\left(\int_{-D_1(y)}^{D_1(y)}dzG_{HP}(x',y',0;\frac{w}{2},y,z)+\int_{-D_2(y)}^{D(y)}dzG_{HP}(x',y',0;\frac{w}{2},y,z)\right)$$

$$+\int_{-\infty}^{\infty_1}dx'\int_{-\infty}^{0}dy'\left(\int_{-D_1(y)}^{D_1(y)}dzG_{HP}(x',y',0;-\frac{w}{2},y,z)+\int_{-D_2(y)}^{D(y)}dzG_{HP}(x',y',0;-\frac{w}{2},y,z)\right)$$

$$+\int_{-\infty}^{\infty_1}dx'\int_{-\infty}^{0}dy'\left(\int_{-D_1(y)}^{D_1(y)}dzG_{HP}(x',y',0;\frac{w}{2},y,z)+\int_{-D(y)}^{D(y)}dzG_{HP}(x',y',0;\frac{w}{2},y,z)\right)$$

$$;+\int_{-\infty}^{\infty}dx'\int_{L_1+L}^{\infty}dy'\left(\int_{-D_1(y)}^{D_1(y)}dzG_{HP}(x',y',0;-\frac{w}{2},y,z)+\int_{-D_2(y)}^{D(y)}dzG_{HP}(-x',y',0;-\frac{w}{2},y,z)\right)\quad(7)$$

$$+\int_{-\infty}^{\infty}dx'\int_{L_1+L}^{\infty}dy'\left(\int_{-D_1(y)}^{D_1(y)}dzG_{HP}(x',y',0;\frac{w}{2},y,z)+\int_{-D(y)}^{D(y)}dzG_{HP}(-x',y',0;\frac{w}{2},y,z)\right)$$

$$+\int_{-\frac{w}{2}}^{\frac{w}{2}}dx'\cdot\int_0^{L_1}dy'\int_{-\frac{w}{2}}^{\frac{w}{2}}dxG_{HP}(x',y',D_1(y');x,y,D_1(y))$$

$$+\int_{-\frac{w}{2}}^{\frac{w}{2}}dx'\cdot\int_{L_1}^{L_1+L}dy'\int_{-\frac{w}{2}}^{\frac{w}{2}}dxG_{HP}(x',y',D_2(y');x,y,D_2(y))$$

$$+\int_{-\infty}^{\infty}dx'\int_{-\infty}^{0}dy'\int_{-\frac{w}{2}}^{\frac{w}{2}}dx\cdot 2G_{HP}(x',y',0;x,y,D_1(y))$$

$$+\int_{-\infty}^{\infty}dx'\int_{L_1+L_2}^{\infty}dy'\int_{-\frac{w}{2}}^{\frac{w}{2}}dx\cdot 2G_{HP}(x',y',0;x,y,D_2(y))$$

We calculated Eq. (7) with a programming language Mathematica working on the workstation of Tohoku Gakuin University (TGU). The PA signal distribution V(y) is a complex number, the amplitude of which is shown in Fig. 3. The result agrees well with experimental data.

Table 1. Combination of heat source planes and observation planes

Heat source	a	b	c	c'
Observation plane	a d e f	b d' e f	c e	c' f

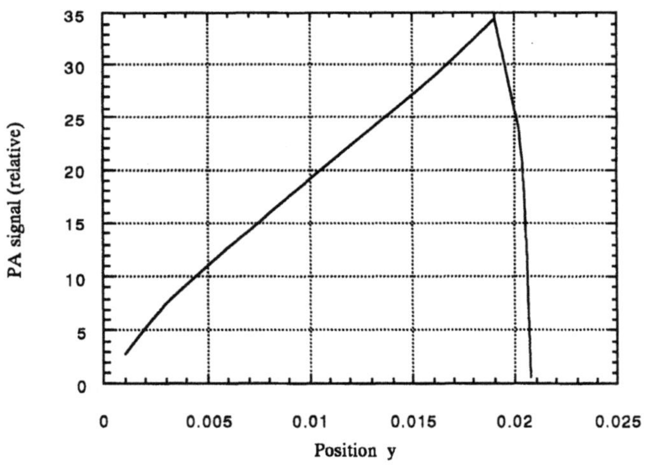

Fig.3 PA signal distribution calculated with Eq. (7).

CONCLUSION

In this paper, we extended the analysis of the PA signal generated for a vertical surface defect, the depth of which is increasing linearly, in order to apply to the NDE of the solid specimen. The theoretical analysis was performed by the use of a simplified heat source model, and an approach based on the Green's function. The obtained results agree well with the experimental data.

ACKNOWLEDGMENTS

The author would like to thank to undergraduate students Ms. Y. Satoh and Mr. T. Sugiyama of TGU for their help to the integration of Eq. (3) with Mathematica .

REFERENCES

1.R. L. Thomas, J. J. Pouch, Y. H. Wong, and P. K. Kuo: J. Appl. Phys., 51: 1152(1980).
2. P. K. Khandelwal , P. W. Heitman, A. J. Silversmith, and T. D. Wakefield: Appl. Phys.Lett., 37: 779 (1980).
3. K. R. Grice, L. J. Ingelhart, L. D. Favro, P. K. Kuo , and R. L. Thomas: J. Appl. Phys., 54: 6245(1983).
4. M. Hangyo, S. Nakashima, S. Sugimoto, T. Yamaguchi, and A. Mitsuishi: Jpn. J. Appl. Phys., 25:376(1986).
5. T. Hoshimiya, H. Endoh, and Y. Hiwatashi: Jpn. J. Appl. Phys.,35; 2916 (1996).
6. P. M. Morse and H. Feshbach: "Methods of Theoretical Physics" (McGrow-Hill Pub.,New York, 1983) , Part 1., Chap.7.

DETERMINATION OF BONDING PROPERTIES IN LAYERED METAL SILICON SYSTEMS USING SEZAWA WAVE MODES

A. Pageler, T. Blum, K. Kosbi, U. Scheer, and S. Boseck

Institute of Materials Science and Structure Research
University of Bremen, Kufsteiner Strasse
28359 Bremen, Germany[1]

ABSTRACT

Scanning acoustic microscopy (SAM) is used to determine the elastic properties of metal single- and multilayered systems. In the work presented here we investigate gold on silicon substrates with and without chromium interlayer, and obtain their experimental dispersion relations and elastic constants by means of V(z)-measurements and recently developed elastic-mapping (ELMA) experiments. These are compared with numerically calculated dispersion relations and turn out to be Sezawa modes. Seeking for a minimum of the errorfunction based on a least-square data fitting the Young's modulus, the shear modulus and Poisson's ratio of the investigated films are determined. The influence of the chromium interlayer is characterized and the bonding properties for these systems are classified.

INTRODUCTION

Thin film technology is of increasing interest for many technical applications. Some of their important properties are their lightness which makes it possible to reduce weight on devices significantly, their high conductivity and their chemical resistance. In order to

[1] Correspondence should be addressed to A. Pageler.
E-mail: pageler@schall.physik.uni-bremen.de

Acoustical Imaging, Volume 24.
Edited by Hua Lee, Kluwer Academic/Plenum Publishers, 2000.

investigate the bonding properties, several destructive testing methods have been developed, e.g. the well known scratch- and pull tests. With these methods one can determine the critical force at which the perfect adhesion breaks down. Unfortunally, it is very difficult to obtain reproducible results in such experiments since the mechanical tests destroy the sample.[1]

Obviously there is a need for non-destructive testing methods. Widely used in this context are ultrasonic techniques which are coast effective and easy to handle.[2]

Another appropriate way to determine the elastic properties of thin single- and multilayered systems is scanning acoustic microscopy (SAM).[3] The method is based on surface acoustic wave (SAW) dispersion measurements, which are very sensitive for the detection of differences in elastic properties of thin layers. SAW dispersion depends on bonding conditions between film and substrate. Generally, the SAW phase velocity v_{SAW} decreases with worse bonding conditions.[4]

MEASURING TECHNIQUES

Reflection-type SAM is used for imaging and detecting the elastic properties of the samples investigated in the present paper. The SAM can be used in two different modes, the V(z)- and the imaging mode.

V(z)-method

A V(z) curve, also known as the acoustic material signature, shows several oscillations depending on the investigated material. Using the V(z)-modus of the SAM we get information about the elastic properties of the layer.

The oscillation of the V(z) curve is an interference effect between two acoustic waves: the specularly reflected wave with normal incidence of the surface of the sample and a surface wave excited in the sample with symmetrical incident and reflecting angle. A transducer records the reflected waves and produces the output voltage V(z). The oscillations in the V(z) curves are due to different ray paths of the two waves and have a spacing Δz.[5]

With known frequency f, sound velocity v_w in the coupling fluid, generally water, and the period Δz of the V(z) curve it is possible to calculate the surface wave velocity v_{SAW}.[5]

ELMA-method

For the elastic-mapping (ELMA) measurements the imaging-modus of the SAM is used. For several constant ultrasonic frequencies a focus-series in steps of 0,1 µm in a range of 60 µm is measured as shown in figure 1. For every measurement we get an image with a characteristical grey tone. The images yield information about an area of 1 mm².

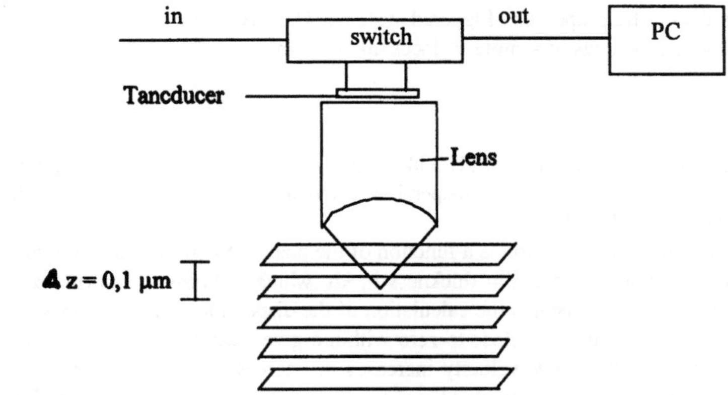

Figure 1: ELMA-measurement

Every image has its own characteristical grey tone depending on the defocus z. The grey tone distribution of the images of the series changes from light grey to dark grey. Any homogeneous image is divided in parts of 20 x 20 pixel areas as shown in figure 2. The white dotted areas have a size of 40 μm and the resulting grey tone at a defined point is the average of the grey tones of all pixels inside these areas. These detected grey tone distributions for the whole series depend on the defocus z can be compared and treated like the interference pattern of a V(z) curve. The SAW velocity is calculated from these detected distributions.

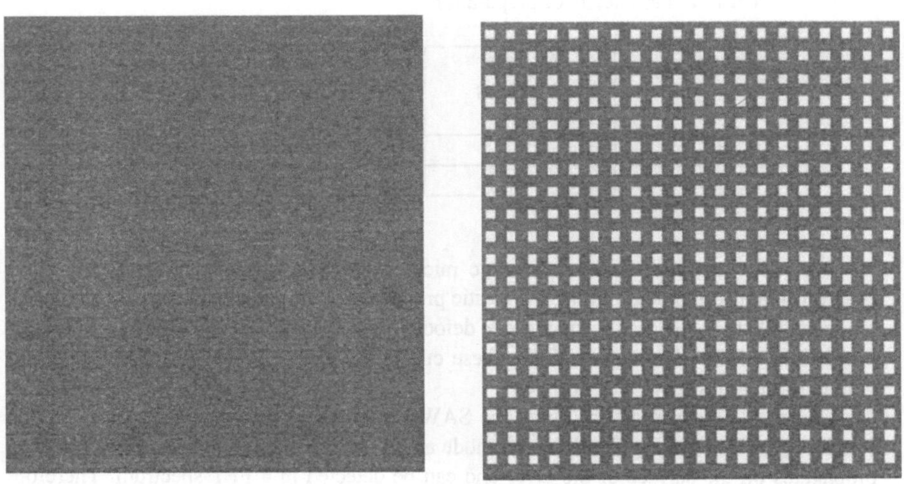

Figure 2: Homogeneous image and the used segmentation for the resulting grey tone distribution

Comparisons with calculated dispersion relations and detected experimental results using the V(z)-method show strong similarity and hence ensure that ELMA is a convenient method to obtain the elastic properties of layered systems. The advantage of ELMA is that it is possible to investigate areas of sample surfaces up to 1 mm².

SAW Dispersion

The elastic properties of layered solids under the assumption of perfect bonding and finite thickness can be determined by dispersion calculations. A detailed review is presented by Farnell and Adler (1972).[6]

The calculated dispersion relation is a function of the phase velocity of the SAW and the product of wave number k and film thickness h, kh, which is known as normalized thickness. For known elastic constants the calculation of the dispersion relation is possible when the boundary determinant gets close to zero within a given accuracy. Depending on the properties of the film the SAW velocity increases or decreases with the normalized thickness. For the films we have investigated the shear wave velocity of the films is smaller than the corresponding shear wave velocity of the silicon substrate. This is known as 'loading case'. Is the shear wave velocity of the film larger than the corresponding shear wave velocity of the substrate it is called 'stiffening' case.

EXPERIMENTAL

The investigated metal layered systems are gold films sputter deposited on oxidized silicon substrates. The samples have a gold layer thickness of 1 μm but only one sample has a chromium interlayer.

Table 1: Parameters of preparation

sample	Au-layer thickness μm	Cr-layer thickness nm
#1	1	none
#2	1	15

A reflection-type scanning acoustic microscope (SAM2000 by KSI, Germany) is used for imaging and determining the elastic properties of the films over a frequency in the range of 0.5 to 2 GHz. The maximum defocus in z-direction is 60 μm. We find four oscillations in our V(z) curves. From these curves the elastic properties of the gold and gold/chromium layer were determined.

From the measured V(z) curve the SAW velocity v_{SAW} can be calculated using the Fourier algorithm (FFT). The Rayleigh mode and higher modes, known as Sezawa modes, propagates on the surface of the layer and can be detected in a FFT spectrum. Therefore some data processing steps have been realized: digital filtering for reducing high frequency components, calculating the derivation of the V(z) curves and removing the geometric effected part of the V(z) data. Finally, the Fourier transform gives the SAW velocity with an accuracy of 3 %.

RESULTS AND DISCUSSION

The elastic constants C_{11}, C_{12} and C_{44}, where $C_{12} = C_{11} - 2C_{44}$ for the investigated layered systems are calculated from $v_{SAW}(kh)$ seeking for a minimum of the errorfunction for the least-square data fitting of theoretical dispersion relations. The used algorithm searched for a minimum of an appropriately chosen errorfunction.[7]

First we plotted the experimental dispersion data for the two measured samples in the numerical calculated dispersion relation of a gold/silicon and a gold/chromium/silicon system shown in figure 3. Sezawa modes are measured. The experimental data for sample #1 measured with the V(z)-method are located on the Sezawa mode calculated for a sputtered thin gold film on silicon substrate using the theory of Farnell and Adler.[6] Therefore the input values for sample #1 are the longitudinal wave velocity $v_l = 3240$ m/s and the shear wave velocity $v_t = 1218$ m/s. The experimental data of sample #2 measured with the V(z)-method are located slightly below the theoretical dispersion relation calculated for the gold/chromium/silicon system. The effective elastic constants of this multilayered system have been derived in terms of the corresponding parameters of the constituent layers by Grimsditch (1985).[8] The chosen input values for sample #2 are $v_l = 3252$ m/s and $v_t = 1248$ m/s. The data for sample #2 detected with the ELMA-method are placed exactly on the calculated dispersion mode and the input values are $v_l = 3240$ m/s and $v_t = 1250$ m/s. This theoretical dispersion relation is calculated using the algorithm of Thomson and Haskell.[9, 10]

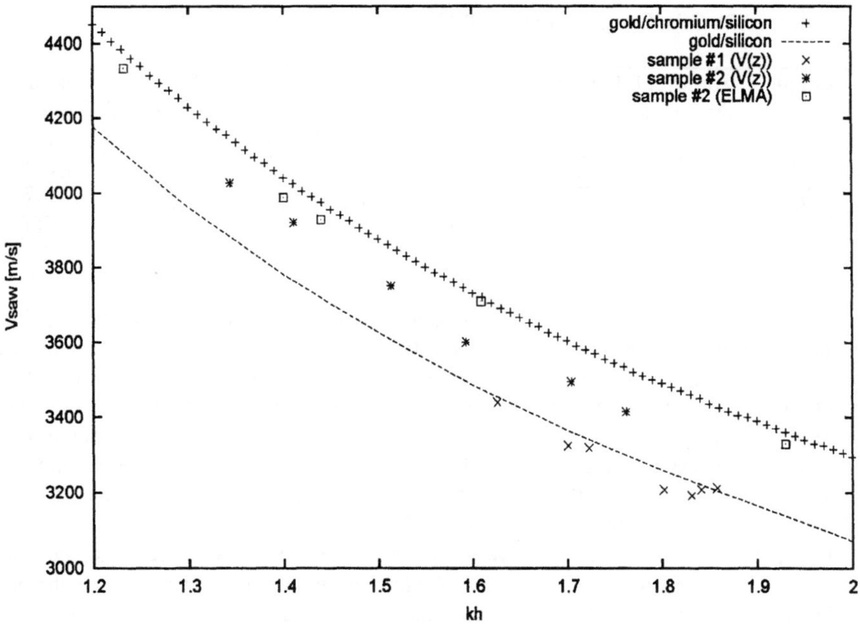

Figure 3: Calculated dispersion relations for a sputter deposited gold thin film on silicon substrate ('----') and a gold/chromium/silicon system ('+') for perfect bonding conditions. 'x' are the experimental data of sample #1 (V(z)-method), '*' are the experimental data of sample #2 (V(z)-method) and '□' are the experimental data of sample #2 (ELMA-method).

The determined elastic constants C_{11} and C_{44} of the investigated films were calculated using[3]

$$v_l = \sqrt{C_{11} / \rho} \qquad v_t = \sqrt{C_{44} / \rho} \tag{1}$$

They are listed in table 2.

Additionally in table 2 the corresponding Young's modulus E, the shear modulus G and Poisson's ratio ν of the measured samples are specified. They can be obtained from following relations[7]

$$E = \frac{C_{44}(3\,C_{11} - 4\,C_{44})}{C_{11} - C_{44}}, \tag{2}$$

$$G = C_{44}, \tag{3}$$

and

$$\nu = \frac{E}{2\,G} \tag{4}$$

Table 2: Experimental results for the elastic constants of the investigated thin films

(Comparison with the theoretical values for solid gold:[3]
E = 79,26 GPa, G = 27,8 GPa, ν = 0,425)

sample	Au-layer thickness µm	Cr-layer thickness nm	measuring technique	C_{11} GPa	$C_{44} = G$ GPa	Young's modulus E GPa	Poisson's ratio ν
#1	1	none	V(z)	202,4	28,6	81,39	0,419
#2	1	15	V(z)	203,8	30,0	105,7	0,355
#2	1	15	ELMA	202,4	30,1	107,1	0,348

We conclude sample #1 is a typical gold layered system with the soft and flexible properties of gold thin layers. The interface between the gold film and the silicon substrate seems to be free of missbonded areas because of the good similarity between the calculated dispersion relation for perfect bonding conditions and experimental data. Sample #2 shows increasing elastic properties. Our interpretation of this result is that the chromium interface causes better adhesion in a layered gold thin film system. Sample #2 is measured with two different techniques: the V(z)- and the ELMA-method. The comparison of the experimental results confirms the usefulness of the ELMA-measurements. The advantage of the ELMA-method is the sensitive detection of the elastic properties of single- and multilayered systems of an area of 1 mm². We have developed a new measuring technique which makes it possible to characterize the bonding properties of thin layered systems.

SUMMARY

Numerical dispersion relations of the layered systems gold/silicon and gold/chromium/silicon have been calculated. The measured data of these systems show an excellent similarity with the numerically calculated dispersion curves. The sample without the chromium interlayer yields a smaller value for the elastic properties. We assume that the thin chromium interlayer increasing the bonding strength of the soft and flexible gold layer on the silicon substrate. It is possible to characterize the bonding quality of the investigated thin films with both methods: V(z)- and ELMA-measurements. The advantage of using the ELMA-method instead of the well known V(z)-method is that ELMA enables us to obtain the bonding quality over an area of 1 mm².

ACKNOWLEDGEMENT

We would like to thank Krämer Scientific Instruments (KSI), Herborn, Germany, for technical support.

REFERENCES

1. P. Cawley, T. Pialucha, M. Lowe, 'A comparison of different methods for the detection of a weak adhesion/adherend interface in bonded joints', in: D.O. Thompson, D. E. Chimenti (Eds.), Review of Progress in Quantitative Nondestructive Evaluation (Vol.12), Plenum Press, New York (1993), pp 1531-1538
2. P. B. Nagy, 'Ultrasonic classification of imperfect interfaces, J. Nondestr. Eval. **11** (1992), pp. 127-140
3. A. Briggs, *Acoustical Microscopy*, Clarendon Press, Oxford (1992)
4. K. Kosbi, 'Quantitative Untersuchungen zur Adhäsion dünner Schichten mittels akustischer Rastermikroskopie, Dissertation, Universität Bremen (1998)
5. Z. Yu, S. Boseck, 'Scanning acoustic microscopy and its applications to material characterization', Rev. Mod. Phys. **67** (1995), pp. 863-891
6. G. W. Farnell and E. L. Adler, 'Elastic wave propagation in thin layers', in: W. P. Mason and R. N. Turston (Eds.), *Physical Acoustics*, Academic Press, New York (1972)
7. J. D. Achenbach, J. O. Kim, Y.-C. Lee, 'Measuring Thin-Film Elastic Constants by Line-Focus Acustic Microscopy', in: A. Briggs (Eds.), *Advances in Acoustic Microscopy* (Vol.1), Polonium Press, New York, London (1995)
8. M. Grimsditch,'Effective Elastic Constants of Superlattices', Phys. Rev. B (Vol.31) **10** (1985), pp. 6818-6819
9. W. T. Thomson, 'Transmission of Elastic Waves through a Stratified Solid Medium', Appl. Phys. **21** (1950, pp. 89-93
10. N. A. Haskell, 'The Dispersion of Surface Waves on Mulitlayered Media', Bull. Seism. Soc. Amer. **43** (1979), pp. 664-672

INTEGRAL APPROXIMATION METHOD FOR CALCULATING ULTRASONIC BEAM PROPAGATION IN ANISOTROPIC MATERIALS

Brian O'Neill[1] and Roman Gr. Maev[1]

[1]Ultrasonic Research Laboratory
Dept. of Physics
University of Windsor
Windsor Ontario, Canada N9B 3P4

INTRODUCTION

The increasing importance of anisotropic theory in acoustic wave theory is evidenced by the growing number of applications and publications in this area. Besides crystals, anisotropy can be found at some level in stressed materials, as well as materials with oriented cracks, pores, or inclusions, textured metals with oriented grains, thinly layered laminates, and lamellar or fibrous composites. Application areas include geophysical applications such as the propagation of siesmic waves and ultrasonic waves in earth and rock, medical applications in the ultrasonic imaging of both hard and soft tissues, as well as a large number of engineering and non-destructive testing applications using ultrasound. All aspects of the wave propagation are further complicated by material anisotropy. Numerous techniques have been developed to calculate the effects of anisotropy on beam propagation, including ray analysis[1], plane wave superposition[2], paraxial and Gauss-Hermite beams[3,4,5], and Green's functions[6,7], as well as other integral techniques[8]. With the exception of the ray analysis, these all suffer in one of two regards, either they are limited in application, or they are time consuming to set up and solve. The approximation used here for the particular case of a convergent beam, allows us to determine the general parameters of a focused acoustic beam inside anisotropic materials in a very efficient manner similar to ray analysis, while retaining most of the amplitude information from the wave picture. Included in this paper are the results of our calculations for the specific case of an austenitic steel weld metal. This is a good example of the intermediate level of anisotropy typically encountered in materials; however, this analysis also works well for highly anisotropic cases. These latter provide the unique possibility of producing high-quality acoustic images of internal structures at some depths. In such cases we can show that for certain specific orientations, the acoustic beam will propagate inside the anisotropic medium as a powerful collimated beam for significant distances. This technique is therefore ideal for imaging the bulk structure of generally anisotropic materials.

Acoustical Imaging, Volume 24.
Edited by Hua Lee, Kluwer Academic/Plenum Publishers, 2000.

THEORY OF PLANE WAVE PROPAGATION IN ANISOTROPIC MATERIALS

Plane wave solutions to the acoustic wave equation in crystals can be found by solving the well-known Christoffel equation:

$$k^2 \mathbf{\Gamma} \hat{\mathbf{e}} = \omega^2 \rho \hat{\mathbf{e}}, \tag{1}$$

where $\mathbf{\Gamma}_{jk} = c_{ijkl} l_i l_l$ is the 3x3 Christoffel matrix, whose elements are found from the stiffness (or compliance) tensor, and are functions of the direction cosines $l_x = k_x/k$, etc. of the wave-vector $\mathbf{k}$. Here $k = |\mathbf{k}|$ is the magnitude of the wavevector, ω is angular frequency, ρ is the density of the medium, and $\hat{\mathbf{e}}$ is the displacement polarization vector. For each direction $\hat{\mathbf{k}}$, we solve equation (1) to find the slowness k/ω and polarization $\hat{\mathbf{e}}$ of the wave with wavefront normal to $\mathbf{k}$. The actual wave (energy) propagation direction is not in the direction of $\mathbf{k}$, however. Instead it is in the group velocity direction, given by:

$$\det\left[\left(\tfrac{k}{\omega}\right)^2 \mathbf{\Gamma} - \rho \mathbf{1}\right] = \Omega(\omega(\mathbf{k}), \mathbf{k}) = 0$$
$$\nabla_k \Omega = -\tfrac{\partial \Omega}{\partial \omega} \nabla_k \omega = -\tfrac{\partial \Omega}{\partial \omega} \mathbf{v}_g, \tag{2}$$

which is always perpendicular to the slowness surfaces defined by Ω. In the geometric ray analysis, it is this group velocity vector which is taken to be the ray, instead of the phase velocity vector, $\mathbf{k}$. Self focusing phenemena can occur when a beam aperture falls in a k-space direction in which the slowness surface is nearly flat or even concave. Since the group velocity direction is perpendicular to this surface, this condition means that even though the k-space extent of the beam is divergent, in coordinate space it is collimated or convergent.

DISPLACEMENT FIELD OF A FOCUSED BEAM

Plane Wave Decomposition

Assume a focused acoustic beam, produced by an ultrasonic lens in a coupling liquid, is incident on the plane surface of an anisotropic specimen. To more accurately describe the propagation of the acoustic beam transmitted through the interface from the liquid coupling medium to the anisotropic solid, we must use wave analysis. The field in a liquid couplant can be represented as a superposition of plane waves and the displacement field of the acoustic wave excited in the liquid couplant is given by[9] (see Figure 1):

$$P(x, y, z) = \int\int A(k_x, k_y) \exp[ik_x x + ik_y y + ik_z(z - \zeta)] dk_x dk_y, \tag{3}$$

where $A(k_x, k_y)$ gives the k-space distribution at the focal plane $z = \zeta$, $k_z = -\sqrt{k^2 - k_x^2 - k_y^2}$, and k is a constant, assuming a single wavelength is excited. At the plane interface, each harmonic independently obeys Snell's law, leaving the tangential components and the frequency unchanged across the boundary. The normal component will be altered to satisfy equation (1), which will also determine the polarization direction. Each harmonic will be partially transmitted and partially reflected, and divided into the three acoustic modes so that the total displacement field must be written:

$$\mathbf{u}(x, y, z) = \textstyle\sum_\alpha \int\int dk_x dk_y T_\alpha \hat{\mathbf{e}}_\alpha A(k_x, k_y) e^{i\Phi(k_x, k_y, z)} \tag{4}$$
$$\Phi = k_x x + k_y y - k_z \zeta + k_z^\alpha z.$$

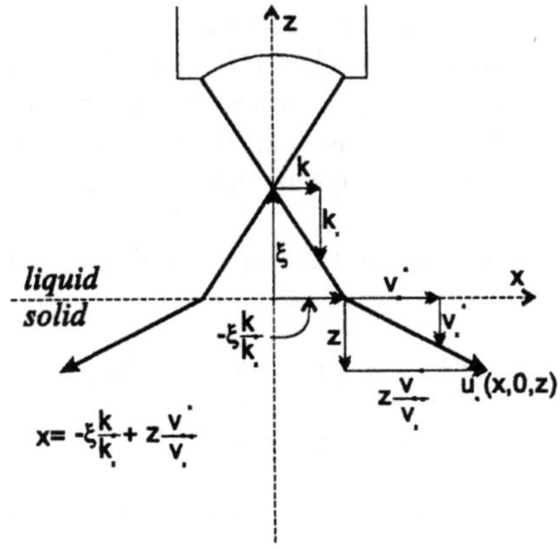

Figure 1: A focusing lens system and the stationary phase approximation.

$T_\alpha(k_x, k_y)$ and $\hat{\mathbf{e}}_\alpha(k_x, k_y)$ are the transmission coefficient and polarization vector of mode α, and $k_z^\alpha(k_x, k_y)$ is the z-component of $\mathbf{k}^\alpha$, the wave vector for mode α as determined by the slowness surface.

Approximation of the Displacement Field Integral

A simple and efficient method of approximating this integral is the stationary phase approximation. We assume that the greatest contribution to the integral comes from points $(\bar{k}_x, \bar{k}_y)$ where the condition for stationary phase holds:

$$\begin{aligned} \frac{\partial\Phi}{\partial k_x} &= x + \zeta\frac{k_x}{k_z} + z\frac{\partial k_z^\alpha}{\partial k_x} = 0 \\ \frac{\partial\Phi}{\partial k_y} &= y + \zeta\frac{k_y}{k_z} + z\frac{\partial k_z^\alpha}{\partial k_y} = 0. \end{aligned} \tag{5}$$

This condition is equivalent to demanding that energy at a field point is due to the rays incident at that point, by virtue of the fact that $\frac{\partial k_z^\alpha}{\partial k_x}$ is equivalent to $\frac{-v_x^\alpha}{v_z^\alpha}$, and similarly for y (Figure 1). The resulting first approximation is then[10]

$$\mathbf{u}\left[x(\bar{k}_x, \bar{k}_y), y(\bar{k}_x, \bar{k}_y), z\right] \approx \sum_\alpha \frac{2\pi \bar{T}_\alpha \bar{\mathbf{e}}_\alpha \bar{A} \exp\left[i\Phi(\bar{k}_x, \bar{k}_y, z) + i\frac{\pi}{2}\right]}{\left[\frac{\partial^2\Phi}{\partial k_x^2}\frac{\partial^2\Phi}{\partial k_y^2} - \left(\frac{\partial^2\Phi}{\partial k_x \partial k_y}\right)^2\right]^{1/2}} \tag{6}$$

where the bar indicates the function evaluated at $(\bar{k}_x, \bar{k}_y)$. By using $\bar{k}_x$ and $\bar{k}_y$ as parameters, and x and y as given by (5), $\mathbf{u}$ can be easily plotted, using complicated or simple transmission and aperture functions as necessary. Decay factors can easily be added, and if necessary, standard methods exist for finding the second term in the expansion around stationary phase points[10,11]. Also, this last expression is equally valid for the case of two anisotropic media, although this would require a suitable modification of the stationary phase condition (5).

Most of the interesting additional information comes from the expression under the square root in the denominator:

$$\frac{\partial^2\Phi}{\partial k_x^2}\frac{\partial^2\Phi}{\partial k_y^2} - \left(\frac{\partial^2\Phi}{\partial k_x \partial k_y}\right)^2 \tag{7}$$

which is a measure of the curvature of the slowness surface at each point. We should expect a stronger than incident beam when (7) is less than unity, and weaker when it is greater than unity. When (7) is equal to zero, the beam amplitude goes to infinity; in some sense, this must indicate a focus. This occurs when there is a "flat" spot in the phase - indicating a large number of rays are contributing to the amplitude at a single point.

In the case of a focused beam and isotropic couplant, (with $k_z = -\sqrt{k^2 - k_x^2 - k_y^2}$) we can easily find:

$$\frac{\partial^2 \Phi}{\partial k_{x,y}^2} = \frac{\zeta}{k_z}\left(1 + \frac{k_{x,y}^2}{k_z^2}\right) + z\frac{\partial^2 k_z^\alpha}{\partial k_{x,y}^2},$$
$$\frac{\partial^2 \Phi}{\partial k_x \partial k_y} = \zeta\frac{k_x k_y}{k_z^3} + z\frac{\partial^2 k_z^\alpha}{\partial k_x \partial k_y}. \tag{8}$$

Substituting into (6) we have the result for a general beam passing from an isotropic to an anisotropic medium

$$z^2\left[\frac{\partial^2 k_z^\alpha}{\partial k_x^2}\frac{\partial^2 k_z^\alpha}{\partial k_y^2} - \left(\frac{\partial^2 k_z^\alpha}{\partial k_x \partial k_y}\right)^2\right]$$
$$+\frac{z\zeta}{k_z}\left[\frac{\partial^2 k_z^\alpha}{\partial k_x^2} + \frac{\partial^2 k_z^\alpha}{\partial k_y^2} + \frac{k_x^2}{k_z^2}\frac{\partial^2 k_z^\alpha}{\partial k_y^2} + \frac{k_y^2}{k_z^2}\frac{\partial^2 k_z^\alpha}{\partial k_x^2} - 2\frac{k_x k_y}{k_z^2}\frac{\partial^2 k_z^\alpha}{\partial k_x \partial k_y}\right] + \frac{\zeta^2 k^2}{k_z^4} \tag{9}$$

From this expression there are a number of deductions that can be made. When we set z to 0, we can find the expected spot size at the interface using the stationary phase conditions (5), with k_x and k_y set to their maximum values as determined by the lens aperture. The displacement amplitude at the interface $z = 0$ is then determined from the last term in (9), which can also be written as r^2/k_z^2 where r is the radial distance from the focal point to the interface, giving the standard form of an outgoing spherical wave. If ζ is also zero (that is, the focus is at the interface), this expression is zero at the focus as predicted, and the amplitude becomes infinite. Also for the theoretical case of a planar slowness surface, all of $\frac{\partial^2 k_z^\alpha}{\partial k_x^2}$, $\frac{\partial^2 k_z^\alpha}{\partial k_y^2}$, $\frac{\partial^2 k_z^\alpha}{\partial k_x \partial k_y}$ are zero, while the first derivatives are constant. It is trivial to show that the beam width at all depths z is equal to the spot size at the interface, and that the displacement amplitude remains constant (recall that we are ignoring attenuation due to effects other than beam divergence). Clearly, directions in which a material's slowness surface is nearly flat provide excellent opportunities for the study of bulk structures using acoustic microscopy. In these directions, the acoustic beam will penetrate deep into the specimen with little or no divergence. Finally, it is expected that expression (9) can be set to zero, and solved for z to give the focal depth. To illustrate, consider the simple case when the second material also is isotropic.

Having $k_z^\alpha = -\sqrt{(k^\alpha)^2 - k_x^2 - k_y^2}$, we obtain:

$$\frac{\partial^2 k_z^\alpha}{\partial k_x^2} = -\left[\frac{k_x^2}{(k_z^\alpha)^2)} + \frac{1}{k_z^\alpha}\right];\ \frac{\partial^2 k_z^\alpha}{\partial k_y^2} = -\left[\frac{k_y^2}{(k_z^\alpha)^2)} + \frac{1}{k_z^\alpha}\right];\ \frac{\partial^2 k_z^\alpha}{\partial k_x \partial k_y} = -\frac{k_x k_y}{(k_z^\alpha)^2} \tag{10}$$

Replacing in (9), expanding and simplifying with the substitutions $k_x^2 + k_y^2 = k_r^2$, $k_r^2 + k_z^2 = k^2$, $k_r^2 + (k_z^\alpha)^2 = (k^\alpha)^2$, we find:

$$z^2\frac{(k^\alpha)^2}{(k_z^\alpha)^4} - z\zeta\frac{k_r^2 k_z^2 + k_r^2(k_z^\alpha)^2 + 2k_z^2(k_z^\alpha)^2}{k_z^3(k_z^\alpha)^3} + \zeta^2\frac{k^2}{k_z^4}. \tag{11}$$

If we trivially set $k^\alpha = k$, $k_z^\alpha = kz$, that is, both media are the same, (11) vanishes only when $z = \zeta$, as one should expect. Otherwise we can let $k^\alpha = k/R$, where R is

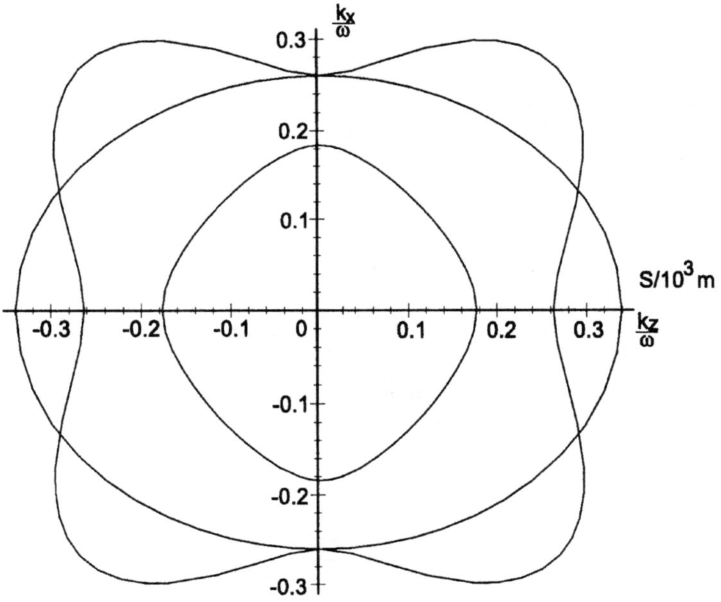

Figure 2. Slowness surfaces for austenitic stainless steel from experimental data by Kupperman and Reimann (1980) [13]

the ratio of the speed in the solid to the speed in the liquid, and setting (11) to zero, find solutions for z:

$$
\begin{aligned}
z &= \frac{\varsigma}{R}\frac{(k^2 - k_r^2/2[(1+R^2) \pm (1-R^2)])\sqrt{k^2 - k_r^2R^2}}{(k^2 - k_r^2)^{3/2}} \\
&= \left\{\frac{\varsigma}{R}\Delta, \frac{\varsigma}{R}\Delta^3\right\};\ \Delta \equiv \frac{(k^2 - k_r^2R^2)^{1/2}}{(k^2 - k_r^2)^{1/2}} = \frac{(1 - R^2\sin^2\theta)^{1/2}}{\cos\theta} \qquad (12)
\end{aligned}
$$

where θ is the angle of incidence. It can be shown that the first solution (+ve) corresponds exactly to the line focus at the beam axis predicted by the geometric ray analysis. The second solution corresponds to a surface formed by the convergence of neighbouring rays at the edge of the beam. This solution forms the outside boundary of the beam in the material and is known as a caustic zone[12].

APPLICATION TO ANISOTROPIC MEDIA - AN EXAMPLE

The real utility of this result is only realized in cases involving anisotropy. In such cases, an accurate ray analysis becomes increasingly difficult. Using simple computational methods, the stationary phase approximation provides an easy way to find a more accurate and flexible solution to the problem. To illustrate this, we can use the simple case of an austenitic stainless steel weld. It is an established fact that this metal exhibits an approximately transversely isotropic (hexagonal) symmetry[1,13–15]. In this case, reasonably compact solutions to the Cristoffel equations exist, in fact, we have shown that for these solutions:

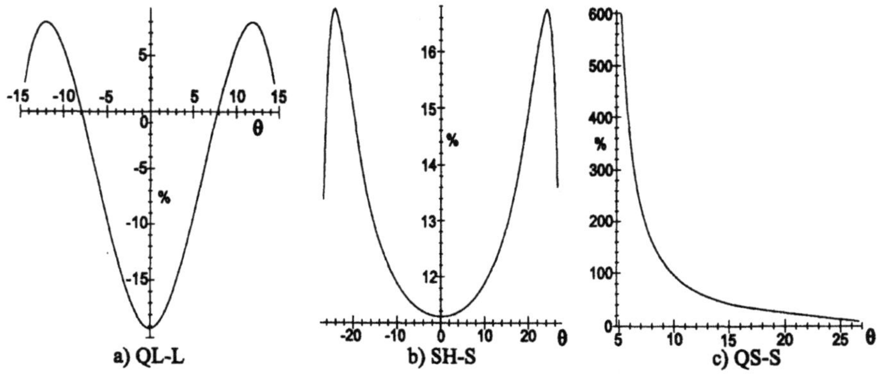

Figure 3: Difference in amplitude between an ultrasonic beam in weld metal and in parent metal, as percent of the peak amplitude in the parent.

$$
\begin{aligned}
c_{SH}^2(\theta) &= A + B\sin^2\theta \\
c_{QL}^2(\theta) &= a + b\sin^2\theta + \sqrt{c + d\sin^2\theta - e\sin^4\theta} \\
c_{QS}^2(\theta) &= a + b\sin^2\theta - \sqrt{c + d\sin^2\theta - e\sin^4\theta}
\end{aligned} \tag{13}
$$

where cSH, cQL, and cQS are the phase velocities which can be found experimentally for the pure horizontal shear (SH) mode, the quasi-longitudinal (QL) mode and the quasi-shear (QS) modes respectively; and A, B, a, b, c, d, and e are parameters to be determined, only three of which are theoretically independent of the others. Using available experimental data[13], we can find the values of these parameters. From this, it is easy to plot the slowness surfaces (Figure 2).

Although this gives all the necessary information, we can more easily use these expressions in (6) and (7) if they are expressed as $k_z^\alpha(k_x, k_y)$, where once again, α represents the modes SH, QL, and QS. This can be arranged by replacing $\sin^2\theta = \frac{k_r^2}{(k_z^\alpha)^2+k_r^2}$, $c_\alpha^2 = \frac{\omega^2}{(k_z^\alpha)^2+k_r^2}$, where $k_r^2 = k_x^2 + k_y^2$, in the relations (17), and solving for k_z^α. The resulting expressions are then substituted directly into (6) and (7), and, using k_x and k_y as parameters, the field strength at any position can be found. In order to investigate only the effects of beam spreading, we can set the transmission coefficient and aperture function to unity, and parametrically plot each mode separately. The difference in amplitude between the weld and the isotropic parent metal beams at the depth $z = -\zeta/10$ are plotted in Figure 3, as a percentage of the maximum isotropic amplitude. The difference is greater than zero in directions of phonon focusing, and less than zero in directions of phonon defocusing. It is easy to see from Figure 3(c) that rays incident at angles greater than about ten degrees will not contribute to the focusing, which occurs as the beam amplitude becomes infinite. This gives a rough limit to the useful size of the aperture needed if we wish to make use of the phonon focusing effects of this mode. The other modes are not focused, but show various degrees of collimation. Predicted beam profiles for the isotropic parent metal are shown in Figures 4(a) and (c), and can be easily compared to the weld metal in Figures 4(b), (d), and (e). While the first-order stationary-phase approximation is not really valid near the focus, the general behaviour of the beam is still easily predicted.

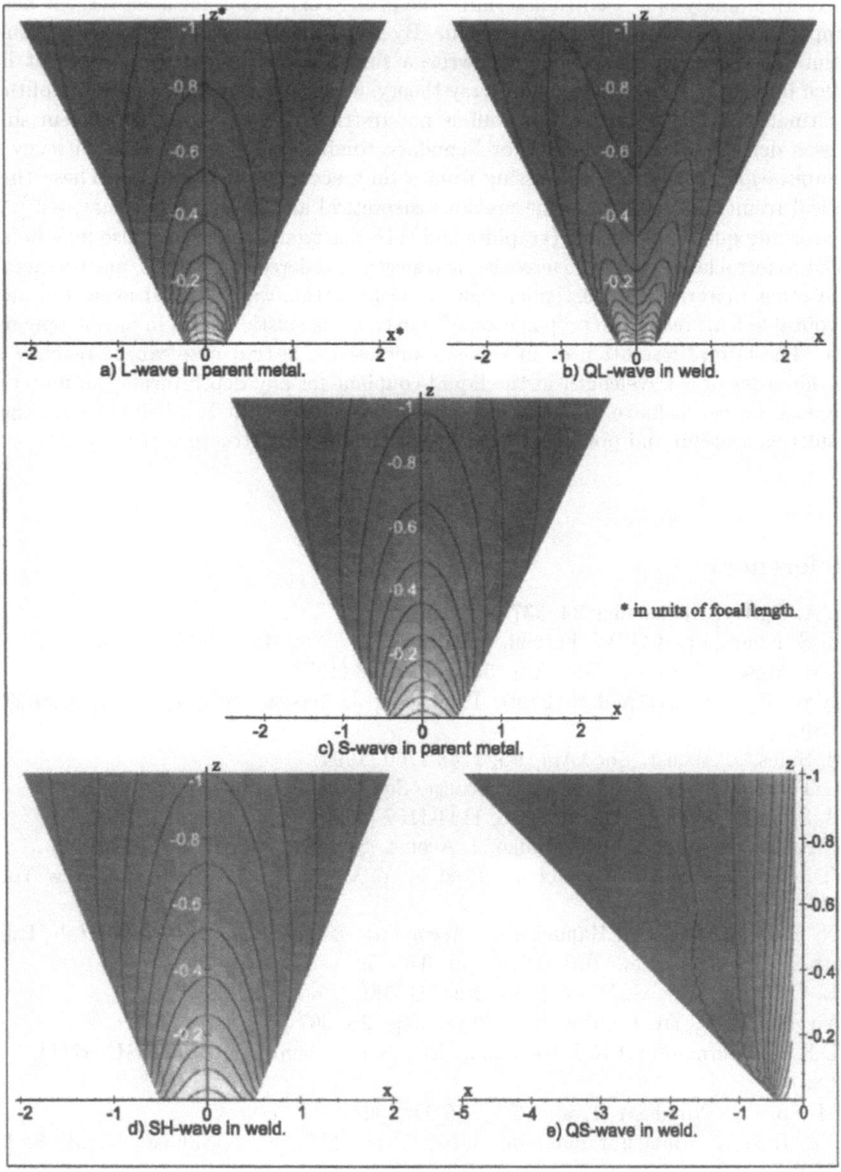

Figure 4: Beam profiles in isotropic parent metal (a, c) and in anisotropic weld metal (b, d, e), calculated using stationary phase.

CONCLUSION

The theory of elastic plane wave propagation has been applied to the specific case of a convergent acoustic beam inside an anisotropic material. The result of this theoretical analysis is used to determine the general expression for a convergent beam propagating in an anisotropic specimen. By using an asymptotic expansion around points of stationary phase, one can write a simple analytic approximation that has much in common with the geometric ray theory, while also containing useful amplitude information. The approximate result is not restricted to any specific problem since it was derived without concern for boundary conditions. The result is efficiency in computation that would be missing from a fully accurate wave theory. These theoretical results also confirm some previous theoretical and experimental analyses[9,16–17] concerning quasi-2D crystals (graphite and HTS materials), and which also may be applied to ferroelastics near the ferroelastic transition, ordered composites, matrix metals, and other materials. Acoustomicroscopical visualisation was achieved using the more favourable features of the propagation of convergent acoustic beams in anisotropic media. The lateral resolution of an acoustic microscope in that case can be reached up to the order of a wavelength in the liquid couplant for any depth inside the material. Because we can achieve high contrast at significant depths, it is possible to use these results as a useful and powerful method to study the bulk structure of materials.

References

[1]J. A. Ogilvy, Ultrasonics **24**, 337-347 (1986).

[2]M. S. Kharusi and G. W. Farnell, J. Acoust. Soc. Am. **48**, 665-670 (1970).

[3]I. M. Mason, J. Acous. Soc. Am. **53**, 1123-1128 (1973).

[4]Byron P. Newberry and R. Bruce Thompson, J. Acoust. Soc. Am. **84**, 2290-2300 (1989).

[5]M. Spies, J. Acoust. Soc. Am. **95**, 1748-1760 (1994).

[6]A. Tverdokhlebov and J. Rose, J. Acoust. Soc. Am. **83**, 118-121 (1988).

[7]M. Spies, J. Acoust. Soc. Am. **96**, 1144-1157 (1994).

[8]V. K. Tewary and C. M. Fortunko, J. Acoust. Soc. Am. **91**, 1888-1896 (1992).

[9]V. Levin, in *Acoustical Imaging*, edited by Y. Wei and B. Gu (Plenum, New York, 1993), Vol. 20, p.233-240.

[10]N. Bleistein and R.A. Handelsman, *Asymptotic Expansions of Integrals* (Holt, Rinehart and Winston, New York, 1975), pp. 340-359.

[11]B. O'Neill and R. G. Maev, Phys. Rev. B, **58**(9), 1-7, (1998).

[12]Yu. Kravtsov, Yu. I. Orlov, Sov. Phys. Usp. **26**, 307 (1983).

[13]D.S. Kupperman and K.J. Reimann, IEEE Trans. Sonics Ultrason. **SU-27**(1), 7-15 (1980).

[14]M. Spies, J. Nondestr. Eval. **13**(2), 85-99 (1994).

[15]J.L. Rose, K. Balasubramaniam, and A. Tverdokhlebov, J. Nondstr. Eval. **8**, 165 (1989).

[16]V.V. Novikov, L.A. Chernozatonskii, Sov. Phys. Acoust. **34**(2), 215 (1988).

[17]V.M. Levin, R.G. Maev, K.I. Maslov et al. in *Acoustical Imaging*, edited by H. Ermert and H. P. Harjes (Plenum, New York, 1992), Vol 19, pp. 651-656.

ELASTIC STRESS INFLUENCE ON THE PROPAGATION OF ELECTROMAGNETIC WAVES THROUGH TWO-LAYERED PERIODIC DIELECTRIC STRUCTURES

Gregory V. Morozov, Roman Gr. Maev, and G. W.F. Drake

Dept. of Physics
University of Windsor
Windsor, Ontario, Canada N9B 3P4

INTRODUCTION

The propagation of waves of any nature through media with one-dimensional periodicity has long been a topic of interest in various areas of physics and technology, beginning with the well-known paper of Kronig and Penney[1]. There is now a vast literature on the subject, including several monographs and reviews[2,3,4,5,6]. Much of this work is devoted to the development of effective numerical and approximate analytical methods, such as the Floquet-Bloch formalism[7], the matrix method[8], the theory of Kogelnik coupled waves[9], and the modified theory of coupled waves[10].

However, even for the simplest case of a two-layered periodic structure, previous expressions for the exact analytical solutions in terms of Floquet-Bloch waves were very complicated and cast in a form involving several parameters whose physical meaning was not clear. As a result the problem of finding the reflection and transmission coefficients for waves incident on multi-layered periodic structures was being usually solved by the use of matrix methods[5,8,10,11]. Although the solutions are exact in form, they are quite cumbersome and difficult to analyze in terms of their dependence on the parameters characterizing each layer. The theory of coupled waves yields relatively simple approximate expressions for the reflection coefficient in the case of two-layered periodic structures. However, depending on the range of parameters involved, the accuracy may be no better than $\pm 50\%$. This is not sufficient for most applications.

In the present paper we developed an improved analytic formulation for the propagation of electromagnetic Floquet-Bloch waves inside a two-layered periodic dielectric structure which allows us to extract exact analytic expressions for the reflection and transmission coefficients in the case of normal incidence of the plane electromagnetic wave on a two-layered periodic dielectric structure in more simple and physically understandable form. As a result we found the extremely high sensitivity of reflection and transmission coefficients of the electromagnetic wave to the parameters of the structure such as refraction indexes and widths of the basic layers in some cases. This makes it possible to influence on the light propagation through such structures by an acoustic wave even if it is only able to change the parameters of the structure by 0.1%. As the simplest example of such effect we consider the constant strain influence on the reflection and transmission of the plane electromagnetic wave.

THE GENERAL THEORY OF ELECTROMAGNETIC WAVES IN TWO-LAYERED PERIODIC DIELECTRIC STRUCTURES

The parameters specifying the problem are as shown in Figure 1. We assume a two-layered periodic transparent (without absorption) nonmagnetic ($\mu = 1$) medium with refractive indexes n_1 and n_2 and thicknesses d_1 and d_2 of the layers such that $d = d_1 + d_2$ is the period of the function $n(z)$ ($n^2(z) = \varepsilon(z)$), and N is the number of the periods of the structure. Denote the index of refraction of the medium on either side of the structure as n_0, i.e. $n = n_0$ if $z < 0$, and $n = n_0$, if $z > Nd$.

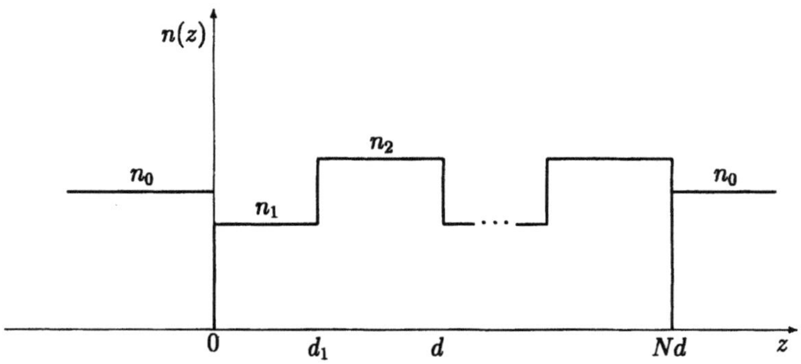

Figure 1: Two-layered periodic dielectric structure.

Monochromatic plane wave with wave vector in vacuum $\mathbf{k_0}$ is assumed to be normally incident upon the two-layered periodic structure from the region $z < 0$. In the case of normal incidence, light polarization does not play a role and the calculations given below are valid for both possible cases of polarization (along y- or x-axis). Taking the amplitude of the incident wave as unity, we can express total electric field in the region $z < 0$ as

$$E(z) = \exp(ik_0n_0z) + A\exp(-ik_0n_0z)\,, \tag{1}$$

where $A\exp(-ik_0n_0z)$ is the reflected wave, and A is its amplitude. In the region $z > Nd$ the total field is just the transmitted wave, which can be written in the form

$$E(z) = B\exp(ik_0n_0(z - Nd))\,, \tag{2}$$

where B is the amplitude. Inside the structure we can represent the total field as a superposition of two travelling Floquet-Bloch waves in the form

$$E(z) = C_1E_1(z) + C_2E_2(z)\,, \tag{3}$$

where

$$E_{1,2}(z) = F_{1,2}(z)\exp\left(i\frac{\xi_{1,2}}{d}z\right), \quad F_{1,2}(z) = F_{1,2}(z+d) \tag{4}$$

The quantities $\xi_{1,2}$ are the so-called characteristic Floquet indices, related by $\xi_2 = -\xi_1$[12]. To find exact analytic expressions for the Floquet-Bloch waves $E_1(z)$ and $E_2(z)$, we represent them inside the n_1 layer of the first period as

$$E_{1,2}(z) = A_{1,2} \sin\left(k_0 n_1 z - \frac{k_0 n_1 d_1}{2} + \varphi_{1,2}\right), \tag{5}$$

Analogously, inside the n_2 layer of the same period,

$$E_{1,2}(z) = B_{1,2} \sin\left(k_0 n_1 (z - d_1) - \frac{k_0 n_2 d_2}{2} + \psi_{1,2}\right), \tag{6}$$

The coefficients A_1 and A_2 in formulae (5) can be set equal to unity (they play the role of normalization constants for the Floquet-Bloch waves). The phases $\varphi_{1,2}$ and $\psi_{1,2}$ are in general complex. Their introduction represents the major point of departure[13] from the usual representation in terms of the exponential function[3,5,10,11,14].

To find the parameters $\varphi_{1,2}$, which define Floquet-Bloch waves $E_{1,2}(z)$ in the first layer of the first period, we use the conditions of continuity for the functions $E_{1,2}(z)$ and its derivative (corresponding to the continuity of the magnetic field) at the points $z = d_1$ and $z = d_1 + d_2 \equiv d$ and then use Floquet theorem in the form $E_{1,2}(z) = \exp(i\xi_{1,2}) E_{1,2}(z - d)$, which means that in the system with periodical coefficients solutions are different by a factor $\exp(i\xi)$ under shifting on period d [2,7,12]. As a result we have

$$\varphi_{1,2} = \pm \frac{1}{2} \arccos\left[\frac{a^{-1} \sin\Omega + a \sin\Delta}{2 \sin(k_0 n_2 d_2)}\right], \tag{7}$$

where

$$a = \frac{n_2 - n_1}{n_2 + n_1} = \left(1 - \frac{n_1}{n_2}\right) \Big/ \left(1 + \frac{n_1}{n_2}\right),$$
$$\Omega = k_0(n_2 d_2 + n_1 d_1), \quad \Delta = k_0(n_2 d_2 - n_1 d_1). \tag{8}$$

For the characteristic Floquet indices $\xi_{1,2}$ ($\xi_2 = -\xi_1 \equiv \xi$) we can obtain the well-known dispersion equation[3,5,10,11,14]. In terms of a, Ω and Δ, it has form

$$\cos\xi = \frac{1}{1 - a^2} \cos\Omega - \frac{a^2}{1 - a^2} \cos\Delta. \tag{9}$$

Now using continuity conditions for the total field $E(z)$ at the points $z = 0$ and $z = Nd$ and Floquet theorem at the form $E(Nd) = C_1 E_1(0) \exp(i\xi_1 N) + C_2 E_2(0) \exp(i\xi_2 N)$ we receive the system with four equations and four unknown coefficients C_1, C_2, A, B. Solving it for coefficients A and B and taking into account the relations $\xi_1 \equiv \xi = -\xi_2$ and $\varphi_2 = -\varphi_1$, we have the final results for reflection coefficient $R = | A^2 |$ and transmission coefficient $T = | B^2 |$ for the electromagnetic wave normally incident on a two-layered periodic dielectric structure

$$R = \frac{[(1 - b)\sin\Omega + a(b + 1)\sin\Delta]^2 + \dfrac{4a^2 b}{1 - a^2}(\cos\Omega - \cos\Delta)^2}{[(1 + b)\sin\Omega + a(1 - b)\sin\Delta]^2 + \dfrac{4b}{1 - a^2}\left|[(\cos\Omega - a^2\cos\Delta)^2 - (1 - a^2)^2]\cot^2(N\xi)\right|}, \tag{10}$$

$$T = \frac{\dfrac{4b}{1 - a^2}\left|[(\cos\Omega - a^2\cos\Delta)^2 - (1 - a^2)^2]\sin^{-2}(N\xi)\right|}{[(1 + b)\sin\Omega + a(1 - b)\sin\Delta]^2 + \dfrac{4b}{1 - a^2}\left|[(\cos\Omega - a^2\cos\Delta)^2 - (1 - a^2)^2]\cot^2(N\xi)\right|}, \tag{11}$$

where $b = n_0^2/(n_1 n_2)$. We can see that the reflection and transmission coefficients depend on five parameters: a, b, Ω, Δ, N (ξ is a function of a, Ω and Δ). Each of them has clear physical meaning. a characterizes the variation in optical properties from one layer to the next (i.e. the amount of optical "modulation"). It depends only on the ratio n_1/n_2, and for common dielectrics (dielectrics with positive ε) $-1 < a < 1$. b characterizes the Fresnel interaction between an electromagnetic wave and the boundaries of the structure, $\Omega = k_0(n_1 d_1 + n_2 d_2)d/d = k_0 n d$ is a period-average dimensionless wave vector of the light inside the structure, $\Delta = k_0(n_2 d_2 - n_1 d_1)$ is the difference in optical paths of the wave inside each layer, and N is the number of periods.

The analysis of the dispersion equation allows one to establish that, there are two physically different regions of parameters for our structure in which behavior of reflection and transmission coefficient is quite different. In the first, the $\xi_{1,2}$ are real ($|\cos(\xi)| \leq 1$), and the function $R = R(\Omega)$ is oscillatory with amplitude depending on b (the increasing of Fresnel interaction on the boundaries makes the amplitude of oscillations higher). Such regions are called allowed regions. In the second, the $\xi_{1,2}$ are complex ($|\cos(\xi)| \geq 1$), and R almost reaches unity if the number of periods N is sufficiently great. Such regions are called forbidden regions. Physically, the accumulated Fresnel reflection from inhomogeneities of the dielectric permittivity $\varepsilon(z)$ with the periodicity of the structure results in an increase of the amplitude of the backward Floquet-Bloch wave at the expense of the forward wave, leading to an increase of the reflection coefficient.

The widths and places of forbidden (or allowed) regions on the scale Ω are determined by a and Δ. With the aid of (9) we have

$$\begin{aligned} \Delta\Omega_{odd} &= 2\arccos(1 - 2a^2\cos^2(\tfrac{\Delta}{2})), \\ \Delta\Omega_{even} &= 2\arccos(1 - 2a^2\sin^2(\tfrac{\Delta}{2})), \end{aligned} \tag{12}$$

where $\Delta\Omega_{odd}$ and $\Delta\Omega_{even}$ are the widths of odd (with the centers $\Omega = (2l+1)\pi, l = 0, 1, 2...$) and even (with the centers $\Omega = 2l\pi$) forbidden regions.

ELASTIC STRESS INFLUENCE ON THE REFLECTION AND TRANSMISSION

The parameters of the two-layered periodic dielectric structure which define the propagation of the electromagnetic wave are the widths of the basic layers d_1 and d_2 and their refraction indexes n_1 and n_2. Each of them can be changed by creating an elastic stress according to formulae

$$\delta d_{1,2} = S_{1,2} d_{1,2}, \quad \delta n_{1,2} = -1/2 n_{1,2}^3 p_{1,2} S_{1,2}, \tag{13}$$

where $S_{1,2}$ are the stresses in the layers $n_{1,2}$, $p_{1,2}$ are the elastooptic coefficients in those layers, which depend on the material and directions of the stress and electromagnetic wave propagation. The variations in $n_{1,2}$ and $d_{1,2}$ cause the changes in Ω, and Δ for the wave with fixed wavelength λ

$$\begin{aligned} \delta\Omega &= k_0(n_2\delta d_2 + n_1\delta d_1 + \delta n_2 d_2 + \delta n_1 d_1), \\ \delta\Delta &= k_0(n_2\delta d_2 - n_1\delta d_1 + \delta n_2 d_2 - \delta n_1 d_1). \end{aligned} \tag{14}$$

Of course, there are also changes in a and b, but they are respectively not as big as $\delta\Omega$ and $\delta\Delta$. As a result, for the fixed wavelength λ_0 ($k_0 = \frac{2\pi}{\lambda_0}$), reflection and transmission coefficients (10, 11) are also changed. In some cases the difference between the unstressed coefficients and the stressed coefficients can be 70%-90%. In order to get such a difference we should have sufficient stress to shift λ_0 from a forbidden to an allowed region. For creation of the stress inside the structure we can use, for example, the modulation of the structure by an acoustic wave, piezoelastic properties of the materials and so on. From a mathematical point of view, the simplest case for analytical description is to create a constant stress in each layer applying a constant compression force to the boundaries of the structure.

Let us demonstrate that even small variations (less than 0.1%) of the structure parameters may cause a great change in the reflection coefficient (10). As the first example we consider the two-layered periodic structure which consists of $N = 15$ periods of polystyrene[11] ($n_1 = 1.59$) with $d_1 = 7.5\mu m$ and chlorotellurite glass[15] ($n_2 = 2.00$) with $d_1 = 5.9\mu m$. Outside the structure there is also polystyrene ($n_0 = 1.59$). Figure 2 represents the reflection coefficient dependence on λ_0. We can see that for $\lambda_0 = 0.6328\mu m$ (He-Ne laser) the reflection coefficient almost equals 100% (no transmission).

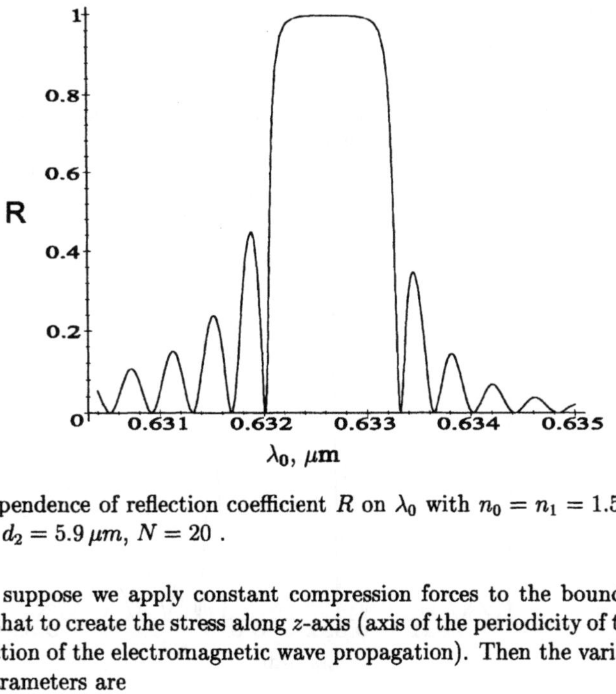

Figure 2: Dependence of reflection coefficient R on λ_0 with $n_0 = n_1 = 1.59$, $n_2 = 2.0$, $d_1 = 7.5\,\mu m$, $d_2 = 5.9\,\mu m$, $N = 20$.

Now suppose we apply constant compression forces to the boundaries of the structure so that to create the stress along z-axis (axis of the periodicity of the structure and the direction of the electromagnetic wave propagation). Then the variations of the structural parameters are

$$\delta d_1 = \frac{T_3}{c_{11}^{(1)}} d_1\,, \quad \delta d_2 = \frac{T_3}{c_{11}^{(2)}} d_2\,, \quad \delta n_1 = -\frac{1}{2} {n_1}^3 p_1 \frac{T_3}{c_{11}^{(1)}}\,, \quad \delta n_2 = -\frac{1}{2} {n_2}^3 p_2 \frac{T_3}{c_{11}^{(2)}}\,, \qquad (15)$$

where $c_{11}^{(1,2)}$ are stress coefficients of polystyrene and chlorotellurite glass, $p_{1,2}$ their elastooptic coefficients, T_3 is a traction force (force per unit area) along z-axis. Let $T_3 = -30\,N/mm^2$, then $\delta d_1 = -0.039\,\mu m$, $\delta d_2 = -0.004\,\mu m$, $\delta n_1 = 0.0032$, $\delta n_2 = 0.0003$ and the reflection coefficient is plotted on Figure 3. We can see now that for $\lambda_0 = 0.6328\mu m$ R is less than 10%, i.e. the reflection is decreased in ten times (in fact we have almost full transmission).

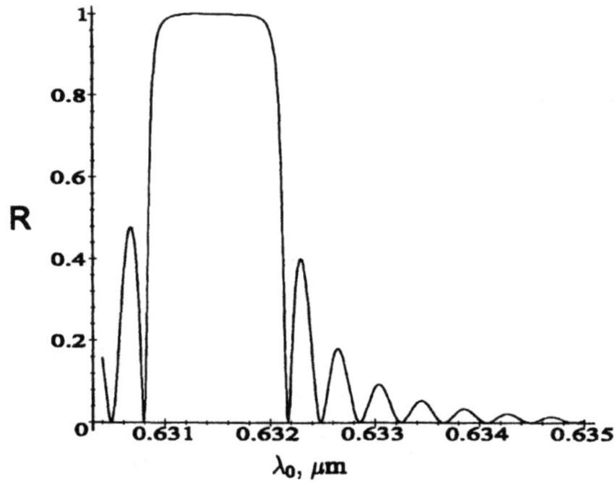

Figure 3: Dependence of reflection coefficient R with λ_0 under $n_0 = 1.59$, $n_1 = 1.59 + 0.0032$, $n_2 = 2.00 + 0.0003$, $d_1 = 7.5 - 0.039\,\mu m$, $d_2 = 5.9 - 0.004\,\mu m$, $N = 20$.

As the second example we consider the structure from $N = 40$ periods of fused silica[16] ($n_1 = 1.457$) and flint glass[16] ($n_2 = 1.616$) which is surrounded by the flint glass itself ($n_0 = 1.616$). Let $d_1 = 5.1\mu m$ and $d_2 = 4.6\mu m$. Then the reflection coefficient dependence on λ_0 is plotted on Figure 4. We can see again that $R(0.6328\mu m) \approx 1$.

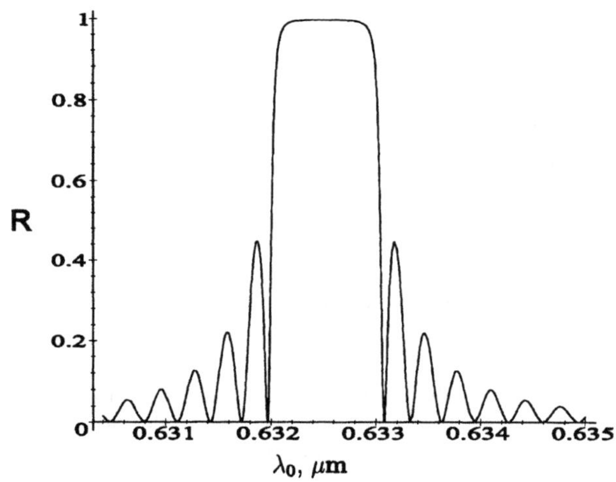

Figure 4: Dependence of reflection coefficient R on λ_0 with $n_0 = n_2 = 1.616$, $n_1 = 1.457$, $d_1 = 5.1\mu m$, $d_2 = 4.6\mu m$, $N = 40$.

If we apply traction force $T = -70\,N/mm^2$ to the boundrary of the structure along z-axis and corresponding forces along x and y-axis (in order to have stress just along z-axis) the variations of the structural parameters will be $\delta d_1 = -0.005\,\mu m$, $\delta d_2 = -0.007\,\mu m$, $\delta n_1 = 0.0004$, $\delta n_2 = 0.0008$ and the reflection coefficient $R\,(0.6328\,\mu m) \approx 0$ (Figure 5) , i.e situation is exactly same as in previous example. Before stress to be applied - no transmission. When stress is applied - almost full transmission.

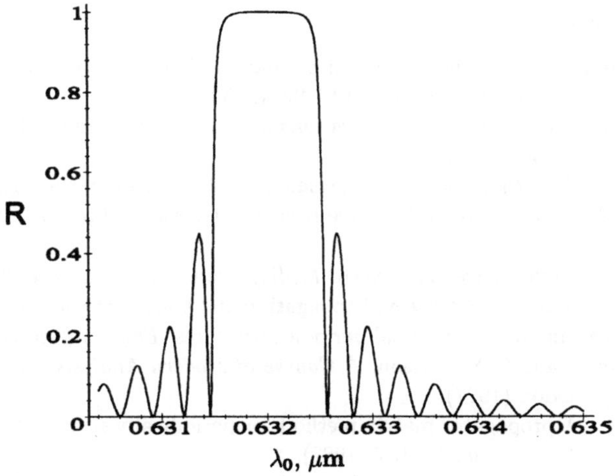

Figure 5: Dependence of reflection coefficient R on λ_0 with $n_0 = 1.616$, $n_1 = 1.457 + 0.0004$, $n_2 = 1.616 + 0.0008$, $d_1 = 5.1 - 0.005\,\mu m$, $d_2 = 4.6 - 0.007\,\mu m$, $N = 40$.

CONCLUSION

We have applied Floquet-Bloch theory to the well-known problem of the electromagnetic wave propagation through a two-layered periodic dielectric structure, using an exact analytical method. The main idea of this method is to represent the solution for each Floquet-Bloch wave inside each basic layer as a sinusoidal function. This allows us to determine general expressions for the reflection and transmission coefficients in a more physically transparent form. Using the results of this analysis we found extremely high sensitivity of the reflection and transmission to the parameters of the structure. It was shown the possibility to change them with the aid of the creation of elastic stress inside the structure. As the simplest example we considered the influence of constant strain along the axis of periodicity on the reflection and transmission of the electromagnetic wave and demonstrated the possibility of changing them up to 90% in some cases.

REFERENCES

1. R.Kronig and W.G. Penney, Quantum mechanics of electrons in crystals, *Proc. Roy. Soc. (London)* **A 130**:499 (1930).
2. N.W. McLachlan, *Theory and Applications of Mathieu Functions,* Oxford Univ. Press, London (1951).
3. L. Brillouin, *Wave Propagation in Periodic Structures,* Dover, New York (1953).
4. C. Elachi, Waves in active and passive periodic structures, *Proc. IEEE* **64**:1666 (1976).
5. L.M. Brekhovskikh, *Waves in Layered Media,* Academic Press, New York (1980).
6. S.Yu. Karpov and S.N. Stolyarov, Propagation and transformation of electromagnetic waves in one-dimensional periodic structures, *Physics-Uspekhi* **36**:1 (1993).
7. E.T. Whitteker and G.N. Watson, *A Course of Modern Analysis,* Cambridge Univ. Press, London (1927).
8. K.E. Gilbert, A propagator matrix method for periodically stratified media, *J. Acoust. Soc. Am.* **73**:137 (1982).
9. H. Kogelnik, Coupled wave theory for thick hologram grating, *Bell Syst. Tech. J.* **48**:2909 (1969).
10. A. Yariv and P. Yeh, *Optical Waves in Crystals,* John Wiley & Sons, New York (1984).
11. M. Born and E. Wolf, *Principles of Optics,* Pergamon Press, Oxford (1975).
12. M.V. Fedoryuk, *Ordinary Differential Equations,* Nauka, Moscow (1985).
13. G.V. Morozov, R.G. Maev, and G.W.F. Drake, Exact analytic expression for the reflection of the electromagnetic wave from a two-layered periodic dielectric structure, *Kvant. Elektron.* **25**:1 (1998).
14. L.M. Levin, Propagation of plane electromagnetic wave through layered periodic medium, *Zh. Tekh. Fiz.* **18**:1399 (1948).
15. I. Abdulhalim, G.N. Pannel, J. Wang, G. Wylangovski, and D.N. Payne, Acousto-optic modulation using a new chlorotellurite glass, *J.Appl.Phys.* **75**:519 (1993).
16. R.W. Dixon, Photoelastic properties of selected materials and their revelance for applications to acoustic light modulators and scanners, *J.Appl.Phys.* **38**:5149 (1967).

A NEW SYSTEM FOR QUANTITATIVE ULTRASONIC BREAST IMAGING OF ACOUSTIC AND ELASTIC PARAMETERS

M. Krueger[1], A. Pesavento[1], H. Ermert[1],
K. M. Hiltawsky[1,2], L. Heuser[2], H. Rosenthal[3], A. Jensen[3]

[1] Dept. of Electrical Engineering,
Ruhr University, 44780 Bochum, Germany
[2] Dept. of Radiology and Nuclear Medicine,
University Hospital 44892 Bochum-Langendreer, Germany
[3] Dept. of Gynaecology and Obstetrics
University Hospital 44892 Bochum-Langendreer, Germany

INTRODUCTION

Early detection of breast cancer is essential: approximately one in ten women in industrialized countries are affected by breast cancer. Although various groups developed concepts and systems for quantitative ultrasonic breast imaging, ultrasound has been only a secondary diagnostic tool for breast cancer detection up to now. Currently, X-ray mammography is used for screening supported by palpation. If suspicious lesions are detected, free-hand ultrasonic examination and needle biopsy provide further diagnostic information. Ultrasonic examination of the breast requires an experienced physician. Results are not always reproducible and the acquired images are not compatible with X-ray mammography. Furthermore, ultrasonic B-mode images provide morphological information about the reflectivity which is not always well correlated to the tissue type.

To support the early detection of breast cancer, a method to measure the spatial distribution of acoustic[3,4,5] and elastic[6] parameters can provide important information. We will introduce an ultrasonic system which measures the spatial distribution of the sound velocity, attenuation, and elasticity using "CARISMA" (Computer Assisted Reconstructive IMAGING by Sonographic and Mechano-Elastic Analysis). The acquired 3D data sets are compatible with X-ray mammography, since the breast is compressed in the same manner and all obtained images have a well defined orientation. In the following we will explain the transmission tomography algorithm, which reconstructs sound velocity c and attenuation α, the strain imaging concept, which provides elasticity information, and the ultrasonic system, which was implemented to acquire the raw data. Phantom and *in vivo* results demonstrate the potentials of the system and the CARISMA algorithms.

Acoustical Imaging, Volume 24.
Edited by Hua Lee, Kluwer Academic/Plenum Publishers, 2000.

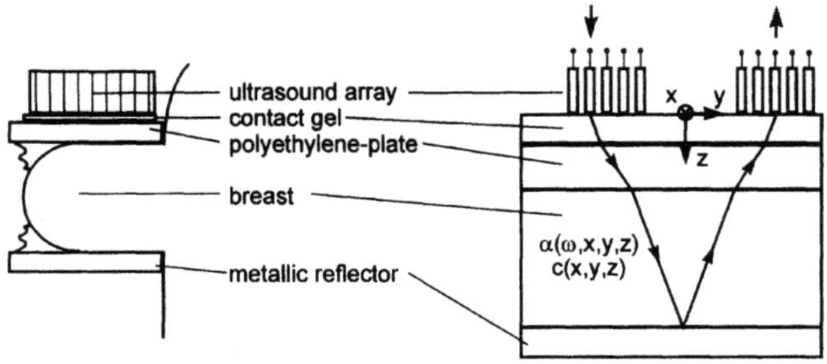

Figure 1. Compression of the breast and time-of-flight tomography concept

TRANSMISSON TOMOGRAPHY

The ultrasonic transmission data are acquired using a linear array and a metallic reflector. As presented in Figure 1 the breast is compressed between two plates of polyethylene and of metal, respectively. The breast is imaged through the upper polyethylene plate in a multistatic single element operating mode of the array. One array element transmits a short ultrasonic pulse and the echoes caused by the lower metallic reflector which are measured by all other elements after penetrating the breast two times provide the tomographic raw data.

In [1,2] we demonstrated, that a ray-tracing approach is adequate to describe the relationship between the acoustic parameters and the echo signals if refraction caused by the planar interfaces of the polyethylene plane is considered. Furthermore, shear wave modes can be neglected, since they are significantly slower than the longitudinal wave modes. This leads to a linear relationship between the attenuation α and the insertion loss A as well as between the "slowness" $s = 1/c$ and the time-of-flight τ. This relationship can be expressed using a matrix equation. Inversion of the equation provides the material parameters slowness and attenuation.

In contrast to earlier concepts[7,8,9], which perform a transmission tomography with a ring aperture around the breast, the directions of the transmission measurements do not cover the complete angular range 0 ... 2π. Consequently, the resolution of our tomographic concept will be anisotropic. Structure with planar horizontal layers cannot be reconstructed, since a measurement parallel to the compression plates is missing. This leads to a singular matrix equation. The equation can be solved using singular value decomposition (SVD)[12] if a sufficient number of measurements is acquired. Even though the SVD provides the optimal reconstruction result, artifacts have to be taken into account.

These problems do not occur in the concepts using a circular aperture. However our concept has some important advantages. It can be applied using modified sonographic standard equipment and the 3D data are compatible with X-ray mammograms.

A more comprehensive description of the tomographic concept has been presented before[1,2].

STRAIN IMAGING CONCEPT

We controlled the compression of the breast with a stepper motor. Thus, raw data for strain imaging concepts could be acquired. The array was moved mechanically in elevation direction. We acquired complex base-band data of several cross-sectional areas of the breast under two different compression levels. Using a correlation analysis the axial displacement and hence the axial strain can be estimated. The resulting strain images provide information about the spatial distribution of the elasticity. However, not the elastic modulus itself but the strain is measured.

Optical Flow

Strain images suffer from noise of the echo data but even more from decorrelation noise. Decorrelation noise is caused by transversal motion[10] and by the strain itself[11]. To address the latter problem, temporal stretching can be used. In a first approach we focus on the problem of transversal displacement.

We used a computational efficient technique know as "optical flow"[13] to estimate the transversal (lateral and out of plane) displacement. This technique is based on the assumption that in case of compression pixels do move but do not change their brightness. By estimating the local gradient of the pre- and post-compression images the spatial distribution of the three-dimensional displacement can be estimated.

With the transversal components of the displacement vectors a new trajectory passing axially and transversally moved scatterers is defined. We used bilinear interpolation to obtain the required echo data. The cross-correlation between the pre-compression echo signal and the corrected post-compression echo signal allows a better estimate of the displacement.

Phase Root Seeking

Furthermore, we developed a time-efficient method to estimate the axial displacement. An approved estimate of the axial displacement is defined by the maximum of the cross-correlation function of the pre- and post-compression echo signals[11]. An alternative approach proposes to evaluate the phase of the cross-correlation function of the complex base-band signals assuming a linear relationship between the phase difference of the signals and the displacement[14]. This approach is faster than finding the maximum of the cross-correlation function. However, the technique suffers from aliasing unless the displacements are very small.

Our method uses the phase of the complex cross-correlation as done in [14]. We found out that for the first window, the method described in [14] gives a rough estimate of the phase root of the cross-correlation function. The phase is zero at the maximum of the cross-correlation function. Consequently, more accurate results are obtained by seeking the exact phase root. We use a modified Newton iteration to find the root. The phase is approximated by a linear function with the slope $2k_0$ (k_0: centroid wave number), as suggested in [14], but more than one iteration is performed. After 2-3 iterations, the difference of the estimation and the limit of the iteration is less than one percent of the sampling time. We use the result as a starting value for the next window. To consider roots between two samples a linear interpolation is used. A more detailed description of the method is provided in [15].

DATA ACQUISITION SYSTEM

Applicator

We implemented an applicator which compresses the breast and moves the ultrasonic array into elevation direction in order to obtain 3D data sets. The lower compression plate consists of polished stainless steel, the upper compression plate is made

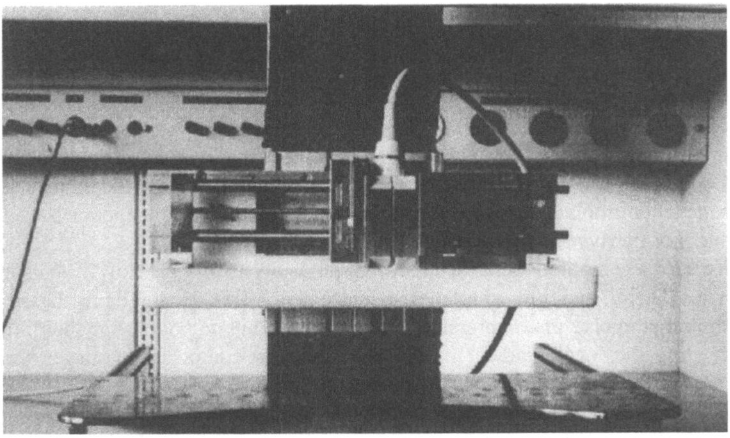

Figure 2: Photo of the applicator (view from the patient's position)

of polyethylene. A stepper motor controls the compression. On the top of the upper compression plate a second stepper motor system moves the transducer array. Figure 2 presents a photo of the applicator.

Ultrasonic System

We used a Siemens Sonoline™ Elegra in the "small parts" operating mode. The probe was a 7.5 MHz linear array. We modified the beam former in order to acquire single element echo data. After a B-scan is recorded, up to 128 frames with different single element transmit and receive apertures are acquired to obtain tomographic raw data. Up to now the acquisition of complex or rf single element data is not possible. Therefore, an estimation of α is not presented in this paper.

For elastographic measurements two levels of compression are applied. The overall strain was 0.5 %. Complex base-band echo signals and the B-scans were acquired. The centroid frequency was 7.2 MHz and the fractional bandwidth of the echo signals was nearly 80 %. The B-scans were stored using a sampling rate of 18 MHz while lateral spacing was 0.16 mm. The data were used for the optical flow estimation The complex base-band echo signals with a 12 MHz sampling rate and a lateral spacing of 0.63 mm were obtained from the color flow signals of the ultrasound system and were used for the axial displacement estimation.

RESULTS

Transmission Tomography

We applied our reconstruction method to several phantom measurements. Our tissue mimicking phantom (CIRS Inc., Norfolk, VA) contains several lesions with higher sound velocity in a homogeneous environment as was found by transmission measurements. A typical reconstruction of a plane with two conspicuous lesions is presented in Figure 3. The most left lesion which was hardly detectable using real time B-scan ultrasound

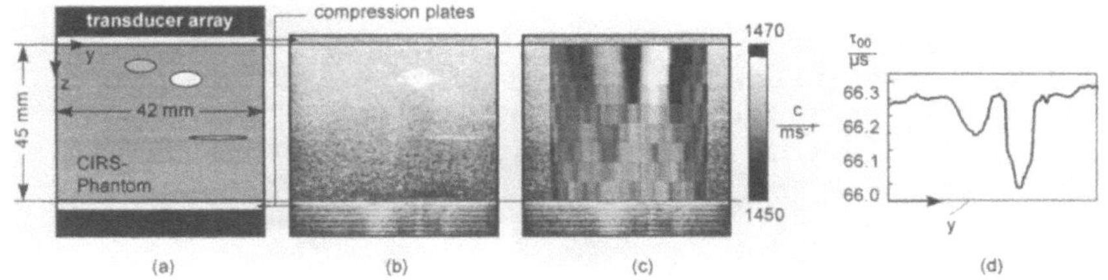

Figure 3: Reconstruction results of a tissue mimicking phantom: (a) sketch of the imaged plane of the phantom (the extension and the gray scale of the lesions provide qualitative information only), (b) B-scan of the phantom, (c) reconstruction of the sound velocity of the phantom (d) axial projection of the sound velocity (presented as time-of-flight).

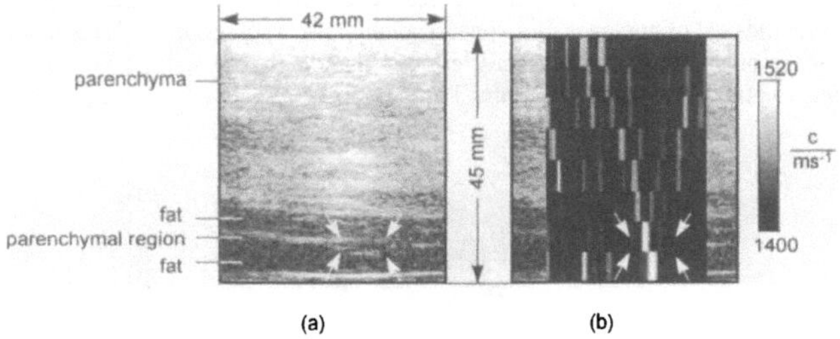

Figure 4: Reconstruction results of a normal breast of a 55 year old women: (a) B-scan of the compressed breast. Inside the fat layer a parenchyma layer with varying thickness is visible. The region with maximal thickness is indicated by arrows. (b) reconstruction of the sound velocity, near the region of interest a velocity peak is found. The arrows are located at the same position as in Figure 4a.

becomes visible in the velocity image. The coarse axial resolution as well as the lateral side lobes are caused by the limited range of projections.

Furthermore, an *in vivo* measurement was performed. The breast of a 55 year old woman had a fat layer with some parenchymal components. The basic structure of the breast tissue is a horizontally layered one (Figure 4). The layers cannot be reconstructed, since a transversal time-of-flight measurement was not performed. However, a spatially isolated parenchymal lump is detected by the tomographic method. The reconstructed velocity distribution looks reasonable. Further clinical evaluation is planned.

A more effective presentation of the results, which is not used in this paper, can be obtained using a B-scan stained by a color-coded representation of the reconstructed velocity.

Strain Imaging

We built an agar agar based phantom to verify our elastography algorithm. A description and a sketch of the phantom are presented in Figure 5. In the three-dimensional strain image the hardness of the spherical lesions appears clearly.

Furthermore, we imaged a breast *in vivo* having a lipoma. Capsule and septa of the lipoma give rise to enhanced strain in the adjacent tissue. However, not all septa can be detected by this method. Using optical flow, the contrast of the lower right septum is enhanced. B-scan and strain images of the phantom are presented in Figure 6.

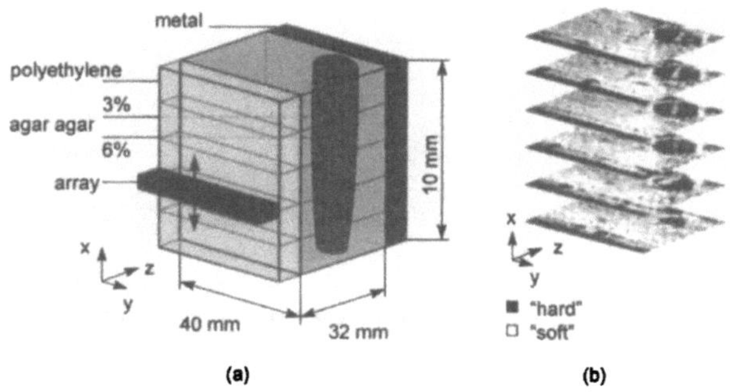

Figure 5: 3D strain image of a phantom (anisotropic scaling). The phantom consisted of agar agar with a concentration of 3 % (30 g per liter of water). Three spherical and "hard" lesions possess a higher concentration (6 %). Silica gel of homogeneous concentration in the whole phantom leads to an isoechoic B-scan. (a) Sketch of the phantom, the largest hard lesion is darker. (b) 3D strain image of the phantom. . Notice the different orientation compared to Figures 1 and 2.

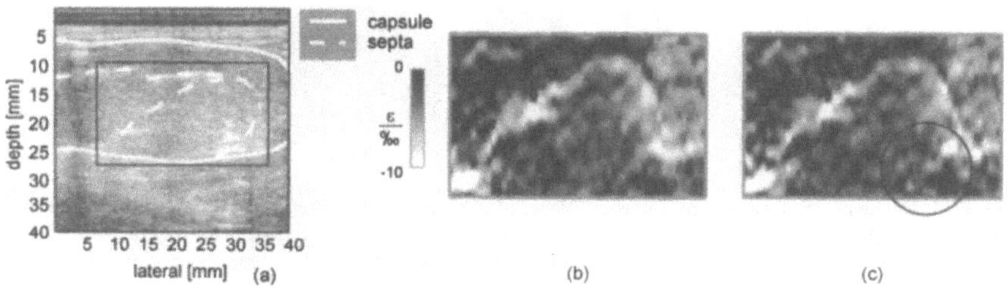

Figure 6: Strain image of a breast with lipoma: (a) B-scan with lines indicating approximate location of capsule and septa in the image plane (drawn by a physician), (b) strain image using phase root seeking, (c) strain image using optical flow and phase root seeking: the contrast of the lower right septum is enhanced.

CONCLUSION

Using a commercially available ultrasound scanner, a PC, and a mechanical applicator, a new system to improve ultrasonic tissue characterization of the female breast was implemented. We introduced a new concept "CARISMA" to obtain the spatial distribution of acoustic and elastic parameters using our system. Limited angle transmission tomography using a linear array and a plane reflector leads to artifacts and to a limited axial resolution but it still delivers relevant information in addition to B-scans. In spite of low temporal and spatial resolution strain images are available using time efficient tools: phase root seeking and 3D optical flow. With our system and CARISMA images of acoustic and elastic parameters can be obtained. Phantom results and preliminary *in vivo* results look encouraging.

Attenuation tomography is will also be implemented, as soon as rf single element data are available. Furthermore, we plan to use a transducer array with a larger aperture. For better strain images we want to improve the signal-to-noise ratio as well as spatial and temporal sampling. We intend to implement adaptive temporal stretching to reduce the decorrelation noise more effectively.

Nevertheless, a clinical evaluation of CARISMA is already possible now and is intended for the near future.

ACKNOWLEDGEMENT

We appreciate the support of Siemens Medical Systems, Ultrasound Group, Issaquah, WA, USA.

REFERENCES

1. M. Krueger, A. Pesavento, and H. Ermert, A modified time-of-flight tomography system for ultrasonic breast imaging, in: 1996 *IEEE Ultrason. Symp. Proc.*, IEEE Press, New York, pp. 1381-1385 (1996).
2. M. Krueger, A. Pesavento, and H. Ermert, Breast ultrasound assisted by acoustic velocity reconstruction, in: *Acoustical Imaging*, vol. 23, S. Lees and L. A. Ferrari, eds., Plenum Press, New York (1996).
3. F.S. Foster, M. Strban, and G. Austin, The ultrasound macroscope: initial studies of breast tissue, *Ultrasonic Imaging* 6:243-261 (1984).
4. K. Richter. Clinical Amplitude/Velocity Reconstructive Imaging (CARI) - A New Sonographic Method for Detecting Breast Lesions. *Brit. J. Radiol.*, 68: 375-384, 1995.
5. K. Richter et al., Detection of malignant and benign breast lesions with an automated us system: results in 120 cases, *Radiology* 205:823-830 (1997).
6. B.S. Garra, E.I. Cespedes, J. Ophir, S.R. Spratt, R.A. Zuurbier, C.M. Magnant, and M.F. Pennanen, Elastography of breast lesions: initial clinical results, *Radiology* 202(1):79-86 (1997).
7. J.F. Greenleaf, S. A. Johnson, W. F. Samoya, and F. A. Duck, Algebraic reconstruction of spatial distributions of acoustic velocities in tissue from their time-of-flight profiles, in: *Acoustic Holography*, vol. 6, Both and Newell,. eds., Plenum Press, New York, pp. 307-326 (1975).
8. J.F. Greenleaf, J. Ylitalo, and J. Gisvold, "Ultrasonic Tomography for Breast Examination," *IEEE Eng. Med. & Biol. Mag.*, pp. 27-32 (Dec. 1987).
9. J.R. Jago, T.A. Whittingham, Experimental studies in transmission ultrasound computed tomography. *Phys. Med. Biol.* 35:1515-1527 (1991).
10. F. Kallel and J. Ophir, Three-dimensional tissue motion and its effects on image noise in elastography, *IEEE Trans. Ultrason. Ferroelect. Freq. Contr.* 44(6):1286-1296 (1997).
11. K. Alam, J. Ophir, and E.E. Konofagou, An adaptive strain estimator for elastography," *IEEE Trans. Ultrason, Ferroelect., Freq. Contr.* 45(2):461-472 (1998).
12. W.H. Press, S.A. Teukolsky, W.T. Vetterling, and B.P. Flannery, *Numerical Recipes in C*, Cambridge University Press, Cambride (1992).
13. M. Bertrand, J. Meunier, M. Doucet, and G. Ferland, Ultrasonic biomechanical strain gauge based on speckle tracking, in: *1989 IEEE Ultrason. Symp. Proc.*, IEEE Press, New York, pp. 859-863 (1989)
14. A. Cohn, S.Y. Emelianov, M.A. Lubinski, and M. O'Donnell, An Elasticity Microscope. Part I: Methods, *IEEE Trans. Ultrason. Ferroelect. Freq. Contr.*, vol. 44(6), pp. 1304-1319 (1997).
15. A. Pesavento, C. Perrey, M. Krueger, and H. Ermert, A time-efficient and accurate strain estimation concept for ultrasonic elastography using iterative phase zero estimation, submitted to *IEEE Trans. Ultrason. Ferroelect. Freq. Contr.* (1998).

DETERMINATION AND EVALUATION OF THE SURFACE REGION OF BREAST TUMORS USING ULTRASONIC ECHOHRAPHY

Xiangyong Cheng, Iwaki Akiyama[1], Syuji Ogawa[1]
Kouichi Itoh[2], Kiyoka Omoto[2], Yi Wang[2], and Nobuyuki Taniguchi[2]

Imaging Systems, Mitani Sangyo Co., Ltd.
1-16-3 Shinkawa, Chuo-ku, Tokyo, 104-0033 Japan
[1]Department of Electrical Engineering, Shonan Institute Of Technology
1-1-25 Nishikaigan, Tsujido, Fujisawa, 251-0046 Japan
[2]Department of Clinical Pathology, Jichi Medical School
3311-1 Yakushiji, Minamikawauchi, Tochigi, 329-0498 Japan

INTRODUCTION

The authors have developed a breast tumor diagnosis system using fuzzy reasoning and three-dimensional ultrasonic imaging techniques[1,2]. In this system, the volume data of the breast tumor is obtained by integrating the probe's position and orientation information with the grayscale cross-sectional images using three-dimensional coordinate transformation and image interpolation. An alternating magnetic sensor attached to the ultrasonic probe tracks its position and orientation. The surface region of the tumor is extracted using fuzzy reasoning, and the tumor is evaluated according to the roughness of the extracted tumor surface. However, as the shape of the tumor surface becomes more complex, it becomes more and more difficult to accurately extract the region. In particular, making the automatically extracted region agree with the manually traced contours is a significant problem.

This paper proposes approaches that use fuzzy reasoning to automatically and consistently extract the breast tumor in the ultrasound data and then evaluate its surface roughness for distinguishing between benign and malignant tumors. The method consists of three stages of processing. In the first stage, the breast volume data is roughly segmented by a three-dimensional Laplacian of Gaussian (LoG) filter. Three types of segmented regions are obtained and labeled as "tumor", "normal tissue" and "boundary". The fuzzy membership functions are then automatically generated by Rician approximation of the histogram of each fuzzy feature parameter in the three-labeled regions. In the second stage of processing, the tumor surface region is extracted by fuzzy reasoning followed by a

Acoustical Imaging, Volume 24.
Edited by Hua Lee, Kluwer Academic/Plenum Publishers, 2000.

defuzzification based on relaxation techniques. In the third stage, the surface roughness of the extracted tumor is evaluated. In this paper, two parameters are investigated; a ratio of the surface area to the volume, and a standard deviation of the distances from the center of gravity of the tumor region to the surface points.

The algorithm has been evaluated using thirty-two cases of malignant and eleven cases of benign tumors, and the results demonstrate the efficient and reliable performance of the proposed scheme. It is shown that the proposed method has the potential to significantly improve breast cancer screening using ultrasound echography.

VOLUME DATA ACQUISITION

Figure 1 shows our three-dimensional ultrasound image acquisition system[2]. It consists of commercially available ultrasound diagnostic equipment (SSD-2000, Aloka, Japan) with a 7.5 MHz, concave-type, electronic scanning probe (Aloka, Japan); a magnetic position sensor (Fastrak, Polhemus, US); PCI Bus Frame Grabber (DT3155, Data Translation, US); and PC (Dell PentiumII 400MHz, RAM 256MB). The position sensor is attached to the probe, and tracks its position and orientation. A volume of ultrasound data is acquired by slowly sweeping (10-15 seconds) the probe over the area. Both the cross sections, and probe's position and orientation data are transferred to PC at a transmission rate of 20 frames per second. The volume data is then constructed using a three-dimensional coordinate transformation and a linear interpolation techniques. Figure 2 shows an original cross section (a) and a slice of the constructed volume data (b) for a breast cancer case.

BREAST TUMOR DETERMINATION AND EVALUATION

A feature of the breast tumor is that it is displayed in the ultrasonic images as being darker than the surrounding normal tissue[3]. Our method focuses on this feature to extract the tumor surface region. In order to consider the context of the images, we classify the breast ultrasound data into three types of regions (tumor, normal tissue and boundary) using a fuzzy logic based approach.

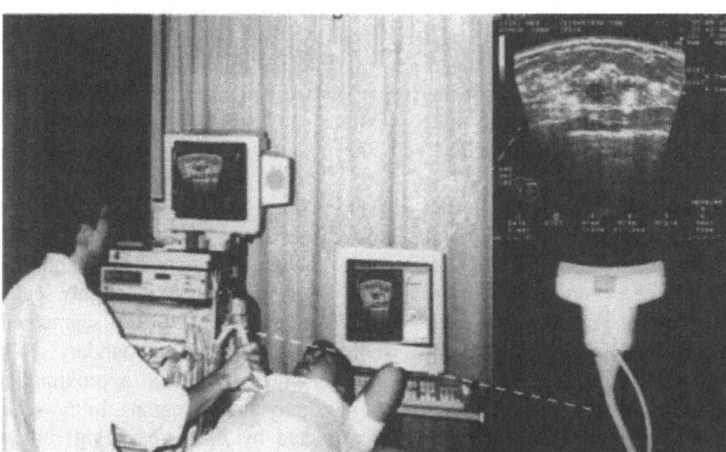

Figure 1. The three-dimensional ultrasound image acquisition system in our approach.

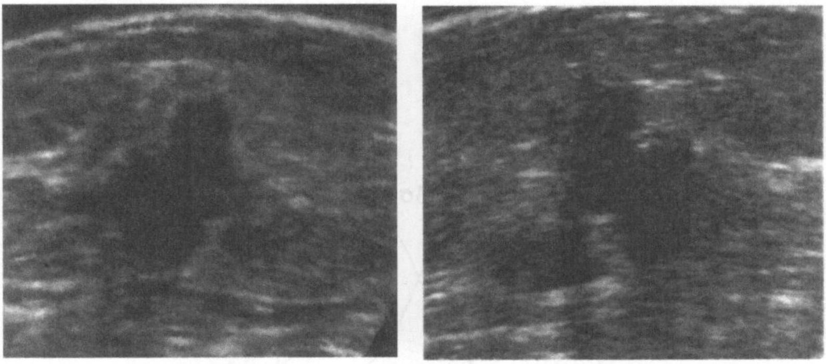

(a) (b)

Figure 2. (a) A B-scan ultrasonic image of breast cancer. (b) A slice of the volume data.

A. Membership Function Generation

Ultrasound images are usually degraded by speckle noise[4,5] and various kinds of artifacts due to the phenomena of reflection, refraction, etc. In addition, the image brightness varies with changes in the circumstances of the patient mammary gland and the ultrasound equipment settings. Because of these reasons, conventional image processing methods[6-14], which are used in CT, MRI or X-ray Mammography, cannot be straightforwardly applied to ultrasound data. Therefore, in order to accurately determine the tumor surface region, we propose a fuzzy logic based approach. In this approach, fuzzy logic membership functions are automatically generated for the ultrasound data by three-dimensional Laplacian of Gaussian (LoG) filtering[10]. During the process of the fuzzy reasoning, three kinds of fuzzy feature parameters for specifying the three classes, "tumor", "normal tissue" and "boundary" are employed. All the feature parameters are computed in a reference volume of 7×7×7 voxels (each voxel is approximately $0.014\times0.014\times0.014$ cm^3). The three parameters are as follows:

(a) u: brightness average,
(b) d: distance from the center of intensity gravity to the center of reference volume,
(c) v: brightness dispersion.

Three types of outputs (positive value, negative value and zero-crossings) can be obtained by applying three-dimensional LoG filtering to the volume data. The three outputs in the breast tumor ultrasound data have the following significance: Regions with the positive values correspond to the cavity, a darker area that has a higher possibility attributable to "tumor"; Regions with the negative values correspond to "normal tissue", a brighter area; and the zero-crossings correspond to "boundary".

A probability density function for the brightness of the ultrasound image can be defined as the Rician density function[15-17] ((1)).

$$P_A(x) = x / \sigma^2 \exp\{ -(x^2 + s^2) / 2\sigma^2 \} I_0(xs /\sigma^2) \quad (1),$$

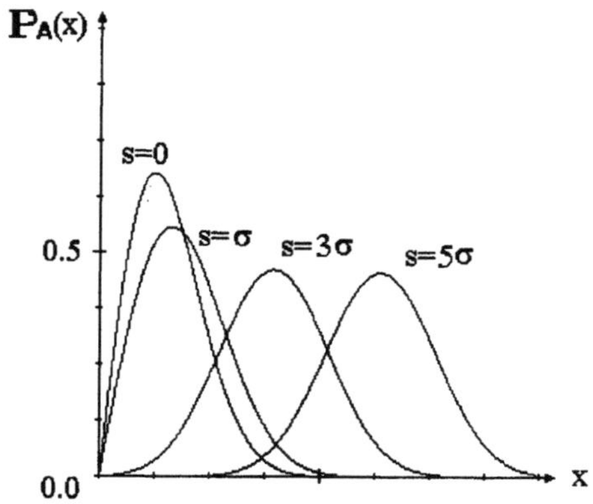

Figure 3. Rician density function.

where, $I_0(x)$ is a modified Bessel function of the first kind, zero order. When s = 0, the Rician is the Rayleigh density function. As s increases, the shape of the Rician changes from that of the Rayleigh density to approximately that of Gaussian density with the mean equal to s (Figure 3). Therefore, we use this characteristic to generate the fuzzy membership function or the characteristic distribution of each feature parameter in the three classes by Rician approximation. That is, Rician approximates each histogram of the feature parameters, u, d, v in the three types of outputs of the three-dimensional LoG filtering.

Figure 4 shows the outline of the fuzzy membership function generation. In the approach, Gaussian approximates the membership function of u for "normal tissue" and "boundary", and Rayleigh approximates the fuzzy membership of u for "tumor" and of d and v for all the three classes.

B. Fuzzy Logic Based Tumor Surface Determination

Figure 5 shows the outline for the fuzzy logic[18,19] based tumor surface determination algorithm[1,2]. First of all, three types of images $\{\mu_t, \mu_n, \mu_b\}$ corresponding to the grade of each voxel attributable to the three classes, "tumor", "normal tissue" and "boundary" are produced by fuzzy reasoning. Those images $\{\mu_t, \mu_n, \mu_b\}$ are then defuzzified into the three classes by a relaxation technique[20-22]. Table 1 shows the fuzzy rule used, where, ∩ and → express the "and" and "then" which are normally used in the fuzzy "if-then" rule, respectively.

Table 1. Fuzzy rule

u		d		v		tumor	normal tissue	boundary
Small	∩	Small	∩	Medium	→	Large	Small	Small
Medium	∩	Large	∩	Large	→	Small	Small	Large
Large	∩	Large	∩	Small	→	Small	Large	Small

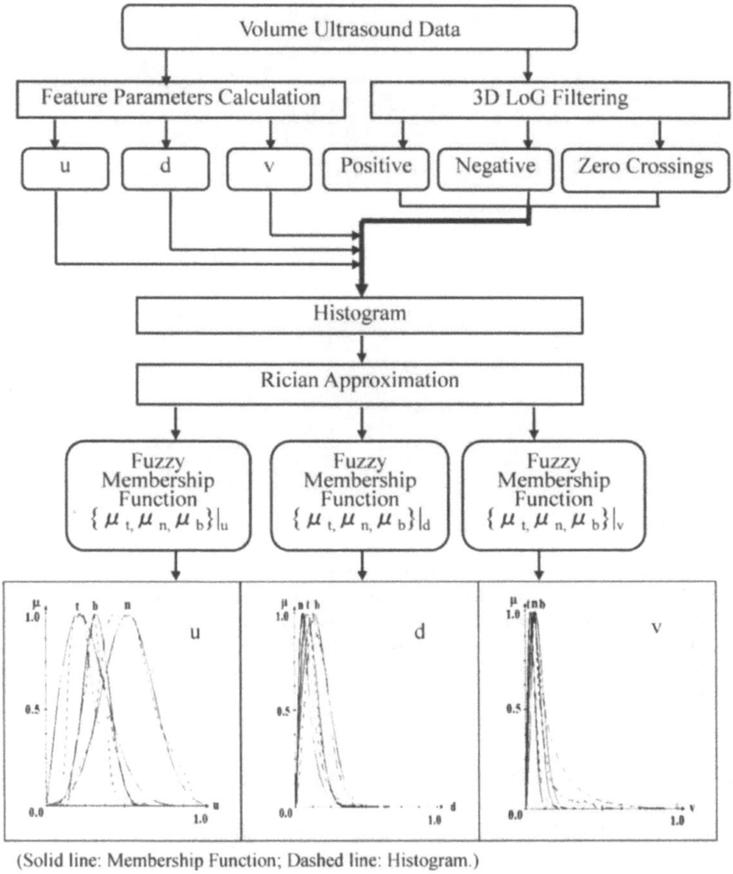

Figure 4. Outline of generation of the fuzzy membership function in the proposed approach.

Since there are gaps and "fuzzy" parts on the fuzzy-reasoning-generated "boundary" region, we proposed a relaxation-based technique to defuzzify the data and determine the final surface region. The constraints used in the iterative operation of the defuzzification are defined as follows.

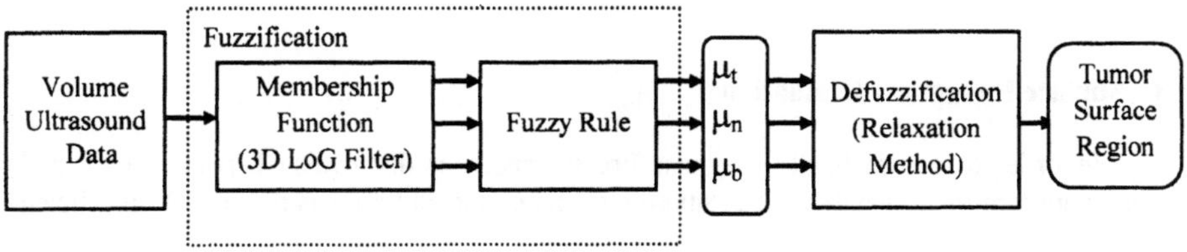

Figure 5. Outline for the fuzzy logic based tumor surface determination algorithm.

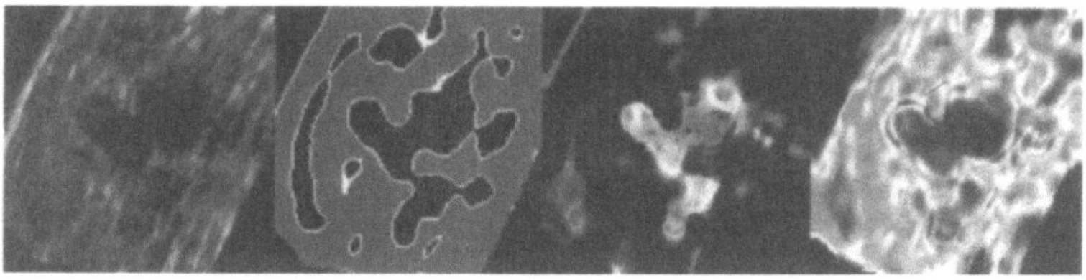

(a) (b) (c) (d)

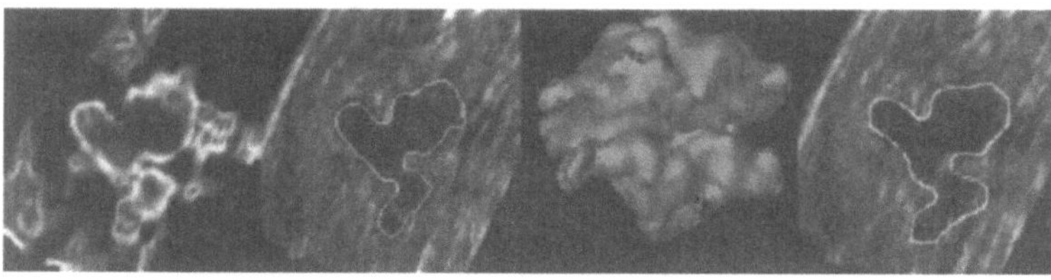

(e) (f) (g) (h)

Figure 6. Procedure of the tumor surface determination method for data in a case of breast cancer. $\sigma = 5$ is used in the three-dimensional LoG filtering, and the number of iterations of the defuzzification is 6. (a) A slice of volume data. (b) Outputs of the three-dimensional LoG filtering. (c),(d),(e) The fuzzy-reasoning-generated images $\{\mu_t, \mu_n, \mu_b\}$. (f) A slice of the determined tumor surface region with the data of (a) superimposed. (g) A rendered tumor surface image. (h) A hand-traced contour for comparison with the results by the proposed method..

(a) The "boundary" voxels (the surface region) must be adjacent to both "tumor" and "normal tissue" voxels.
(b) The "tumor" and "normal tissue" voxels are not adjacent.

The grade of each voxel attributable to the three classes is updated gradually by the above constraints. The operation terminates when the total updated grade is over a given threshold. Figure 6 shows the procedure of the proposed approach for data in a case of breast cancer. (a) is a slice of the volume data. (b) shows the three types of outputs (positive value (dark area), negative value (gray area) and zero-crossings (white lines)) of the three-dimensional LoG filtering. (c),(d),(e) are the fuzzy-reasoning-generated images $\{\mu_t, \mu_n, \mu_b\}$. (f) is a slice of the determined tumor surface region with the data of (a) superimposed. (g) shows a computer rendered surface image of the tumor. (h) was a hand-traced contour, for comparison purposes. It is shown that the results obtained by our method are in good agreement with the hand-traced contour.

C. Surface Roughness Evaluation

In order to automatically diagnose breast cancer or distinguish between begin and malignant tumors, a quantitative evaluation of the extracted tumor is desired. A significant characteristic of malignant tumors is that the surface appears as uneven complex shapes. We define two parameters for evaluating the tumor surface roughness. One is a ratio of the surface area to the volume (*SV ratio*) which is defined as,

$$SV = (\text{surface area}^3 / \text{volume}^2) / 36\pi \qquad (2),$$

where the coefficient $1/36\pi$ is used to normalize the value so that the *SV ratio* equals one when a tumor is a sphere. Another parameter is the standard deviation (*SD*) of the distances from the center of gravity of the tumor region to the surface points.

As the roughness of tumor surface increases, the values of the parameters become larger. The larger the parameter, the higher the possibility of the tumor being malignant.

RESULTS

The proposed algorithm is evaluated using thirty-two cases of malignant and eleven cases of benign tumors. Figure 7 shows a sample of the results. In each sample, the left shows a cross section of the determined surface with the volume data superimposed, and the right is a computer rendered surface image. As shown in the figure 7, both types of tumor ((a)~(c):malignant, (d):benign) surfaces are successfully determined, and the difference between the two types of tumor can be observed in the rendered surface images. As a typical malignant cancer (a), an uneven roughness of the tumor surface can be observed, and the appearance of the cancer invasion into the surrounding tissue is understood to be true. An uneven appearance of the malignant tumor surface ((b),(c)) can also be observed in the rendered surface images. A characteristic of a benign tumor is that the surface is relatively smooth compared to a malignant tumor. This appearance is presented clearly in the result ((d)). The values of the parameter *SV ratio* for (a)~(d) are 7.78, 8.65, 4.0, 1.89, and the *SD* are 0.15cm, 0.21cm, 0.16cm, 0.06cm, respectively. Figures 8 (a) and (b) show the results for the two evaluation parameters.

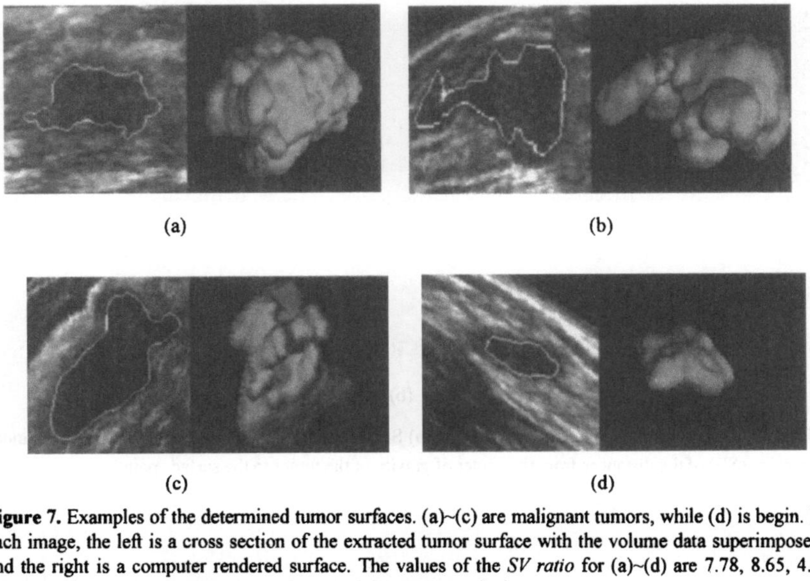

(a) (b) (c) (d)

Figure 7. Examples of the determined tumor surfaces. (a)~(c) are malignant tumors, while (d) is begin. In each image, the left is a cross section of the extracted tumor surface with the volume data superimposed, and the right is a computer rendered surface. The values of the *SV ratio* for (a)~(d) are 7.78, 8.65, 4.0, 1.89, and the *SD* are 0.15cm, 0.21cm, 0.16cm, 0.06cm, respectively.

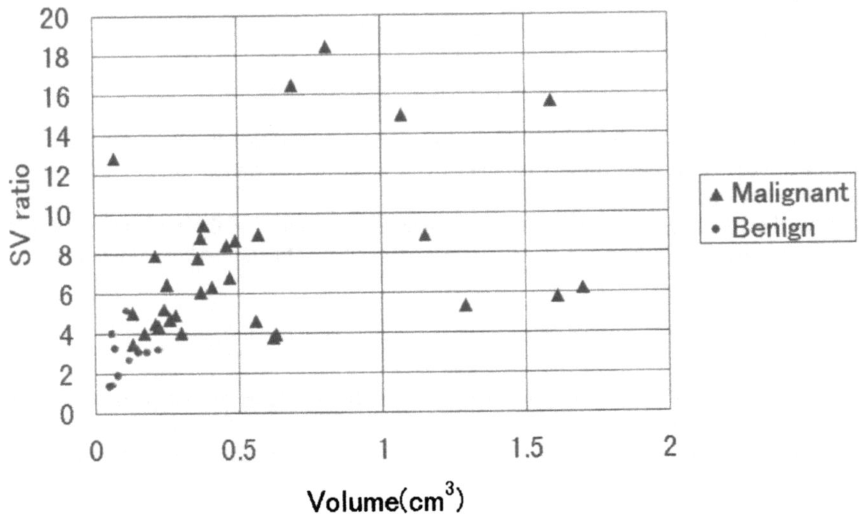

(a)

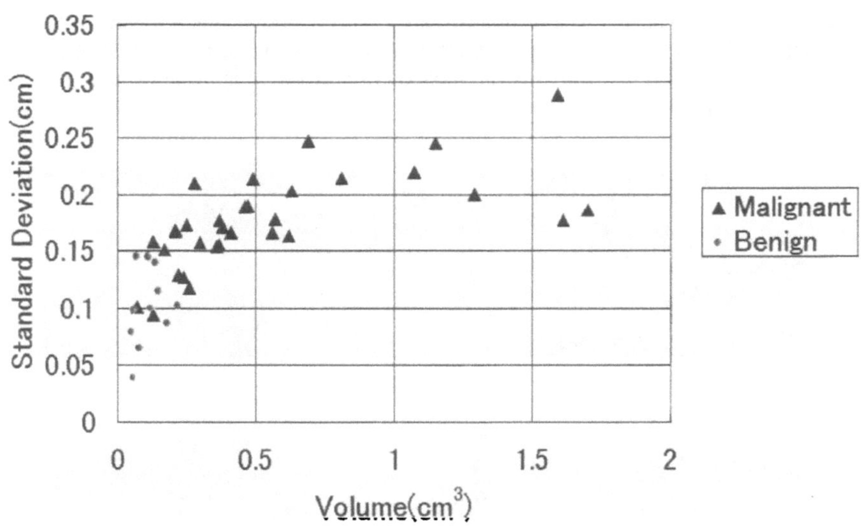

(b)

Figure 8. Results for the evaluation parameters. (a) Surface area to volume ratio (*SV ratio*). (b) Standard deviation (*SD*) of the distances from the center of gravity of the tumor to the surface points.

DISCUSSION

In this paper, we propose the use of a three-dimensional LoG filter to generate fuzzy membership functions. These functions are then applied to the image data in order to consistently and automatically extract the surface region of breast tumor. The membership functions change as variations are made to the size of σ used in the LoG filter. These variations affect the results of the extracted surface region. Figure 9 shows the determined boundaries and rendered surface images as σ varies from 3 to 11. The membership function is generated using the Rician density function to approximate the histogram. It is shown that the result (Figure 9) is in most agreement with the hand-traced contour, when $\sigma = 5$ (Figure 6 (h)).

On the other hand, Figure 10 shows the results obtained by using the histogram (with σ varying from 3 to 11) as the fuzzy membership function. It can be seen that the result obtained from using the proposed Rician approximated membership function (with $\sigma = 5$) is in better agreement with the hand-traced contour than results obtained by using the histogram alone.

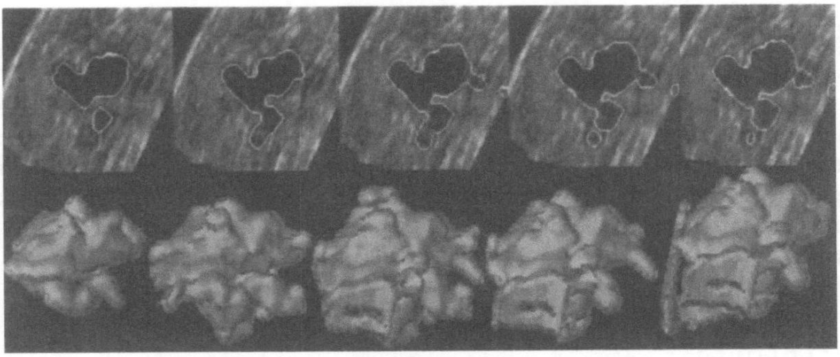

Figure 9. The determined boundaries and rendered surface images by the Rician approximated fuzzy membership function. σ is 3,5,7,9 and 11 (from left to right).

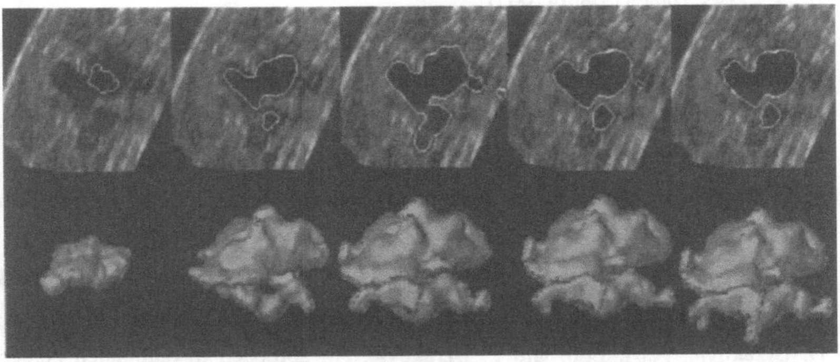

Figure 10. The determined boundaries and rendered surface images by using the histogram as fuzzy membership function. σ is 3,5,7,9 and 11 (from left to right).

CONCLUSIONS

In this paper, we proposed a method for the determination and evaluation of the surface region of breast tumors using ultrasound echography. The breast tumor volume data is obtained using conventional ultrasound diagnostic equipment with a magnetic position sensor attached to the probe for tracking its position and orientation. The surface region of breast tumor is determined using fuzzy reasoning whose fuzzy membership functions are generated by a three-dimensional Laplacian of Gaussian filter. This deals with the brightness variation of the ultrasonic images, which occurs with the changes in circumstances of the patient's mammary gland and the ultrasound diagnostic equipment settings. Two parameters, which are the surface area to volume ratio and the standard deviation of the distances from the center of gravity of the tumor region to the surface points, are investigated to evaluate the surface roughness of the tumor for quantitatively diagnosing the breast cancer.

The experimental results for thirty-two cases of malignant and eleven cases of benign tumors demonstrate the efficient and reliable performance of the proposed method. It is shown that the method has the potential to significantly improve breast cancer screening using ultrasound echography.

ACKNOWLEDGMENTS

The authors would like to thank Professor Masato Nakajima, Department of Electrical Engineering, Keio University and Dr. Akihisa Ohya, Institute of Information Sciences and Electronics, University of Tsukuba for their suggestions and discussions during the beginning of this work. The authors are also very grateful to Dr. Shin Ohtsuka, Dr. Hirobumi Mizunuma and Professor Kyotaro Kanazawa, Department of Surgery, Jichi Medical School, for their cooperation in the acquisition of the clinical data.

REFERENCES

1. X.-Y. Cheng, I. Akiyama, K. Itoh, Y. Wang, N. Taniguchi, S. Otsuka, and H. Mizunuma, Breast tumor diagnosis system using three dimensional ultrasonic echography, *Proc. IEEE EMBS-1997* 517-520(1997).
2. X.-Y. Cheng, I. Akiyama, K. Itoh, Y. Wang, N. Taniguchi, and M. Nakajima, Automated detection of breast tumors in ultrasonic images using fuzzy reasoning, *Proc. IEEE ICIP-1997* III:420-423(1997).
3. D. Leucht and H. Madjar. Teaching Atlas of Breast Ultrasound, Thieme(1995).
4. C.B. Burckhardt, Speckle in ultrasound B-mode scans, *IEEE Trans. Sonics & Ultrasonics* 25(1):1-6(1978).
5. R.F. Wagner, S.W. Smith, J.M. Sandrik, and H. Lopez, Statistics of speckle in ultrasound B-scans, *IEEE Trans. Sonics & Ultrasonics* 30(3):156-163(1983).
6. D. Brzakovic, X.M. Luo, and P. Bzrakovic, An approach to automated detection of tumors im mammography, *IEEE Trans. Med. Imag.* 9(3):233-241(1990).
7. S. M. Lai, X. Li, and W.F. Bischof, On techniques for detecting circumscribed masses in

mammograms, IEEE Trans. Med. Imag. 8(4):377-386(1989).

8. D. Wei, H.-P. Chan, M.A. Helvie, B. Sahiner, N. Petrick, D.D. Adler, and M.M. Goodsitt, Multiresolution texture analysis for classification of mass and normal breast tissue on digital mammograms, *Proc. SPIE* 2434:606-611(996).
9. M. Kamber, R. Shinghal, D.L. Collins, G.S. Francis, and A.C. Evans, Model-based 3D segmentation of multiple sclerosis lessions in MR brain images, *IEEE Trans. Med. Imag.* 14(3):442-453(1995).
10. M. Bomans, K.-H. Hoehne, U. Tiede, and M. Riemer, 3D segmentation of MR images of the head for 3D display, *IEEE Trans. Med. Imag.* 9(2):177-183(1990).
11. S. K. Pal and R.A. King, On edge detection of X-ray images using fuzzy sets, *IEEE Trans.* PAMI 5(1):66-77(1983).
12. D. Marr and E. Hildreth, Theory of edge detection, *Proc. the Royal Society* B207:187-217(1980).
13. J.F. Canny, A computational approach to edge detection, *IEEE Trans. PAMI* 8(6):679-698(1986).
14. T. Law, H. Itoh, and H. Seki, Image filtering, edge detection, and edge tracing using fuzzy reasoning, *IEEE Trans. PAMI* 18(5):481-491(1996).
15. Joseph W. Goodman, Statistical Optics, *John Wiley & Sons* (1985).
16. Julius S. Bendat, Principles and Applications of Random Noise Theory, *John Wiley & Sons* (1958).
17. A. Papoulis, Probability, Random Variables, and Stochastic Processes, *McGraw-Hill Book Company* (1984).
18. L.A. Zadeh, Fuzzy algorithm, *Info. Control* 12:94-102(1968).
19. E.H. Mamdani, Application of fuzzy logic to approximate reasoning using linguistic synthesis, *IEEE Trans. Computer* C-26:1182-1191(1977).
20. A. Ronsenfeld, R.A. Hummel, and S.W. Zucker, Scene labeling by relaxation operation, *IEEE Trans. Syst., Man., and Cybern.* SMC-6:420-433(1976).
21. S.W. Zucker, R.A. Hummel, and A. Ronsenfeld, An application of relaxation labeling to line and cave enhancement, *IEEE Trans. Computer* C-26(4):394-403(1977).
22. R.A. Hummel and S.W. Zucker, On the foundation of relaxation labeling processes, *IEEE Trans. PAMI* 5(3):259-288(1983).

DETERMINATION OF ULTRASOUND BACKSCATTER LEVEL OF VASCULAR STRUCTURES, WITH APPLICATION TO ARTERIAL PLAQUE CLASSIFICATION

Peder C. Pedersen and Zeljko Cakareski

Dept. of Electrical and Computer Engineering
Worcester Polytechnic Institute, Worcester, MA 01609

INTRODUCTION

Stroke is the third leading cause of death and long term disability among middle-age and older patients, afflicting roughly 500,000 Americans each year. Stroke survivors, numbering more than 3 millions, often face years of debilitating physical and mental impairment, emotional distress and overwhelming medical costs. Strokes have been estimated to cost this nation an estimated $19.7 billions in 1994, and over $3 billions in lost productivity[1].

Some studies indicate that 60% of strokes are directly due to rupture or ulceration of atherosclerotic plaques in the carotid artery[2], although other studies place this figure much lower. Recent studies done on carotid arteries suggest that the morphology and composition of atherosclerotic plaque is predictive of stroke risk[3,4,5]. In fact, most plaque types, such as calcified and fibrous plaques, can be considered stable and have a low risk of inducing strokes. To consider whether ultrasound is an appropriate modality for assessing stroke risk, one must consider to which extent plaque types with different stroke risks have different acoustic properties.

Acoustic characterization *in vitro* of normal arterial wall and four different categories of atherosclerotic lesions (fatty, fibrofatty, fibrotic and calcified) has been carried out in terms of the backscatter level and its angle dependence at 10 MHz[6]. The results show a dramatic variation in backscatter coefficient at normal incidence between the different plaque types, with as much as a factor 100 (40 dB) change from normal to calcified. In addition, the angle dependence of the scattered signal varies significantly across plaque types.

Despite these *in vitro* results, *in vivo* characterization of plaques has so far been only moderately successful, in contrast to assessing the degree of arterial stenosis which can be done accurately, using either ultrasound color Doppler flow mapping or contrast arteriography. The limitations in plaque characterization lie to a great extent in the phase aberrating effects of the intervening soft tissue. The ultrasound B-mode appearance of a carotid plaque has been shown to be related to the presence of brain infarcts, and to the risk of developing new neurological symptoms[7,8]. A recent study showed[9] that carotid plaques with a low gray scale value on the ultrasound B-mode display are associated with cerebral infarctions much more frequently than plaques with a high gray scale value. By visually inspecting the B-mode image whether the plaque is echolucent, hypoechogenic, hyper-echogenic or inhomogeneous, some degree of plaque classification is possible[10].

Acoustical Imaging, Volume 24.
Edited by Hua Lee, Kluwer Academic/Plenum Publishers, 2000.

If classification of atherosclerotic plaques *in vivo* in terms of stroke risk could be achieved, the decision whether to remove the plaque surgically (by carotid endarterectomy) could be made with much greater confidence than is possible today. This surgical procedure is expensive and carries a certain risk of inducing stroke. Even so, the benefit of carotid endarterectomy has been clearly demonstrated in the NASCET study (North American Symptomatic Carotid Endarterectomy Trial)[11].

BASIC CONCEPTS FOR DETERMINING BACKSCATTER LEVEL

This paper presents a new ultrasound approach for determining backscatter level, intended for *in vivo* classification of carotid artery plaque. Direct assessment of absolute ultrasound backscatter level is difficult, due to the phase aberrating effects of the intervening soft tissue. To correct for this aberration, an appropriate reference backscatterer is needed which must meet two requirements: i) the backscatterer must be located adjacent to the plaque or vessel wall region of interest, and ii) the backscatterer must have nearly omnidirectional scattering characteristics. Moving arterial blood adjacent to the region of interest fulfills these requirements. Thus, the plaque classification concept is based on determining the absolute ultrasound backscatter level, from regions within the atherosclerotic lesion and from the blood-lesion interface, by using *the backscatter level of an adjacent range cell in blood as a reference*. The clinical utility of such a technique lies with the fact that different plaque types have different stroke risk as well as dramatically different ultrasound backscatter level, as discussed earlier.

The concept of using moving (arterial) blood as a backscatter reference in order to determine the absolute value of integrated backscatter of an unknown tissue interface was proposed in 1988 by Nakayama[12] and has successfully been used for *in vivo* myocardial tissue characterization[13,14]. The absolute backscatter power was measured for the purpose of estimating the size of microemboli[15]. Recently, an approach[16] for plaque classification was published where the echogenecity of blood in the selected vessel was tagged to the color green, and the echogenecity of nearby tissues were divided into 9 ranges, and a specific color assigned to each range.

Application of the Concept to Vessel Walls and Plaques

The technique for determining the Integrated Back-Scatter (IBS) level from a vessel wall or a sclerotic lesion is specifically applicable for measurements through an intervening inhomogeneous soft tissue layer; the IBS represents a normalized measure of signal energy, as will be discussed later. The concept is illustrated in Fig. 1(a) where the integrated backscatter is measured with a focused transducer from two selected range cells: one range cell placed in the middle of the blood vessel and one range cell placed at the interface between blood and the lesion. The measured backscatter level from the range cell in blood provides a reference with which to compare the backscatter level from the lesion. If the true backscatter coefficient for moving (arterial) blood has been determined *a priori*, the backscatter level of the selected plaque region can be determined in an absolute sense. As an extension of the method, the scattering from different angles may be determined. This concept is illustrated in Fig. 1(b).

Unfortunately, the backscattered signal from the vessel lumen contains both the backscattered signal from the blood mimicking fluid as well as clutter echoes originating from the vessel wall. The presence of the intervening inhomogeneous tissue may further increase the clutter level. Thus, processing of the lumen signal is necessary in order to determine the energy of the signal component due to blood or blood mimicking fluid.

Broadband Integrated Backscatter (IBS)

The rationale behind using IBS is that the integration operation removes much of the random fluctuation in the RF signal which is observed when the backscatter transfer function is measured; IBS still preserves the spatial resolution. IBS measures the normalized energy of the RF backscatter signal for a given sample volume of the plaque specimen being tested.

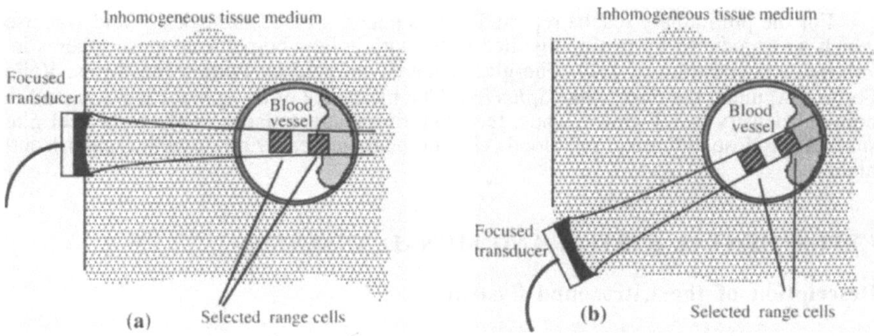

Figure 1. Concept of the measurement of the integrated backscatter (or angle scatter) level of plaque interfaces. (a) Measurement under normal incidence; (b) measurement under non-normal incidence. The value of the integrated backscatter from the range cell in the middle of the blood vessel provides the reference level with which to normalize the integrated backscatter level from the range cell, placed at the blood/plaque interface.

The definition of IBS used in this work is closely related to the form presented by O'Donnell et al[17] and is given in (1):

$$IBS = \frac{\int_{window} \left|V_{sample}(f)\right|^2 df}{\int_{-\infty}^{+\infty} |P(f)|^2 df} = \frac{\int_{window} \left|v_{sample}(t)\right|^2 dt}{\int_{-\infty}^{+\infty} |p(t)|^2 dt} \tag{1}$$

$P(f)$ is the Fourier transform of $p(t)$, and $p(t)$ is the impulse response of the transducer in pulse-echo mode, obtained with a perfect reflector placed at the location of the sample volume, and the medium between the transducer and the reflector assumed homogeneous and lossless. $v_{sample}(t)$ is defined as the *windowed* backscattered signal from the sample volume of interest, and $V_{sample}(f)$ is the corresponding Fourier transform.

The integrated backscatter level of the arterial wall or arterial plaque can be obtained in an absolute sense by means of a calibration procedure which uses the received signal from a reference reflector in blood at the intended range, obtained with the given transducer. For the results in this paper, a *normalized* backscatter level, σ_{norm}, is determined instead as

$$\sigma_{norm} = \frac{IBS_{wall}}{IBS_{scatt}} = \frac{E_{wall}}{\overline{E}_{scatt}} \tag{2}$$

In (2), E_{wall} is the energy of the echo signal from the vessel wall, or alternatively, energy from the plaque or the plaque-blood interface, while $\overline{E}_{scatt}$ is the mean energy of the echo signal from the lumen of the vessel, with clutter echoes removed. If $v_{sample}(t)$ in (1) represents the echo signal, the numerator in (1) yields either E_{wall} or $\overline{E}_{scatt}$.

Blood as a Normalizing Backscatterer

It is well documented[18,19] that ultrasound backscatter from blood is highly shear-rate dependent, due to the aggregation of red blood cells (RBCs). The effect can be dramatic, with slow moving blood producing up to 40 dB higher backscatter level than fast moving blood. Fortunately, arterial blood exhibits minimal aggregation, and the backscatter level from arterial blood can be considered nearly constant. Specifically, it has been found[15] that mean shear rate exceeding 50 s^{-1} minimizes RBC aggregation.

For the preliminary results reported in this paper, a blood mimicking fluid was used which was produced by mixing distilled water with hollow borosilicate glass spheres in a volume concentration of 1:40. The glass spheres are sold by Potters Industries, Valley Forge, PA, under the trade name *Sphericel.* The mean size of the spheres is 8 μm, and the nominal density is 1.1 gm/cc. Thus, the scattering characteristics of the Sphericel glass spheres are close to those of red blood cells, although the scattering strength is significantly higher.

EXPERIMENTAL SYSTEM AND SIGNAL ANALYSIS

Description of the Ultrasound System

The complete experimental system is shown in Figure 2. It is built around the Hewlett-Packard ImagePoint ultrasound scanner and includes a flow system and a scanning tank. The tank contains an artery mimicking tube with blood mimicking fluid flowing through it; the ultrasound measurements are carried out through a layer of soft tissue, typically a layer of beef, approx. 2 cm thick. A linear array transducer, operating at 5 MHz, is used.

The scanner has been modified to make the RF signal (after array summation) externally available in the form of a 16 bit digital signal. The digital RF signal is sent to an HP 16500B logic analysis system which functions as a fast, real-time data storage unit from where the data is transferred via GPIB to a Pentium PC for analysis in Matlab. A number of the scanner parameters are controllable (overriding the preset parameters) via keyboard, such as amplitude and length of the excitation signal, the pulse repetition frequency (PRF), receive gain, gate width, and scan line selection.

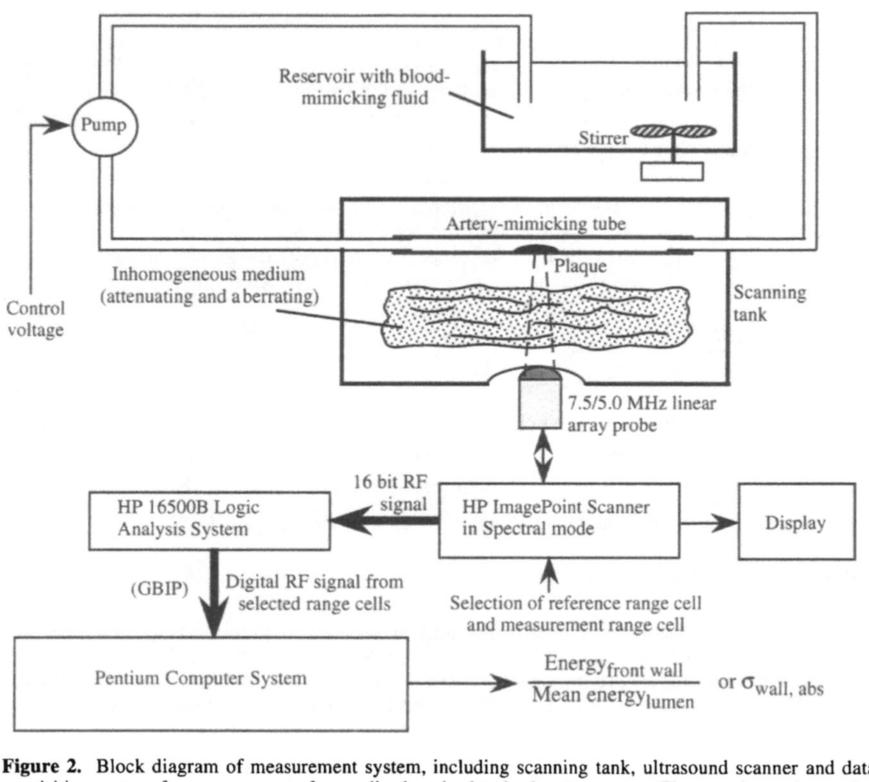

Figure 2. Block diagram of measurement system, including scanning tank, ultrasound scanner and data acquisition system, for measurement of normalized or absolute backscatter energy. The scanning tank contains water, and the intervening soft tissue can be translated laterally.

The scanner is operated in spectral mode and the acquired RF signal is determined by the location and width of the Doppler gate, adjusted to either include the wall of the tube, the lumen of the tube, or the whole tube. A PRF of approx. 2 KHz is used.

Description of Flow System

The scanning tank has an acoustic window at the side of the tank and is lined with 1/4" butyl rubber in order to minimize reverberations. A tissue holder system (not shown in Figure 2) allows accurate placement and lateral translation of the soft tissue. For the measurements, the scanning tank is filled with degassed water. The artery-mimicking tube is a silicon rubber penrose tube with 1/4" ID (6.35 mm) and 15 mil (0.38 mm) wall thickness.

Blood mimicking fluid, as previously described, is pumped through the tube at a specified flow rate. Mean flow rates from 1.4 m/s to 2.8 m/s were used. The pumping action is created by a rapidly rotating wheel with impeller blades; this pump type introduces a oscillatory pressure function, in the frequency range of 100 to 400 Hz, into the moving fluid which in turn introduces non-stationary clutter echoes in the lumen with a corresponding frequency content. The oscillatory pressure component is effectively removed by a closed reservoir placed between the pump and the scanning tank (not shown in Figure 2). Magnetic stirrers in both reservoirs are used to maintain a uniform scatterer concentration.

Signal Analysis

Assume that a 3μs range cell (or window) is placed in the lumen of the tube and a 20 MHz sampling rate is used for the RF signal. A sampled RF signal from the range cell will then consist of N = 60 samples. The first RF signal is contained in a 1 x N column matrix with the elements $v_{1,1}$, $v_{2,1}$, ..., $v_{N,1}$ which represent the N samples. Assume that M consecutive measurements of the backscattered RF signal from the blood lumen are made. The samples of the jth measurements will then be represented by the values $v_{1,j}$, $v_{2,j}$, ..., $v_{N,j}$. M is typically chosen to be 60. For a PRF of 2 KHz, the time to acquire the M signals will be 30 ms which represents only a short fraction of a cardiac cycle. The signals can be represented in a N x M data matrix, **V**, as shown in (3).

$$\mathbf{V} = \begin{bmatrix} v_{1,1} & v_{1,2} & \cdot\cdot & v_{1,j} & \cdot\cdot & v_{1,M} \\ v_{2,1} & v_{2,2} & \cdot\cdot & v_{2,j} & \cdot\cdot & v_{2,M} \\ \cdot\cdot & \cdot\cdot & \cdot\cdot & \cdot\cdot & \cdot\cdot & \cdot\cdot \\ v_{N,1} & v_{N,2} & \cdot\cdot & v_{N,j} & \cdot\cdot & v_{N,M} \end{bmatrix} \quad (3)$$

The rows in **V** represent the scattering amplitude of the echo signal from a specific range, sampled at the PRF. When the signals in the data matrix represent echoes from completely stationary structures, the elements of a given row in the data matrix all have the same value, corresponding to a sampled dc signal. This situation is approximately observed with the pump turned off.

As mentioned earlier, the lumen signal contains contributions from both clutter echoes and scatterer echoes. A data matrix containing only clutter echoes can be obtained by measurements with only circulating distilled water through the tube. In this case, the rows contain a strong dc component, corresponding to the stationary clutter, and a time varying component, corresponding to the non-stationary clutter. An illustrative magnitude spectrum of the non-stationary clutter component is shown in Figure 3. Note that the spectral shape is determined by the pump speed.

If a data matrix from the lumen were filled only with signals from the moving scatterers, and no clutter echoes were present, the magnitude row spectra would be significantly different, as illustrated in Figure 3. The row spectra due to scatterer are principally determined by the flow rate across the ultrasound beam and by the ultrasound beam width in the flow direction. To determine the energy from only the scatterers (= blood or blood mimicking fluid), the clutter energy needs to be removed from the lumen The present approach is a two-step method: First the dc component of each row is removed, followed by high pass filtering, either with a 50 Hz, a 100 Hz, or a 200 Hz cut-off frequency. As can be seen from Figure 3, neither of the filter cut-off frequencies can achieve the task of removing all the clutter energy, while retaining all the energy from the scatterers.

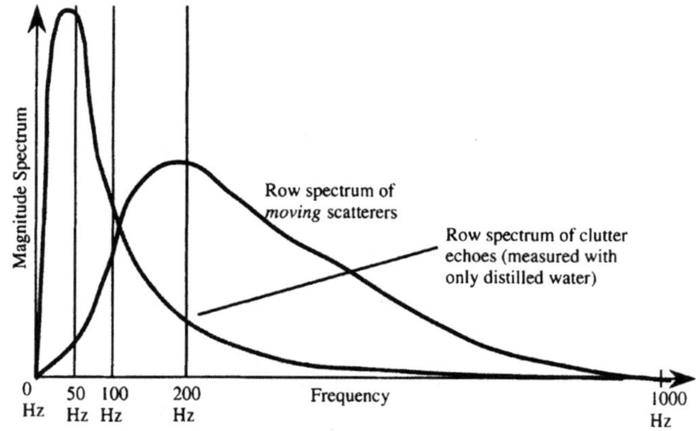

Figure 3. Mean magnitude spectra of the lumen row signals, due to moving scatterers and due to clutter, obtained from the data matrix

Given the short duration of the row signals (30 ms), the high pass filtering requires special attention. The rows were symmetrically zero-padded, and a 15 order, zero-phase IIR filter was used which maintains the length, M, of the rows. The result is an N x M filtered data matrix, **U**, shown in (4).

$$\mathbf{U} = \begin{bmatrix} u_{1,1} & u_{1,2} & .. & u_{1,j} & .. & u_{1,M} \\ u_{2,1} & u_{2,2} & .. & u_{2,j} & .. & u_{2,M} \\ .. & .. & .. & .. & .. & .. \\ u_{N,1} & u_{N,2} & .. & u_{N,j} & .. & u_{N,M} \end{bmatrix} \tag{4}$$

Next, the energy of each of the M row-filtered RF signals (columns in the data matrix) is determined. The energy of the jth signal will be denoted $E_{filt,j}$ and is given in (5).

$$E_{filt,j} = \sum_{i=1}^{N} (u_{i,j})^2 \Delta t \tag{5}$$

where Δt is the sampling interval, here 50 ns. Finally, in (6), $\overline{E}_{scatt}$ is determined as the mean energy of the M filtered signals.

$$\overline{E}_{scatt} = \frac{1}{M} \sum_{j=1}^{M} E_{filt,j} \tag{6}$$

To determine E_{wall}, required to find σ_{norm} in (2), a data matrix, **W**, is generated. The matrix contains M consecutive measurements of the echo of the front wall of the tube, in the form of sampled RF signals, with P samples in each RF signal. The data matrix **W** is of the same format as **V** in (3) except that P is lower than N, the number of samples in the RF signal from the lumen. If the samples of the jth measurements are represented by the sample values $w_{1,j}$, $w_{2,j}$, ..., $w_{P,j}$, E_{wall} is calculated analogously with $\overline{E}_{scatt}$ as follows:

$$E_{wall} = \frac{1}{M} \sum_{j=1}^{M} \sum_{i=1}^{P} (w_{i,j})^2 \Delta t \tag{7}$$

With the results in (6) and (7) thus obtained, σ_{norm} can be calculated.

RESULTS

Measurements have been carried out with the experimental system in Figure 2, using a layer of fresh beef, warmed to room temperature. First, the energy of the front wall echo from the silicon rubber tube, E_{wall}, was found using eq. (7), followed by calculation of $\overline{E}_{scatt}$ using (6) and determination of σ_{norm}, as given in (2).

The measurement variables are the pump speed and the tissue position. For the results presented here, the pump speed was held constant at setting 4, corresponding to a mean velocity through the tube of 2.2 m/s. First, measurements of E_{wall} were carried out at 10 different lateral positions of the 2 cm thick beef tissue, for the purpose of observing how much the echo energy varies for different types of overlying tissues. The results shown in Figure 4 (a) reveal an energy spread of roughly a factor 10:1. If a more inhomogeneous tissue were used, or if one tissue were replaced with another for each measurement, a far greater variation in E_{wall} would be observed. Hence, classification of plaque type from direct observation of the energy or the signal amplitude is not possible.

In Figure 4 (b), the normalized energy σ_{norm} is shown for the same 10 tissue positions. The 3 curves are obtained for $\overline{E}_{scatt}$ determined with the 3 different high pass filter cut-off frequencies. It can be seen that the variation of a factor 10 for the directly measured front wall energy has been reduced to a variation of a factor 4. This is only a modest reduction, and further improvement is required. Possible reasons for the inadequate improvement and means to overcome this are given in the next section of the paper.

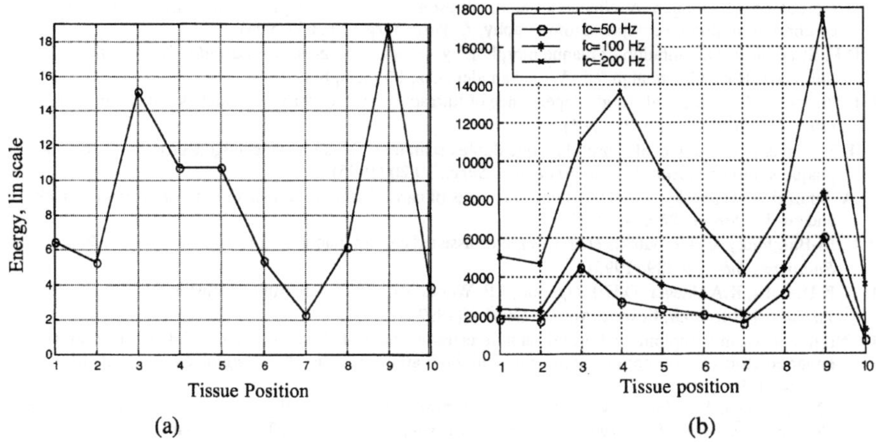

Figure 4. (a). Variation in energy of echo from front wall for 10 lateral positions of 2 cm thick layer of beef, placed between the transducer and the silicon rubber tube. The normalized energy, σ_{norm}, of echo from front wall for the same 10 position of the tissue layer.

DISCUSSION AND CONCLUSION

The paper has described a new technique for *in vivo* determination of the normalized backscatter level from arterial structures, such as vessel walls or atherosclerotic lesions. The technique makes use of the energy of the backscattered signal from blood as a normalizing parameter, and may with further improvements make accurate classification of plaque types possible.

The measurements were performed with a silicon rubber tube, representing an artery, through which flowed a blood mimicking fluid in the form of distilled water with borosilicate glass spheres. Intervening inhomogeneous soft tissues layers of varying structure and composition were imitated by placing a beef layer at various lateral positions. Only a moderate reduction in the spread of front wall echo energy with tissue position was achieved when normalizing with the mean energy of the filtered echoes from the lumen. The complicating factor is the high clutter level in the lumen with a high degree of non-stationarity which makes the removal of the clutter energy by means of high pass filtering difficult. It is possible that turbulence widens the clutter component of the lumen spectra, and increasing the viscosity of the blood mimicking fluid may somewhat alleviate this. Other ways to improve performance may include different signal processing approaches and/or a the use of a single element focused transducer (sector scanner), to reduce beam width in both lateral directions.

Acknowledgements

The authors greatly appreciate the ultrasound scanner and logic analyzer, made available by Hewlett Packard, and the technical assistance from Ronald Gatzke and Ted Fazioli, HP.

REFERENCES

1. A.J. Labowitz. Data presented at the 67th Scientific Sessions of the American Heart Assoc. Mtg., Dallas, Nov. 13, 1994, in program entitled "Non Invasive Evaluation of Patients with Embolic Stroke".
2. R.W. Bock, A.C. Gray-Weale, P.A. Mock, D.A. Robinson, L. Irwig, and R.J. Lusby, The natural history of asymptomatic carotid disease, *J. Vasc. Surg.*, 17:160 (1993)
3. H. Van Damme, M. Vivario, Pathologic aspects of carotid plaques: surgical and clinical significance, *Intl. Angiology*, 12:299 (1993).
4. G. Geroulakos, J. Domjan, A. Nicolaides et al, Ultrasonic carotid artery plaque structure and the risk of cerebral infarction on computed tomography, *J. Vasc. Surg.*, 20:263 (1994)
5. D.E. Strandness, Quantitation of plaque morphology – work in progress, in: *Vascular Imaging for Surgeons*, R.M. Greenhalgh, ed., W.B. Saunders Co., Philadelphia (1995).
6. E. Picano, L. Landini, et al, Angle dependence of ultrasonic backscatter in arterial tissues: A study *in vitro*, *Circulation*, 72:572 (1985).
7. E.M. Cave, N.D. Pugh et al, Carotid plaque duplex scanning: Does plaque echogenecity correlate with plaque symptoms, *Eur. J. Vasc. Endovasc. Surg.*, 10:20 (1995).
8. A. Ianuzzi, T. Wilcosky et al, Ultrasonic correlates of carotid atherosclerosis in transient iscemic attack and stroke, *Stroke*, 26:614 (1995).
9. N. El-Barghouty, G. Geroulakis et al, Computer-assisted carotid plaque characterization, *Eur. J. Vasc. Endovasc. Surg.*, 9:389 (1995).
10. K.H. Labs, K.A. Jaeger, D.E. Fitzgerald, J.P. Woodcock and D. Neuerburg-Heusler, ed., *Diagnostic Vascular Ultrasound*, Edward Arnold, London (1992).
11. North American Symptomatic Carotid Endarterectomy Trial Collaborators. Beneficial effect of carotid endarterectomy in asymptomatic patients with high grade stenosis. *New England J. of Medicine*, 325:445 (1991).
12. K. Nakayama and S. Yagi, In vivo tissue characterization using blood flow Doppler signal as a reference, *Proc. of 52nd Mtg. of The Japanese Soc. of Ultrasonics in Medicine*, Tokyo, Japan, 399 (1988).
13. A. Shiba, I. Yamada et al, A measurement method for absolute value of integrated backscatter, *1991 IEEE Ultrasonics Symp. Proc.*, Lake Buena Vista, FL, 1089 (1991).
14. J. Naito, T. Masuyama et al, Validation of transthoracic myocardial ultrasonic tissue characterization: Comparison of transthoracic and open-chest measurement of integrated backscatter, *Ultrasound in Med. & Biol.*, 21:33 (1995).
15. M.A. Moehring and J.R. Klepper, Pulse doppler ultrasound detection, characterization and size estimation of emboli in flowing blood, *IEEE Trans. Biom. Eng.*, 41:35 (1994).
16. K. Beach, J.F. Primozich, and D.E. Strandness, Pseudocolor B-mode arterial images to quantify echogenecity of atherosclerotic plaque, *Ultrasound in Med. and Biol*, 20:731 (1994).
17. M. O'Donnell et at, Broadband integrated backscatter: An approach to spatially localized tissue characterization in vivo, *1979 Ultrasonics Symposium Proceedings*, Ed. B.R. McAvoy, IEEE, New York, 175 (1979).
18. Y.W. Yuan and K.K. Shung, Ultrasound Backscatter from Flowing Whole Blood. I: Dependence on Shear Rate and Hematocrit, *JASA*, 84:52 (1988).
19. K.K. Shung, Ultrasound characterization of blood, in: *Tissue Characterization with Ultrasound, Vol. II*, J.F. Greenleaf, ed., CRC Press, Boca Raton, FL (1986).

INVESTIGATION OF THE MICRO BUBBLE SIZE DISTRIBUTION IN THE EXTRACORPOREAL BLOOD CIRCULATION

Georg Dietrich,[1] Klaus V. Jenderka,[1] Ulrich Cobet,[1]Bernhard Kopsch,[1] Albrecht Klemenz[1], Petr Urbanek[2]

[1]Institute for Medical Physics and Biophysics
Martin-Luther-University Halle-Wittenberg
Medical Faculty, D-06097 Halle, Germany
[2]HP-medica GmbH
Bahnhofstrasse 30, D-86150 Augsburg, Germany

INTRODUCTION

During open-heart surgeries gas micro bubbles of different sizes and numbers are generated in the blood due to leaks in the extracorporeal circulation or because of the oxygenation. The micro bubbles may cause gas embolism with severe complications for the patients - even death. Despite of the used filters, gas micro bubbles still occur in the arterial area which increase the risk of embolism. Although the present measuring systems allow to detect larger gas micro bubbles and to roughly classify the bubbles size, an exact detection of the size distribution, in particular in the clinically interesting range of 10 - 40 µm, was not possible. Important factors, e.g. absorption of blood and tube material, sound field distribution and probe coupling have not been considered so far. The presented measuring system shows a high sensitivity to bubbles up to 5 µm. The special ultrasound probe construction secures an optimal connection of the transmitters to the tube. The control and evaluation software allows the clinical application for monitoring during operations as well as laboratory interventions with regard to special questions.

METHOD

Theoretical basis: Ultrasound waves are scattered at small gas micro bubbles. The part of the backscattering energy, resp. the scattering section is proportional to the square of the gas micro bubble radius[1,2,3,4]:

Acoustical Imaging, Volume 24.
Edited by Hua Lee, Kluwer Academic/Plenum Publishers, 2000.

$$I_S \sim I_0 \, e^{-axf} \, r_B^2 \qquad (1)$$

were I_S is the backscattered energy, I_0 is the energy of the incident wave, e^{-axf} is the absorption of the penetration path and r_B is the bubble radius. However, the backscattering energy measured at the transducer also depends on other parameters: transmitting intensity, tube absorption (including the coupling conditions), acoustical features of the suspension and resonance conditions for gas micro bubbles.

Using a Doppler system, only the scattering signals of flowing scatters can be detected, this means that the essentially larger wall echos are being extracted. The scattering signal of the blood particles can be considered to be constant, since the transmitter frequency is far below the resonance frequency of the cells[5]. To consider the absorption features of various tube materials or the different haematocrit values of blood, which are strongly affecting the scattering amplitude, since they are damping the irradiated as well as backscattered ultrasound wave, the total absorption between transducer and gas micro bubbles must be known[6,7].

The resonance conditions for free-swinging gas micro bubbles at a transmitter frequency of 2 MHz are within the measuring area (approx. 3 - 5 µm). But in comparison to the effects in water, the amplitude increase of the resonance vibration of the gas micro bubbles in blood can consider to be small.

Device description: The measuring system is a two-channel pulse Doppler construction. Basically, you differ between two measuring methods: The definition of the backscattered Doppler signal for the detection of the gas micro bubble size and the measuring of the reflector echo for the definition of the absorption features.

The ultrasound probe is covered by a collar which is pressed around the tube. The 2 MHz transducer transmits the ultrasound beam under a defined angle through a polystyrene window and through the tube. The collar deforms the tube in such a way that the inner section is placed in close distance to the transducer, which provides a nearly homogeneous field distribution. On the opposite side of the tube a reference reflector is placed. At a repeated time interval of a few seconds during the measurements a reference ultrasound beam is transmitted, which echo is continuously used for the correction of absorption properties of the tube material and the suspension.

The envelope of the signal is determined from the Doppler signal by rectification with the

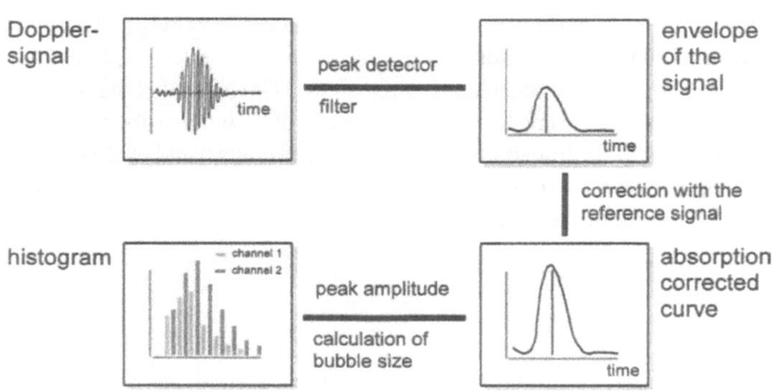

Figure 1. Block scheme of the bubble sizing process.

peak detector as well as different filter systems. This results in a signal progression with peaks on a certain background noise which height correlates with the gas micro bubble size. This signal is then being digitalised and transferred to a data processing system. The gas micro bubble distribution is calculated from the signal after the absorption correction and the correction of double interpretations with the calibration data (Figure 1).

RESULTS AND DISCUSSION

The device was tested with suspension giving ultrasound reflections over a wide size range of gas micro bubbles. Both measuring channels were calibrated with a defined gas micro bubble distribution. The solution used for the calibration contains gas micro bubbles which are surrounded by an albumin cover, which means that they have a sufficient life-time. The suspension was measured with our ultrasound Doppler system and with a normal cell counter. The estimated size distributions of both measurements are well comparable (Figure 2).

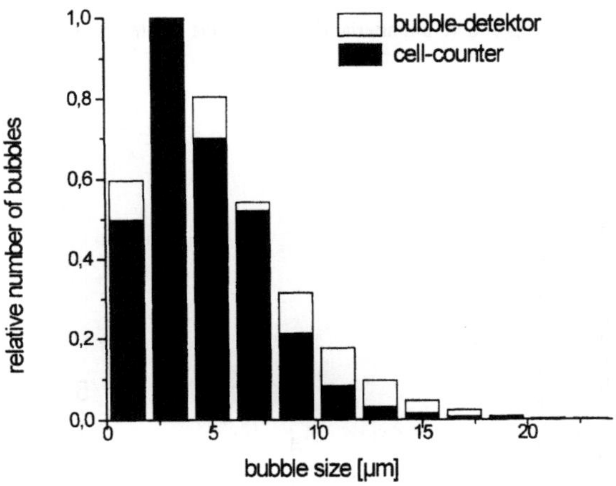

Figure 2. Comparison of the size distribution of albumin covered gas micro bubbles measured with the bubble-detector and a cell-counter.

The two-channel construction of the system allows the investigation of changes in the number and size distribution of bubbles passing filter systems, pumps or other devices generating turbulence. The estimation of the effectiveness of an arterial filter in the extracorporeal circulation is possible by arranging the sensors before and behind the filter system (Figure 3).

During the clinical tests the micro bubble distribution in the extracorporeal circulation was monitored in several open-heart surgeries for the whole operation period. A plot of the counted number of bubbles versus time show, that inevitable manipulations of the operating team causes peaks in the number of gas bubbles. During an open heart surgery not only at the start and the removal phase of the heart-lung-machine a lot amount of bubbles occur - also the drawing of blood and the application of drugs generate bubbles (Figure 4).

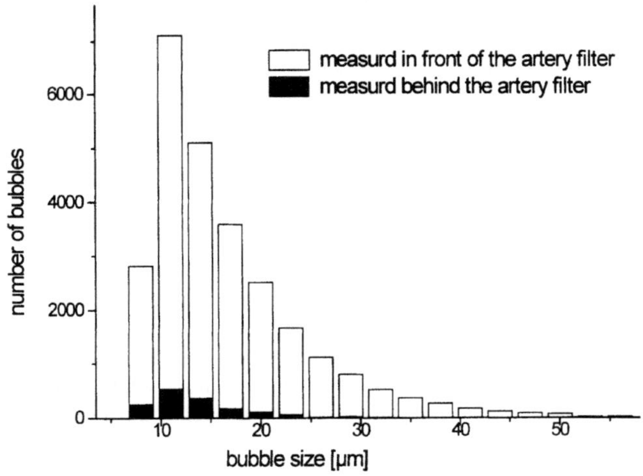

Figure 3. Effectiveness of a 40 µm arterial filter during the whole time of an open heart surgery.

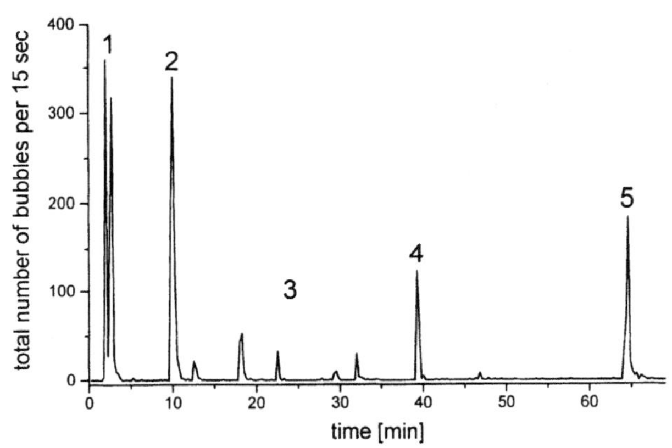

Figure 4. Total number of micro bubbles during an open-heart surgery in correlation with special actions of the operating team (1-start of the heart-lung-machine, 2-venous cut, 3-manipulations on the heart-lung-machine (drawing of blood, application of drugs), 4-venous cut, 5-removal of the heart-lung-machine).

CONCLUSION

A useful tool for the investigation of the number and size distribution of air bubbles in the extracorporeal circulation based on a 2 Mhz pulse Doppler system was developed and adapted to automated control and continuously measurement. The reliable measurement range of booth channels covering the size range of 10 µm to 120 µm. The behaviour in the resonance effected critical size range under 5 µm will be investigate in future work.

REFERENCES

1. C. Devin, Survey of thermal, radiation and viscous damping of pulsating air bubbles in water, *J. Acoust. Soc. Am.*, 31: 1654-1667 (1959)
2. H. Medwin, Counting bubbles acoustically: a review, *Ultrasonics*, 15: 7-12 (1977)
3. N. de Jong, R. Cornet and C.T. Lancée, Higher harmonics of vibrating gas-filled microspheres. Part one: simulations, *Ultrasonics*, 32: 447-453 (1994)
4. P.H. Chang, K.K. Shung, and H.B. Levene, Quantitative measurements of second harmonic Doppler using ultrasound contrast agents, *Ultrasound Med.Biol.*, 22: 1205-1214 (1996)
5. K.K. Shung, R.A. Sigelmann and J.M. Reid, Scattering of ultrasound by blood, *IEEE Trans. Biomed. Eng.*, 23: 460-467 (1976)
6. N. de Jong and L. Hoff, Ultrasound scattering properties of Albunex microspheres, *Ultrasonics*, 31: 175-181 (1993)
7. S. Holm, M. Myhrum and L. Hoff, Modelling of the ultrasound return from Albunex microspheres, *Ultrasonics*, 32: 123-130 (1994)

A METHOD FOR DETECTING ECHOES FROM MICROBUBBLE CONTRAST AGENTS BASED ON TIME VARIANCE

Wilko Wilkening[1], John Lazenby[2], Helmut Ermert[1]

[1] Dept. of Electrical Engineering
Ruhr-University
D-44780 Bochum, GERMANY
[2] Siemens Medical Systems
Ultrasound Group
P.O. Box 7002
Issaquah WA 98027, USA

INTRODUCTION

By increasing the echogenicity of blood[1], microbubble contrast agents improve the sensitivity of Doppler ultrasound[2,3,4]. Nevertheless, there are still limitations to the imaging of small blood vessels and to the quantification of perfusion. Compared to B-mode images, color mode and power mode images suffer from poor spatial resolution and sensitivity to motion artifacts.

The acoustic properties of microbubbles differ significantly from those of scatterers in tissue. The size of a microbubble and the composition and structure of its shell determine the backscatter coefficient, the attenuation coefficient, and the strength of the non-linear scattering from the bubble[5]. Insonification can change the acoustic properties of microbubbles significantly, since it may cause shrinking, growing, splitting, combination, and disintegration of microbubbles. Furthermore, acoustical streaming and the formation of clusters may occur[6,7,8].

Contrast Agent Specific Imaging Techniques

New imaging techniques, which are based on the non-linearity and time-variance of microbubbles, can overcome the limitations of Doppler based imaging techniques.

Methods commonly used for tissue characterization may be expanded to contrast agent imaging, since microbubbles exhibit extraordinary acoustic properties[9]. Due to the large size of the resolution cells, which is needed for the estimation of relevant parameters, such methods will probably not offer an alternative to Doppler ultrasound as far as the axial resolution is concerned.

Second harmonic imaging uses the non-linearity of scattering from microbubbles, since the generation of harmonics is a property of non-linear scatterers[10,11]. Second harmonic B-mode imaging shows an improved contrast between blood and tissue[12,13,14]. However, the improved contrast of that technique is limited because non-linear effects in propagation through tissue and system non-linearity also generate harmonics[15]. If the

Acoustical Imaging, Volume 24.
Edited by Hua Lee, Kluwer Academic/Plenum Publishers, 2000.

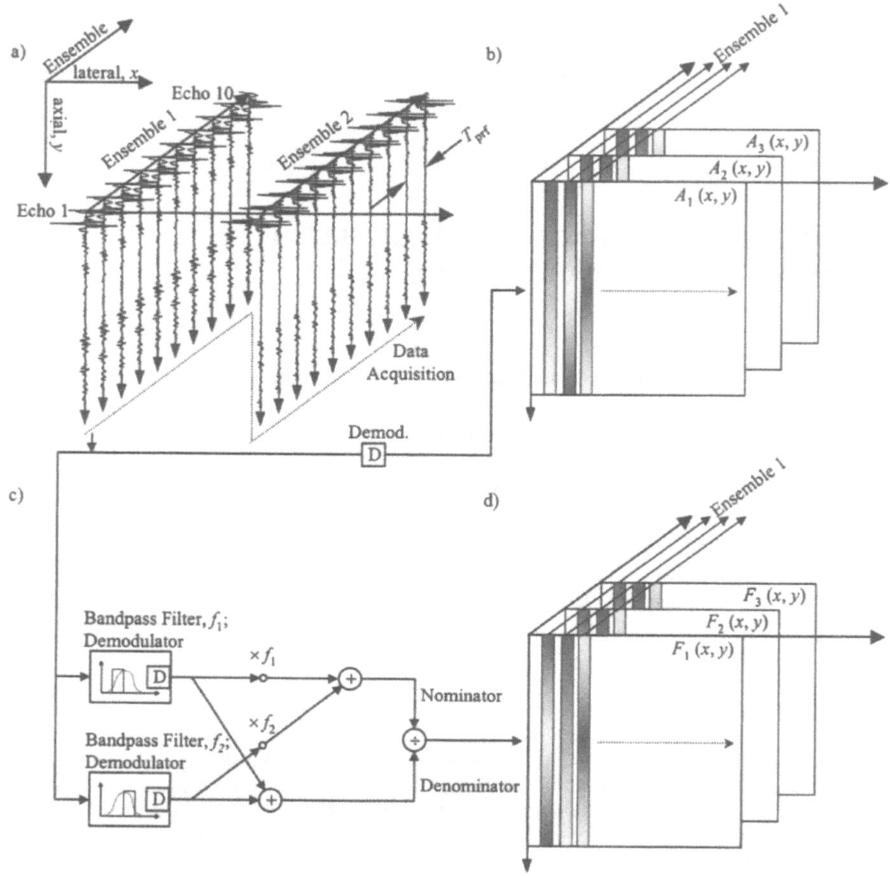

Figure 1. a) Data acquisition. b) Representation of the amplitudes. c) Estimation of centroid frequencies. d) Representation of the centroid frequencies.

second harmonic is separated from the first harmonic by a bandpass filter, the axial resolution will be reduced according to the loss of bandwidth[16]. Second harmonic power mode allows the imaging of low blood velocities due to the better signal-to-clutter ratio, but resolution and minimal detectable velocity are still limitations for some applications.

In this presentation, we propose a new method for contrast agent specific imaging. Time-variance imaging (TVI) evaluates the time-variant properties of the observed scatterers in order to distinguish microbubbles and time-invariant scatters, which are typical for tissue.

TIME-VARIANCE IMAGING

Data Acquisition

In order to detect and characterize time-variance of acoustic parameters, TVI performs multiple pulse-echo measurements, similar to the sequences that are commonly used for color mode and power mode. Figure 1 illustrates the data acquisition. At a given lateral position, denoted by the lateral coordinate x, a sequence of N broadband transmit pulses is

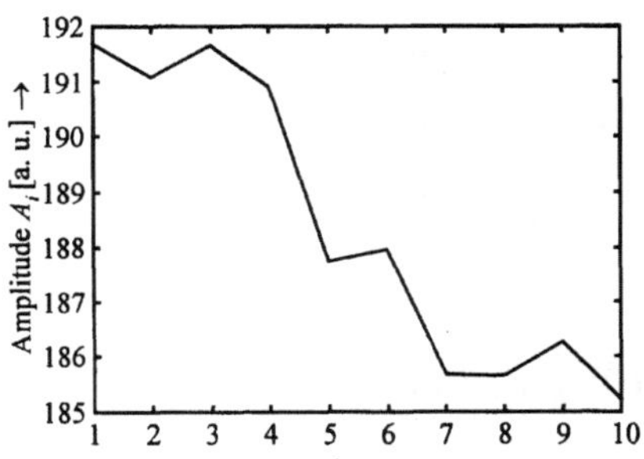

Figure 2. Time course of the echo amplitude measured from Levovist® within a sponge at an arbitrary point (x, y).

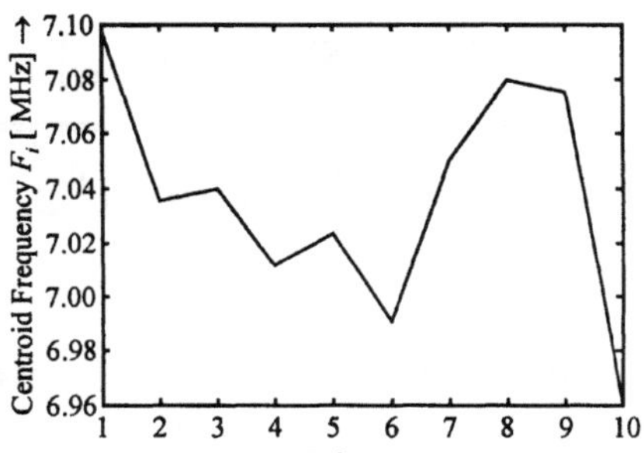

Figure 3. Time course of the centroid frequency measured from Levovist® within a sponge at an arbitrary point (x, y).

transmitted. The group of corresponding receive echoes, which will be referred to as an ensemble, reflects the time-variance of observed scatterers as far as it is measurable by these pulse-echo measurements. The amplitude A and the centroid frequency F are extracted from the echo signals.

In order to reduce motion artifacts and to separate between time-variant effects resulting from normal dissolution and time-variant effects that are due to the insonification, the interval T_{prf} between pulses is minimized. The time between sequences, i. e. the interval between frames, should allow some reperfusion of the insonified area and the approach of a new equilibrium.

We have investigated several parameters which are derived either from the amplitudes or from the spectra of the echo signals.

Amplitude

The elements of a vector **A** defined as

$$\mathbf{A}(x, y) = [A_1(x, y), A_2(x, y), \ldots A_N(x, y)] \tag{1}$$

represent the time course of the amplitude at a given location throughout an ensemble. A typical amplitude curve measured from contrast agent within a reflective material, i. e. a sponge, is shown in Figure 2. For the first three insonifications, the amplitude remains almost constant. Then, the amplitude drops down to a value that is mainly due to the reflectivity of the sponge. We hypothesize that the first pulses break the shells of bubbles. The free air bubble is then subject to all kinds of disintegration processes.

Two parameters that quantify the changes in amplitude are the standard deviation σ_A and the amplitude swing Δ_A which are defined as

$$\sigma_A(x, y) = \sqrt{\frac{1}{N-1}\sum_{i=1}^{N}(A_i(x, y) - \mu_A(x, y))^2}, \qquad \mu_A(x, y) = \frac{1}{N}\sum_{i=1}^{N} A_i(x, y)$$

and (2)

$$\Delta_A(x, y) = \max(\mathbf{A}(x, y)) - \min(\mathbf{A}(x, y)),$$

respectively.

These parameters do not consider the actual course in time of the amplitudes. An approach that does consider the time course correlates the $\mathbf{A}(x, y)$ with the result of a reference measurement $\mathbf{A}_{\text{ref}}$, e. g. the data shown in Figure 2. The correlation coefficient

$$\rho_A(x,y) = \frac{\sum_{i=1}^{N}\left[A_{i,\mathrm{ref}} - \mu_{A,\mathrm{ref}}\right]\cdot\left[A_i(x,y) - \mu_A(x,y)\right]}{\sqrt{\sum_{i=1}^{N}\left(A_{i,\mathrm{ref}} - \mu_{A,\mathrm{ref}}\right)^2 \sum_{i=1}^{N}\left(A_i(x,y) - \mu_A(x,y)\right)^2}}, \tag{3}$$

$$\mathbf{A}_{\mathrm{ref}} = \left[A_{1,\mathrm{ref}}, A_{2,\mathrm{ref}}, \ldots A_{N,\mathrm{ref}}\right], \quad \mu_{A,\mathrm{ref}} = \frac{1}{N}\sum_{i=1}^{N} A_{i,\mathrm{ref}}$$

quantifies the similarity of the two time courses represented by $\mathbf{A}(x, y)$ and $\mathbf{A}_{\mathrm{ref}}$ independent of the means μ_A, $\mu_{A,\mathrm{ref}}$ and standard deviations of the two samples.

Centroid Frequency

The fact that microbubbles have a resonance frequency and non-linear properties, which depend on the size of the bubble and the structure of the shell, suggests that the centroid frequency of the receive spectrum will be shifted if the insonification changes the structure of the bubbles. The centroid frequency at a depth y is calculated from a local estimate of the receive spectrum for all positions i within an ensemble and for all lateral positions x.

$$F_i(x,y) = \frac{\int_{f_{\min}}^{f_{\max}} f \cdot \mathbf{S}_i(f,x,y)\,df}{\int_{f_{\min}}^{f_{\max}} \mathbf{S}_i(f,x,y)\,df},$$

$$\mathbf{F}(x,y) = \left[F_1(x,y), F_2(x,y), \ldots F_N(x,y)\right], \tag{4}$$

where $f_{\min}$, $f_{\max}$ limit the bandwidth to e. g. the –20 dB bandwidth of the transducer in order to suppress noise. The local spectrum is either estimated by a FFT, which is based on a short window centered around y, or by dividing the receive signal into two or more frequency bands and calculating the amplitudes for the frequency bands, which leads to a coarse representation of the receive spectrum at discrete frequencies given by the center frequencies of the filters. The latter method, which is illustrated in Figure 1 c), is used exclusively for all spectrum-based TVI images included in this presentation. The data format of the ultrasound machine we used, a Siemens Sonoline® Elegra, allowed a very fast implementation of this method.

Similar to the evaluation of the amplitude, the standard deviation and frequency swing are defined as

$$\sigma_F(x,y) = \sqrt{\frac{1}{N-1}\sum_{i=1}^{N}\left(F_i(x,y) - \mu_F(x,y)\right)^2}, \qquad \mu_F(x,y) = \frac{1}{N}\sum_{i=1}^{N} F_i(x,y)$$

and (5)

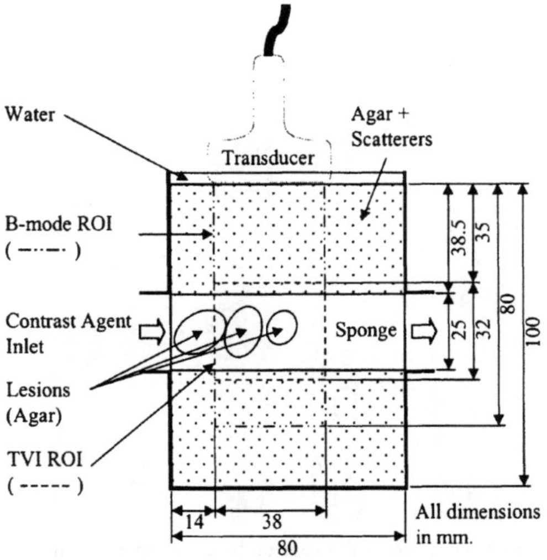

Figure 4. Perfusion phantom containing unperfusable lesions in a perfusable sponge.

$$\Delta_F(x,y) = \max(\mathbf{F}(x,y)) - \min(\mathbf{F}(x,y)).$$

However, a parameter that considers the course in time of the centroid frequency could not be found as easily as for the amplitude, since in experiments both down-shifts and up-shifts of the centroid frequency were observed. Figure 3 shows a result from a measurement. Experimental data suggest that the time course of the centroid frequency measured from microbubbles exhibits a linear trend which might be interrupted by a sudden rise or fall, whereas no linear trend was observed if changes in the centroid frequency were due to noise. Therefore, σ_F weighted by the magnitude of the linear trend of F_i in ensemble direction was chosen as a robust parameter for spectrum-based TVI (S-TVI).

Perfusion Measurement

Color and power mode are important tools for vascular and perfusion imaging but are limited to applications where the flow velocity is significantly greater than the velocity of tissue motion. Contrast agents can improve the performance of these imaging modes. However, the visualization of perfusion remains a problem in many cases. Contrast specific imaging techniques, like TVI, offer another way of performing perfusion measurements.

The concentration of the microbubbles within that region of interest will approach an equilibrium determined by the destruction rate and the reperfusion rate. If the frame rate, the amount of contrast agent injected into the blood stream, and the acoustic power are set appropriately, a sequence of TVI images will show the different regions of the image approaching different equilibrium intensities. The equilibrium intensities will be higher in regions with more perfusion and lower in regions with less perfusion. Time-intensity curves calculated from regions of interest within TVI images offer an additional tool for quantifying perfusion. The information of the B-mode image and the TVI image can be combined by a 2-dimensional colormap.

EXPERIMENTAL RESULTS

Experiments were conducted using a perfusion phantom shown in Figure 4. The phantom consists of a perfusable cylindrical sponge within an agar block. The concentration of agar was 30 g/l. The reflectivity of the agar results from silica gel at a

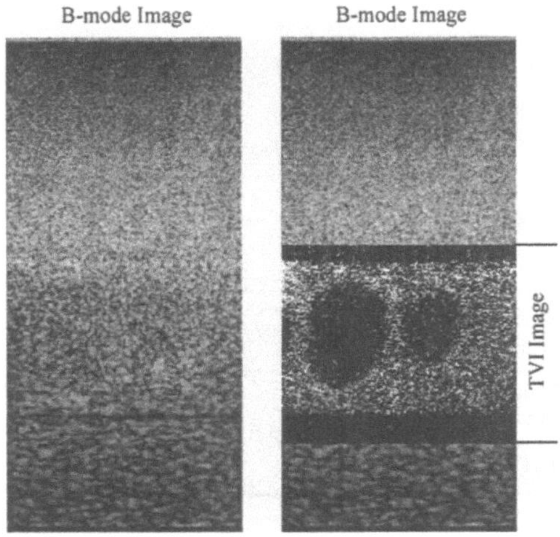

Figure 5. B-mode image and TVI image of the perfusion phantom. Flow velocity: 20 mm/s.

concentration of 10 g/l, which provides solid particles having a mean diameter of 15 μm. The sponge, which has a reflectivity similar to the surrounding agar, contains unperfusable lesions. The lesions were made by injecting agar without additional scatterers into the sponge. Levovist®, ~0.05 g/l, was pumped through the sponge at mean flow velocities of 0 – 20 mm/s. In the B-mode image, this concentration did not change the intensity by more than 1 dB.

We used a Siemens Sonoline® Elegra with a 7.5 MHz linear probe for the acquisition of digital data. The pulse repetition frequency f_{prf}, which is the reciprocal of T_{prf}, was 6499 Hz.

Time-Variance Imaging

Figure 5 gives a comparison of a B-mode image with a dynamic range of 55 dB with the corresponding S-TVI image of the perfusion phantom. In the B-mode image, the sponge-agar interface is visible because of some trapped air bubbles and because of a small, water-filled gap. However, the intensities of the agar and the sponge are very similar. The lesions cannot be clearly identified. In the TVI overlay, the agar outside the sponge and the lesions appear black, since they cannot be perfused by the contrast agent while the perfusable sponge is bright.

S-TVI performed best in the experiments. The visualization of σ_A and Δ_A yielded good results, too, as long as the SNR is sufficiently high, which was the case for a depth of up to 8 cm in our *in vitro* experiments. The visualization of ρ_A is less sensitive to noise but the perfused regions have a very granular appearance, which indicates that the reference curve representing bubble destruction only covers some destructive processes. Attempts to make the algorithm more flexible resulted in an increased sensitivity to noise.

Perfusion Measurement

We calculated time-intensity curves over 16 frames from an ROI within the sponge at different flow velocities, ranging from 0 – 20 mm/s. As Figure 6 shows, the intensity of the S-TVI images decreases for the first 3 frames and then, it reaches an equilibrium. For the measurement at 0 mm/s, the equilibrium is equivalent to the noise floor, which is reasonable, since there is no reperfusion of the insonified region. It is interesting to note that the initial intensities are not identical although an equal amount of contrast agent was used for all experiments. Due to the overlap of the acoustic fields of adjacent beam lines,

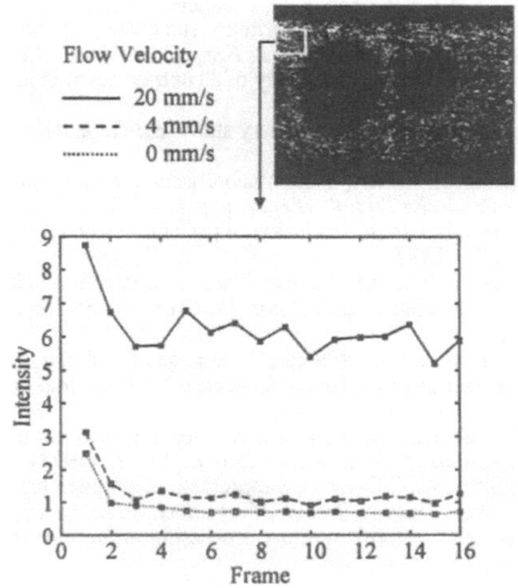

Figure 6. Time-intensity curve calculated from 16 S-TVI frames for 3 different flow velocities.

microbubbles are being destroyed by a beam before their time-variant acoustic properties are measured by the next beam.

CONCLUSIONS

Time-variance imaging is a feasible method for detecting microbubble contrast agents. Compared to Doppler modes, TVI provides better resolution and does not require a minimum velocity. The lateral resolution equals the resolution of the corresponding B-mode image, the axial resolution for e. g. S-TVI is half the resolution of the corresponding B-mode image, since the bandwidth is halved. Series of TVI images will allow the qualitative evaluation of perfusion.

ACKNOWLEDGEMENT

We would like to thank Schering AG, Berlin, Germany, for providing us with the contrast agent SHU 508 A (Levovist®).

REFERENCES

1. A. Boukaz, N. De Jong, C. Cachard, Standard properties of ultrasound contrast agents, *Ultrasound in Med. & Biol.* 24(3) 469-472 (1998).
2. P. Burns, J. Powers, D. Simpson, A. Brezina, A. Kolin, C. Chin, V. Uhlendorf, T. Fritsch, Harmonic power mode Doppler using microbubble contrast agents: an improved method for small vessel flow imaging, *1994 IEEE Ultrasonics Symposium Proc.*, pp. 1548-1550 (1994).
3. Y. Kono, F. Moriyasu, T. Nada, Y. Suginoshita, T. Matsumura, K. Kobayashi, T. Nakamura, T. Chiba, Gray scale second harmonic imaging of the liver: A preliminary animal study, *Ultrasound in Med. & Biol.*, 23(5): 719-726 (1997).
4. B. Schrope, V.L. Newhouse, Second harmonic ultrasonic blood perfusion measurement, *Ultrasound in Med. & Biol.* 19(7): 567-579 (1993).
5. P.J.A. Frinking, N. De Jong, Acoustic modeling of shell-encapsulated gas bubbles, *Ultrasound in Med. & Biol.* 24(4): 523-533 (1998).
6. S. Meerbaum, Principles of echo contrast: *Advances in Echo Imaging using Contrast Enhancement*, N. Nanda, R. Schlief, ed., Kluwer Academic Publishers (1993).

7. P.A. Dayton, K. Morgan, A.L.S. Klibanov, G. Brandenburger, K.R. Nightingale, K.W. Ferrara, A preliminary evaluation of the effects of primary and secondary radiation forces on acoustic contrast agents, *IEEE Trans. Ultrason., Ferroelect., Freq. Contr.* 44(6): 1264-1277 November (1997).
8. J. Wu, F. Tong, Experimental study of stability of a contrast agent in an ultrasound field, *Ultrasound in Med. & Biol.* 24(2): 257-265 (1998).
9. M. Arditi, T. Breiner, M. Schneider, Preliminary study in differential contrast echography, *Ultrasound in Med. & Biol.* 23(8): 1185-1194 (1997).
10. M. Krüger, W. Wilkening, H. Ermert, Systemtheoretische Analyse des Ultraschallkontrastmittels Levovist®, *Biomedizinische Technik (Ergänzungsband)* 40: 271-272 (1995).
11. A.O. Maksimov, On the subharmonic emission of gas bubbles under two-frequency excitation, *Ultrasonics* 35: 79-86 (1997).
12. P.H. Chang, K.K. Shung, S. Wu, H.B. Levene, Second harmonic imaging and harmonic Doppler measurements with Albunex®, *IEEE Trans. Ultrason., Ferroelect., Freq. Contr.* 42(6): 1020-1027 November (1995).
13. P.M. Shankar, P.D. Krishna, V.L. Newhouse, Advantages of subharmonic over second harmonic backscatter for contrast-to-tissue echo enhancement, *Ultrasound in Med. & Biol.* 24(3) 395-399 (1998).
14. W. Zheng, V.L. Newhouse, Onset delay of acoustic second harmonic backscatter from bubbles or microspheres, *Ultrasound in Med. & Biol.* 24(4): 513-522 (1998).
15. S. Krishnan, J.D. Hamilton, M. O'Donnell, Suppression of propagating second harmonic in ultrasound contrast imaging, *IEEE Trans. Ultrason., Ferroelect., Freq. Contr.* 45(3): 704-711 May (1998).
16. S. Krishnan, M. O'Donnell, Transmit aperture processing for nonlinear contrast agent imaging, *Ultrasonic Imaging* 18: 77-105 (1996).

ANALYSIS OF INTRAVASCULAR ULTRASOUND (IVUS) ECHO SIGNALS FOR CHARACTERIZATION OF VESSEL WALL PROPERTIES

Wolfram Schmidt[1], Mathias Niendorf[1], Detlef Maschke[1], Detlef Behrend[1], Klaus-Peter Schmitz[1], Wilhelm Urbaszek[2]

[1]Institute for Biomedical Engineering
[2]Clinic for Internal Medicine, Department of Cardiology
University of Rostock
D-18055 Rostock, Germany

INTRODUCTION

It is established that intravascular ultrasonic imaging provides best results in quantifying atherosclerotic changes in vessels. One of the current development areas in Intravascular ultrasound (IVUS) is to improve methods for characterizing tissue properties[1]. There are numerous different approaches, using highly sophisticated methods of image analysis to obtain geometrical data such as plaque extensions (thickness) and tissue composition. These investigations are usually made in terms of gray scaled video images, which are processed from the amplitude of ultrasonic echo signals.

However, there is much more information in the radio-frequency (RF) echo signal which can be used for tissue characterization and identification[2,3].

It is the goal of our work to determine the material and structure properties of normal and pathologic vessel walls by analyzing the RF echo signals from intravascular ultrasound (IVUS) catheters. For this purpose signal analyzing techniques were adapted that provided parameters from IVUS echo signals in the time and frequency domain[4,5,6,7].

MATERIALS AND METHODS

Measurement setup: These studies were carried out with an IVUS device (CVIS Insight, Cardiovascular Imaging Systems, Sunnyvale, CA) using ultrasound catheters MicroView or MicroRail (2.9 or 3.2 F distal diameter, respectively). The transducers of both types are identical. They operate with a rotating crystal of nominal 30 MHz center frequency. IVUS images are obtained from 256 single A-scans, one scan for each 1.4° sector of the image. The maximum speed of the rotation is 30 cycles per second. This enables real time images.

The control device provides a triggered RF output where the analog echo signals can be measured after passing the patient module, which contains potential isolation and some filtering and preamplification. The signal was digitized by a digital oscilloscope (LeCroy 9350AL) at a sample frequency of 250 MHz. Every scan was sampled with 10k points of 8 bit data. Measurements of a whole image were triggered by a frame trigger signal which was provided by the control device to obtain the starting point for the image. To record all shots of the image, this frame trigger had to be added by using a trigger logic circuit because the recording of the single A-scans could only be triggered by the amplitude of the scans themselves. Therefore the phase stability of the single scans is limited. The gain adjustment of the scope was fixed at a high level (50 mV/div.) in order to obtain sufficient resolution of the backscatter amplitude. Effects of overmodulation during the transmission pulse could not be avoided, but they were reduced by using an additional high pass filter. A block diagram of the setup is provided in figure 1.

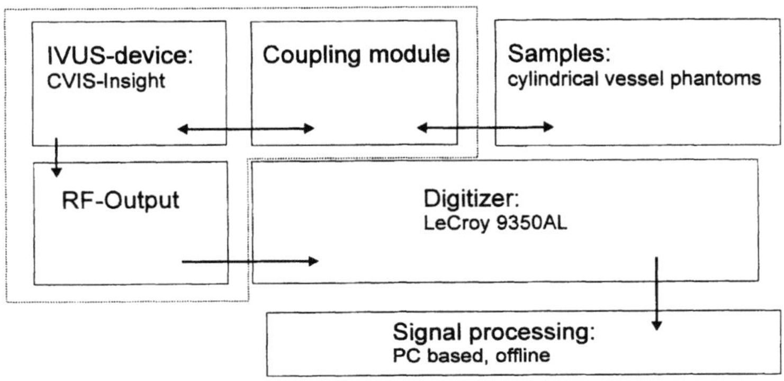

Figure 1. Test setup for IVUS RF signal measurements

IVUS catheter specifics: The IVUS catheter is the heart of the measurement but it is recommended for single use only. To minimize cost, each catheter was used several times which can be done since there were no human patients involved. Nevertheless, the characteristics of the single unit have to be taken into account. It was observed that specific parameters such as resonance frequency or electrical impedance as well as the point spread function (PSF) vary with different catheter transducers[8]. This may be caused basically by the multiple use in the lab, and it cannot be neglected.

On the other hand it is not possible to measure the catheter characteristics for each catheter before its use. For that reason a reference procedure was introduced. The echo signals of a whole image were first obtained from the catheter in a pure water bath without any other surrounding material, in order to determine the specific characteristics of the single unit. The resulting image scans were taken to compare with echoes from special phantoms and single objects in the following studies.

Test samples: Morphology, structure and components of normal as well as of atherosclerotic arterial vessels are well known from histologic studies. In depth reviews have been published elsewhere[9,10]. The earliest lesion in atherosclerotic progression is the fatty streak, which is a lipid-rich region. Advanced atherosclerosis is characterized by fibrous plaques surrounded by connective tissue and containing intracellular and extracellular lipids. Underneath this cellular structure there may be an area of necrotic tissue, cholesterol and calcification[9].

To avoid the variability of living objects, ultrasonic phantoms were used to mimic typical and constant material properties. All phantoms are based on cylindrical tubes as carriers for additional materials. The latter were chosen to simulate three types of vessel walls: Normal tissue, soft plaques and hard plaques. The procedures to produce such phantoms were described earlier[3]. The acoustical properties of the model layers were investigated in terms of IVUS and scanning acoustic microscopy (SAM) techniques in the frequency range of 25 to 30 MHz. Results of specific parameters such as the speed of sound and attenuation coefficient are given for plane films and IVUS phantoms[11].

Signal processing: The digitized echoes were used to analyze, in the time domain, the pulse shape (described by the slope of the signal envelope in V/ns), pulse duration (ns) and energy content (mW). These parameters can be calculated according to their definition in figure 2.

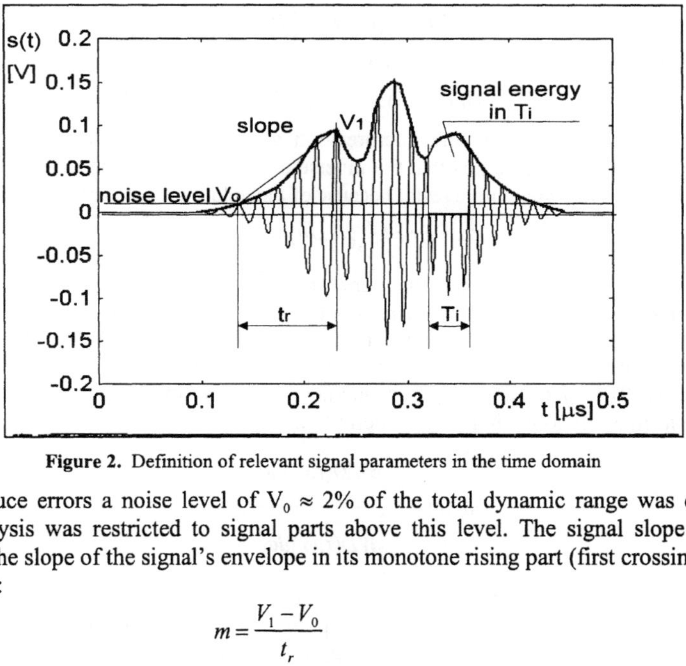

Figure 2. Definition of relevant signal parameters in the time domain

To reduce errors a noise level of $V_0 \approx 2\%$ of the total dynamic range was defined. Signal analysis was restricted to signal parts above this level. The signal slope m was defined as the slope of the signal's envelope in its monotone rising part (first crossing of the noise level):

$$m = \frac{V_1 - V_0}{t_r} \tag{1}$$

where t_r denotes the rise time of the envelope. The signal energy was calculated by integrating s(t) over t within the inspection interval T_i:

$$E(t) = \int_{t-T_i/2}^{t+T_i/2} s^2(t)dt \tag{2}$$

The interval T_i was shifted along the signal s(t) to obtain an instantaneous value of E(t) for each sample value.

Furthermore the Wigner-Ville distribution (WVD) was calculated from single A-Scans providing a time-frequency distribution of the signal. From this distribution the frequency parameters, i.e. the difference between the center frequency of the received pulse and the center frequency of the catheter transducer, as measured as the reference, were obtained.

The Wigner distribution (WD) for signals is given by:

$$\begin{aligned} WD(t,\omega) &= \int_{-\infty}^{\infty} s(t+\tau/2)s^*(t-\tau/2)e^{-j\omega\tau}d\tau \\ &= F_\tau\left\{s(t+\tau/2)s^*(t-\tau/2)\right\} \end{aligned} \tag{3}$$

Thus, the WD can be seen as the Fourier transform (FT) of the inner product. That product results from the time signal and its conjugated part, both shifted by the time variable τ. It is one of the most important properties of the WD to be considered as a time-varying power spectrum of stationary as well as of instationary signals. The WD provides instantaneous spectral information. This is in contrast to the FT which describes the spectrum of a time window.

Some further properties of the WD shall be given to prove these statements: Calculating the integral of the WD over t provides the power density spectrum $P(\omega)$:

$$\int WD(t,\omega)dt = F(\omega)F^*(\omega) = P(\omega). \tag{4}$$

The total signal power P can be obtained by integrating over t and ω:

$$\iint WD(t,\omega)dtd\omega = \int P(\omega)d\omega = P, \tag{5}$$

and the instantaneous signal power P(t) is

$$\int WD(t,\omega)d\omega = s(t)s^*(t) = |s(t)|^2 = P(t) \tag{6}$$

These relations facilitate the interpretation of the WD as a time dependent power spectrum. It must be mentioned that the inverse transformation of the time signal from the WD is restricted to a qualitative result because of the unknown constant s*(0):

$$\frac{1}{2\pi}\int WD(t/2,\omega)e^{j\omega t}dt = s(t)s^*(0) \tag{7}$$

The latter is not of special importance for the signal analysis in the current work. Further basics and interesting applications of the WD are given in the literature[4,5,6,7].

The extraction of characteristic parameters of ultrasonic backscatter from vessel or phantom structures requires a special postprocessing of the WD time-frequency distribution. The first order frequency moment of every time slice was calculated by a discrete form of the following analytic definition:

$$\bar{\omega}(t) = \frac{\int_{-\infty}^{\infty}\omega WD(t,\omega)d\omega}{\int_{-\infty}^{\infty}WD(t,\omega)d\omega}. \tag{8}$$

For this calculation, the integral limits had to be changed to finite record length and integration was performed by a discrete approximation. The width of frequency steps depends on the length of the time window, as known from Fourier transformation.

In order to make the mean frequency parameter more sensitive and comparable to the measurements with different catheter transducers, the first order moment $\bar{\omega}(t)$ was compared to the mean frequency of the catheter reference signal $\bar{\omega}(cath)$ (region of ring-down artifacts) by calculating the difference:

$$\bar{\omega}_r(t) = \bar{\omega}(t) - \bar{\omega}(cath) \tag{9}$$

In this way the catheter specific characteristics could be considered.

RESULTS

IVUS measurements were done with several different phantoms. In order to find signal parameters, which can be based on acoustical effects, only two types of phantoms shall be discussed in the following: soft and hard plaque phantoms. Typical echo signals are shown in figure 3 and figure 4 for the two types.

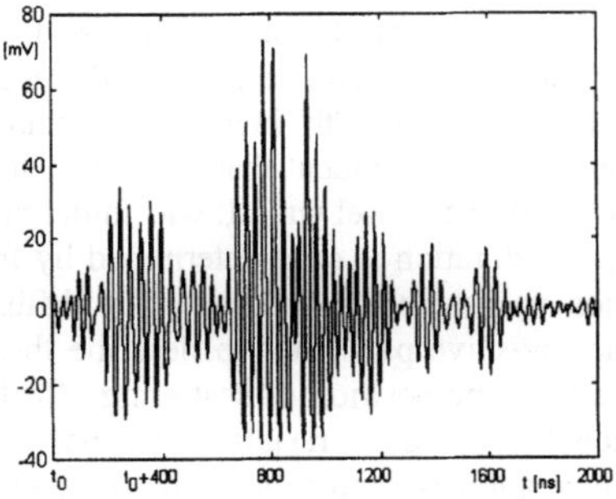

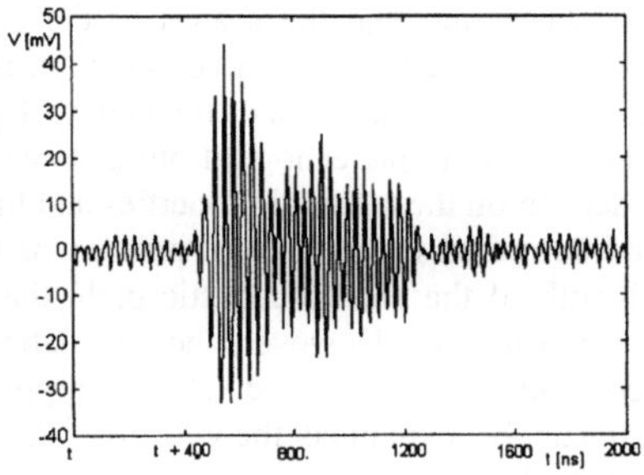

Figure 3. Echo signal from soft plaque phantom (detail)

Figure 4. Echo signal from hard plaque phantom (detail)

It can be seen that there are differences in the echo signals, caused by the different structure of the insonated material and the distance between the transducer and the backscattering structure. For quantitative analysis and for obtaining objective parameters the calculation of signal parameters was performed, both in the time as well as in the frequency domain.

Extracting the mean frequency parameter, the WD was calculated for each echo containing typical structures. In figure 5 and 6, the WD for soft and hard plaque phantom signal is shown as an example. Note, that the figure is zoomed to the pulse region. This way the catheter artifacts in the beginning of each pulse are removed.

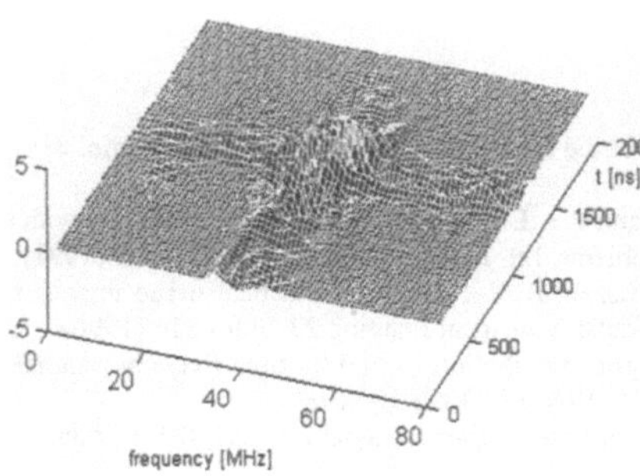

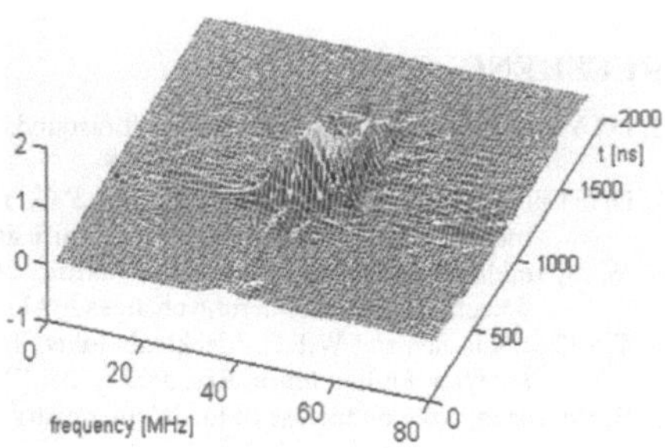

Figure 5. WD of the soft plaque phantom signal from fig. 3 (zoomed into the plaque echo region)

Figure 6. WD of the hard plaque phantom signal (plaque echo region from fig. 4)

The relative shift of the mean frequency within the pulse region was derived from the IVUS echo signal of both phantom types.

DISCUSSION

In order to classify the constitution of the vessel wall several parameters of the ultrasonic echo signal have to be included. The use of the WD for time-frequency distributed analysis provides additional information in the frequency domain without loosing the time resolution. The local position can be determined by the time information assuming a specific speed of sound in the IVUS system. Changes in frequency content caused by various effects, e.g. frequency dependent attenuation or scattering, can be observed and quantified.

It was found that the shapes of the echoes, represented by the slope and rise time parameters as well as the energy content, are determined by several complex influences. Effects of the distance and frequency dependent attenuation with increasing distance overlap with multiple echoing from different surfaces. But it is evident that the parameters also depend on the material properties and the structure of the vessel wall. It was found that in addition to an ideal reflected pulse, where the pulse duration is only determined by the bandwidth of the electro-acoustic path, the duration is increased significantly by diffuse echoing structures. However, the calculation of an objective parameter to describe these effects, could not be derived. The main problem is the superposition of scattering effects, which can be explained theoretically[12,13]. Additional disturbance results from the non-stationary location of the rotating transducer. The rotation is accompanied by a varying distance between the transducer and the catheter tube, which is the first echoing object in the scan. This configuration provides multiple echoes and interferences.

Analyzing the WVD, a shift of the center frequency to lower values was observed at the surface of weak phantoms. This should be caused by the rough and irregular structure of these phantom materials even at the surface. The effect could not be shown with homogenous harder materials such as pure polyurethane or metal surfaces.

Further studies will have do deal with the verification of the extracted parameters. Up to now no single parameter was found that could be used alone for classification of the vessel tissue.

ACKNOWLEDGEMENT

This work is supported by the German Research Foundation (DFG, INK27/A1-1).

REFERENCES

1. PG Yock, PJ Fitzgerald: Intravascular Ultrasound: State of the art and future directions, Am J Cardiol 81: 27E-32E (1998)
2. D.T. Linker, A. Kleven, A. Gronningsaeter, P.G. Yock, and B.A.J. Angelsen: Tissue characterization with intra-arterial ultrasound: special promise and problems, Int. J. of Cardiac Imag. 6: 225-263 (1991)
3. W. Schmidt; D. Behrend, O. Skerl, H. Martin, W. Urbaszek, K.-P. Schmitz: Acoustical tissue images for detection of atherosclerotic changes in blood vessels; Acoustical Imaging 22: 209 - 214 (1996)
4. T.A.C.M. Claasen and W.F.G. Mecklenbräuker, The Wigner distribution - a tool for time-frequency signal analysis, Philips Journ. Res. 35:217-250(I), 276-300(II), 372-389 (III) (1980)
5. B. Boashash, Note on the use of the Wigner distribution for time-frequency signal analysis; IEEE Trans. ASSP 36:1518-1521 (1988)
6. O. Skerl, and I. Hartmann, The Wigner distribution in Doppler sonography, Acoust. Imag. 19:335-339 (1992)
7. O. Skerl, W. Schmidt, and O. Specht, Wigner-Verteilung als Werkzeug zur Zeit-Frequenz-Analyse nicht-stationärer Signale; Technisches Messen 61:7-15 (1994)
8. W. Schmidt, M. Niendorf, D. Behrend, K.-P. Schmitz, W. Urbaszek: Elimination of Specific Transducer Influence on Signals from Intravascular Ultrasonic Catheters; Proc. of the 18th Annual Conference of IEEE Engineering in Medicine and Biology Society, Amsterdam, 31.10.-3.11.1996
9. R. Ross, The pathogenesis of atherosclerosis - an update; The New England J. of Med. 314:488-500 (1986)
10. J. Honye, D.J. Mahon, A. Jain, C.J. White, S.R. Ramee, J.B. Wallis, A. Al-Zarka, and J.M. Tobis, Morphological effects of coronary balloon angioplasty in vivo assessed by intravascular ultrasound imaging, Circulation 85:1012 - 1025 (1992)
11. W. Schmidt, M. Niendorf, D. Behrend, K.-P. Schmitz: Ultrasonic Investigation of normal and pathologic arterial walls; Medical & Biological Engineering & Computing 35: 404 (1997) Suppl. I
12. J.A. Jensen, A model for propagation and scattering of ultrasound in tissue, JASA 89:182-190 (1991)
13. B.A.J. Angelsen: Wave propagation and scattering in inhomogeneous tissue; in:Waves, signals and signal processing in medical ultrasonics, vol. II, Trondheim, Norway (1996)
14. M. Niendorf, W. Schmidt, D. Maschke, D. Behrend, K.-P. Schmitz, W. Urbaszek: Zeit- und Frequenzparameter zur Analyse von IVUS-Echosignalen, Biomedizinische Technik 43: 104 - 105 (1998) Suppl. I

IN VIVO STUDY OF THE INFLUENCE OF GRAVITY ON CORTICAL AND CANCELLOUS BONE VELOCITY

PP Antich[12], S Mehta[12], M Daphtary[1], M Lewis[1], B Smith[1] and CYC Pak[2]
[1]Advanced Radiological Sciences, Department of Radiology
[2] Center for Mineral Metabolism and Clinical Research
University of Texas Southwestern Medical Center at Dallas
5323 Harry Hines Blvd, Dallas, Texas 75235

ABSTRACT

This study presents new information obtained using ultrasound on the elasticity and anisotropy of human bone tissue in normal conditions and on their changes in the absence of gravity. Strict bedrest maintained for twelve weeks was used to simulate the absence of gravity encountered in space flight. The elastic properties of bone were assessed via the ultrasound velocity, measured noninvasively and nondestructively using a novel device developed in our laboratory for these studies, based on the principles of UCR (Ultrasound Critical-angle Reflectometry).

INTRODUCTION

Weight-bearing physical activity is essential for the development and maintenance of a healthy musculoskeletal system: conversely, prolonged bedrest and inactivity are detrimental and increase the risk of bone fractures.[1] The loss of both muscle and bone mass has been documented. [1-4] Just 4-6 weeks of bedrest result in a substantial loss of muscle mass and a concomitant decrease of up to 40% in muscle strength; in parallel with these, bone mineral density decreases throughout the skeleton.[2] Based on physiological principles, equally significant changes are expected to occur in the the tissue themselves: these remain, however, largely unexplored. Our research is directed at obtaining new information on the elasticity and anisotropy of human bone tissue and on its changes in space flight.

The normal organization of structural and material properties of bone represents an optimal adaptation to gravity, as normal gravity determines the preferential direction of loading for bone, and bone tissue is modeled and remodeled to accommodate loads. It is then natural to test the hypothesis that removal of gravity will alter this organization. This hypothesis has not been extensively studied heretofore in humans, and this work presents new evidence that i) under normal conditions there are substantial differences in bone elasticity and anisotropy in different bones, ii) the absence of gravity produces rapid, site- and orientation-dependent changes in bone elasticity and iii) these changes are different in cancellous and cortical bone.

For this study we employed the prolonged bedrest model of space flight, relying on its ability to mimic the absence of gravity, and measured the ultrasound velocity in bone in vivo, noninvasively and nondestructively. The velocity is an important parameter, as a deterministic relation exists between it and elasticity or stiffness ($E = \rho \cdot V^2$); furthermore, various studies have shown a strong positive correlation between velocity and mechanical strength in bone [3-5]. It should be noted that the measurement of elasticity through ultrasound velocity and density is routinely used in mechanical engineering and in material science, and has been tested in bone studies by several researchers. [6-13] Although various methods and devices exist for measuring ultrasound velocity in bone [14], a device based on the principles of UCR (Ultrasound Critical-angle Reflectometry) [15] did not previously exist was specifically developed in our laboratory for these studies.

MATERIALS AND METHODS

In UCR, velocities are measured by detecting the angle at which total internal reflection occurs in the medium in which the ultrasound beam originates, upon its arrival at a bone surface. The usefulness of UCR in bone studies is due to the fact that it permits i) the absolute measurement of both pressure and shear wave velocity at multiple orientations, ii) the separate measurement of cortical and trabecular bone, iii) can be applied both *in vivo* and *in vitro* [15-19]. Here we summarize its salient points. A pressure wave (velocity *c*) propagating in water, upon arriving at the surface of a bone sample, gives rise to a reflected wave and two waves propagating in bone, pressure waves of velocity v_p and shear waves of velocity v_s. As the angle of incidence θ (angle between the direction of propagation of the wave and the normal to the bone surface) increases, it attains two critical angles of incidence for the pressure and the shear wave respectively. At these two angles, recognized as discontinuities in the reflected amplitude and phase, the pressure and shear wave velocities are unambiguously obtained from their values and the velocity of sound in water c, using $v_i = c/\sin\theta_i$. As the velocity of sound in water is accurately known, the measurement of critical angle of incidence in water constitutes an absolute measurement of ultrasound velocity in bone. As indicated in Fig.1, cortical and trabecular bone volumes can be distinguished at the same site exploiting the physical separation of the beams following

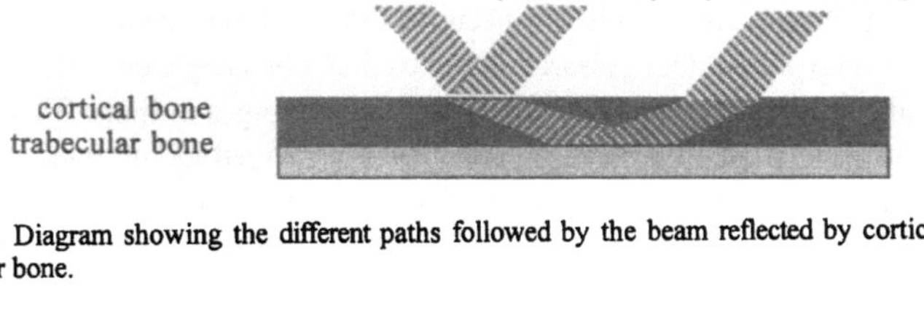

Figure 1. Diagram showing the different paths followed by the beam reflected by cortical bone and trabecular bone.

reflection from, and refraction in, the two bone volumes. The two reflected beams are separately detected at different locations and times (phases), allowing measurement of the properties of the two bone tissues: the lateral displacement of the beam can be used to infer the depth of measurement. The formal justification for this method for obtaining trabecular bone properties is briefly outlined in the Appendix. Experimentally, the values measured at the surface are indeed different from those measured at depth when the sample is heterogeneous. This is the case of bone, which consists of three structurally and functionally distinct regions: cortical, endosteal and trabecular. By mechanically removing successive surface layers of a bone sample, the "at depth" measurement has been validated by direct comparison with the surface measurement.

Using these principles, a device was built in our laboratory, in which a latex membrane separates the applicator from the body and which allows the observation of bone properties at multiple sites in the upper, lower and middle skeleton for this study. The applicator, which houses the transducers and the water calibration medium, is placed in contact with the skin over the bone of interest via this membrane. Movements of the applicator are controlled by the operator via computer to permit velocity measurement at any selected point and orientation in the area of interest, as well as to select reflections from different depths.

RESULTS

In the present analysis, completed in ten patients, we utilized an average of over 25 individual determinations per site per patient for cortical and trabecular bone along different orientations, pre- and post- bedrest. For bone material the velocity is expected to have a linear-quadratic dependence upon the square of the cosine of the orientation, based on an assumed hexagonal symmetry for its matrix[16]: in vivo measurements closely conform to this expectation as shown in Fig.2

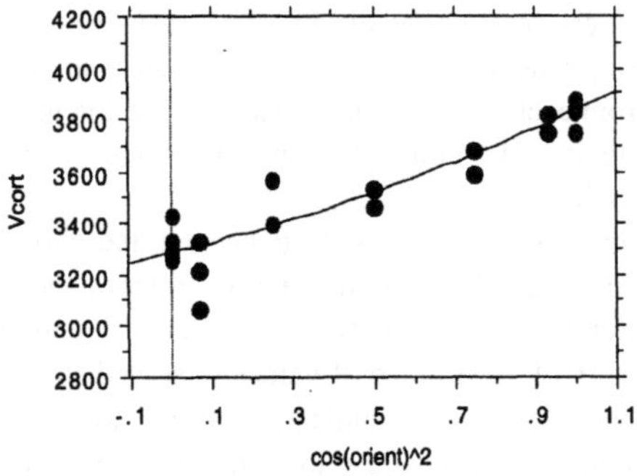

Fig. 2. Cortical bone velocity as a function of the angle between the plane of measurement and the long axis of the tibia. The data for an individual subject follows the expected linear-quadratic relationship, with an R^2 of 0.89, P=.0001.

To minimize the time necessary to complete a study, the angular dependence was taken into account in most subjects by measuring velocities in a restricted range of angles at two orthogonal orientations, parallel (Longitudinal) and perpendicular (Transverse) to the axis of the body. Multiple measurements within a small area were averaged to result in a precision ≈ 1%.[16] The velocities were derived from 258 spectra obtained from the calcaneus, 194 from the tibia, 244 from the trochanter, 287 altogether from the left forearm (1/2 in the radius and 1/2 in the ulna) and 223 from the forehead. At each site, pre- and post-bedrest values of the trabecular and cortical bone velocity were obtained and separately averaged for the two orientations. Significant differences exist between both cortical and cancellous bone velocities at the different sites as shown in Table I for the trabecular bone velocity in the longitudinal direction at various sites before bedrest.

TABLE I -

TRABECULAR BONE VELOCITY AT FIVE DIFFERENT SITES BEFORE BEDREST

a-longitudinal orientation

	calcaneus	trochanter	radius	ulna	forehead
calcaneus	2930 (30)	.95	.0001	.0001	.0016
trochanter		2933 (29)	.0001	.0001	.0019
radius			3144 (27)	.85	.29
ulna				3152 (29)	.22
forehead					3090 (35)

b-transverse orientation

	calcaneus	trochanter	forehead
calcaneus	3126 (32)	.0001	.0001
trochanter		2973 (32)	.24
forehead			3094 (35)

c-dependence upon orientation

	longitudinal	transverse	P
calcaneus	2930 (30)	3126 (32)	.0001
trochanter	2933 (29)	2973 (32)	.33
forehead	3090 (35)	3094 (35)	.95

TABLE I: Mean values and standard errors of the trabecular bone velocity for five bones at two orientations prior to bedrest, and P values for the difference between values.

In the longitudinal direction, the values at the calcaneus and trochanter were close to each other and significantly different from those at the radius, ulna and forehead; these in turn were substantially identical to each other. In the transverse direction, the value at the calcaneus was substantially higher than at the trochanter and forehead, which were not significantly different from each other. At the calcaneus, the longitudinal velocity was lower than the transverse velocity. By contrast, trabecular bone at the trochanter and at the forehead appeared isotropic.

No definite relationship was found between trabecular and cortical bone velocities at the different sites, or between longitudinal and transverse values of the trabecular bone velocity, and only a weak correlation is found between longitudinal and transverse cortical bone velocities with R=.36, p=.014. The range of values observed in the same individual under normal conditions for the different bones indicates that the biomechanical proeprties of bone tissue have high plasticity, as necessary to achieve optimal adaptation to functional requirements, which are influenced by metabolic and physiological processes as well as gravity.

The effects of gravity alone are better revealed by the changes from baseline values induced by bedrest. The pre- and post- values both at all sites and for the lower skeleton (calcaneus, tibia), the middle skeleton (forearm and trochanter) and the forehead are shown in Table II for all patient data combined. Overall, there was a significant decrease in the trabecular bone velocity along the transverse axis (orientation T), while the cortical bone velocity increased significantly in both directions. In addition, the values of the velocities were strongly site-dependent, as were their responses to the absence of gravity. Significant differences between the lower, middle and upper skeleton are shown in Table II.

TABLE II - BONE VELOCITY BEFORE AND AFTER BEDREST

orientation	sites	trabecular				cortical			
		Vpre	Vpost	dV	p	Vpre	Vpost	dV	p
L	all	3042	3006	-37	.08	3804	3854	**50**	**.009**
T	all	3045	2939	**-105**	**.002**	3514	3637	**123**	**.0008**
L	lower	2930	2973	43	.30	3757	3918	**166**	**.0001**
T	lower	3126	3004	**-122**	**.04**	3413	3659	**246**	**.0001**
L	middle	3075	2997	**-78**	**.003**	3889	3895	6	.81
T	middle	2973	2931	-42	.41	3830	3781	-49	.55
L	head	3090	3110	20	.69	3670	3574	**-95**	**.02**
T	head	3094	2867	**-227**	**.0006**	3477	3498	28	.28

As already seen for the baseline values, the values of trabecular and cortical bone and their changes were markedly site-dependent and uncorrelated to each other. Indeed, there were substantial differences between changes in the two bone types at the two orientations and among sites. The changes tended to have opposite sign in trabecular and cortical bone, so that bedrest resulted in a redistribution of bone elasticity and strength among the two bone types. These observations demonstrate that it is not possible to assume that equal changes occur at different sites and for the two bone types.

Further evidence for a substantial rearrangement of structural and material properties was seen in the changes in the dependence of velocity upon orientation. The bedrest-induced changes at the different sites for the anisotropy (with sign) are shown in figure 3 below.

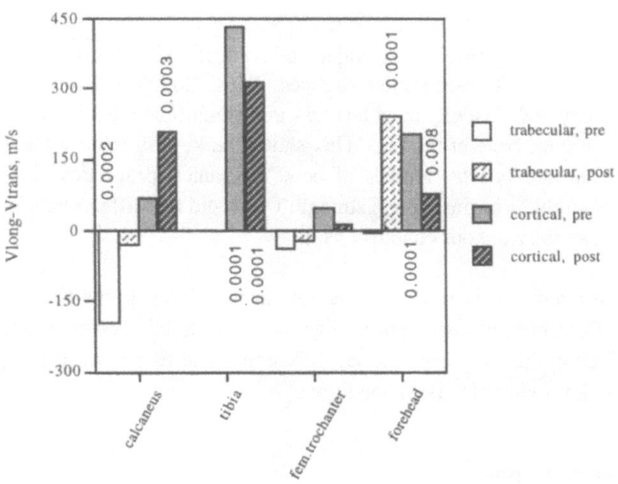

Figure 3. Anisotropy (V_{long}-V_{trans}) of the trabecular and cortical bone velocity before and after bedrest at the calcaneus, tibia, femural trochanter and forehead. (significance as shown)

The anisotropy at the calcaneus decreased in trabecular and increased in cortical bone; at the forehead these trends were reversed. At the tibia the anisotropy of cortical bone decreased while at the greater trochanter both trabecular and cortical bone velocities remained substantially unchanged.

DISCUSSION

Cortical and trabecular bone velocities were separately measured *in vivo* at different skeletal sites in ten subjects using UCR before and after prolonged bedrest, a well accepted clinical model of the effects of absence of gravity and therefore of space flight.

The bone velocities before bedrest were markedly different among sites, both in their value and dependence upon orientation. It is natural to hypothesize that the initial intersite differences delineate the range of values available to normal physiological processes for the remodeling of bone material subjected to different loading patterns, and that the changes observed after bedrest indicate the extent to which gravity (as distinguished from metabolic or genetic causes) affects these values.

After twelve weeks of bedrest there were significant changes in the relations between the elastic properties of trabecular and cortical bone. In general the trabecular bone velocity decreased, indicating a loss in trabecular bone strength, but this trend did not hold for cortical bone, accentuating the differences between the two bone materials. In the lower appendicular skeleton the anisotropy decreased in trabecular, but increased in cortical bone at the calcaneus, whereas at the tibia the cortical bone anisotropy remained high but decreased after bedrest. In the middle skeleton the changes were less pronounced: at the greater trochanter both trabecular and cortical bone velocities remained substantially unchanged. Changes at the forehead were opposite to those at the calcaneus.

There was no correlation between values of cortical and trabecular bone velocity, nor were there correlations between their changes. Thus, both the baseline values and the changes in velocity and elasticity must be measured at multiple sites in order to disclose the effects of an agent on bone properties. This should be kept in mind when interpreting the results of *in vivo* and *in vitro* studies of bone mechanical properties; in particular, it is apparent that a study of intrinsic bone strength which did not differentiate between the two tissues would lead to erroneous conclusions.

Thus, bedrest resulted in profound changes in the elastic properties of trabecular and cortical bone. The implications for space flight are that effective countermeasures should be capable of affecting the two bone dissues differently, and to restore the dependence upon site and orientation found in normal conditions.

ACKNOWLEDGMENTS
This study was supported by NASA (NAGW 3582).

APPENDIX

Studies of trabecular bone properties *in vivo* are complicated by the fact that this tissue lies under a cortical bone layer. A partial wave analysis of multiple scattering and reflection shows that while the first reflected wave is influenced exclusively by cortical bone properties, scattering from trabecular bone contributes to subsequent reflections, the order of which is indicated below with the superscript i. Let water be characterized by its density ρ, velocity c; cortical bone by density ρ', pressure wave velocity v and shear wave velocity w; trabecular bone by ρ", u, ç, , while the angles with the normal are θ, β and γ, η and χ.for the incident and refracted waves respectively. Using Snell's law $\text{Sin}\theta/c=\text{Sin}\beta/v = \text{Sin}\gamma/w=\text{Sin}\eta/u=\text{Sin}\chi/ç$, and defining the phase terms $F_s=e^{-i\omega x\text{Cos}\gamma/w}$ and $F_p= e^{-i\omega x\text{Cos}\beta/v}$ it may be seen that the amplitudes of the reflected (W^i), refracted (T^i), internally reflected (R^i) and transmitted (T'^i) waves are related to each other and to the incident wave amplitude (J) via the equations of continuity of the normal components of the displacements (eq. 1 below) and of the stress tensor at the first interface (eq. 2,3) and of the normal (4) and tangential (5) components of the displacements and the normal components of the stress tensor (6,7) at second interface.

1. $(J-W^i)\text{Cos}\theta=(T_p^i-F^2_pR_p^{i-1})\text{Cos}\beta-(T_s^i-F^2_sR_s^{i-1})\text{Sin}\gamma$
2. $J+W^i = \rho'v(T_p^i+F^2_pR_p^{i-1})\text{Cos}2\gamma- \rho'w(T_s^i+F^2_sR_s^{i-1})\text{Sin}2\gamma$
3. $(T_p^i-F^2_pR_p^i) = -(T_s^i-F^2_sR_s^{i-1})\text{Cos}2\gamma/2\text{Sin}\gamma\text{Cos}\beta$
4. $F_p(R_p^i-T_p^i)\text{Cos}\beta-F_s(R_s^i-T_s^i)\text{Sin}\gamma=-T'^i_p\text{Cos}\eta+T'^i_s\text{Sin}\chi$
5. $F_p(R_p^i+T_p^i)\text{Sin}\beta+ F_s(R_s^i+T_s^i)\text{Cos}\gamma= T'^i_p\text{Sin}\eta+T'^i_s\text{Cos}\chi$
6. $\rho'v\text{Cos}2\gamma F_p(R_p^i+T_p^i)- \rho'w\text{Sin}2\gamma F_s(R_s^i+T_s^i)= \rho''uT'^i_p\text{Cos}2\chi- \rho''çT'^i_s\text{Sin}2\chi$
7. $\rho'w[F_p(R_p^i-T_p^i)\text{Sin}\gamma\text{Cos}\beta+F_s(R_s^i-T_s^i)\text{Cos}2\gamma/2] = - \rho''ç[T'^i_p\text{Cos}\eta\text{Sin}\chi+T'^i_s\text{Cos}2\chi/2]$

These equations constitute a system of 7 independent equations which can be solved iteratively for the 7 unknowns R .... T'_s . The superscript i=0 indicates the first reflection of an incoming wave front (for which $R_p^{-1}= R_s^{-1}=0$); i=1 indicates the next reflection (for which $R_p^{-1}\neq R_s^{-1}\neq 0$). Without developing here the most general solution, it can be readily seen that for a sufficiently thin cortical bone layer, the problem is formally identical to that for reflection from a homogeneous sample of cancellous bone. With $F_p\approx F_s\approx 1$, if we use the first 3 equations in the second 4 we find that

4. $(J-W^i)\text{Cos}\theta =T'^i_p\text{Cos}\eta-T'^i_s\text{Sin}\chi$
6. $J+W^i=\rho''uT'^i_p\text{Cos}2\chi- \rho''çT'^i_s\text{Sin}2\chi$
7. $0= - \rho''ç[T'^i_p\text{Cos}\eta\text{Sin}\chi+T'^i_s\text{Cos}2\chi/2]$

These 3 equations in the 3 unknowns W^i, T'^i_p and T'^i_s in which the details of refraction and reflection in cortical bone are entirely absent, can be solved to give:

- $W^i=J(1-P/\text{Sin}\chi\text{Cos}\theta)$

with $P=1/(\rho''u\text{Cos}^22\chi/2\text{Sin}\chi\text{Cos}\eta+\rho''ç\text{Sin}2\chi+1/2\text{Sin}\chi\text{Cos}\theta)$.

The solution is considerably more complicated when F_p and F_s are significantly different from unity, but the main conclusion, that a solution dependent only upon trabecular values exists, holds for cortical bone thicknesses not exceeding ≈ 1 mm at 3.5 MHz. The device is accordingly designed so that the receiver either accepts only reflections from the first surface, or avoids these reflections while accepting reflections generated from internal reflection in the cortical bone layer. The trabecular bone parameters can be then derived as in the case of cortical bone. It should be noted that the procedure has been validated experimentally.[20]

BIBLIOGRAPHY

1. Anon, American College of Sports Medicine position stand: Osteoporosis and exercise Med Sci Sports Exerc 27: I-VII (1995)
2. SA Bloomfield: Changes in musculoskeletal structure and function with prolonged bedrest Med Sci Sports Exerc 29:197-206 (1997)
3. I Hvid, N Jensen, C Bunger, K Solund , J Djurhuus Bone mineral assay: its relation to the mechanical strength of trabecular bone Eng Med 14:79-83 (1985)
4. J Vahey, J Lewis, R Vanderby: Elastic moduli, yield stress, and ultimate stress of trabecular bone from the proximal tibial epiphysis J Biomech 20:29-33 (1987)
5. C Turner, M Eich 1991 Ultrasonic velocity as a predictor of strength in bovine trabecular bone Calcif Tiss Int 49(2):116-119
6. R Ashman, S Cowin, W Van Buskirk, J Rice: A continuous wave technique for the measurement of the elastic properties of cortical bone J Biomech 17(5):349-361 (1984)
7. W Bonfield, A Tully Ultrasonic analysis of the Young's modulus of cortical bone J Biomedical Engineering 4(1):23-27 (1982)
8. J Katz Anisotropy of Young's modulus of bone Nature 283:106-107 (1980)
9. H Kim, W Walsh Mechanical and ultrasonic characterization of cortical bone Biomimetics 1:293-370 (1993)
10. S Lang Ultrasonic method for measuring elastic coefficients of bone and results of fresh and dried bones IEEE Trans, Biomedical Engineering 17:101-105 (1970)
11. S Lees, J Heeley, P Cleary A study of some properties of bovine cortical bone using ultrasound Calcif Tiss Intnl 29:107-117 (1979)
12. J Rho, R Ashman, C Turner Young's modulus of trabecular and cortical bone material: ultrasonic and microtensile measurements J of Biomechanics 26(2):111-119 (1993)
13. W Van Buskirk, S Cowin, R Ward Ultrasonic measurement of orthotropic elastic constants of bovine femoral bone J of Biomechanical Eng 103:67-72 (1981)
14. J Kaufman, T Einhorn Perspectives: Ultrasound assessment of bone J Bone and Min Res 8(5):517-525 (1993)
15. P Antich, S Mehta, M Daphtary, B Smith, T Nguyen, V Vaguine Bone elastometric imaging using ultrasound critical-angle reflectometry Ultrasonic Imaging Vol 23 145-150, (1997)
16. P Antich, J Anderson, R Ashman, JDowdey, J Gonzales, R Murry, J Zerwekh, CYC Pak Measurement of mechanical properties of bone material in vitro by ultrasound reflection: methodology and comparison with ultrasound transmission J Bone and Min Res 6(4):417-426 (1991)
17. P Antich: Ultrasound Study of Bone in Vitro: Calcif Tissue Int, 53 (Suppl 1):S157-S161, (1993)
18. R Ashman, P Antich, J Gonzales, J Anderson, J Rho A comparison of reflection and transmission ultrasonic techniques for measurement of trabecular bone elasticity J Biomechanics 27(9):1195-1199 (1994)
19. P Antich, S Mehta Ultrasound Critical-angle Reflectometry (UCR): A new modality for functional elastometric imaging Physics Med Biol 42, 1763-1777 (1997)
20. J Gonzales, Measurement of Cancellous bone velocity by reflection ultrasound: application to the assessment of bone response to treatment in osteoporotic patients, PhD Dissertation, Southwestern Graduate School of Biomedical Sciences, The University of Texas Southwestern Medical Center at Dallas (1993).

ULTRASOUND CONTRAST IMAGING OF PROSTATE TUMORS

F. Forsberg,[1] M. T. Ismail,[2] E. K. Hagen,[3] D. A. Merton,[1] J. B. Liu,[1]
L. Gomella,[2] D. K. Johnson,[3] P. E. Losco,[3] G. J. Miller,[4] P. N. Werahera,[4]
R. deCampo,[4] J. S. Stewart,[4] A. K. Aksnes,[3] A. Tornes,[3] B. B. Goldberg[1]

[1] Department of Radiology
[2] Department of Urology
Thomas Jefferson University
Philadelphia, PA 19107, USA
[3] Nycomed-Amersham
Oslo, N-0401, Norway
[4] Department of Pathology
University of Colorado Health Sciences Center
Denver, CO 80262, USA

INTRODUCTION

Prostate cancer was the second-leading cause of cancer related deaths among men in the United States in 1993.[1] The detection of prostate cancer can be difficult with most physicians relying on a combination of digital rectal examination, prostate specific antigen level and transrectal ultrasound (TRUS).[2,3] However, the standard gray scale detection with TRUS of prostate cancer has limitations, with many investigators reporting low predictive values ranging from 18 % to 60 %.[4,5,6] These low predictive values reflect the nonspecificity of the sonographic features utilized to predict the presence of cancer.

Color Doppler imaging (CDI) has been suggested to identify foci of increased flow (i.e., tumor neovascularity) and, thus, improve the detection of lesions that warrant biopsy. Brawer et al. microscopically demonstrated increased angiogenesis in adenocarcinoma compared with benign prostate tissue and that a stepwise increase in angiogenesis was seen toward the center of a prostate tumor.[7,8] However, pathology does not allow assessment of the dynamic flow abnormalities present in prostate cancer. Hence, the need for CDI which has produced somewhat encouraging results for the detection of tumor blood flow [9,10] and for determining the cancer prognosis.[11]

Acoustical Imaging, Volume 24.
Edited by Hua Lee, Kluwer Academic/Plenum Publishers, 2000.

However, conventional CDI may not be sufficiently sensitive to detect flow in some prostate cancers, due to the very slow flow involved as well as the small vessel size and tortuousity associated with tumor angiogenesis. Ultrasound contrast agents may alleviate these problems given the up to 25 dB enhancement of Doppler signals produced by injection of such gas microbubble based agents.[12] The potential for improving diagnostic outcomes by using an ultrasound contrast agent in conjunction with TRUS to detect tumors earlier is, therefore, of utmost importance. A contrast agent which can be seen in the small vessels and provides marked vascular enhancement is highly desirable. Prostate cancers should be easier to detect with such an agent. This in turn should lead to a more accurate diagnosis — in the short term by enabling better biopsy guidance and in the long turn by providing the means for tumor characterization based on the tumor's contrast uptake and washout characteristics.

To investigate the potential of contrast enhanced ultrasound imaging of prostate lesions a novel contrast agent (NC100100; Nycomed-Amersham, Oslo, Norway) was evaluated in canines. The dog was chosen as the animal model to ensure a prostate of a size comparable to the ones found in humans.

METHODS

Twenty-two mongrel dogs were used in this study. Sedation of the dogs was achieved by intramuscular administration of a mixture of 0.04 mg/kg atropine sulfate (Anthony Products, Arcadia, CA), 0.75 mg/kg acepromazine (Promace, Aveco, Fort Dodge, IA), and 23 mg/kg ketamine hydrochloride (Ketaset). Anesthesia was maintained with continuous infusion of 15 mg/kg/hour of Diprivan (10 mg/ml concentration; Veneca Pharmaceuticals, Wilmington, DE). The dogs received intravenous injections of a 1 % solution of NC100100 (at a rate of 1 - 2 ml/s) in dosages of 0.00625 - 0.05 μl microbubbles/kg bodyweight through an 18 gauge angiocatheter placed in a fore limb vein. This contrast agent consists of encapsulated microbubbles of a perfluorocarbon gas with a relatively narrow size distribution and a mean diameter of 3 μm. Each animal received a Fleets enema to facilitate endorectal ultrasound scanning.

The dogs were studied in three phases. In phase I, 8 normal dogs were examined to establish the vascular patterns of the normal prostate when imaged with ultrasound contrast. In 4 of the normal dogs a radio-opaque silicone microfil (Flow Tech, Carver, MA) was injected immediately prior to sacrifice producing a cast of the vascularity, which was then x-rayed. In the remaining 4 dogs colored microspheres were injected to allow absolute perfusion in ml/min/g to be determined (for further details, see Chaudhari et al.[13]). Next, phase II simulated prostate cancer by inducing lesions (i.e., pseudo-tumors) in 6 dogs. These lesions were created by placing a needle-based laser probe directly into the gland and inducing tissue ablation using a CL60 laser system (Surgical Laser Technologies, Oaks, PA) on a 10 W setting for 3 minutes. Finally, phase III of our studies focused on the development of a unique new animal model for investigating prostate cancer. Following subcutaneous growth in a host dog, 8 dogs were injected with a Canine Transmissible Venereal Sarcoma (CTVS) cell line (50 million cells/ml) directly into the prostate. The dogs were mildly immunosuppressed and no clinical effects nor abnormal hematological parameters were found. The CTVS cell line is capable of growing and producing a cancer-like mass in the dog prostate, and tumor growth was noted within 10 days after the cell line injection. By 20-30 days post injection, multi-nodular tumors were evident in the prostate (as established by digital rectal examination). After the lesions had grown to a suitable size the dogs underwent ultrasound contrast studies.

Transrectal CDI and color amplitude imaging (CAI; a.k.a. power Doppler) of the prostate were performed in 2D and 3D with a C9-5 end-fire probe and an HDI 3000 scanner (ATL, Bothell, WA). The 3D imaging technique is based on free hand data acquisition and reconstruction of CAI images stored in the scanner's cine-loop. To limit the color blooming artifact associated with ultrasound contrast agent studies[14] the pulse repetition frequency (PRF) was adjusted to display only limited flow information prior to injection. All imaging parameters were kept constant during injections. To avoid any cumulative effect of contrast sufficient time was allowed between injections (10 -15 minutes) to ensure a return to baseline conditions. After the ultrasound examinations the dogs were sacrificed and pathology performed.

The animal studies reported here were carried out in an ethical and humane fashion under the direct supervision of a veterinarian. All protocols were approved by the University's Animal Use and Care Committee.

RESULTS

In the 8 normal dogs, injection of NC100100 demonstrated a radial, spoke-like pattern of vascularity (Figure 1). First the peripheral arteries enhanced (within 10 - 15 seconds) in the arterial phase followed by venous flow drainage (mainly in the urethral area) in the venous phase some 5 to 6 seconds later. Overall vascular enhancement was observed in the normal glands for more than 10 minutes even for dosages as low as 0.00625 μl microbubbles/kg (see Figure 1). Both 2D and 3D CAI images were found to correspond well with the vascularity seen on the microfil x-ray images and by pathology. The 3D CAI display produced better visualization of the vascularity than did 2D CAI (compare Figures 1B and 1D). Finally, the continuity of the spoke-like vascular pattern was better visualized with CAI than with CDI (images omitted for the sake of brevity) and, therefore, only CAI was performed in Phase II and III.

Five pseudo-tumors were created in five dogs (the sixth study failed due to a laser malfunction) ranging in size from 5 x 6 mm to 10 x 14 mm as seen by ultrasound with contrast. All lesions were more clearly delineated in both 2D and 3D after administration of NC100100 and matched well with pathology as shown in Table 1. An example of a tissue ablation study is presented in Figure 2. Pre contrast injection the lesion was thought to be relatively small (9 x 3 mm; Figure 2A), but after administration of NC100100 the true extent of the lesion was better appreciated (cf., Figures 2C and 2D). Although this pseudo-tumor appeared to disrupt normal vascularity within an entire lobe of the prostate (Figure 2C) it was estimated by post-contrast ultrasound to be approximately 8 x 12 mm in size (cf., Table 1).

The CTVS tumor dogs of Phase III were studied on average 28 days post CTVS implantation (range: 17 - 44 days). The ultrasound contrast studies showed tumors inside the prostate and outside the capsule as well as in the rectal wall and adjacent lymph nodes. NC100100 improved visualization of the tumor vascularity and delineation of tumor size and shape. Intra-tumoral vessels could be seen in most cases with the larger CTVS tumors (> 25 mm in diameter) demonstrating avascular central regions associated with necrosis. The tumors varied in size from 4 x 5 mm to 28 x 45 mm (Table 2). Histopathology confirmed the ultrasound findings and revealed typical CTVS cells infiltrating the prostate with moderate neovascularity. Neoplasms with lymphocytic infiltrates distorted the normal

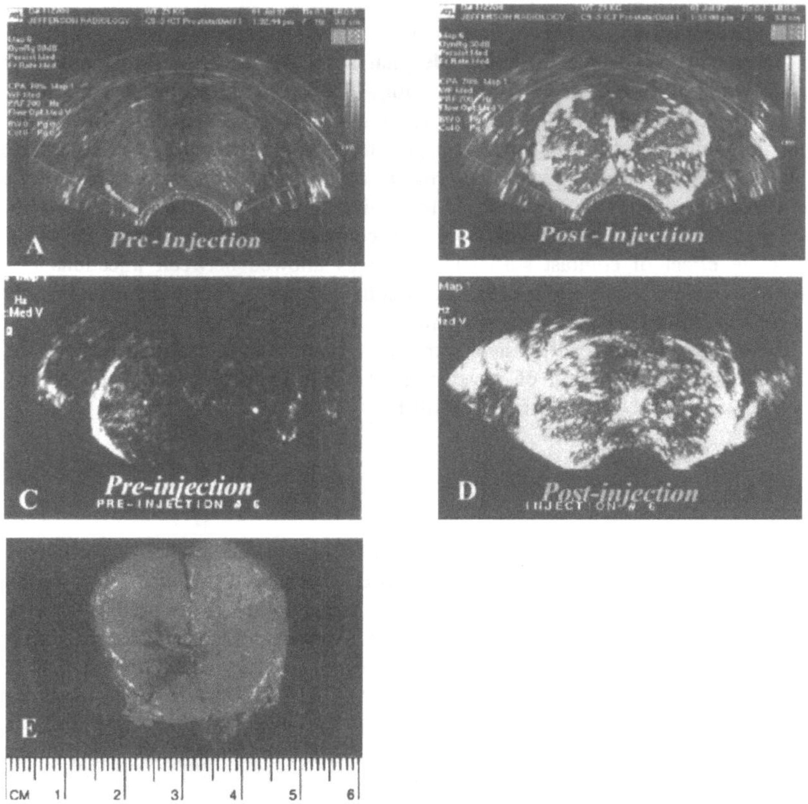

Figure 1. CAI of normal dog prostate. Images reproduced in B/W. (A) 2D image pre injection; (B) 2D image post injection of 0.00625 μl/kg of NC100100; (C) 3D image pre injection; (D) 3D image post injection of NC100100 (same dose as B). Notice, the improved depiction of the vasculature throughout the gland; (E) pathological specimen demonstrating the spoke-like vasculature seen with CAI.

prostate parenchyma leading to necrotic centers as the tumor outgrew its vascular supply, exactly as observed with contrast-enhanced TRUS. Pathology confirmed that the CTVS had metastasized to the iliac and para-lumbar lymph nodes as well as, on one occasion, to the liver.

In dogs no. 2 and 3 contrast enhanced ultrasound identified several smaller tumor nodules, which pathology established were all part of larger multinodular CTVS tumors (8 x 9 mm and 40 x 50 mm, respectively; Table 2). The former tumor contained a 4 x 5 mm focal nodule in agreement with the ultrasound findings, which was assumed to be an independent tumor at the time of the ultrasound study. When the ultrasound images were re-evaluated, retrospectively, the smaller tumors measured as one mass matched the size of

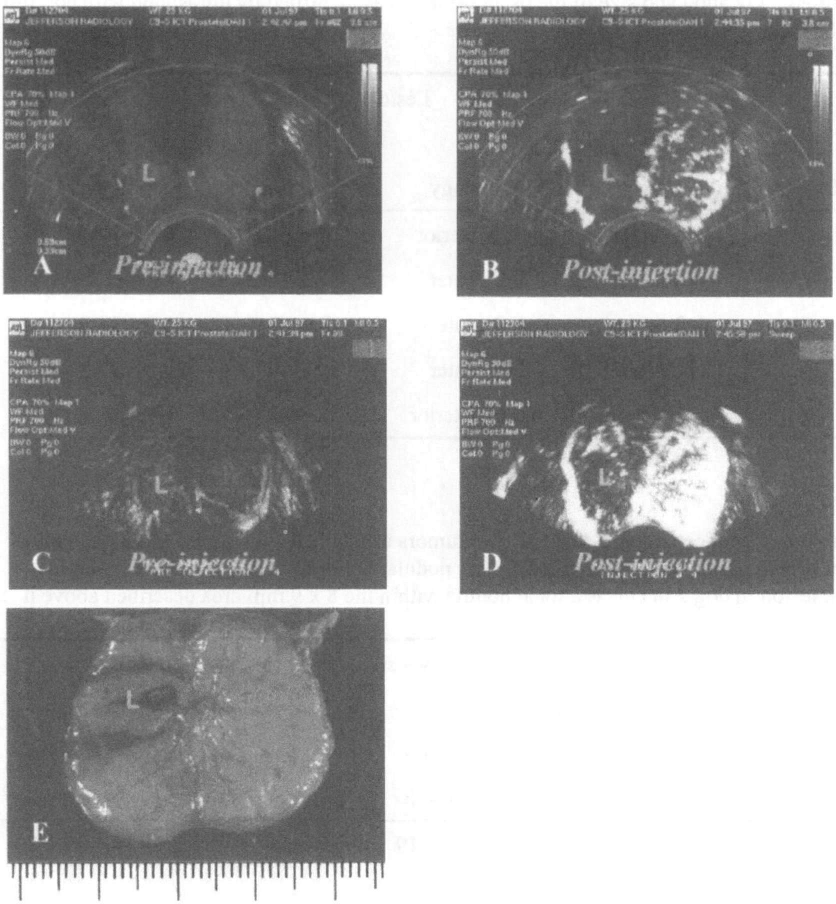

Figure 2. CAI of dog prostate after laser ablation has created an avascular lesion (L) or pseudo tumor. Images reproduced in B/W. (A) 2D image pre injection; (B) 2D image post injection of 0.00625 μl/kg of NC100100. An entire lobe of the gland appear avascular; (C) 3D image pre injection; (D) 3D image post injection of NC100100 (same dose as B). Notice, the improved depiction of the 10 x 14.9 mm lesion; (E) pathological specimen demonstrating the spoke-like vasculature seen with CAI.

the larger tumor identified by pathology. On two occasions, lesions seen by ultrasound corresponded to areas of hemorrhaging due to trauma at the injection site (marked by * in Table 2) rather than to CTVS tumors. However, as these lesions still represent abnormalities they have been included in Table 2. No tumors were found by ultrasound with contrast nor by pathology in dog no. 8.

In one case (in dog 5) the administration of NC100100 enabled the detection of a prostate lesion, which was not seen prior to injection (Figure 3). Attention pre contrast administration was focused on a small 4 x 6 mm extra-capsular tumor, but post contrast a

Table 1. Location and size of the pseudo-tumors as identified by ultrasound with contrast (US) and by pathology

	Lesion			
	location		size in mm	
Dog no	US	pathology	US	pathology
1	left center	left posterior	5 x 6	6.7 x 7.8
2	left center	left center	10 x 14	7.8 x 11.1
3	left center	left center	8 x 12	10.0 x 14.9
4	left anterior	left center	10 x 13	10.0 x 12.2
5	right posterior	right posterior	11 x 12	6.7 x 13.3

Table 2. Location and size of the CTVS tumors as identified by ultrasound with contrast and by pathology. Lesions marked * were nodular abnormalities but not CTVS tumors. The lesion in dog 2 in () was a focal nodule within the 8 x 9 mm area described above it

	CTVS size [mm] by					
	ultrasound			pathology		
Dog no	intra-prostatic	capsular	extra-capsular	intra-prostatic	capsular	extra-capsular
1	33 x 35	18 x 16	10 x 19	30 x 35	15 x 15	12 x 20
		18 x 19	22 x 30		15 x 20	20 x 35
2	5 x 7		14 x 25	8 x 9		20 x 25
	4 x 5		18 x 29	(4 x 5)		23 x 27
3	15 x 19		23 x 36	40 x 50		20 x 25
	11 x 13		28 x 45			25 x 40
	18 x 26					
4	27 x 27		14 x 14	25 x 30		15 x 15
5	10 x 11	4 x 6		4 x 6*	4 x 6	
		3 x 8			5 x 8	
6			6 x 17			6 x 17
7	5 x 7			10 x 10*		

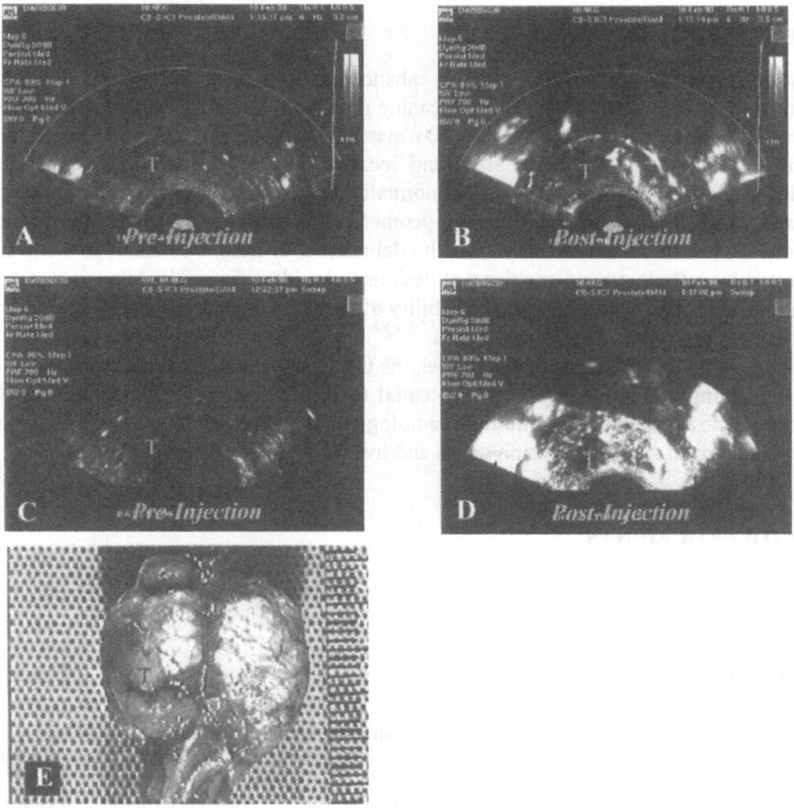

Figure 3. CAI of dog prostate tumor (T) 19 days after CTVS implantation. Images reproduced in B/W. (A) 2D image pre injection; (B) 2D image post injection of 0.025 μl/kg of NC100100 showing a 10 x 11 mm tumor inside the prostate; (C) 3D image pre injection; (D) 3D image post injection of NC100100 (same dose as B). A better depiction of the tumor blood supply is seen relative to B; (E) pathological specimen demonstrating the extent of the lesion within the prostate parenchyma.

second 10 x 11 mm hypovascular focal nodule was detected within the left lobe of the prostate (Figure 3D). At the time of the contrast study, the second lesion was assumed to be a CTVS tumor. However, pathology revealed this area to be a 4 x 6 mm non-tumor lesion (fibrosis with a few atrophied glands) probably associated with trauma incurred at the time of the CTVS cell-line injection, as mentioned above. Nonetheless, the size of the area and its lack of vascularity corresponds relatively well with the ultrasound contrast findings (10 x 11 mm; cf. Table 2) confirming the ability of contrast enhanced TRUS to identify unsuspected focal abnormalities.

CONCLUSIONS

NC100100 produces marked vascular enhancement (for over 10 minutes) and better visualization of the normal blood flow in canine prostates. Contrast-enhanced ultrasound imaging in CAI mode (2D as well as 3D) matched the pathological measurements of ablation lesion and CTVS tumor sizes and location well. While free hand 3D CAI provided the best visualization of focal abnormalities, measuring the actual tumor size in this mode was limited by the inherent geometric uncertainties of the reconstruction technique. Ultrasound contrast improves the delineation of lesions as small as 4 x 5 mm (confirmed by pathology) and enables new lesions to be identified. The latter is clinically very important, since it opens up the possibility of reducing the currently very high rate of blind biopsies performed in prostate studies.[2,3,4,5,6]

Finally, a completely new animal model, the CTVS in canines, has been developed for prostate imaging. This model has the potential to be applicable for evaluation of other diagnostic contrast agents using multiple radiological modalities in the prostate as well as in other anatomical sites (e.g., lymph nodes and liver).

ACKNOWLEDGMENTS

This project was supported by a grant from Nycomed-Amersham, Oslo, Norway.

REFERENCES

1. C.C. Boring, T.S. Squires, and T. Tong. Cancer statistics, 1993. *CA Cancer J Clin* 43:7-26, (1993).
2. F. Labrie, A. Dupont, E. Suburu, et al. Serum prostate-specific antigen as prescreening test for prostate cancer. *J Urol* 147:846-852, (1992).
3. W.H. Cooper, B.R. Mosley, C.T. Rutherford, et al. Prostate cancer detection in a clinical urological practice by ultrasonography, digital rectal examination and prostate specific antigen. *J Urol* 143:1146-1154, (1990).
4. M.D. Rifkin, and H. Choi. Endorectal prostate ultrasound: implications of the small peripherally placed hypoechoic lesion. *Radiology* 166:619-622, (1988).
5. F. Lee, S. Trop-Pedersen, P.J. Littrup, R.D. McLeary, T.A. McHugh, A.P. Smid, P.J. Stella, and G.S. Borlaza. Hypoechoic lesions of the prostate: clinical relevance of tumor size, digital rectal examination and prostate specific antigen. *Radiology* 170:29-32, (1989).
6. Radge H, Bagely CM, Aldape HC, Blasko JC: Prostate cancer screening with high resolution transrectal ultrasound. *J Endourol* 3:115-123, 1989.
7. Brawer MK, Bigler SA, Deering RE: Quantitative morphometric analysis of the microcirculation in prostate carcinoma. *J Cell Bioch* 16H:62-64, 1992.
8. Siegal JA, Enyou Yu AB, Brawer MK: Topography of neovascularity in human prostate carcinoma. *Cancer* 75:2545-2551, 1995.
9. Rifkin MD, Sudakoff GS, Alexander AA: Prostate: techniques, results, and potential applications of color Doppler US scanning. *Radiology* 186:509-513, 1993.
10. Kelly IM, Lees WR, Rickards D: Prostate cancer and the role of color Doppler US. *Radiology* 189:153-156, (1993).
11. Ismail MT, Petersen RO, Alexander AA, Newschaffer C, Gomella LG: Color Doppler imaging in predicting the biological behavior of prostate cancer: correlation with disease free survival. *Urology* 50:906-912, (1997).
12. Goldberg. *Ultrasound Contrast Agents*, Martin Dunitz, London, (1997).
13. Chaudhari, F. Forsberg, D.A. Merton, M. Magno, J.B. Liu, E.K. Hagen, D. Johnson and B.B. Goldberg: Fractional moving blood volume estimates compared to absolute perfusion (abstract). *Ultrasonic Imaging*, 1998. In press.
14. Forsberg, J.B. Liu, P.N. Burns, D.A. Merton and B.B. Goldberg. Artifacts in ultrasonic contrast Agent Studies. *J. Ultrasound Med.* 13:357-365, (1994).

HIGH RESOLUTION ESTIMATION OF AXIAL AND TRANSVERSAL BLOODFLOW WITH A 50 MHZ PULSED WAVE DOPPLER SYSTEM FOR DERMATOLOGY

M. Vogt[1], H. Ermert[1], S. el Gammal[2], K. Kaspar[2], K. Hoffmann[2], M. Stücker[2], P. Altmeyer[2]

[1]Dept. of Electrical Engineering
[2]Dermatologic University Hospital
Ruhr-University
D-44780 Bochum
Germany

INTRODUCTION

Dermatologic questions regarding the upper skin, namely the epidermis and the dermis, require high spatial resolution and noninvasive imaging techniques. Imaging systems based on high resolution and broadband ultrasound in the range from 30 to 150 MHz turned out to solve various diagnostic problems.

In addition to morphology visualization, imaging of blood flow in the skin is of growing interest. Sensitive detection techniques with sufficiently high spatial resolution are necessary to account for the small vessels and for the low flow velocities inside the microcirculatory system of the skin. In the horizontal subpapillary plexus and inside the capillaries, which branch vertically from this plexus, various vessel structures and flow directions are encountered. Conventional Pulsed-Wave-Doppler-Systems only estimate the axial flow velocity component with respect to the direction of sound propagation. An estimate for the true flow velocity inside the vessel can be obtained from the axial velocity by numerical correction if the angle between the flow direction and the direction of sound propagation is known. Due to the complex structure of the microcirculatory system, it is difficult to determine the orientation of the vessels and, therefore, to determine the incident angle from ultrasonic images. Also, the method fails at incident angles near 90°, i.e. at flow directions perpendicular to the direction of sound propagation. Thus, it is desirable to get an estimate of the true flow velocity directly from the ultrasonic echoes. If the geometry of the sound beam is known, it is possible to estimate the transversal velocity component from the echo data in addition to the Doppler related axial velocity component by application of appropriate estimation strategies[1,2,3]. The true velocity can be calculated from the estimates for both perpendicular directions.

We have extended an earlier published strategy for high resolution axial velocity processing[4] to estimate both, the axial and the transversal flow velocity. High spatial and

Acoustical Imaging, Volume 24.
Edited by Hua Lee, Kluwer Academic/Plenum Publishers, 2000.

high velocity resolution is emphasized. Results from simulations and flow phantom measurements with a 50 MHz Pulsed-Wave-Doppler-System for dermatology are presented.

HIGH FREQUENCY PULSED WAVE DOPPLER SYSTEM

The spatial resolution of an ultrasonic imaging system with respect to the lateral direction is determined by the sound beam geometry. For the axial direction the resolution is determined by the length of the ultrasonic pulse, i.e. the bandwidth of the transducer[5]. To meet the requirements for diagnostic imaging of the skin high resolution and broadband ultrasound has to be applied. The single element transducer used in our Pulsed-Wave-Doppler-System consists of a piezoelectric polymer (PVDF) foil with a concave curved focusing aperture in order to increase the sensitivity and the lateral resolution. The center frequency of the transducer is $f_0 = 57$ MHz, the -6 dB bandwidth is $B = 50$ MHz, the axial resolution (-6 dB) $\delta_{axial} = 13.3$ µm and the lateral resolution (-6 dB) in the focus $\delta_{lateral} = 58$ µm. The length of the focal zone is approximately $d_{axial} = 600$ µm.

Because of the rotational symmetry of the transducer the sound beam is rotationally symmetric, too. Let a vessel under observation be insonified under an incident angle α between the flow direction and the direction of sound propagation. Then the movement of the blood particles inside the vessel, mainly the erythrocytes, can be described by two velocity components with respect to the sound beam of the transducer which are perpendicular to each other. The axial velocity component describes the movement of the particles parallel to the direction of sound propagation. The movement perpendicular to the sound propagation direction is described by the transversal velocity component.

A single transducer is used for transmission and reception of the acoustic waves (monostatic approach). The acoustic waves which are scattered from the moving blood particles are determined by two effects:

- First, due to the axial movement of the scatterers, the backscattered waves are temporarily stretched or compressed compared to the incident waves in the sense of the classical Doppler effect.
- Second, if the scatterers pass the sound beam in the transversal direction, the echo signal is weighted by the sound beam characteristic of the transducer. This weighting is influenced by the transversal velocity of the scatterers.

From the backscattered signal it is possible to estimate both, the magnitude of the axial velocity component and the axial flow direction. But, due to the rotational symmetry of the setup, only the magnitude of the transversal velocity component is determinable. The flow direction in the plane perpendicular to the sound propagation direction is generally not accessible by the echo signal of a monostatic system. The magnitude of the true velocity in the vessel and the axial direction of the true velocity can be calculated from the velocity estimates for both directions.

Pulsed wave transmit signals are applied in order to resolve different flow conditions in different depths. However, it is difficult to track the scatterer movements adequately if only one single pulse is applied. To increase the observation time, a train of successive pulsed wave signals with the pulse repetition period T_{PR} is transmitted. In order to distinguish moving scatterers from stationary scatterers at different depths, the corresponding sequence of echo signals (echo sequence) is recorded. To obtain a two-dimensional flow image this procedure is performed for each lateral transducer position, see Figure 1.

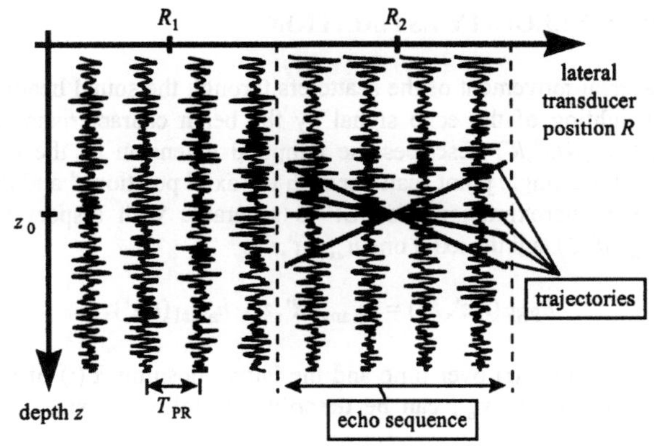

Figure 1. Sequence of echo signals of a Pulsed Wave-Doppler-System, response to a train of transmitted pulses with the pulse repetition period T_{PR} applied at every lateral transducer position R_n, trajectories of constant velocities ('Butterfly-lines')[4].

AXIAL VELOCITY ESTIMATION

In the past many methods have been proposed to estimate the axial velocity from the echo sequence of a Pulsed-Wave-Doppler-System. The simplest approach is to sample the quadrature demodulated echo sequence (I/Q: inphase / quadrature component) at constant time lags after transmission of the consecutive pulses[6], i.e. at constant depth $z = z_0$ and various R, see Figure 1. From this samples usually the Doppler spectrum is analyzed applying the Fast Fourier Transform (FFT) to obtain the velocity distribution for a single gated depth range. Alternatively, the mean axial velocity can be calculated in a very fast and effective manner using an autocorrelation approach[7]. However, by sampling at constant time lags the axial scatterer movement is not taken into account. Applying broadband pulses, the spatial resolution and the velocity resolution are worse and depend on the axial velocity[4,8,9].

The axial movement of the consecutive echo pulses has to be tracked to maintain high spatial resolution with broadband transmit pulses. Different processing schemes can be performed in order to reach this goal. One approach is to process the quadrature demodulated echo signal inside range gated windows along trajectories of constant velocities (Wideband maximum likelihood estimation, WMLE[9]). With shrinking window length a similar approach is obtained, where the echo signal is sampled along lines corresponding to trajectories of constant velocities (Butterfly Search Technique, BST[4]). Since the actual axial velocity is not known, the echo data from a set of trajectories corresponding to predetermined discrete axial velocities has to be processed.

In the WMLE approach the windowed quadrature demodulated echo signal is processed in a matched filter sense using a Doppler shifted quadrature demodulated echo model that is expected to be obtained from the corresponding depth[9]. In the BST approach the quadrature demodulated echo signal along the trajectories is analyzed for the frequency component which corresponds to that trajectory (BST on quadrature components[4]).

The Butterfly Search on quadrature components approach is used below to estimate the axial velocity component. Furthermore, this approach is extended to estimate the transversal velocity component.

TRANSVERSAL VELOCITY ESTIMATION

The transversal movement of the scatterers through the sound beam of the transducer results in a weighting of the echo signal by the beam characteristic. The point spread function (PSF) $h_{PSF}(t,z',R')$ describes the temporal extension of the echo signal of the imaging system for a single point scatterer with the axial position z' and the lateral position R'. With a good approximation the PSF is separable with respect to the transversal component $h_{trans}(R',z')$ and the axial one $h_{axial}(t',z')$:

$$h_{PSF}(t,z',R') = h_{trans}(R',z') \cdot h_{axial}(t,z') \quad (1)$$

The axial position $z'(t)$ over time and the lateral position $R'(t)$ of a scatterer moving with the constant velocity v_{true} can be described as follows (axial coordinate z, lateral coordinate R) :

$$z'(t) = z_0 + v_{true} \cdot \cos\alpha \cdot t \quad , \quad R'(t) = R_0 + v_{true} \cdot \sin\alpha \cdot t \quad (2)$$

From equations (1) and (2) the weighting of the echo signal caused by the sound beam (described by the PSF) due to the transversal scatterer movement can be calculated. In the following the axial scatterer movement is neglected ($z'(t) = z_0 = \text{const}$), i.e. it is assumed that the observation takes place for only a short observation time. In this case the envelope (magnitude) of the I/Q-demodulated echo signal $x_{echo+}(t)$ is proportional to the envelope of the transversal part of the PSF. With v_{true} denoting the true velocity and v_{trans} denoting the transversal velocity component of a single point scatterer, the following proportionality is obtained:

$$\left| x_{echo+}(t) \right| \sim \left| h_{trans}(R_0 + v_{trans} \cdot t, z_0) \right| \quad , \quad v_{trans} = v_{true} \cdot \sin\alpha \quad (3)$$

The transversal velocity can be estimated from this weighting using the measured PSF of the transducer. However, the scatterers in a given depth do not pass the sound beam separately but in groups of scatterers with the same velocity. Therefore, the resulting echo signal is given by the superposition of the point spread functions of all scatterers inside the sound beam. A typical statistically varying speckle structure results from this superposition and is determined by the spatial resolution of the imaging system.

For example Figure 2 shows the simulated envelope of the echo sequence resulting from the superposition of the echo signals which are backscattered at moving point scatterers inside a vessel. In this simulation the three dimensional geometry of the vessel was taken into account. The vessel diameter was assumed to be $D = 500\ \mu m$ and the incident angle was $\alpha = 65°$. The scatterers pass through the sound beam in the transversal direction with different velocities at different depths due to the parabolic flow profile. As is visible in Figure 2 this leads to a characteristic depth dependent speckle texture, which is determined by the PSF of the imaging system and the velocity distribution.

Because of this texture, the transversal velocity cannot be estimated directly from the envelope of the echo signal. However, an independent representation of the weighting with respect to the speckle texture is given by the autocovariance function (ACVF) of the envelope of the signal[10].

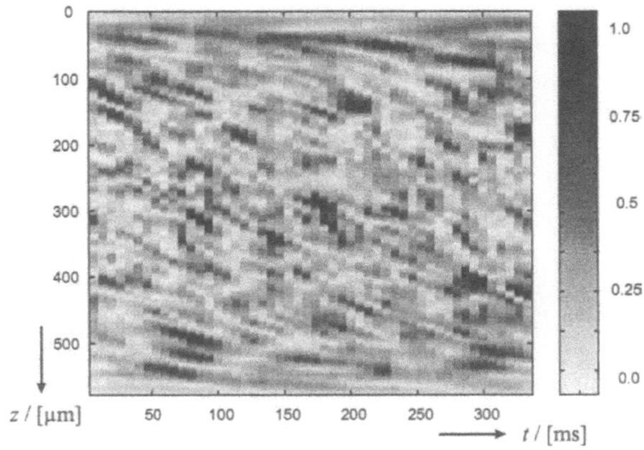

Figure 2. Simulation data, envelope of echo signal for a single transducer position, train of 50 transmitted pulses with pulse repetition period $T_{PR} = 6.7$ ms, incident angle $\alpha = 65°$, vessel diameter D = 500 µm, parabolic flow profile with $v_{true,max} = 2.22$ mm/s.

Using this representation a method for the estimation of the transversal velocity from the envelope of the echo signal was developed.

As a reference in our approach the autocovariance function of the transversal part $h_{trans}(R',z')$ of the point spread function is calculated for every constant depth $z = z_0$. The corresponding ACVF $c_{trans}(R')$ is only a function of the lateral coordinate R for a constant depth. The PSF of our high frequency imaging system was obtained from a thin wire target (tungsten wire with a diameter of 6.8 µm) at different depths.

With our approach the estimation of the transversal velocity component is performed as follows. First, the ACVF $c_{echo}(t)$ of the envelope of the echo signal along the trajectory corresponding to the estimated axial velocity is calculated for all depths. The slope of the reference $c_{trans}(R')$ is then normalized to the slope of the calculated function $c_{echo}(t)$. With this approach the relationship between the arguments of the terms in equation (3) is estimated. From this relationship the transversal velocity component is calculated.

With the above described extension of the Butterfly Search Technique a method for the combined estimation of both velocities with high spatial and velocity resolution is available. To estimate the transversal velocity the envelope of the echo signal is used. Therefore, compared to the axial velocity estimate using the phase sensitive I/Q-demodulated echo signal, a higher variance is expected.

VALIDATION OF THE METHOD

In order to test the above introduced approach simulation data and measurements from a flow phantom were used. The echo data in Figure 2 from the simulation for a constant transducer position shows that the successive echo signals decorrelate very rapidly at the center of the vessel because of the relatively high velocity of the scatterers. At the border of the vessel the echo signals decorrelate very slowly and the effect of the weighting by the beam characteristic is of less significance because of the lower velocities. The pulse repetition period was $T_{PR} = 6.7$ ms and the maximum true velocity at the center of the

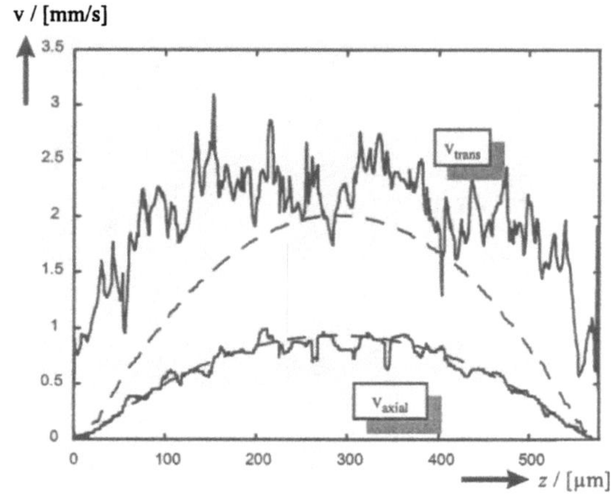

Figure 3. Estimate for simulation data (solid curves), simulation input (dashed curves), compare Figure 2.

vessel was $v_{true,max} = 2.22$ mm/s. With the incident angle $\alpha = 65°$, the maximum axial velocity is $v_{axial,max} = 0.94$ mm/s and the maximum transversal velocity $v_{trans,max} = 2.01$ mm/s.

In Figure 3 the estimate for the axial and the transversal velocity is compared with the simulation input. The estimate for the axial velocity component reflects the parabolic flow profile over the vessel diameter with good agreement. For the transversal flow component a good agreement between simulation input and estimate is only found at the center of the vessel. At the border the transversal velocity is significantly overestimated. This may be caused by the fact that the weighting of the echo signal by the sound beam is same significant for different velocities, compare Figure 2.

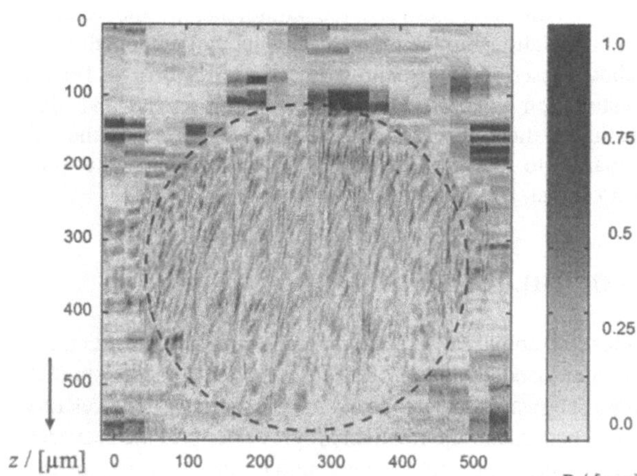

Figure 4. Flow phantom, envelope of echo signal for 20 equidistant lateral transducer positions, train of 50 transmitted pulses with pulse repetition period $T_{PR} = 6.7$ ms at each position, incident angle $\alpha = 65°$, vessel diameter $D = 500$ μm, flow rate FR = 0.177 μl/s, mean true velocity over vessel cross section (marked by circle) $v_{true,mean} = 1.11$ mm/s, maximum true velocity $v_{true,max} = 2.22$ mm/s for parabolic flow profile.

We built a flow phantom with a small vessel in an agar block as tissue mimicking material. Silica gel was added to the agar to model stationary scatterers. A motor driven syringe pump was used to feed the vessel with a blood mimicking fluid (42% glycerol, 58% water, 180 g yeast per liter fluid). The vessel diameter was equivalent to the simulation $D = 500$ µm, the flow rate was FR = 0.18 µl/s resulting in a mean true velocity $v_{true,mean} = 1.11$ mm/s. With the small vessel diameter, the low mean velocity and the mechanical properties of the fluid, a parabolic flow profile is expected inside the vessel. Under this assumption the maximum true velocity at the center of the vessel was $v_{true,max} = 2.22$ mm/s. The incident angle in the setup was adjusted to $\alpha = 65°$, therefore a maximum axial velocity $v_{axial,max} = 0.94$ mm/s and a maximum transversal velocity $v_{trans,max} = 2.01$ mm/s are expected at the center of the vessel.

The echo signals from the flow phantom are shown in Figure 4. A train of 50 pulses with the pulse repetition period $T_{PR} = 6.7$ ms was applied at each of the 20 equidistant lateral transducer positions with a distance of 30 µm from each other. Echoes due to stationary scatterers are clearly distinguishable from echo signals that result from backscattering at moving scatterers inside the vessel. The largest echo movement appears at the center of the vessel corresponding to the highest velocity.

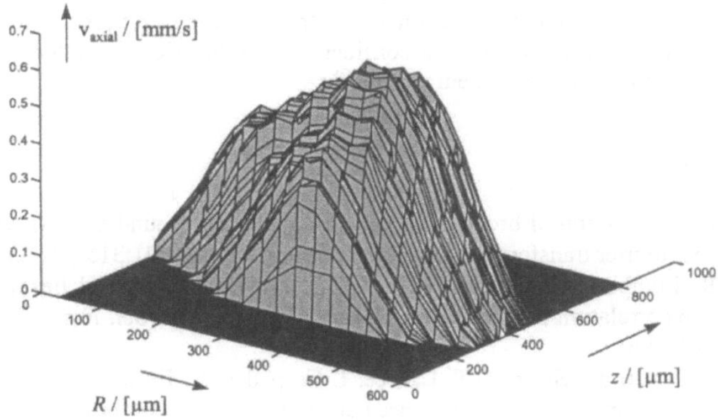

Figure 5. Estimate of axial velocity profile over vessel cross section, pulse repetition period $T_{PR} = 6.7$ ms.

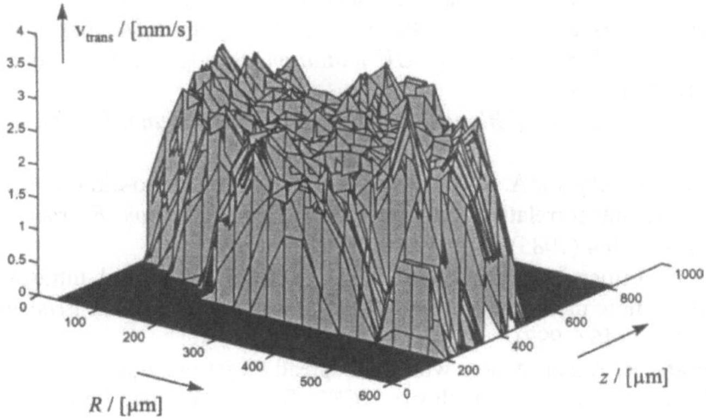

Figure 6. Estimate of transversal velocity profile over vessel cross section, pulse repetition period $T_{PR} = 10$ ms.

In Figure 5 the estimated axial velocity profile over the vessel cross section is shown. As expected a parabolic flow profile is obtained and the maximum axial velocity is estimated with $v_{axial,max} = 0.6$ mm/s. The result for the transversal velocity distribution is shown in Figure 6. In accordance with the simulation the velocity distribution at the border of the vessel is overestimated, while the velocity at the center is adequately estimated with $v_{trans,max} = 2.2$ mm/s. One of the reasons for the insufficient axial velocity estimate could be a false assumption regarding the adjusted incident angle α. Aberrations of the angle are of great influence on the axial velocity component in the cases at angles around 90°.

CONCLUSIONS

In this paper an approach for the combined estimation of axial and transversal velocity with special emphasis on high resolution is presented. Our proposed approach was tested using simulated data and measurements obtained from a flow phantom. With the estimated flow profile for the axial velocity it was shown that the flow profile inside the vessel with a diameter of 500 µm and with a flow rate of 0.18 µl/s was parabolic. The proposed method gives adequate results at the center of the vessel, while the low velocities at the border are over estimated. With the combined estimation strategy complex vessel structures with unknown flow directions are taken into consideration. Subject of our future work is the imaging of the microcirculatory system of the skin.

REFERENCES

1. L.S. Wilson, Description of broad-band pulsed doppler ultrasound processing using the two-dimensional fourier transform, *Ultrasonic Imaging*, No. 13, 301:315 (1991)

2. D. Dotti, R. Lombardi, Estimation of the angle between ultrasound beam and blood velocity through correlation functions, *IEEE Trans. Ultrason. Ferroel. Freq. Contr.*, UFFC 43(5), 864:869 (1996)

3. W. Li, A.F.W. van der Steen, C.T. Lancee, I. Cespedes, N. Bom, Blood flow imaging and volume flow quantitation with intravascular ultrasound, *Ultras. in Med. & Biol.*, Vol. 24, No. 2, 203:214 (1998)

4. S.K. Alam, K.J. Parker, The butterfly search technique for estimation of blood velocity, *Ultras. in Med. & Biol.*, Vol. 21, No .5, 657:670 (1995)

5. H. Ermert; M. Vogt; C. Paßmann; S. el Gammal; K. Kaspar; K. Hoffmann; P. Altmeyer. High-frequency ultrasound (50-150 MHz) in dermatology, in: P. Altmeyer; K. Hoffmann; M. Stücker (Eds.), *Skin Cancer and UV Radiation*, Springer, Berlin, Heidelberg, New York, 1023:1051 (1997)

6. J.A. Jensen. *Estimation of Blood Velocities Using Ultrasound*, Cambridge University Press (1996)

7. C. Kasai, K. Namekawa, A. Koyano, R. Omoto, Real-time two-dimensional blood flow imaging using an autocorrelation technique, *IEEE Trans. Ultrason. Ferroel. Freq. Contr.*, UFFC SU-32, 458:464 (1985)

8. M. Vogt, H. Ermert, Application of high frequency broadband ultrasound for high resolution blood flow measurement, in: *IEEE Ultrasonics Symposium Proceedings*, IEEE, New York, 1243:1246 (1997)

9. K.W. Ferrara, R. Agazi, A new wideband spread target maximum likelihood estimator for blood flow estimation-part I: theory, *IEEE Trans. Ultrason. Ferroel. Freq. Contr.*, UFFC 38(1), 1:16 (1991)

10. R.F: Wagner, S.W Smith, J.M. Sandrik, H. Lopez, Statistics of speckle in ultrasound B-scans, *IEEE Trans. Sonics and Ultras.*, Vol. 30, No.3, 156:163 (1983)

DIFFRACTION TOMOGRAPHY BREAST IMAGING SYSTEM: PATIENT IMAGE RECONSTRUCTION AND ANALYSIS

Helmar S. Janée,[1,3] Michael P. André,[1,2] Mariana Z. Ysrael[1,2] Peter J. Martin,[4]

Department of Radiology, [1]Veterans Affairs Medical Center and [2]University of California School of Medicine, La Jolla, CA, [3]University of Hertfordshire, Hatfield, UK, [4]ThermoTrex Corporation, San Diego, CA.

INTRODUCTION

Ultrasound is the primary adjunctive tool to mammography in the United States but its role is mainly limited to distinguishing solid from cystic masses. Some studies have concluded that modern high-resolution ultrasound in experienced hands can yield greater sensitivity than mammography for the classification of palpable breast tumors, although there is not significant difference in specificity between the two modalities.[1,2] The suggestion is that ultrasound can play a wider role in the management of breast masses and may reduce unnecessary biopsies.

Ultrasound tissue characterization of breast masses has been researched for many years. These studies suggest that a diagnostic imaging device which is sensitive to the subtle differences in speed of sound and attenuation of breast tissues may be useful for enhancing the detection and characterization of these lesions. In particular, it holds promise of improved image contrast of lesions over x-ray mammography and pulse-echo ultrasound. Although there is considerable overlap of sound speeds for benign and malignant lesions, there is evidence that the slope of the attenuation versus frequency curve is different for malignant and benign masses. Elasticity or hardness, which is related to the sound speed and density of breast masses, is now being investigated by non-invasive means and shows potential for improvement of detection and classification of nodules.

The approach to ultrasound CT reported here utilizes a diffraction tomography technique that addresses many of the short-comings of previous work and present breast ultrasound. It provides a very large field of view (20 cm diameter) at high resolution (<1 mm), it is not strongly dependent on operator expertise, it provides a standardized sequential tomographic approach to surveying the entire breast, and it has the potential to measure tissue properties *in vivo*. The methods of image reconstruction that we employ are similar to previous diffraction tomography work in which the wave equation for the propagation of sound through a spatially variant medium is solved via simplifying approximations. One significant difference is that this method acquires the entire scatter field around the object in a very short time using sophisticated electronic multiplexing techniques. We have developed several imaging approaches which are unique to our transducer configuration and data acquisition.

The breast imaging system and its operational principles have been described in detail

Acoustical Imaging, Volume 24.
Edited by Hua Lee. Kluwer Academic/Plenum Publishers, 2000.

elsewhere[3,4]. This study combines the discussion of image reconstruction methods with an analysis of the scatter fields from both phantom objects and clinical breast images from 25 patients. A hybrid image reconstruction method is employed that is able to minimize aberrations induced by large-sized objects and is computationally efficient--an important consideration for clinical use. The entire ultrasound scatter field (2π) was measured from widely varying breast types in a series of clinical trials. The amplitude distributions of these data are important for classification of breast tissue type in terms of scattering properties. Conventional mammograms were evaluated alongside the corresponding ultrasound tomographic images in order to determine the distribution of breast tissue types for a particular patient. The relative contrast of tissue types was used to develop a tissue classification index and the predominant tissue type was correlated with the measured scattering amplitude distribution. Finally, three image reconstruction approaches and presentations were evaluated on the basis of their ability to quantify breast tissue scattering in relation to composition.

The assessment of clinical images obtained in these preliminary trials is intended to address three key concerns in the development of this new imaging modality: 1—To what extent are simplifying assumption inherent in the image reconstruction algorithms satisfied in a clinical application? 2—Is it possible to interpret reconstructed clinical images in terms of (breast) tissue types and properties? And 3—Are observed tissue property contrast values clinically relevant?

These questions are addressed in part in this paper though the assessment of clinical utility is most certainly preliminary. The patient data base is much too small to draw definitive conclusions about clinical utility and the value of the current clinical images is that they, in conjunction with phantom data, can be used to establish guidelines for experimental system and image reconstruction improvements.

HYBRID IMAGE RECONSTRUCTION

This is an abbreviated discussion of particular aspects of the image reconstruction, the general procedure has been described in a number of recent papers[3,4]. Figure 1 shows a diagram of the experimental geometry for scatter measurements of cw, single-frequency sound waves. Each element in turn acts as a transmitter while the remaining transducers record the signal emanating from the object. Two separate arrays 20-cm in diameter were developed: 512 elements with center frequency at 0.5 MHz, and 1024 elements with center frequency at 1.0 MHz. The entire measurement procedure is normally repeated at A=10 frequencies, ω_α (α=1,2,...,A), spaced at 62.5 kHz intervals from 687 kHz to 1.250 MHz for the 1024 array, and for A=20 frequencies spaced at 31 kHz from 300 to 600 kHz for the 512 array.

Our general approach to image reconstruction is based on the Born Approximation which is used to linearize the inhomogeneous Helmholtz wave equation that describes the interaction of the insonifying wave with the scattering potential. This takes the form:

$$\nabla^2\psi(r,\theta)+K(r,\theta)\psi(r,\theta)=0 \qquad \text{where} \qquad K(r,\theta)\equiv k^2(r,\theta)-\rho(r,\theta)^{1/2}\,\nabla^2\rho(r,\theta)^{-1/2} \qquad (1)$$

The scattering potential which represents the object to be imaged (and reconstructed) is given by:

$S(r,\theta)=K(r,\theta)-k_0^2$ where k_0 is the propagation constant for the water medium. (2)

Transformation to the integral form of equation 1 for the solution to the scattered wave function yields for our experimental system:

$$\psi_{\alpha}^{sc}(r_0,\theta) = m_{j,k}(\alpha) = \sum_{m,n} H'_m(k_0 r_0) H'_n(k_0 r_0) e^{-i(m\theta_j + n\theta_k)} \int\int J_m(k_0 r) J_n(k_0 r) e^{i\theta(m+n)} S^{\alpha}(r,\theta) r dr d\theta \tag{3}$$

where r_0 is the radius of the ring, j,k are transmit and receive transducer indeces, their antenna patters are given by the first derivative of the Hankel function and the incident wave is represented by Bessel functions of the first kind.

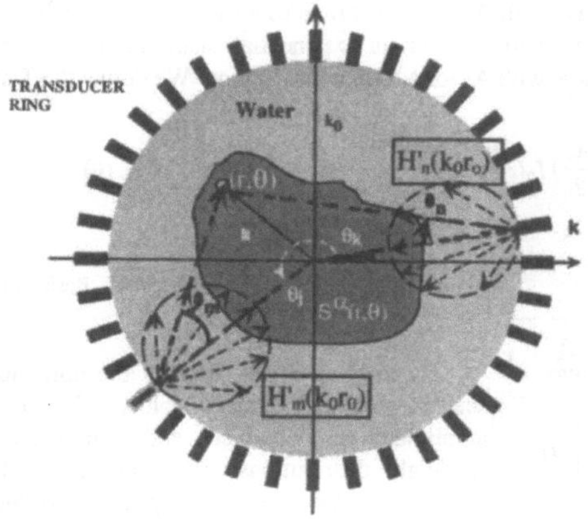

Figure 1. Measurement Geometry of Diffraction Tomography Scanner. j is the Transmit transducer, k the Receive Transducer. The scatterer is at (r,θ). Two rings have 512 or 1024 transducers, respectively. Hankel functions H'_m and H'_n describe the transducer transmit and receive antenna patterns. S^{α} (r,θ) is the scattering potential at frequency α one desires to recover.

The major steps leading toward solution for the object function $S^{\alpha}(r,\theta)$ in (3) are outlined in Figure 2. The raw data acquisition refers to the recording of the complex scatter data of amplitude and phase in quadrature. Following this stage a number of calibrations and gain settings are performed and one of several optional reconstruction techniques are selected which yield the estimate of the object function, S_{ij}. In series of phantom experiments, a number of approaches have been developed to reduce artifacts and to extend the range of linearity of the reconstructed images. These include computation of magnitude images (i.e. the modulus of the complex image), combinations of single-frequency images, variation of the transducer center frequency and variation of the propagation constant of the coupling medium. Iterative reconstruction schemes have been developed that that use synthesized multi-frequency pulse data to develop a time-of-flight map which is used to phase-correct single frequency images.

The assessment of the utility of these various "hybrid" reconstruction techniques was performed with phantoms of controlled sound speed and, to some extent, attenuation. From the recovered object function S(r,θ) of equation (2) , the known constant k_0 of the coupling medium and the relationship for the object propagation constant k:

$$k^2(r,\theta) = \frac{\omega^2}{c^2(r,\theta)} + i\omega\mu(r,\theta) \tag{4}$$

it is, in principle, possible to recover the object speed of sound and attenuation distributions.

TISSUE SCATTER AMPLITUDE DISTRIBUTIONS

The analysis of scatter amplitude distributions and their interpretations were reported recently[5,6] and we summarize here only the relevant formalisms. The relative strength of scattering is reflected in the nature of the amplitude distributions (modulus of the complex probability density function) of scatter signals: In the case of weak scattering, the observed distribution will be Rician (Rice Nakagami) and ultimately Gaussian; the transition from weak to strong scattering is indicated by the change to a Rayleigh distribution. Amplitude distortion, however, is observed when scattering becomes strong and its appearance can be thought to define the limit of applicability of weak-scattering theories.

The general form of the distribution to which our data were fitted is given by the expression:

$$p(A) = \frac{A}{\sigma^2}\exp[\frac{-A^2 + A_0^2}{2\sigma^2}]\, I_0(\frac{AA_0}{\sigma^2}) \qquad (5)$$

where p(A) is the probability density for amplitude A, the random term. A_0 is a constant phasor, has integral values and when A_0=0, equation 3 tends toward a Rayleigh distribution. I_0 is the modified Bessel function. For large values of A_0, p(A) becomes Gaussian.. The ratio of the mean value of the distribution p(A) to its standard deviation is found to increase with A_0 and thus provides a useful index of whether a scatterer is mostly Rayleighian or Rician. This ratio is 1 for an exponential distribution, 1.24 for a Rayleigh distribution and becomes arbitrarily large with A_0 for a Rician distribution. We define the Tissue Index, $\langle a\rangle/\sigma_a$, by:

$$TI \equiv \frac{\bar{a}}{\sigma_a} = \sqrt{\frac{\pi}{2}}\exp(-\frac{A_0^2}{4})[(1+\frac{A_0^2}{2})\, I_0(\frac{A_0^2}{2}) + \frac{A_0^2}{2}\, I_1(\frac{A_0^2}{2})] \qquad (6)$$

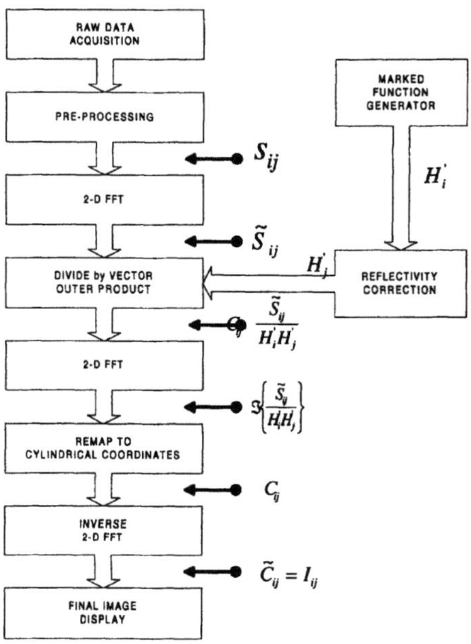

Figure 2. Outline of key steps of Basic Image Reconstruction Procedures.

By forming a histogram of the normalized scatter amplitude values, plots of pdf vs. amplitude are obtained and compared to Rayleigh distributions of identical standard deviation. If there is close agreement between the two, that is, if the means of the computed Rayleigh and the experimentally determined distributions are close, then the scatter field is considered to be fully developed. An example of a weak scatterer with a significant phasor contribution is provided by Figure 3, the scatter amplitude distribution of 3 270 micron nylon filaments. Both the shape of the distribution and the high Tissue Index are indicative of a weak scatter field. This will be compared to patient data in the next section.

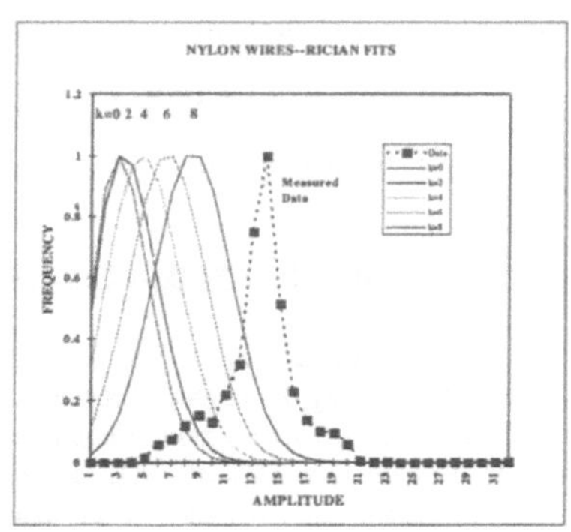

Figure 3. Scatter Amplitude Distribution from Wire Phantom Array of 270 micron Nylon Filaments. TI = 4.58

TISSUE TEXTURE ANALYSIS

The evaluation procedures discussed here deal not with the raw, time-dependent data but with the reconstructed images and the pixel values of image elements of clinical significance. One of us (radiologist Dr. Ysrael) viewed images of coronal breast slices on the computer monitor and, with the aid of corresponding mammograms (usually orthogonal CC and MLO views were available), identified the locations and spatial extents of significant tissue structures such as : **Skin, Subcutaneous Fat, Parenchyma, Ligaments, Cysts, Lesions** and **Unidentified Features.** Examples of the last item would be bright dots or specks that might represent calcifications or some unidentified strongly reflecting (or scattering) small object. Although the pixel value of the skin layer was recorded for many cases, it was found to have rather arbitrary magnitudes and is not believed to be of diagnostic value. We suspect that this is in part due to as yet poorly understood edge artifacts.

Following this identification of key image features each was evaluated by recording mean pixel values for regions of interest (ROI's) that encompassed the feature. For each image slice, the mean pixel value for water was obtained by defining an annular region that excluded the breast object as well as artifact-filled regions adjacent to the transducer ring. The gray-scale value for water was used to compute the contrast for each image element from the relationship: $\mathbf{C = (p_{object} - p_{water})/ |p_{water}|}$. The use of the absolute value for the water pixel value assures that the sign of the contrast value unambiguously indicates the relationship to the pixel value of water which quite often happens to be less than zero.

Image slices from all patients, some examined with the two transducer rings, were analyzed in this fashion. Generally around 10 tomographic slices were available for each patient but not all slices have as yet been analyzed. We concentrated on those slices that included interesting anatomical features or diagnosed structures such as cysts, fiber adenomas or malignant lesions.

RESULTS AND DISCUSSION

We have extended the scatter amplitude analyses to frequencies lower than previously reported in our studies of patients who were part of a clinical trials series at the UCSD Center for Women's Health. The initial analyses used only those receivers of the transducer ring that were in the shadow of the object, using the sinograms of the data to select the minimum range of receivers subtended by the object. This is comparable to the technique used by Steinberg et al.[7] who used a single transmitter and a linear receiving array to record only pulses transmitted through the object. We compare this subset of the scatter data with the amplitude distributions for the whole scatter field, which we obtain in a single measurement.

Our intent in these measurements is to assess the strength of the scatter field and thus obtain an estimate of the magnitude of the speed of sound, density and attenuation fluctuations with respect to the coupling medium. A goal of this analysis is to correlate the scatter amplitude distributions with the object contrast properties. The mean object contrast (speed of sound) and the object size establish the expected overall phase shift and thus permit an assessment of the validity conditions of the Born approximation. Ultimately, it is hoped that the dominant nature of the distribution and the Tissue Index will provide an a priori tool for selecting an optimal reconstruction technique for the particular object. This assessment would be part of the Pre-Processing step of the image reconstruction procedure outlined in Figure 2.

Figure 4 shows the scatter data over all angles for both the scan tank alone and for the case with a breast object. Data shown are for Patient 14, classified as an average breast and were obtained with the 0.5 MHz SCT ring. Though smoothed, the data reveal high spatial frequency fluctuations that are believed to be acoustical signals possibly due to diffraction effects and multiple reflections. The breast object data show significant attenuation and energy scattered at angles larger than those subtended by the object. This suggests that such large-angle scattering is due to compressibility fluctuations which gives rise to monopole terms.

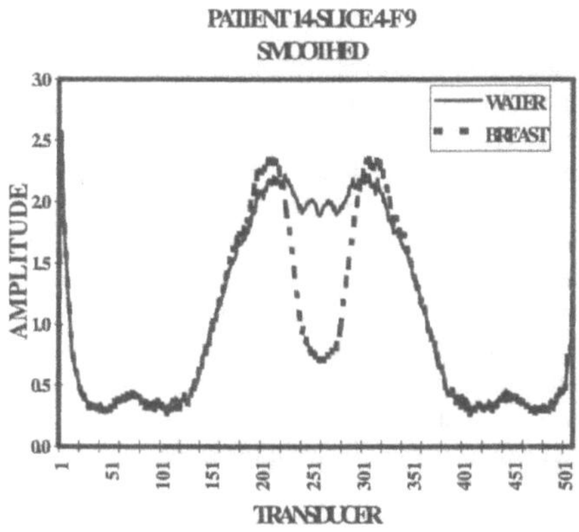

Figure 4. Raw Scatter Data for Patient 14, Slice 4 at 1 MHz. Solid Line are Data for the Empty Scanner Tank. Dashed Line is the Scatter Distribution with a Breast Object.

An example of a scatter amplitude distribution obtained from data such as these is presented in Figure 5. The breast scatter data were normalized by the empty tank data and the resultant amplitude histogram data are to be contrasted with the example of Figure 3. In this case the data are from a tomographic slice of Patient 24, classified as a dense, glandular breast with scattered microcalcifications. Both the shape of the distribution and the Tissue Index are indicative of a Rayleigh distribution and thus suggest strong scattering.

A summary of the scatter distribution analysis of five patients is presented in Table 1. These represent three distinct breast classifications and include also a patient with a history of radiation and chemo therapy following lumpectomy. Data of the Tissue Index are presented for both restricted angular views (TI-2) and the complete angular scatter field (TI-6). These data were obtained at the transducer center frequency and generally for a slice in the center of the breast. In all instances, the full angular distribution data exhibit stronger scattering than the data limited to the transducers in the shadow of the object. The observation reported previously[1] that fatty breasts exhibit strong scattering is also born out as is the observation that average breasts exhibit relatively weak scattering.

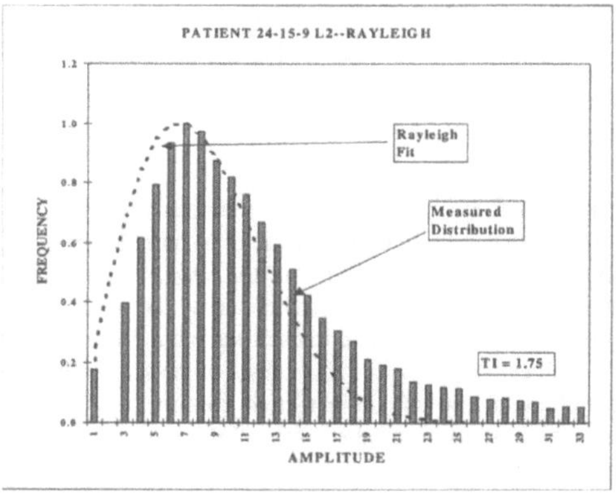

Figure 5. Scatter Amplitude Distribution for Patient 24, Slice 15. Restricted Angular View. TI = 1.75

Studies of the contrast of breast tissue elements were initially performed for a few patients on images from several reconstruction approaches. These included the real, imaginary and magnitude images from back-propagation reconstructions as well as reconstructions incorporating various aberration correction and frequency summing techniques. On the basis of these initial investigations, the magnitude images were found to provide the clearest separation of tissue type and thus have been utilized for the rest of the studies. The first concern was to establish that contrast values associated with various tissue classifications be relatively consistent throughout the breast volume. Only if that was found to be the case could the next step be taken, the comparison of tissue contrast values for several patients. Figure 6 shows the tissue contrast values for fat and glandular tissue (parenchyma) as well as the mean value of the whole slice for nine 8 mm slices. The separation of contrast values are fairly constant throughout the breast volume and show that fat, as expected, generally has a negative contrast value while parenchyma is significantly above the mean object level. The latter, however, is disturbingly high and suggests that clinical images are well outside the Born approximation validity conditions of total phase shift.

Table 1. Tissue Index Summary

PT. ID	MHz	TI-2	TI-6	TYPE	COMMENTS
5	1.0	2.04	1.35	DENSE	Micro CA++
8	1.0	1.77	1.65	FATTY/ AVERAGE	Fibrous, Rad/Chemo Tx, Micro CA++
14	0.5	2.61	1.97	AVERAGE	
15	0.5	1.97	1.48	FATTY	
24	0.5	1.51	1.40	DENSE	MICRO CA++

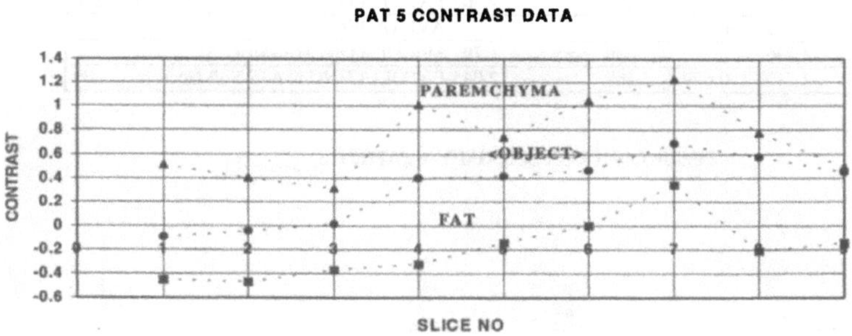

Figure 6. Intra-Breast Contrast Values of Breast Tissue Types. Dense Breast scanned with 1 MHz Transducer Ring.

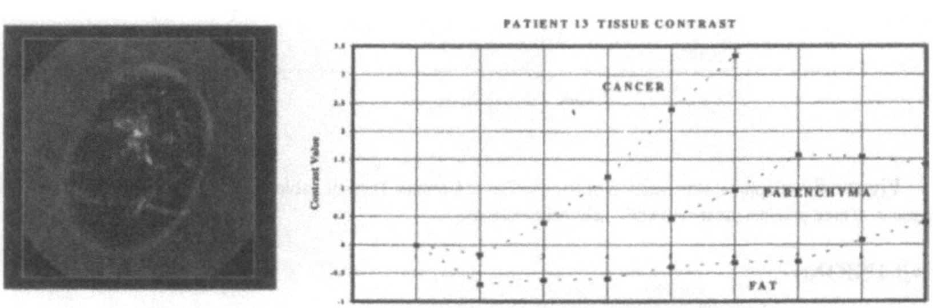

Figure 7. Image and Contrast Values of Patient 13. Fatty Breast with Invasive Lobular Carcinoma. Spiculated Mass of 2.5 x 3 cm.

A further demonstration of the relative consistency of contrast values for tissue elements within the breast is provided in Figure 7. Patient 13 was diagnosed with invasive lobular carcinoma, a spiculated mass of 2.5 x 3 cm roughly 5 cm behind the nipple. Since this mass is within glandular tissue, its markedly higher contrast value in that region provides encouragement that under favorable conditions, image contrast analysis may be able to diagnose malignancies and thus confirm mammography findings.

To demonstrate that the tissue texture analysis of our clinical diffraction tomography breast images provided reasonable discrimination of tissue type for different patients, we have selected images from 8 patients for analysis. The results are summarized in Table 2 and graphically presented in Figure 7.

Table 2. Tissue Contrast Data for Eight Patients

CONTRAST DATA SUMMARY: EIGHT PATIENTS							
PATIEN	WATER	OBJECT	SQ FAT	PAREN	CYST	CANCE	BREAST TYPE
5	5.414	0.321	-0.194	0.725			DENSE/Scattered Micro Ca++
7	0.178	0.044	-0.266	0.636	-0.082		DENSE/SubAreolar Cyst
9	4.759	0.263	-0.162	0.856			FATTY/ Large Breast
12	5.687	0.812	-0.152	1.887			DENSE/Calc. Fibroademomas
13	6.513	0.281	-0.302	1.196		2.307	FATTY/AVERAGE/ Invasive Lobular Carcinc
14	0.182	0.145	-0.565	0.639			AVERAGE/
19	-0.152	1.236		1.36		1.795	SPICULATED MASS/DCIS
21	-0.134	1.219	-0.629	1.169		2.045	EXFOLIATING MASS/Soft Necrotic Center

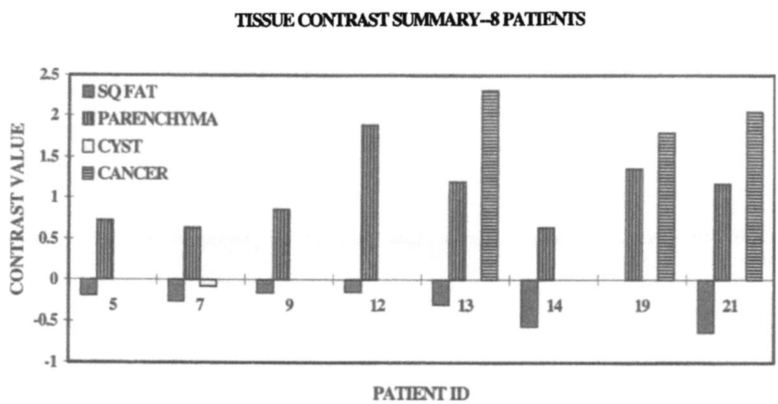

Figure 8. Graphical Presentation of Patient Tissue Contrast Data in Table 2. Tissue types are grouped with each patient. Three patients presented with malignant lesions.

CONCLUSIONS

Analysis of patient breast images from a clinical trials series of 25 patients suggest that ultrasound diffraction tomography has the potential for discriminating between different tissue types.

On the basis of very limited data, it further appears that image contrast values of malignant lesions provide sufficient discrimination from normal breast tissue to be potentially of diagnostic value. An analysis of tissue scatter amplitude distributions shows some correlation between the nature of the distributions, characterized by a Tissue Index, and predominant breast tissue classifications. Since this is reflective of the strength of scattering, the main utility of this analysis is potentially as an a priori tool for the selection of optimal reconstruction algorithms. The data are not sufficiently specific nor interpretable to be of diagnostic value.

The observed strong scattering of breast tissue and the object dimensions lead to phase shifts that are well beyond the validity conditions for the Born approximation. We conclude that our hybrid reconstruction techniques extend the linear range of the propagation constant that can be reconstructed but lose the ability to separate it into components of speed of sound and attenuation.

REFERENCES

1. S.A. Feig, "The role of ultrasound in a breast imaging center", Seminars in US, CT and MR 10, 90-105 (1989).
2. V.P. Jackson, "Sonography of malignant breast disease," Seminars in US, CT and MR 10, 119-131 (1989).
3. P. J. Martin, M. P. André, B. A. Spivey, D. A. Palmer, "Computed tomography using multi-hologram reconstruction of low-frequency ultrasound," J. Ultrasound Med. 12, S26 (1993).
4. M.P. André, P.J. Martin, H.S. Janée, G.P. Otto, et al: "Investigation of contrast sensitivity and phase aberration in low-frequency ultrasound diffraction tomography." Radiology 193 (P), 308 (1994).
5. H.S. Janée, J.P. Jones, M.P. André, "Analysis of scatter fields in diffraction tomography experiments." *Acoustical Imaging* 22 (1996).
6. H.S. Janée, M.P. André, M.Z. Ysrael, L.K. Olson, G.P. Otto, G.R. Leopold, "Amplitude distributions of ultrasound scatter fields: *In vivo* results in the breast. ", presented at AAPM Annual Conference, Philadelphia, PA, 1996.
7. B.D. Steinberg, A discussion of two wavefront aberration correction procedures, Ultrason. Imag. 14:398 (1992).

CALIBRATION OF THE *URTURIP* TECHNIQUE

Bruno Migeon, Philippe Deforge, and Pierre Marché

Laboratoire Vision et Robotique
63, avenue de Lattre de Tassigny
18020 Bourges Cedex - France

INTRODUCTION

As a part of our project concerned with the development of an ultrasound scanner dedicated to limb study, the URTURIP Technique (Ultrasound Reflection-mode Tomography Using Radial Image Processing) has been developed[1]. It consists of using classical B-scan images instead of projections[2-5] and gives qualitative images instead of quantitative images. The final goal of this project is the 2D and 3D reconstruction of anatomical structures at limb level by using echographic image processing. The developed process consists of several successive steps like : multiple reflection removing[6], 2D reconstruction[1], segmentation[7], contour association, contour interpolation[8], 3D reconstruction and visualization[9]. It has been validated by in vitro experiments on anatomical pieces of limbs of new-borns using a simple acquisition system prototype[9].

To validate the URTURIP Technique by in vivo experiments with the help of a new acquisition system, the subpixel coordinates of the rotation center must be accurately determined. The calibration consists in solving this problem of rotation center determination and is the purpose of this paper. A reference calibration method has recently been developed[10] and gives the coordinates of the rotation centre with a subpixel accuracy.

But, this reference method is very time-consuming and cannot be used systematically for each routine exam so that faster methods are developed. At present, the more sophisticated one is based on a least squares fitting technique, which requires a good first estimation close to the real solution in order to satisfy a condition[11] (hypothesis of small displacements). This first estimation is given by a basic method that is noise sensitive, and is therefore often unreliable. This paper presents a new method based on a least squares fitting but which does not require a first estimation of the solution and is more accurate and less time consuming.

Acoustical Imaging, Volume 24.
Edited by Hua Lee, Kluwer Academic/Plenum Publishers, 2000.

2D RECONSTRUCTION

The Ultrasound Reflection-mode Tomography Using Radial Image Processing (URTURIP Technique) principle is to utilize radial B-scan images instead of projections as most other methods do[2-5]. In comparison, fewer radial directions are needed, it is less time-consuming, but qualitative images instead of quantitative ones are computed.

Let L_i^* i=1,..,N be N radial images obtained from N angulary equidistant directions around the rotation center where $L_i^*(k, l)$ denotes the luminance of the (k, l)-pixel on the L_i^* image.

An adjustment step consists in turning each L_i^* image around the rotation center with its own acquisition angle, to construct N adjusted images L_i in such a way that a pixel (x, y) on each L_i image corresponds to the same real point of the cross-section.

Then, a method based on the URTURIP Technique reconstructs an image L by a combination of the N adjusted images L_i , i.e. $L(x, y) = f(L_i(x,y))$, i=1..N, where f denotes the reconstruction method.

CALIBRATION PROBLEM

Introduction

Before applying a method of the URTURIP Technique, the radial images must be adjusted according to the rotation center of the probe. The object of the calibration stage is to determine the rotation center, the coordinates of which are identical on each radial image.

The reference method

A reference method based on an energy minimization technique has been developed[10]. It allows one to obtain the subpixel coordinates of the rotation center with an accuracy as much as one could wish for by using several radial images of a very hyperechogenic solid object.

The basic idea is that the real center is such that the reconstruction of a cross section of this object is perfect i.e. all radial information of the external contour overlaps perfectly with the others. Then, if an energy is able to represent the non-overlapping, the real rotation center minimizes this energy.

Let $E(x_c,y_c)$ be the energy of a reconstructed image after an adjustment of the radial images by considering the rotation center (x_c,y_c). $E(x_c,y_c)$ is defined by the number of the occupied pixels of the image reconstructed by the *maxima* method[1]. In other words :

$$E(x_c,y_c) = \sum_{(x,y)\,\in L^{(x_c,y_c)}} \delta^{(x_c,y_c)}(x,y) \quad \text{with} \quad \delta^{(x_c,y_c)} = \begin{cases} 0 & \text{if } L^{(x_c,y_c)}(x,y) = 0 \\ 1 & \text{if } L^{(x_c,y_c)}(x,y) \neq 0 \end{cases}$$

$$\text{and} \qquad L^{(x_c,y_c)}(x,y) = \operatorname*{Max}_{i} L_i^{(x_c,y_c)}(x,y)$$

where $L^{(x_c,y_c)}$ is the reconstructed image after an adjustment according to the (x_c,y_c) rotation center, i is the index of the radial direction of investigation, $L_i^{(x_c,y_c)}$ is the ith radial image adjusted according to the rotation center (x_c,y_c).

To achieve the desired accuracy, each pixel of the images is subdivided into an adequate number of subpixels. Then, by minimizing this energy with a solid reference object, the subpixel coordinates of the rotation center are determined.

This reference method is very time-consuming and it is difficult to apply it in a routine exam if one often wants to change the probe (for each change, the calibration problem has to be solved). For this reason we must develop new methods, which are faster and comparable with the reference method in terms of accuracy.

Others methods

The reference method being very time-consuming, fast methods must be developed in order to be easily used in routine.

Superimposing N radial images of a point P comes to express the different rotations of P in the same referential, and theses N images Pi of P belong to a circle whose rotation center is the demanded one.

So, a very simple and extremely fast method consists in computing the gravity center of N points Pi corresponding to P viewed from N angulary equidistant directions of investigation. But, in practice, the points Pi are segmented from noisy spots[12] blurred in function of several factors (distance from the probe, incidence,...) so that the results are not very accurate.

Another method is based on a least squares fitting technique but it requires a first estimation close to the solution in order to satisfy a condition of small displacements. In practice, the first estimation is given by the method of gravity center and the results are limited.

Influence of the water temperature

The interface used in all ultrasound scanner systems is water. But, the ultrasound speed in water depends of the temperature and may be different from the speed taken into account by the echograph device.

Using the model developed by Medwin[13], it can be shown that the difference of water temperature may lead to significant errors in terms of position.

Due to this problem, in our system, all radial images are firstly corrected before calibration and 2D reconstruction. This temperature correction consists in computing the speed of ultrasound in water knowing its temperature (with the help of Medwin's model) and then in applying the linear adequate transformation of the water area on each radial image.

THE PROPOSED METHOD BASED ON THE LEAST SQUARES TECHNIQUE

Principle

Let us consider a point P belonging to the exploration plane, and Pi, i=1..N, corresponding to P on the N radial images. Superimposing the N radial images (which comes to express the different rotation of P in the same referential), the N points Pi belong to a circle whose rotation center is unknown and demanded.

In practice, a vertical taut nylon thread of 0.3 mm diameter is used and N radial images taken according to N angulary equidistant radial directions give the N points Pi. A segmentation step is required to determine the points Pi which appear as noisy spots on the images[12].

The rotation center has to be determined from these extracted points Pi. A simple method consists of computing the centre of gravity of these points[12] if they are angulary equidistant. It is very fast and gives an estimation of the rotation center of the probe. However, the accuracy is insufficient due to the uncertainty on the Pi position after segmentation.

The method presented here consists of determining the center of the circle which is best fitted by the points Pi to a least squares sense.

Theoretical description

Considering the N extracted points Pi (figure 1) and the rotation center C, two angles can be defined as follows :

$$-\alpha_j = (CP_j, CP_{j+1}) \quad j \in [1..N] \qquad \text{and} \qquad -\beta_j = (CP_1, CP_{j+1}) = -\sum_{i=1}^{j-1} \alpha_i$$

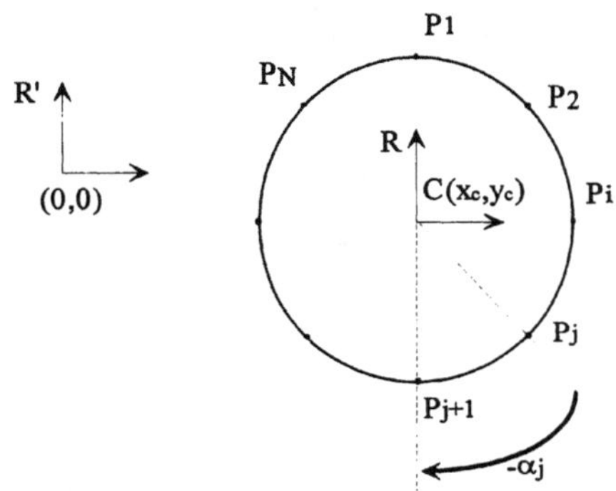

Figure 1: Extracted points and rotation centre.

Let $P_i^{'}$ be the transformed point of Pi by a rotation of angle β_i around C. Let (x_i, y_i) be respectively the coordinates of $P_i^{'}$ and P_i . The points $P_i^{'}$ must correspond to P_1 so that :

$$\begin{cases} x_1 = x_i \cos\beta_i - y_i \sin\beta_i - x_c \cos\beta_i + y_c \sin\beta_i + x_c \\ y_1 = x_i \sin\beta_i + y_i \cos\beta_i - x_c \sin\beta_i - y_c \cos\beta_i + y_c \end{cases} \qquad i = 2..N$$

This system of (N-1) equations with 2 unknowns can be solved by minimizing the function W defined by :

$$W = \sum_i \left\| \overrightarrow{P_1 P_i^{'}} \right\|^2$$

After some simplifications, the partial derivatives $\partial W / \partial x_c$ and $\partial W / \partial y_c$ are given by :

$$\frac{\partial W}{\partial x_c} = 2\left\{2x_c\sum_i(1-\cos\beta_i) - \sum_i\left[(1-\cos\beta_i)(x_i+x_1)+\sin\beta_i(y_i-y_1)\right]\right\}$$

and

$$\frac{\partial W}{\partial y_c} = 2\left\{2y_c\sum_i(1-\cos\beta_i) - \sum_i\left[\sin\beta_i(x_1-x_i)+(1-\cos\beta_i)(y_1+y_i)\right]\right\}$$

Then, the solving of the system $\begin{cases}\dfrac{\partial W}{\partial x_c}=0\\ \dfrac{\partial W}{\partial y_c}=0\end{cases}$ leads to the solution :

$$x_c = \frac{\sum_i\left[(1-\cos\beta_i)(x_i+x_1)+\sin\beta_i(y_i-y_1)\right]}{2\sum_i(1-\cos\beta_i)}$$

$$y_c = \frac{\sum_i\left[\sin\beta_i(x_1-x_i)+(1-\cos\beta_i)(y_1+y_i)\right]}{2\sum_i(1-\cos\beta_i)}$$

RESULTS

This new method and the previous one[11] have been applied with 4, 8, 16 and 64 points Pi (corresponding to 4, 8, 16 and 64 radial directions of investigation) and compared to the reference method[10] with respect to accuracy and execution time. In all cases, a temperature correction has been held.

In terms of execution time, they are very similar and give immediately the result (less than 1 second) when the reference method takes requires about 30 minutes[10].

In terms of accuracy, the presented method gives better results as illustrated in table I. Moreover, contrary to the previous one, the accuracy increase with the number of points Pi.

Table 1 . Comparison beetween the previous and presented method with the reference method for a required accuracy of 1mm and after temperature correction.

Number of points Pi	4	8	16	32	64
Previous method	+/- 1.5mm	+/- 0.6 mm	+/- 0.4mm	+/- 0.3 mm	+/- 0.3 mm
Presented method	+/- 1.2mm	+/- 0.5 mm	+/- 0.3mm	+/- 0.2 mm	+/- 0.1 mm

CONCLUSION

As a part of our development project of an ultrasound scanner based on the URTURIP Technique, a reference calibration method has been developed. It allows determining the subpixel coordinates of the rotation center which is required, to use the URTURIP Technique, with an accuracy as one could wish. This method is in return very time-consuming, so that faster methods providing a result close to the one given by the reference method must be developed .

After a brief reminder of the different existing methods and the temperature correction, this paper presents a new least squares based calibration method. It does not require a first estimation of the solution, and takes into account the knowledge of the different radial directions of investigation. The comparison with the reference method shows that it allows obtaining immediately very good results.

This calibration method is now currently used in our application and allows good results to be obtained immediately.

REFERENCES

1. B. Migeon and P. Marché, Ultrasound tomography by radial image processing, *Innov. Tech. Biol. Med.*, vol. 13, n°3, pp. 292-304, (1992).
2. C. M. Sehgal, et al., Ultrasound transmission and reflection computerized tomography for imaging bones and adjoining soft tissues, *IEEE Ultrasonic Symp.* Chicago, IL, vol. 2, pp. 849-852, (1988).
3. H. Hiller, H. Hermert, System analysis of ultrasound reflection mode computerized tomography, *IEEE Trans. Sonics Ultrason.*, vol SU 31, pp. 240-250, (1984).
4. M. Friedrich , et al., Computerized ultrasound echo tomography of the breast, *Europ. J. Radiol.*, vol 2, pp. 78-87, (1982).
5. J. Ylitalo, et al., Ultrasonic reflection mode computed tomography through a skullbone, *IEEE Trans. Biomedical Engineering*, Vol. 37, N° 11, Nov. (1990).
6. B. Migeon, P. Vieyres, P. Marché, A simple solution for removing echo bars for URTURIP Technique, Acoustical Imaging, *Acoustical Imaging*, Vol. 22, pp. 543-548, (1995).
7. B. Migeon, V. Serfaty, M. Gorkani, P. Marché, An Adaptive Smoothing Filter for URTURIP Images Applying the Maximum Entropy Principle, *IEEE Engineering in Medicine and Biology*, pp. 762-765, Nov/Dec (1995).
8. B. Migeon, P. Vieyres, P. Marché, Interpolation of star-shaped contours for the creation of lists of voxels : application to 3D visualisation of long bones, *Int. J. of CADCAM and Computer Graphics*, Vol.9, n°4, pp. 579-587, (1994).
9. B. Migeon, P. Marché, Echographic Image Processing for Reconstructing Long Bones, *Proc. of 17th Annual International Conference IEEE-EMBS*, Montreal, (1995).
10. B. Migeon, P. Deforge, P. Marché, Calibration for the URTURIP Technique using an energy minimization method, *Proc. of the 23rd Int. Symp. On Acoustical Imaging*, Boston, April, (1997).
11. B. Migeon, P. Deforge, P. Marché, Calibration of an ultrasound scanner based on the URTURIP Technique using a least square method, *Proc. of the Int. Symp. of Computer Methods in Biomechanics & Biomedical Engineering*, Barcelona, 1997.
12. P. Deforge, B. Migeon, P. Marché, Calibration du système d'acquisition d'un scanner à ultrasons médical basé sur la technique URTURIP, *Proc. of the 4th French Congress on Acoustics*, Marseille, April, (1997).
13. H. Medwin, Speed of sound in water: A simple equation for realistic parameters, *J. Acoust. Soc. Am.*, Vol. 58 (6), pp. 1318-1319 (1975).

STUDIES OF BONE BIOPHYSICS USING ULTRASOUND VELOCITY

S Mehta[12], PP Antich[12], M Daphtary[1], M Lewis[1], B Smith[1] and WJ Landis[3]

[1]Advanced Radiological Sciences, Department of Radiology
[2] Center for Mineral Metabolism and Clinical Research
UT Southwestern Medical Center at Dallas,
5323 Harry Hines Blvd, Dallas, Texas 75235
[3] Department of Biochemistry and Molecular Pathology,
Northeastern Ohio Universities College of Medicine, Rootstown, OH 44272

INTRODUCTION

The biological processes in bone are expressed in its final structure, achieved to fulfill specific physical functions. Thus a biophysical study of bone is a study of the relationship between its organization (at various scales) and its functions. Bone tissue primarily provides structural support and mineral homeostasis. Its microstructure has a hierarchical organization and a symmetry (transverse isotropy) optimized to meet a variety of structural needs over the entire skeleton; the material is thus heterogeneous and anisotropic with its mechanical properties varying with site and orientation. The two compartments (types) of bone, cortical and cancellous, are made of the same material but vary in the composition of the components. Bone material is constituted from organic matter which consists mostly of proteins (primarily the fibrous protein Type I collagen), glycoproteins and proteoglycans and from inorganic matter, the bone mineral crystals. The microstructural organization of these two phases is not very well described but it has been demonstrated that the mineral crystals are normally preferentially oriented with their major axis along the long fibrous collagen proteins [1]. The properties of these components also change with location and age. This complexity in the microarchitecture of bone matrix makes it extremely difficult to diagnose causes of mechanical or functional failure of the tissue.

The strength of bone is a clinically relevant biophysical property, the assessment of which is currently carried out by calculating the amount of mineral present in bone from a projected X-ray image of the entire bone. Biochemical markers such as collagen degradation products have also been used to assess bone physiology but give no insight into location or compartment-specific responses. Biopsies of bone can provide valuable information about bone status but this is an invasive technique and repeated measurements are generally not possible. Magnetic resonance microscopy has been used at peripheral sites to quantitate bone structure, through which the functional status of bone can be inferred [2]. However, all of the above techniques provide indirect measures of the mechanical properties of bone.

Acoustical Imaging, Volume 24,
Edited by Hua Lee, Kluwer Academic/Plenum Publishers, 2000.

Ultrasound is the only noninvasive technique that can make a direct measurement of bone mechanical properties. The ultrasound velocity in bone (v) is related to the stiffness of the material (as $\mu=\rho \cdot v^2$; where μ is longitudinal modulus, ρ is density) [3]. Measurements of bone velocity using the UCR (Ultrasound Critical-angle Reflectometry) technique have revealed several previously unknown properties of bone material [4-6].

The UCR technique measures all five independent parameters of the stiffness tensor of the bone matrix [7]. These material constants of stiffness describe fully the elastic mechanical behavior of bone and are related in a complex but predictable fashion to its microstructure and biophysical interactions among individual elements of the bone extracellular matrix [8]. The mechanical strength of bone depends on these biophysical relationships. Thus the strong empirical relationship shown to exist between the velocity of sound and the strength of bone can be exploited by UCR and the roles of individual components of the bone matrix in this interaction, now not well characterized [9, 10], can be better understood.

MATERIALS AND METHODS

The UCR technique may be used to determine changes in mechanical properties of biochemically modified bone material and this information may in turn be applied to analyze the biophysical causes of alterations in bone behavior.

Ultrasound Critical-angle Reflectometry (UCR)

A brief description of the UCR technique is provided here; details are available elsewhere [7]. An analytical formulation of the phenomena of reflection and refraction of ultrasound through a range of incident angles at the interface of water (soft-tissue) and bone has been developed and is used to recognize the critical angles of pressure and shear waves at total internal reflection in a calibration medium, water, in which the velocity is known accurately. Therefore, the velocities of these waves can be calculated from these critical angles by Snell's law within the plane of scattering in bone. This scattering plane is defined by the incident beam and the normal to the surface at the point of incidence. Rotation of the scattering plane at a point gives an orientation-dependent distribution of these velocities as depicted in Figure 1 below (shear velocity data similar; not shown). The velocities measured at any orientation are related to the rotated stiffness tensorial elements at that angle and the entire distribution, when fit to this well-known relationship, yields the elements of the stiffness tensor of the material at that point. This point can then be scanned over an area and the measured velocity or stiffness values assigned using a color-table to produce images [7].

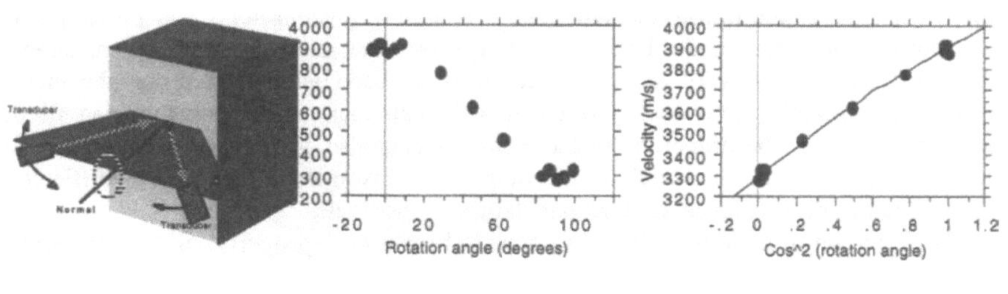

Figure 1 (L to R): i) Schematic of UCR setup depicting rotation of scattering plane around a normal to the material surface. ii) Example of directional dependence (anisotropy) of measured pressure wave velocities in a cortical bovine bone sample (vs. rotation angle) iii) When the data are fit to a quadratic function with $\cos^2$(rotation angle), the coefficients of that function yield the stiffness constants.

The transverse isotropy symmetry of bone material reduces to 5 the number of independent constants in its stiffness matrix [7]. These stiffness constants are, in turn, formally related to the technical constants (elastic moduli, Poisson's ratios), which are the mechanical properties measured during bulk mechanical testing. The technical constants for bone, determined through an inversion of the stiffness matrix (and the precision with which they are determined), are the principal (Young's) moduli in the major and minor directions, E_{33} (3%) and E_{11} (2%), respectively, the Poisson's ratios ν_{13} (10%), ν_{12} (8%) and ν_{31} (13%), and the shear moduli, G_{12} (6%) and G_{23} (3%). Standard abbreviated notations for the technical constant suffixes are used here [7]. The directions 3 and 1 correspond to the long axis of the cylindrical long bone and its transverse axis, respectively, and the direction 2 corresponds to the radial direction (Figure 2). Thus V_{33} and V_{11} are pressure wave velocities along the 3 and 1 directions, and V_{44} and V_{66} are shear wave velocities in the same directions respectively. Therefore in Figure 1 and all subsequent figures, V_{33} corresponds to the measurement at rotation angle=0° and V_{11} to the measurement at rotation angle=90°.

Modification of Bone Material Properties

In order to assess the biophysical properties of its extracellular matrix, four physico-chemical treatments reproducibly altering specific organic or inorganic components were applied to samples of bovine cortical bone obtained from the femoral diaphysis of two animals from a local slaughterhouse. Samples were divided into treatment groups [5] as described briefly below :

1) To examine the effects of a limited denaturation of the organic matrix, a group of bone samples (n=12) were placed in a 3M (pH 7.5) solution of urea at room temperature for 72 hours. After completion of the UCR measurements, the urea was removed using the original storage solution and the measurements were repeated.

2) To examine the effects of total removal of the organic matrix, the organic components were completely dissolved using NaOCl (sodium hypochlorite) solution (Sigma Chemical Co., >5% available chloride), leaving behind the virtually undisturbed mineral phase [5]. Bone samples (n=12) were placed in the solution for 72 hours and remeasured by UCR.

3) To degrade specifically the primary structural component of the organic matrix, Type I collagen, the enzyme Type I Collagenase was used. Bone samples (n=14) were immersed in a solution of Type I Collagenase (Sigma Chemical Co.; 1% w/v; pH 7.5) for up to 30 hours at 32°C, with mild agitation. The mineral crystalline state was not affected [5].

4) To simulate the physiological process of resorption of the mineral component, a calcium-chelating (EDTA; 0.5M, pH 7.4) solution was used. A discrete demineralization process, in which the solution was sprayed on for a few minutes and removed each time, was used in repeated steps (18 steps) to create local areas of mineral dissolution. The organic phase was not significantly altered [11]. The mineral phase has been shown to have a major contribution to the magnitude and anisotropy of bone mechanical properties [12].

Studies of Osteogenesis Imperfecta (OI) in a Mutant Mouse Model of the Disease

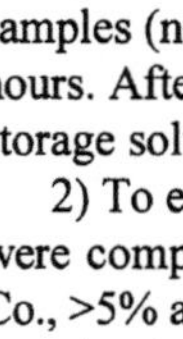

Figure 2 Schematic of femur with coordinate axes

OI arises primarily through a defect in Type I collagen production. In humans, the disease is characterized by spontaneous fracture of bone under normal loading but the disease is variable in its severity, the effects ranging from mild disability to death in utero. Femurs from homozygous mice (*oim*; both gene copies with mutation in Type I collagen gene and displaying the same types of fragility fractures as found in humans) were compared to those from wild type (normal) mice by UCR measurements of cortical bone properties in distal diaphyses. The axes were assigned as shown in Figure 2.

Analysis of Statistically Significant Changes

The changes in measured properties were evaluated by the nonparametric Wilcoxon signed ranks paired analysis with significance assigned at $p<0.05$.

RESULTS

The results of the material modification experiments are shown below in Table 1 as percent changes from original (pre-treatment) measured values. Statistically significant differences (*) are shown in bold.

TABLE 1 Results of Material Modification Experiments

	Urea % Change	Demineral-ization % Change	Collagenase % Change	NaOCl % Change
Density (g/cm^3)	- 0.2	**- 0.7** *	**– 0.4** *	**- 3** *
V_{33} (m/s)	**2** *	**- 3** *	**– 3** *	**- 16** *
V_{11} (m/s)	**3** *	**- 3** *	**– 1** *	**- 17** *
$V_{44}=V_{55}$ (m/s)	**11** *	**- 5** *	**– 11** *	
V_{66} (m/s)	**15** *	**- 5** *	– 3	
E_{33} (GPa)	**7** *	**- 20** *	**– 10** *	
E_{11} (GPa)	**21** *	**- 19** *	– 5	
G_{12} (GPa)	**34** *	**- 16** *	– 3	
G_{23} (GPa)	**27** *	**- 11** *	**– 10** *	
ν_{12}	**– 41 (urea)*** **– 34 (post)***	26	– 5	
ν_{13}	27 (urea) **56 (post)***	5	5	
ν_{31}	9 (urea) **30 (post)***	6	4	

* indicates signficance at $p<0.05$.

1) Urea altered the physico-chemical properties of the organic matrix and increased the shear and pressure wave velocities. The shear velocities showed much greater change than pressure wave velocities (Figure 3). Remarkably, this change was reversible with all velocities returning to baseline values when urea was removed. Principal and shear moduli increased reversibly but the Poisson's ratios showed some irreversible changes when urea was removed (shown as "urea" and "post" values in Table 1). The matrix anisotropy, marked by the squared ratio of V_{33} and V_{11}, decreased because of an increased V_{11} value.

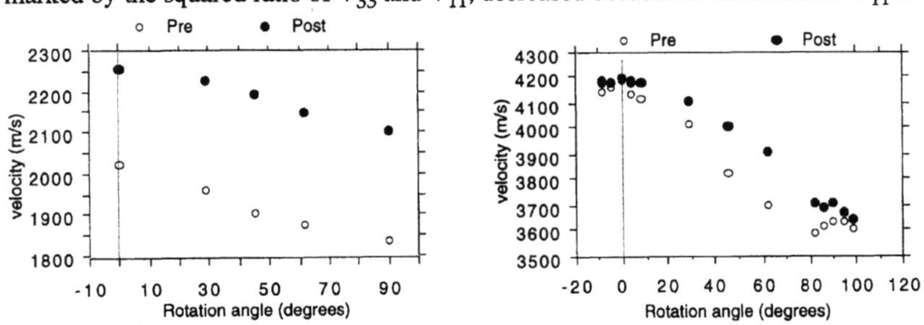

Figure 3 Changes in shear wave (left) and pressure wave (right) velocity distributions with urea treatment.

2) Complete degradation of the organic matrix by NaOCl resulted in pressure wave velocities decreasing significantly with similar losses along both principal axes (Figure 4). Shear velocities fell below the threshold for detection by the UCR technique. The anisotropy did not change substantially (as seen from V_{33}, V_{11} changes in Table 1).

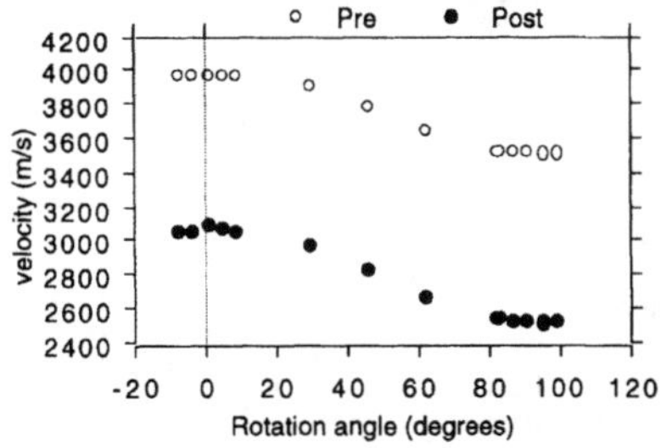

Figure 4 Changes in pressure wave velocity distribution with complete removal of organic matrix (protocol 2, NaOCl,24 hrs).

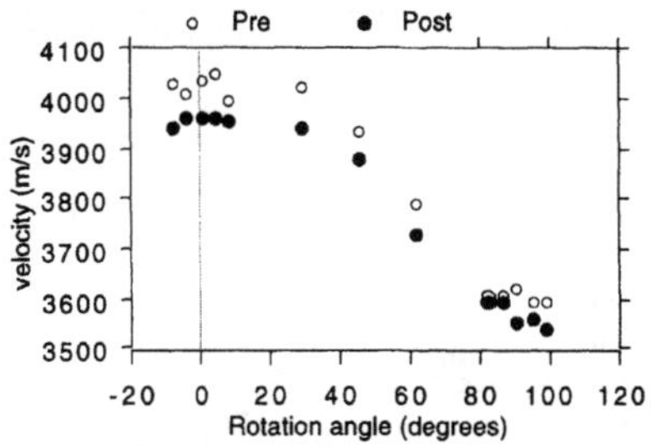

Figure 5 Changes in pressure wave velocity distribution with selective degradation of collagen by collagenase.

3) Enzymatic degradation of the collagen protein network led to a decrease in values of V_{33} and in V_{11} (Figure 5). The modulus in the longitudinal direction E_{33} decreased whereas the decrease in the transverse modulus, E_{11}, did not reach statistical significance. This decreased the anisotropy of the material properties (seen also in changes in V_{33}, V_{11} in Table 1). The shear velocity and shear modulus along the long axis (V_{44}, G_{23} respectively) also showed a significant decrease (Table 1).

4) The discrete demineralization process produced a gradual reduction (with increasing steps of demineralization; Figure 6a) in pressure and shear velocities at all orientations. The losses were equal in all directions and the anisotropy ratio did not change (Figure 6b). The principal and shear moduli also decreased significantly with no change in the Poisson's ratio.

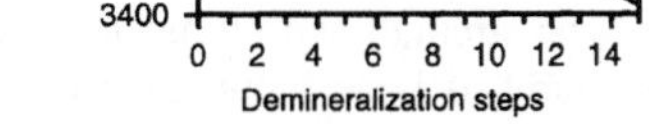

Figure 6a A gradual decrease in the pressure wave velocity (both V_{33} and V_{11}) with increasing steps of discrete demineralization process Representative sample A.

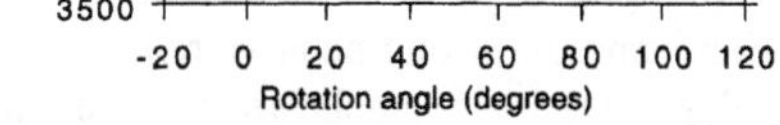

Figure 6b Pressure wave velocity is reduced at all orientations after 18 steps of demineralization. Representative sample B.

TABLE 2: Differences between *Oim* and Wild Type Animals

	V_{33}	V_{11}	V_{44}	V_{66}	E_{33}	E_{11}	G_{12}	G_{23}
% Difference	-8	**23 ***	1	2	**-19 ***	19	-9	-10

* indicates statistical significance with $p<0.05$ (difference = *oim* - wild type)

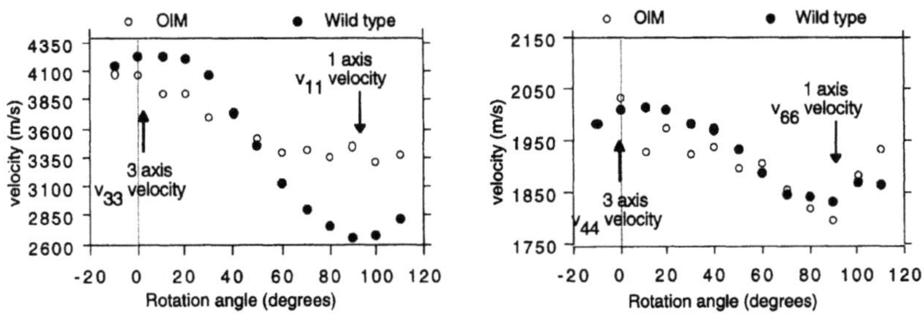

Figure 7 Orientation-dependence of pressure (left) and shear (right) wave velocities shown for two representative bone samples of wild type (filled circles) and *oim* mice (open circles).

The variation of the pressure and shear wave velocities with angle of rotation of the scattering plane is shown graphically in Figure 7 for representative bone samples from homozygous (*oim*) and wild-type mice. There was no change in the velocities along the long axis (V_{33}) but a significant increase ($p<0.05$) was measured in V_{11} (23%) compared to the wild type (Figure 7; Table 2). The *oim* mouse bones had a significantly reduced pressure velocity anisotropy ratio $((V_{33}/V_{11})^2$; mean of 1.65) compared to the wild type specimens (mean 2.8), a result mainly attributable to the increased V_{11} velocity (Figure 7) in the *oim* specimens. The bones from the *oim* mice showed no difference in shear velocity values from the wild type.

Elastic moduli and Poisson's ratios were compared for *oim* and wild type animals (Table 2). The principal elastic moduli for the *oim* specimens increased in E_{11} (19%) and decreased a statistically significant 19% in E_{33} compared to wild type tissues (Table 2). Mean values for the shear moduli, G_{12} and G_{23}, did not show a significant change. Of the Poisson's ratios, ν_{12} was increased (data not shown here).

DISCUSSION AND SUMMARY OF RESULTS

The agreement between measurement of elastic modulus by UCR and by mechanical testing [5] and the strong positive relationship between bone strength and ultrasound velocity (or elasticity) [13,14] emphasize the importance of the study of bone biophysical properties through ultrasound velocity. The high precision and accuracy of the nondestructive UCR measurements make it possible to follow progressive changes in material anisotropy and elasticity in the same sample and at the same point [5]. In the work here, the utility of ultrasound velocity measurement (using UCR) is shown in bone biophysical investigations and in providing a novel understanding of the effects of a human disease on bone material properties.

The biophysical character of specific extracellular matrix components is discussed below, followed by comments on UCR results for mice with brittle bones: The data obtained here quantitatively demonstrate that degradation of collagen affects the relationships among the elasticity constants by decreasing the anisotropy ratio and changing the shear moduli.

Collagen thus contributes to the elasticity of bovine bone material at all orientations with its greatest contribution along the major axis of bone (which coincides with the preferred orientation of collagen fibers[1, 15]).

The results of the NaOCl treatment show that a significant anisotropy ratio is retained after the organic matter is removed. However, without the organic matrix, the mineral component cannot support tensile or compressive loads, a fact reflected in the collapse of the elastic moduli of the material observed here.

Denaturation using urea is expected to lead to altered interfacial bonding and/or packing between organic matter and the bone mineral. This maybe attributable to the altered protein conformations induced by urea [5]. The large increases in shear moduli and velocities seen with this treatment show that the mechanical characteristics of bone, reflected in the physical properties assessed by UCR, are significantly affected by changes in the organic matrix interfacial bonding or packing.

Discrete demineralization maintained the relationship between moduli with no change observed in anisotropy or Poisson's ratios measured by UCR. This result indicates that changes in the inorganic component of bone induced by local demineralization processes, perhaps as during focal bone resorption by cells, do not affect the distribution of overall mechanical stresses and strains in the tissue.

In all, these results confirm that the mineral component and the collagen component of the organic matrix contribute to the anisotropy of the bone matrix properties.

Discussion of the results from the *oim* bones: UCR results from Table 1 and corresponding figures above suggest the small decrease in E_{33} seen in *oim* bones may be consistent with effects simulated on the bone organic matrix, specifically collagen, by collagenase. This interpretation would appear to correlate with the fact that an alteration in collagen structure is known to be the cause of britlle bones in the *oim* mice. The increase in V_{11} (reflected in E_{11}) is consistent with changes in matrix components induced by urea denaturation, again possibly related to collagen alterations. Since collagen provides a framework for mineral organization in vertebrates, any modification would directly affect mineral alignment, organization, distribution or other parameter [16]. Such changes in mineral orientation would be consistent with the observed decrease in *oim* anisotropy. Indeed, studies by other laboratories, utilizing electron microscopic tomography and small angle X-ray scattering analysis, have suggested that the mineral component arrangement is altered with more crystals oriented along the 1-direction (V_{11}) than in the wild type bone [17].

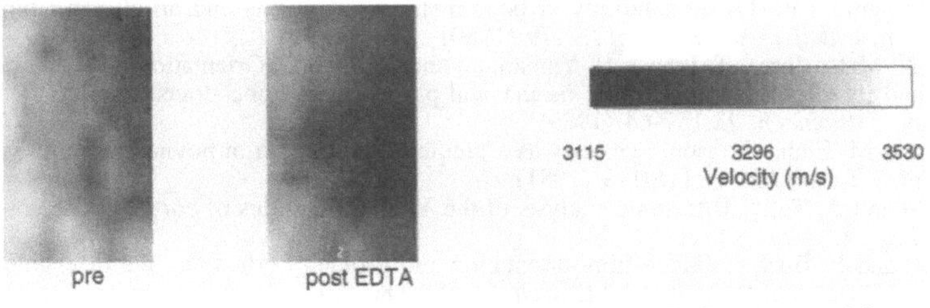

Figure 8 A normal bovine bone sample was imaged by UCR, first untreated (image on left) and then subjected to several steps of discrete demineralization over its right side (image on right). The area where the demineralization process was focused can easily be visualized and velocities assigned. The dimensions of the image area are 7 X 12 mm.

Finally, it should be noted that the components of the matrix of elasticity do not maintain a fixed relationship with each other or with bone density. Changes in bone quality have different effects on these various elastic constants. The above discussion clearly indicates that an assessment of bone functional status is incomplete without the measurement of multiple components of bone mechanical properties as done here by UCR.

An approach to measuring spatial and temporal changes such as these can also be derived from UCR since it is sensitive to location within a sample, is noninvasive and can be used repeatedly over time. An example of this is presented in Figure 8, where the same bone area is imaged before and after discrete demineralization. A precise location is imaged here at different times by a physical property of the material. Thus, UCR has a range of applied uses from studying samples for basic biomechanical research or clinical problems, as described above, to detailed evaluation of bone status in vivo. Such biophysical analysis could also be valuable in leading to alternative therapies for many disorders.

ACKNOWLEDGMENTS

The authors thank Mrs. Karen Hodgens (Children's Hospital, Boston) for technical assistance. WJ Landis acknowledges support from the National Institutes of Health (NIH), grant AR41452. The other authors acknowledge support from Vitel, Inc., and NIH grant AR16061.

REFERENCES

1. F. Evans and R. Vincentelli, Relation of collagen fiber orientation to some mechanical properties of human cortical bone, *J Biomech.* 2:63-71 (1969).
2. H. Chung, F. Wehrli, J. Williams and S. Wehrli, Three-dimensional Nuclear Magnetic Resonance microimaging of trabecular bone, *J Bone Miner Res.* 10:1452-1461 (1995).
3. P. Antich, Ultrasound study of bone in vitro, *Calcif Tissue Int.* 53:S157-S161 (1993).
4. P. Antich, S. Mehta, M. Daphtary, B. Smith and V. Vaguine, Bone elastometric imaging using ultrasound critical-angle reflectometry (UCR), in: *Acoustical Imaging V. 23*, L. a. Ferrari, ed., Plenum Press, NY (1997).
5. S. Mehta, O. Öz and P. Antich, Bone elasticity and ultrasound velocity are affected by subtle changes in organic matrix, *J Bone Miner Res.* 13:114-119 (1998).
6. J. Zerwekh, P. Antich, S. Mehta, K. Sakhaee, F. Gottschalk and C. Pak, Reflection ultrasound velocities and histomorphometric and connectivity analyses: Correlations and effect of slow-release sodium fluoride., *J Bone Miner Res.* 12:2068-2075 (1997).
7. P. Antich and S. Mehta, UCR (ultrasound critical-angle reflectometry): A new modality for functional elastometric imaging, *Physics Med Biol.* 42:1763-1777 (1997).
8. J. L. Katz and A. Meunier, The elastic anisotropy of bone, *J Biomech.* 20:1063-70 (1987).
9. E. Sedlin and C. Hirsch, Factors affecting the determination of the physical properties of femoral cortical bone, *Acta Orthop Scand.* 37:29-48 (1966).
10. J. Mammone and S. Hudson, Micromechanics of bone strength and fracture, *J Biomech.* 26:439-446 (1993).
11. I. Dickson and S. Sohan, Effects of demineralization in an ethanolic solution of triethylammonium EDTA on solubility of bone matrix components and on ultrastructural preservation, *Calcif Tissue Int.* 32:175-179 (1980).
12. N. Sasaki, N. Matsushima, T. Ikawa, H. Yamamura and A. Fukuda, Orientation of bone mineral and its role in the anisotropic mechanical properties of bone--transverse anisotropy, *J Biomech.* 22:157-64 (1989).
13. C. Turner and M. Eich, Ultrasonic velocity as a predictor of strength in bovine cancellous bone., *Calcif Tissue Int.* 49:116-119 (1991).
14. W. Bonfield and A. Tully, Ultrasonic analysis of the Young's modulus of cortical bone, *J Biomed Eng.* 4:23-27 (1982).
15. R. Pidaparti and B. Burr, Collagen fibre orientation and geometry effects on the mechanical properties of secondary osteons, *J Biomech.* 25:869-880 (1992).
16. W. J. Landis, The strength of a calcified tissue depends in part on the molecular structure and organization of its constituent mineral crystals in their organic matrix, *Bone.* 16:533-44 (1995).
17. P. Fratzl, O. Paris, K. Klaushofer and W. J. Landis, Bone mineralization in an osteogenesis imperfecta mouse model studied by small-angle x-ray scattering, *J Clin Invest.* 97:396-402 (1996).

IMAGING OF THE TISSUE ELASTICITY BASED ON ULTRASONIC DISPLACEMENT AND STRAIN MEASUREMENTS

Yasuo Yamashita[1] and Mitsuhiro Kubota[2]

[1]Department of Industrial Engineering and Management,
Nihon University College of Industrial Technology,
Narashino 275-8575, Japan
[2]Department of Surgery, Tokai University School of Medicine,
Isehara 259-0011, Japan

ABSTRACT

A variational formulation is presented for the reconstruction of the elasticity of an isotropic, inhomogeneous elastic medium subject to external mechanical forces. Given a knowledge of the displacements within the medium, a finite-element based model for static deformation is proposed for solving the distribution of the shear modulus of the tissue. The feasibility of the proposed method is demonstrated using the simulated deformation data of the simple inclusion problem. The results show that the relative shear modulus may be reconstructed from the displacement data measured locally in the region of interest within an inhomogeneous medium, and that the relative shear modulus can be recovered to some degree of accuracy in the presence of measurement noise.

INTRODUCTION

Palpation has been the clinical diagnostic modality to detect the changes in soft tissue elasticity, because the elasticity is usually related to some abnormal, pathological process. It is, however, effective only on the lesion near the skin surface and information obtained by palpation is inherently subjective. Therefore, imaging of tissue elasticity will become a new modality of significant diagnostic value (Skovoroda et al., 1995; Kallel et al., 1996; Raghavam and Yagle, 1994).

Internal tissue deformation induced by externally applied mechanical sources or by the primary cardiac pulsation has been evaluated to characterize tissue elasticity. Under the assumption that the tissue is elastic, isotropic, and subject to a constant uniform stress field, the strain field could be interpreted as a relative measure of elasticity distribution, since the strain is small in a hard tissue. This strain field was visualized as a gray level image named *elastogram* and the technique itself *elastography* (Ophir et al., 1991). Although the tissue elasticity is ultimately correlated with strain field of internal deformation, deformational geometry as well as the pattern of external force/displacement

Acoustical Imaging, Volume 24.
Edited by Hua Lee, Kluwer Academic/Plenum Publishers, 2000.

sources can greatly affect the strain distribution as well. Consequently, for a quatitative elasticity imaging, material parameter such as the shear modulus, or equivalently the Young's modulus must be reconstructed from estimates of internal displacement and strain (Emelianov et al., 1995).

O'Donnell et al. (1994) have shown that quantitative elasticity imaging consists of three main components: using real time ultrasound imaging devices, speckle tracking and evaluation of tissue deformation, estimation of strain tensor components, and reconstruction of the elastic modulus based on strain fields. In this paper, we present a formulation and numerical demonstration of the determination of material elasticity of an isotropic, inhomogeneous elastic medium subject to external forces.

In the theory section, the general theoretical approach to determine the deformation inside tissue based on the common model of a linear, elastic, isotropic medium is presented. A variational formulation for the forward problem is presented in the next section. Practical method to estimate the elastic modulus using displacement data is formulated by inversely solving the forward problem. The feasibility of the proposed method is demonstrated using the simulated deformation data of a simple 2-D inclusion problem. This is followed by a result of elasticity reconstruction based on the internal displacement field measured from a gelatin-based phantom to compare with theoretical predictions of a linear elastic model. Finally, the paper concludes with a discussion of the results.

THEORY

Governing Equations for Static Deformation

Biological soft tissue can be considered as an isotropic, continuous incompressible medium. When the soft tissue is deformed very slowly under externally applied mechanical forces, the static equilibrium equation is given by

$$\sum_{j=1}^{3} \frac{\partial \sigma_{ij}}{\partial x_j} + g_i = 0 \qquad i = 1,2,3, \tag{1}$$

where σ_{ij} is the symmetric stress tensor and $x_i (i = 1,2,3)$ are the rectangular Cartesian coordinate (x, y, z), and g_i is the body force which can be neglected for small tissue deformation. Assuming linear elasticity, the constitutive equation in an isotropic, continuous medium under static deformation is

$$\sigma_{ij} = \lambda \nabla \cdot \boldsymbol{u} \delta_{ij} + 2\mu\epsilon_{ij} \qquad i,j = 1,2,3. \tag{2}$$

The strain tensor ϵ_{ij} can be defined using the displacement vector u_i's as

$$\epsilon_{ij} = \frac{1}{2}\left(\frac{\partial u_i}{\partial x_j} + \frac{\partial u_j}{\partial x_i}\right) \qquad i,j = 1,2,3, \tag{3}$$

where, u_i is one component of displacement vector $\boldsymbol{u} = (u_1, u_2, u_3)$. In Eq. (2) the parameters $\lambda = E\nu/(1+\nu)(1-2\nu)$ and $\mu = E/2(1+\nu)$ are the Lamè coefficients in which E is the Young's modulus and ν is the Poisson's ratio. Thus, the static deformation for the isotropic continuous medium is characterized by two elastic parameters.

Since, the biological soft tissues is incompressible (Sarvazyan, 1975) (i.e., the Poisson's ratio approaches 0.5), the volume change due to deformation must be zero:

$$\nabla \cdot \boldsymbol{u} = div \boldsymbol{u} = \epsilon_{11} + \epsilon_{22} + \epsilon_{33} \tag{4}$$

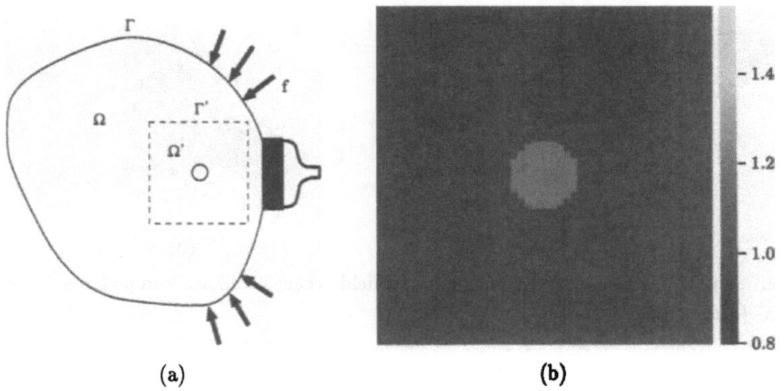

(a) (b)

Figure 1. Biological tissue model being subject to external forces (a) and a gray level image of the shear modulus distribution (b) used for numerical experiments. The tissue elasticity distribution contains a circular hard inclusion in the center. The inclusion is 1.2 harder than its surrounding medium.

Thus, for incompressible medium, the longitudinal Lamè constant λ approaches infinity. Under these conditions, the stress-strain relation for static deformation reduces to

$$\sigma_{ij} = p\delta_{ij} + 2\mu\epsilon_{ij} \qquad i,j = 1,2,3, \tag{5}$$

where p is the static internal pressure, defined as

$$\lim_{\lambda/\mu\to\infty,\ div\boldsymbol{u}\to 0} [\lambda(div\boldsymbol{u})] = p. \tag{6}$$

which, from Eqs. (4) and (5), is equivalent to the mean normal stress:

$$p = \frac{1}{3}\sum_{i=1}^{3}\sigma_{ii}. \tag{7}$$

Therefore, the static deformation of an isotropic, incompressible medium can be completely characterized by a single elastic parameter, μ or E. The static internal pressure p, however, cannot be definitely related to the strain.

Eq. (5) can be solved for ϵ_{ij}; thus, the stress-strain relation is usually written as

$$\epsilon_{ii} = \frac{1}{E}\Big(\sigma_{ii} - \frac{1}{2}\sum_{k=1,k\neq i}^{3}\sigma_{kk}\Big) \qquad i = 1,2,3 \tag{8}$$

$$\epsilon_{ij} = \frac{3}{2E}\sigma_{ij} \qquad i,j = 1,2,3;\ i \neq j \tag{9}$$

where E is the Young's modulus which equals 3μ for an incompressible medium. Since the stress tensor is not determined from the strain tensor, we cannot obtain the constitutive equation for an incompressible medium in the usual sense.

2-Dimensional Case

Let us consider the estimation of the 2-D spatial distribution of the shear modulus. Assuming that the shear modulus μ has spatial distribution only in the (x, y) plane

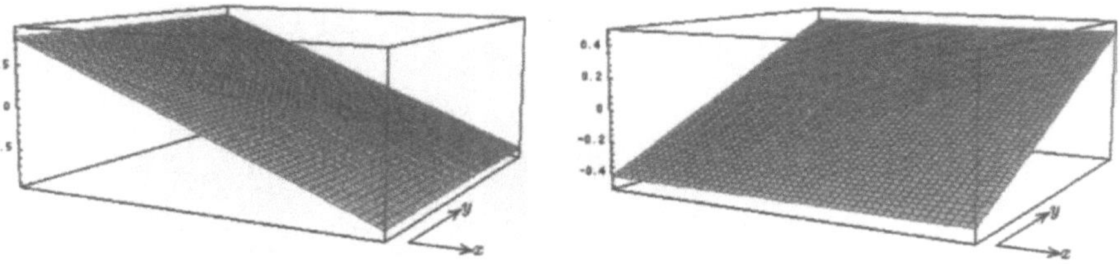

(a) (b)

Figure 2. Analytic solution of the displacement field when a uniform compression is applied in x-direction. (a)u_x. (b)u_y.

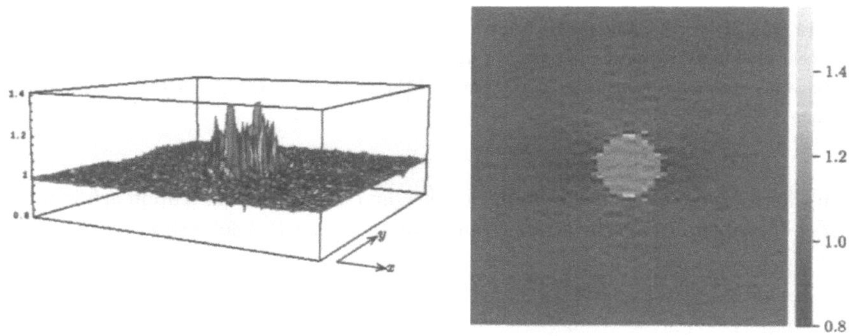

Figure 3. Elasticity distribution reconstructed from the displacement fields (u_x,u_y) given in Fig. 2.

and also that the force is free in the direction of the z-axis, we can obtain (Sumi et al., 1995)

$$\sigma_{xz} = \sigma_{yz} = \sigma_{zz} = 0. \tag{10}$$

The mean normal stress is

$$p = \frac{\sigma_{xx} + \sigma_{yy}}{3} = 2\mu\epsilon_{zz} = 2\mu(\epsilon_{xx} + \epsilon_{yy}) \tag{11}$$

under the assumption of incompressibility. Substituting this into Eq. (5) gives us the stress/strain constitutive equations as follows:

$$2\mu \begin{pmatrix} 2 & 1 & 0 \\ 1 & 2 & 0 \\ 0 & 0 & 1 \end{pmatrix} \begin{pmatrix} \epsilon_{xx} \\ \epsilon_{yy} \\ \epsilon_{xy} \end{pmatrix} = \begin{pmatrix} \sigma_{xx} \\ \sigma_{yy} \\ \sigma_{xy} \end{pmatrix}. \tag{12}$$

Thus, the stresses σ_{xx}, σ_{xy}, and σ_{yy} are definitely related to the strains ϵ_{xx}, ϵ_{xy}, and ϵ_{yy}.

Variational Formulation of the Forward Problem

In the forward problem, the equilibrium equation (1) along with the constitutive equation (5) is solved for the unknown displacement, when the material elasticity and

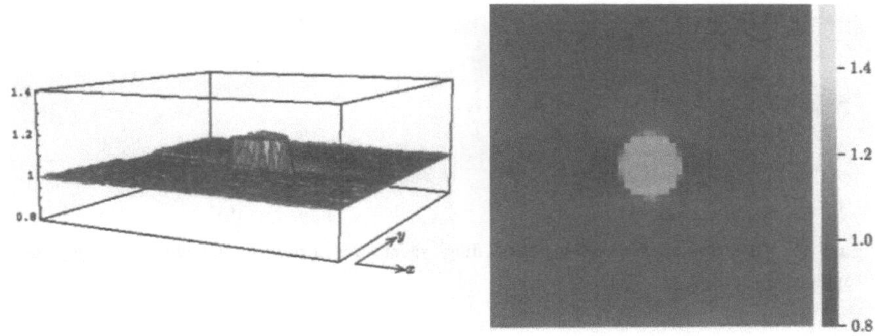

Figure 4. Elasticity distribution reconstructed from the displacement field u_x given in Fig. 2.

the boundary conditions are known (see Fig. 1(a)). The mechanical boundary conditions are generally given by

$$(\sigma_{ij}n_j - f_i)\delta(u_i - u_i^0) = 0 \tag{13}$$

where n_j is the jth component of the unit normal vector at the surface of the medium, f_i ia a force per unit area at the surface acting in direction, x_i, u_i^0 is the initial surface displacement, and δ is a variation symbol (Skovoroda et al., 1994). At a same part of the boundary either external force or displacement should be imposed. The variational principle states that the energy functional

$$\Pi \equiv \frac{1}{2}\iiint_{\Omega}\sum_{ij}\sigma_{ij}\epsilon_{ij}d\Omega - \int_{\Gamma}\sum_{i}u_i f_i d\Gamma \tag{14}$$

being subject to the constraints of $u_i = u_i^0$ (the Dirichlet boundary condition), gives, on minimization, the satisfaction of the equilibrium equation along with the constitutive equation as well as the boundary condition $\sigma_{ij}n_j = f_i$. (The continuity conditions of displacement and normal stress at each internal interface between region of different matrial parameters are also satisfied.)

In applying the finite element method, the region of interest, Ω', over which the equilibrium equation is defined, is divided into a number of subdivision called elements. The true displacement function within each element is approximated using the nodal displacements and the interpolation function. The interpolation function will be chosen to be linear in the nodal displacements but may be nonlinear with respect to space coordinates. The integral in Eq. (14) is evaluated element by element, and then they are all summed up to give the total energy of Eq. (14). Since the energy expression of Eq. (14) is usually quadratic in the nodal displacements, the minimization of Eq. (14), with respect to each and every nodal displacement, results in a set of linear equations for the unknown displacement values

$$W[\vec{\mu}]\,\vec{u} = \vec{f} \tag{15}$$

where a sparse symmetric matrix $W[\vec{\mu}]$ is known as the stiffness matrix which is a function of the shear modulus, $\vec{\mu}$, of each element, $\vec{u}$ is a column vector consisting of all nodal displacements, and $\vec{f}$ is a constant column vector defining the applied external forces. The boundary condition of $u_i = u_i^0$ is included in Eq. (15) by setting the displacement of those boundary nodes to the prescribed values, u_i^0. Thus, the

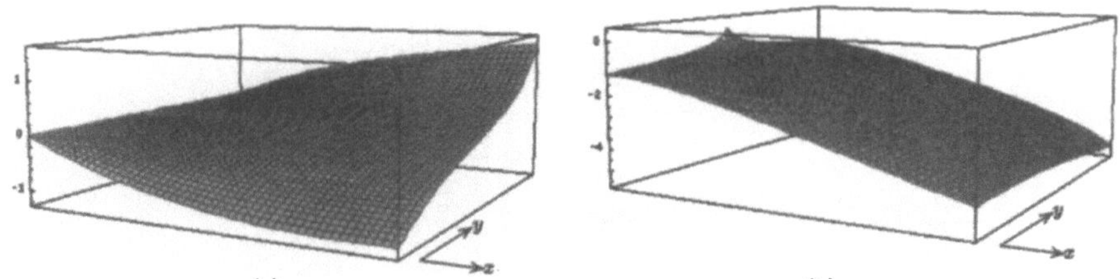

(a) (b)

Figure 5. Displacement vector field when a linear shear force in addition to nonuniform compression is applied. (a)u_x. (b)u_y.

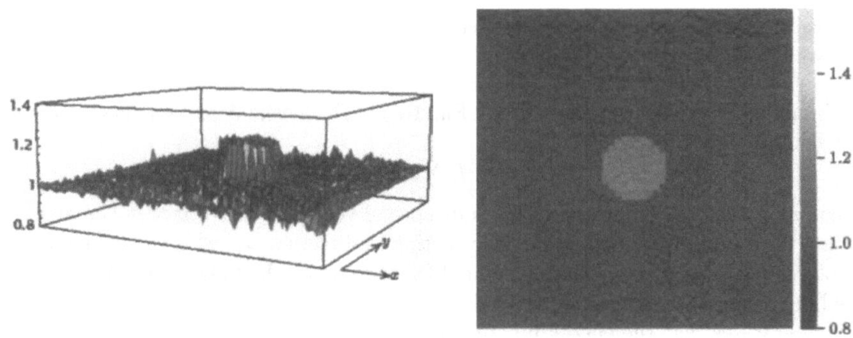

Figure 6. Elasticity distribution reconstructed from the displacement fields (u_x,u_y) given in Fig. 5.

displacements of all nodes are determined from Eq. (15). This procedure gives the Reyleigh-Ritz approximate solution to the partial differential equation (1) along with the constitutive equation (2) in the sense of a least-squares fit to the actual solution in terms of an energy norm (Zienkiewicz, 1977).

Shear Modulus Reconstruction Based on Displacement Fields

The variational formulation for the forward problem leads to a system of equations of the form $W[\vec{\mu}]\,\vec{u} = \vec{f}$, where $\vec{u}$ is a vector of the unknown displacements at the nodal points. Since the left-hand side of Eq. (15) is also linear for the shear modulus $\vec{\mu}$ of each element, Eq. (15) can be rearranged with the shear modulus as the unknown. This yields a different linear system of equations

$$C\,[\vec{u}]\,\vec{\mu} = \vec{f} \tag{16}$$

where $\vec{\mu}$ is a column vector consisting of shear mudulus of each element, and $\vec{f}$ is the same as that imposed in the right-hand side of Eq. (15). The element of the matrix $C\,[\vec{u}]$ is now a function of the measured displacements $\vec{u}$.

Since the matrix $C\,[\vec{u}]$ tends to be ill-conditioned, the solution for Eq. (16) is generally obscured by wildly oscillating noise and is usually very unstable. The regularized least squares technique to stabilize the solution yields

$$\vec{\mu} = (C^tC + \gamma I)^{-1}C^t\vec{f} \tag{17}$$

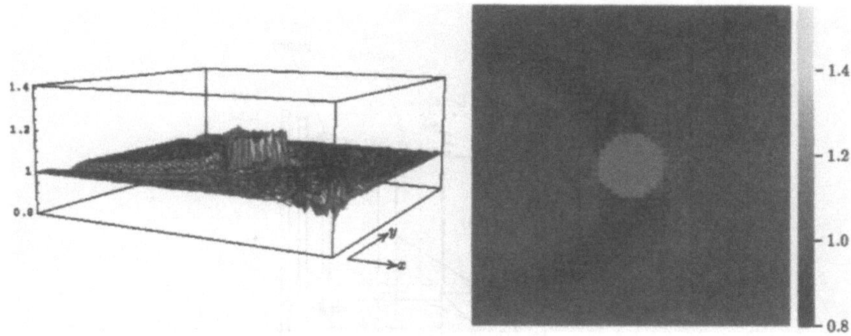

Figure 7. Elasticity distribution reconstructed from the displacement field u_x given in Fig. 5.

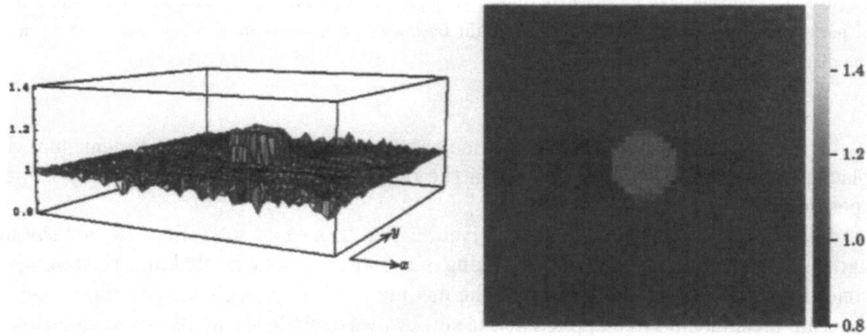

Figure 8. Elasticity distribution reconstructed from the displacement field u_y given in Fig. 5.

where the superscript t denotes the transpose of a matrix, I is an identity matrix, and the scalar quantity γ is the regularizing parameter whose optimal value depends on the noise level included in the measured displacements.

NUMERICAL SIMULATION

The biological soft tissue model consists of a 2-dimensional infinite incompressible elastic medium in the (x,y) plane, which contains a bonded circular inclusion whose shear modulus differs from that of the surrounding medium. In this study, the region of interest is a size of 500 $\times$ 500 which has a circular hard inclusion of radius 50 whose shear modulus is 1.2 whereas the one at the surrounding media is 1.0 (see Fig. 1(b)). Here, we consider two cases: one is that the model is subject to the uniform compression in the x-direction at infinite distance and the other is that the model is subject to a linear shear force in addition to nonuniform compression.

For the first case where the model is subject to the uniform compression in the x-direction, the displacement fields can be analytically given (Sumi et al., 1995; Honein and Herman, 1990), although the analytic solution for the displacement is not presented here. Fig. 2 shows a bird's-eye view of the analytically obtained displacement u_x and

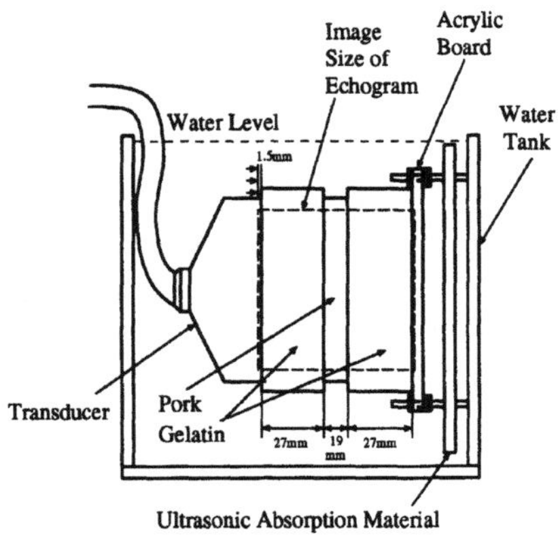

Figure 9. Schematic of *ex vivo* experiment. The phantom consisting of a kind of gelatin and a piece of pork was compressed by 1.5mm to the right by moving the tranceducer which was used to image the displacement field.

u_y in the region of interest. We assume that we can measure the 2-dimensional displacement vector at arbitrary points in the region of interest ultrasonically by using a speckle tracking method.

Based on the displacement fields given in Fig. 2, we can solve Eq. (16) for the unknown shear modulus. Generally speaking, since we may not exactly know the boundary conditions imposed at the surface of the medium or it is difficult to measure the exact external mechanical force applied at the surface, we exclude the equilibrium equation at the surface node at which the unknown external force is acting. The resulting equation become

$$D[\vec{u}]\,\vec{\mu} = 0 \tag{18}$$

so that we can only reconstruct the relative value of the shear modulus. In the reconstruction process we assigned 1.0 to the shear modulus of some elements in the right edge of the region of interest as the reference element.

Fig. 3 shows the relative shear modulus reconstructed from the displacement given in Fig. 2. The reconstructed shear modulus seems reasonable, but the relative shear modulus has some erroneous spikes along the boundary of the circular hard inclusion. The relative shear modulus at the central part of the inclusion has the mean value of 1.14.

The reconstruction was carried out for the case in which only the displacement data in x-direction was measured. Since 1-D displacement measurement can be accomplished more easily by ultrasonic means than 2-D displacement measurement, this reconstruction problem seems to have practical interest (Sumi et al., 1995). We reconstructed the relative shear modulus only from the displacement in the x-direction, in which, since only the equilibrium equation in x-direction is used to solve for the unknown relative shear modulus, the number of the matrix equation of (18) reduces to one half. Fig. 4 shows the relative shear modulus reconstructed using only the x-component of the displacement field, u_x, in which the relative shear modulus gives better estimates compared to that shown in Fig. 3. But the relative shear modulus at the central part of the inclusion still has the mean value of 1.14.

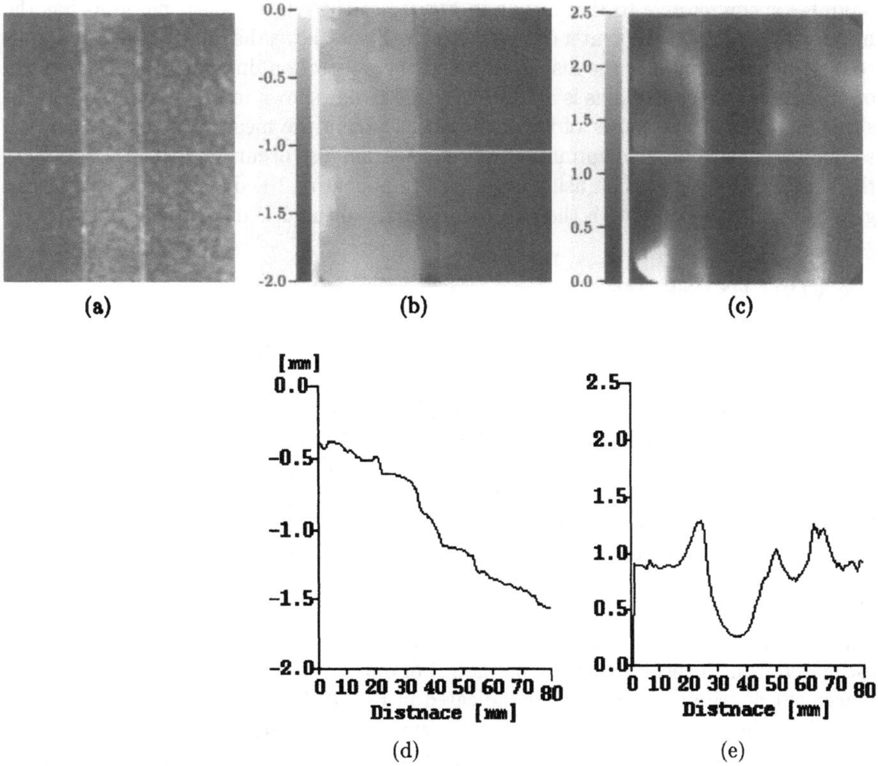

Figure 10. Images of the phantom when the external force was applied. (a) Conventional B-scan image before compression. (b) Image of the component of the displacement, u_x, using a quantitative gray scale. (c) Quantitative image of the relative shear modulus, μ. (d)(e) Profiles of the magnitude along the line of respective images shown above.

For the second case, the tissue model is the same as that shown in Fig. 1(b) but a linear shear force in addition to a nonuniform compression are applied at the surface of the medium. The displacement fields were computed by solving the forward problem using the finite element method. Fig. 5 shows a bird's-eye view of the computed displacements u_x and u_y in the region of interest. The displacement fields are different from one obtained for the first case in which the uniform compression is applied.

The effect of noisy displacement data was also considered. We assume that the measurement noise is white Gaussian noise with mean zero and variance α^2 (Raghavan and Yagle, 1994) and the signal-to-noise ratio (the ratio of the absolute magnitude of the strain to the square root of the variance of the noise) is assumed to 1000, namely 60dB. Using this noise model, white Gaussian noise was added to the computed displacements. A noisy estimate of the relative shear modulus was obtained and is shown in Fig. 6, in which the relative shear modulus at the center of the inclusion has the mean value of 1.19, so that the relative shear modulus distribution agrees fairly with the true distribution.

The reconstruction was also carried out for the case in which only the displacement data in x-direction is measured. Fig. 7 shows the relative shear modulus reconstructed using only the displacement in the x-direction. The distribution of the relative shear modulus is comparable to that shown in Fig. 6 but the relative shear modulus has the mean value of 1.14 at the center of the inclusion. Fig. 8 shows the relative shear modulus reconstructed using only the displacement in the y-direction, in which the distribution of the relative shear modulus is almost the same as one shown in Fig. 6 and the relative shear modulus at the center of the inclusion has again the mean value of 1.14. In the second case, since tissue deformation as well as strain also occurs greatly in y-direction, the elasticity reconstruction using only one component of 2-D displacement vector can give comparable results with the one reconstructed using 2-D displacement fields.

EX VIVO EXPERIMENT

An *ex vivo* experiment has been performed to demonstrate the usefullness of the proposed method to reconstruct the elasticity distribution. The experimental setup is shown in Fig. 9. The tissue-equivalent phantom is composed of a kind of gelatin and a piece of pork. The transducer was displaced to the right by 1.5mm in order to compress the phantom. Measurements were performed using a commercially available ultrasound scanner, with a 3.5MHz focused phased array. Two RF envelope B-scan images of region of interest before and after compression were digitized and transfered to the computer to estimate the local tissue displacement and strain (Yamashita and Kubota, 1994).

Fig. 10(a) shows the B-scan image of the phantom. The image of phantom is displayed over 77mm×96mm. Fig. 10(b) shows the x-component of the displacement fields estimated using a speckle tracking method. For quantitative evaluation, a profile of displacement distribution along a line indicated in Fig. 10(b) is shown in Fig. 10(d). The spatial resolution to obtain a reasonable accuracy in ultrasonic displacement will depend on various imaging paramenters such as ultrasonic frequency, single to noise ratio of digitized image, and so on. It should be noted that the local displacement less than 0.2mm can be detected with spatial resolution of 4mm×4mm. Fig. 10(c) and (e) give the reconstructed distribution of relative shear modulus and its profile along the line indicated in the image. The relative shear modulus in the region of pork has the mean value of 0.4, which means that pork is more soft than gelatin.

DISCUSSION

We have developed a finite-element based mathematical model to solve the inverse problem of elasticity imaging. A system of equations, $D\,[\vec{u}]\,\vec{\mu} = \vec{0}$, for reconstructing the relative shear modulus, $\vec{\mu}$, from the measured displacement, $\vec{u}$, was derived by rearranging the forward equation obtained from the variational formulation for the static deformation. Numerical experiments were carried out to validate the proposed method using the simulated deformation data of the simple inclusion problem. Simulation results demonstrated that the inversion algorithm can be applied to reconstruct the tissue elasticity in the presence of measurement noise.

In the present study, only one component of the displacement vector was also used to reconstruct the relative shear modulus. The result showed good estimates comparable to those obtained from two components of the displacement fields. It should be pointed out that the relative shear modulus recovered from only the displacement in one direction gives a good estimate.

In our *ex vivo* experiment with a tissue-equivalent phantom, the relative shear modulus was clearly small in more soft region.

The degree of accuracy attainable in the inverse estimation of the tissue elasticity was demonstrated. Due to the ill-posedness of this inverse problem, it was relatively difficult to recover fine structures of the elasticity distribution. The accuracy of the reconstructed elasticity distribution depends greatly upon both the quantity and quality of the measured displacement data. The signal-to-noise ratio of the measured displacement as well as the number of measurement points have a strong influence on the accuracy of reconstructed elasticity distribution. The numerical experiments carried out here would suggest that, in order to reconstruct the relative shear modulus with satisfactory accuracy, the signal-to-noise ratio should be at least 60dB.

Although these results demonstrate the feasibility of the proposed method, there are many issues still to be addressed. Of primary impotance are the accuracy of the finite element model for static deformation and the quantification of the sensitivity of the invesion to noise in the displacement data. These will ultimately determine its practical implementation in real, physical situations.

ACKNOWLEGEMENTS

This work was supported in part by the Ministry of Education under Grant 05680766. The authors wish to thank N. Yamane for developing the computer programs during this work, and T. Sai for preparing the manuscript.

REFERENCES

Emelianov, S.Y., Skovoroda, A.R., Lubinski, M.A., and O'Donnell, M., 1995, Reconstructive elasticity imaging, *Acoustical Imag.*, 21:241.

Honein, T., and Herman, G., 1990, On bonded inclusions with circular or strait boundaries in plane elaststatics, *Trans. ASME J. Appl. Mech.*, 57:850.

Kallel, F., Bertrand, M., Ophir, J. and Cespedes, I., 1995, Determination of elasticity distribution in tissue from spatio-temporal changes in ultrasound signals," *Acoustical Imag.*, 21:433.

Kallel, F., and Bertrand, M., 1996, Tissue elasticity reconstruction using linear perturbation method, *IEEE Trans. Med. Imag.*, 15:299.

Ophir, J., Cespedes, I., Ponnekanti, H., Yazdi, Y., and Li, X., 1991, Elastography: A quantiative method for imaging the elasticity of biological tissues, *Ultrasonic Imaging* 13:111.

O'Donnell, M., Skovoroda, A.R., Shapo, B.M., and Emelianov, S.Y., 1994, Internal displacement and strain imaging using ultrasonic speckle tracking, *IEEE Trans. Ultrason. Ferroelect. Freg. Contr.*, 41:314.

Raghavam, K.R., and Yagle, A., 1994, Forward and inverse problems in imaging the elasticity of soft tissue," *IEEE Trans. Nucl. Sci.*, 41:1639.

Sarvazyan, A.P., 1975, Low frequency acoustic characteristics of biological tissues, *Mechnics of Polymers*, 4:691.

Skovoroda, A.R., Emelianov, S.Y.,Lubinski, M.A., Sarvazyan, A.P., and O'Donnell, M., 1994, Theoretical analysis and verification of ultrasound displacement and strain imaging, *IEEE Trans. Ultrason. Ferroelect. Freq. Contr.*, 41:314.

Skovoroda, A.R., Emelianov, S.Y., and O'Donnell, M., 1995, Tissue elasticity reconstruction based on ultrasonic displacement and strain images, *IEEE Trans. Ultrason. Ferroelect. Freq. Contr.*, 42:747.

Sumi, C., Suzuki, A., and Nakayama, K., 1995, Estimation of shear modulus distribution in soft tissue from strain distribution, *IEEE Trans. Biomed. Eng.*, 42:193.

Yamashita, Y., and Kubota, M., 1994, Ultrasonic characterization of tissue hardness in the *in vivo* human liver, *Proc. IEEE Ultrason. Symp.*, 1449.

Zienkiewicz, O.C., 1977, *The Finite Element Method*, MaGraw-Hill, London.

SYSTEM INDEPENDENT IN VIVO ESTIMATION OF ACOUSTICAL ATTENUATION AND RELATIVE BACKSCATTERING COEFFICIENT OF HUMAN TISSUE

Tilo Gaertner,[1] Klaus Vitold Jenderka,[1] Hans Heynemann,[2] Mario Zacharias,[2] and Fank Heinicke[2]

[1]Institute for Medical Physics and Biophysics
[2]Urological Clinic
Martin-Luther-University Halle-Wittenberg, Medical Faculty
D-06097 Halle, Germany

INTRODUCTION

The aim of this work is to improve the ultrasonic diagnosis of cancerous tissue. Therefore the unprocessed rf data of B-mode images are investigated. After corrections for system dependent influences on the data a spectral analysis is applied to derive additional information about the acoustical properties of the tissue.

Normal B-mode images of regions of the human body contain precise information about the organ topography and anatomy. For more information physicians investigate the texture of organ regions in the image. The texture of a grey scaled image is affected by both the parameters of the used B-mode system and the acoustical properties of the tissue itself. Because of that the interpretation of texture parameters is strongly dependent on the used device and tissue parameters derived by different systems are normally not comparable. Furthermore, only a fraction of the information contained in rf echo signals, namely the envelope of the rf signal is used for imaging in B-mode systems.

The soft tissue of some organs like testis and prostate can be assumed to be more or less a random distribution of acoustic scatterers. B-mode images of those media appear as homogeneous images with a characteristic texture. Therefore it is hypothesized, that acoustic parameters like attenuation, backscattering and scatterer distribution may be estimated as a mean value for regions of interest within these organs. All these parameters are frequency dependent and might be estimated by spectral analysis within the bandwidth of the

Acoustical Imaging, Volume 24.
Edited by Hua Lee, Kluwer Academic/Plenum Publishers, 2000.

transducer, as opposed to conventional envelope analysis. In some cases changes of the acoustical parameters due to morphological alterations in the tissue are invisible or not clearly to be seen in B-mode images, whereas it can be shown by spectral analysis.

METHOD

The requirement for the calculation of acoustical parameters is the access to the rf echo signals of a B-mode system without any preprocessing except for time gain compensation (tgc). The application of tgc is necessary to provide the optimal exploitation of the dynamic range of the system and is corrected prior to the spectral analysis. Two different types of B-mode systems were used for the rf data acquisition of the backscattered signals from phantoms, testis, and prostate. The signal of a DORNIER AI 5200 (7.5 MHz, curved array) was digitized with an oscilloscope (LeCroy 9430, 100 MHz, 10 bit). For the second device - Ultramark 9 HDI (ATL) - the digitized data (20 MHz, 16 bit) from the B-mode system were transferred to a pci-bus interface of a pc. In both cases the tgc of the device was used and a corresponding signal was stored in addition to the rf data. Tgc was corrected before any kind of spectral analysis.

A procedure was developed to correct for system specific effects and to determine tissue or phantom specific acoustical parameters in terms of ultrasound spectroscopy. For several time windows (gaussian) along the scan lines in a region of interest (roi) the spectrum is calculated by fft. By employing the method of cepstral smoothing the spectral ultrasound intensity distribution can be determined independently of structural influences. After diffraction correction the attenuation is estimated from the slope of the cepstral smoothed spectra by linear regression for a set of frequencies within the bandwidth of the transducer.

DIFFRACTION CORRECTION

The spectral analysis of the rf data requires the knowledge of the sound pressure distribution in the penetrated medium. This distribution arises from diffraction effects and the different modes of focusing in the transducer control. Because of the great variety of different transducer types and the variable numbers and positions of focal zones, distinctly structured sound fields can appear [1,2,3]. The direct measurement of the sound pressure distribution in the investigated medium is not possible nowadays, so commonly measurements in water are carried out [1,4,5]. However, water is not a scattering medium and its attenuation is much lower than that of biological tissue. Another way of diffraction correction bases on an amount of data of in vivo measurements [6]. This should estimate the sound distribution in tissue best, but a lot of measurements on patients is neccessary.

For this reason we used tissue mimicking reference phantoms made of graphite powder immersed in agar, to build up a set of correction functions for two different B-mode systems and every applied transducer, respectively. The attenuation of the phantoms is measured in transmission mode with a set of single element transducers with different frequencies. Then the phantoms were investigated with the B-mode systems and the spectral analysis decribed above without diffraction correction was carried out. The deviation of the received spectrum in dependence from depth from the ideal course (i.e. only the known attenuation was taken into account) gives the diffraction correction function. The diffraction correction is implemented in the spectral analysis.

Phantoms only give an estimation of the diffraction effects in human tissue. The quality of this kind of diffraction correction strongly depends on wheter the acoustic properties of the phantoms are similar to those of tissue or not. The phantom we used for correction has an attenuation of about 0.5 dB/(MHz cm) in the frequency range of 4-10 MHz what is the standard for normal soft tissue. The speed of sound in the phantoms is 1501 m/s (23°C) and only slightly lower than in testis (1570 m/s). However, in the measurements on testis with the ATL linear array after diffraction correction remains a sligth deviation from the linear course of the frequency dependent data versus depth. Since the amplitude and the shape of this deviation seem to be similar for all data from testis with this trancducer we applied a second correction in this case. This correction was obtained by an estimation of the attenuation in the decribed way. After that the remaining deviation from the linear course was stored. This data were avaraged for 3 healthy persons and 12 measurement with each person. Then this data were applied as a second step correction.

The further improvement of the phantoms should yield a better adaption of the phantom features to specific tissue. That wouldt avoid the need of a set of in vivo data from patients for correction. As discussed below the medium between transducer and roi may affect the transmission of ultrasound in several ways. The skin is the one layer that have to be considered in most in vivo measurements, also with testis. Figure 1 shows an experiment with the same phantom with nonscattering liquid intermediate layers (4 mm, within latex glove) with a different speed of sound. The data were plotted without diffraction correction. The shift of the focal point is higher than an offset due to a correction of the time of flight for the 4 mm. If water and phantom as intermediate layers (only different in attenuation) are compared, no explicit change was found.

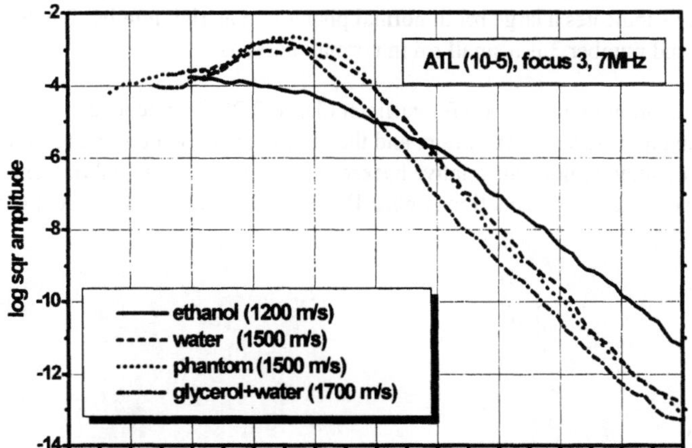

Figure 1

On the other hand measurements of the in vivo speed of sound in testis (transmission, relativ to water) yielded a value of 1570 m/s and no differences wheter the testis and the skin or the skin itself was investigated. Since the speed of sound in phantoms depends on temperature (20°C - 1491 m/s, 35°C - 1526 m/s) the phantom was investigated with this temperatures, but no changes were found.

ESTIMATION OF ACOUSTICAL PARAMETERS

The estimation of the attenuation is independent of the intermediate layer (that is the media between transducer and the ROI). For this reason its application is convenient, especially under in vivo conditions. However, the estimation of the attenuation becomes difficult for small areas (diameter less 1 cm), since there are not sufficient data for the regression and thus the value for the attenuation might not be reliable. For such small ROI's the estimation of the backscatter coefficient may be the better choice, because it is estimated as a mean value, rather than by regression. Both parameters have been shown to be correlated to the tissue state. The relative backscatter coefficient is calculated from the attenuation corrected backscattered data relative to an external reference spectrum obtained from a reflector or phantom. For the attenuation commonly a standard value for normal tissue (0.5 dB/cm/MHz) is applied. This might not be correct for all cases and do not take different types of tissue in the intermediate layer into account.

RESULTS

Various tests of the estimation of the attenuation and the relative backscatter coefficient have been carried out in a phantom with a small (0.5 cm diameter) inclusion with scattering properties different to those of the rest of the phantom. Different transducers and intermediate layers were used. Figures 2, 3, 5, 6 show B-mode images of the measurements. The roi number 1 indicates a large roi in normal phantom material, number 2 is for the roi in the inclusion, and number 3 is a small roi in normal material.

The estimation of attenuation for sufficient large ROI's is independent of the used B-mode device and transducer combination, and the intermediate medium, respectively (figure 4). In the case of small regions the relative backscatter coefficient (external reference) is more reliable than the estimated attenuation (figure 4), because it is also independent of the used

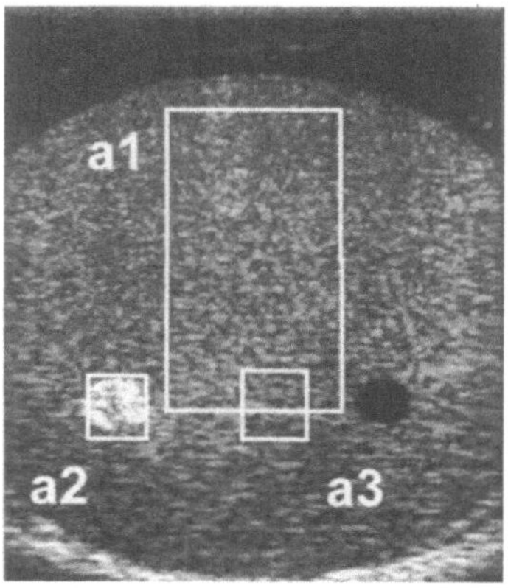

Figure 2 (10-5 MHz linear array, intermediate layer water)

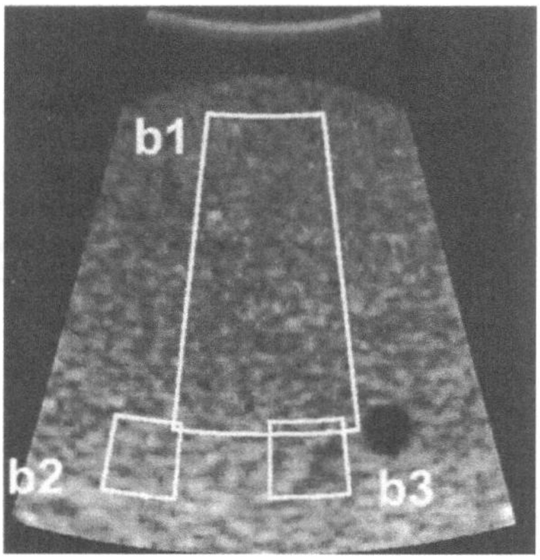

Figure 3 (7-4MHz curved array, intermediate layer water)

ransducer. As reference for the relative backscatter coefficient the backscattered signal from the phantom for the diffraction correction was used.

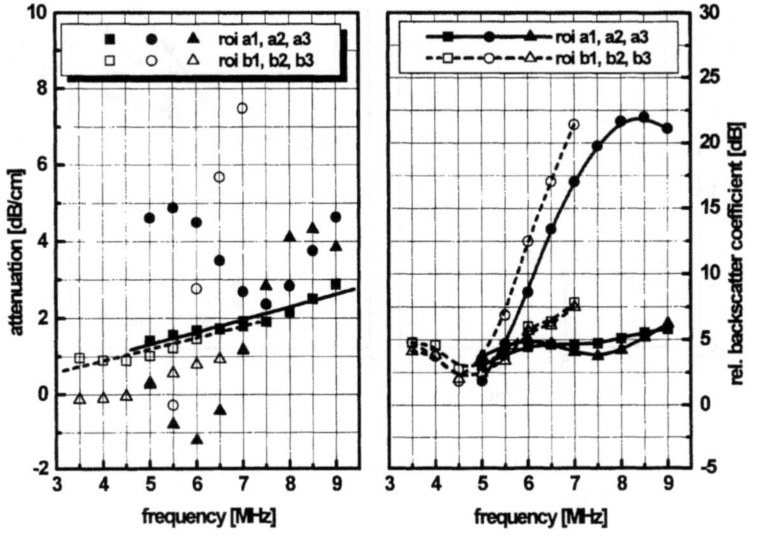

Figure 4 attenuation and rel. backscatter coefficient (extern. reference) with different transducers and the same intermediate layer (water)

But on the other hand the relative backscatter coefficient depends on the intermediate layer, even if the same transducer is used (figure 4). For this reason a combination of both parameters is used. The attenuation is calculated for a large homogeneous region (a1, b1, c1, d1) of interest in the phantom. Then this attenuation was supposed to be constant for the phantom and it was corrected for the way between transducer and the small regions (number 2 and 3).

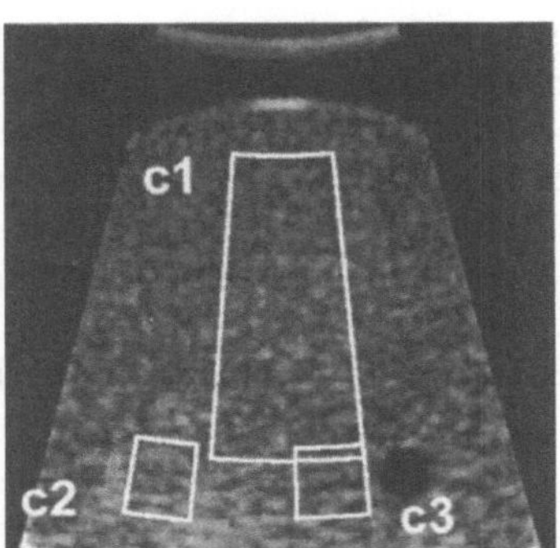

Figure 5 (7-4 MHz curved array, intermediate layer with low attenuation)

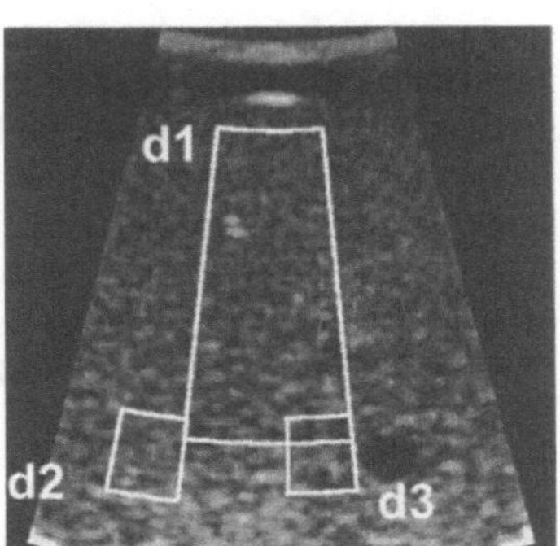

Figure 6 (7-4 MHz curved array, intermediate layer with high attenuation)

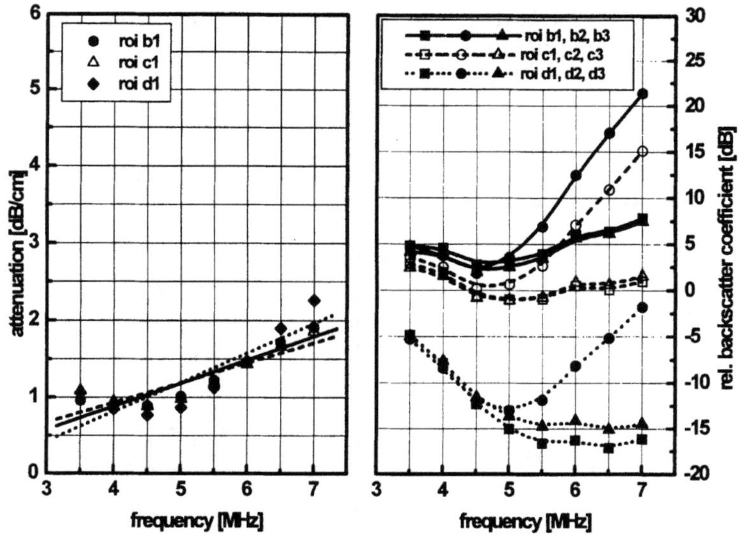

Figure 7 attenuation and rel. backscatter coefficient (external reference) for the same transducers and different intermediate layers

The differences between the relative backscatter coefficient of the inclusion and the rest of the phantom then only depends on the intermediate layer. This can be corrected by using a reference spectrum calculated from a region within the phantom during the same measurement instead of an external reference spectrum. Therefore the same large roi (number 1) is used. In this way the relative backscatter coefficient becomes independent of the used transducer and the intermediate layer (figure 8).

With this method also images of the relative backscatter coefficient may be constructed for separate frequencies. That allows to image changes in the scattering properties, that are not seen in common B-mode devices [7].

Furthermore, the procedure has been applied to in vivo measurements of human testis and prostate. In both cases an intermediate layer (skin) exists for in vivo measurements. Its properties differ from those of the investigated tissue, and the properties of the skin are normally not avaiable, because the high signal near the transducer surface overmodulates the amplifiers. So the attenuation is the better choice for the investigation of testis tumors, especially in large homogeneous changes of the testis (seminoma). For the prostate in some cases the roi in the cancerous region are to small, so that mainly the backscatter coefficient was measured. The measurements showed an increased attenuation (>1dB for testis) and decreased relative backscatter coefficient (5-15 dB for prostate) in cancerous tissue relative to normal tissue.

ACKNOWLEDGMENTS

This work was supported by the German Research Community (He 2156/2-2).

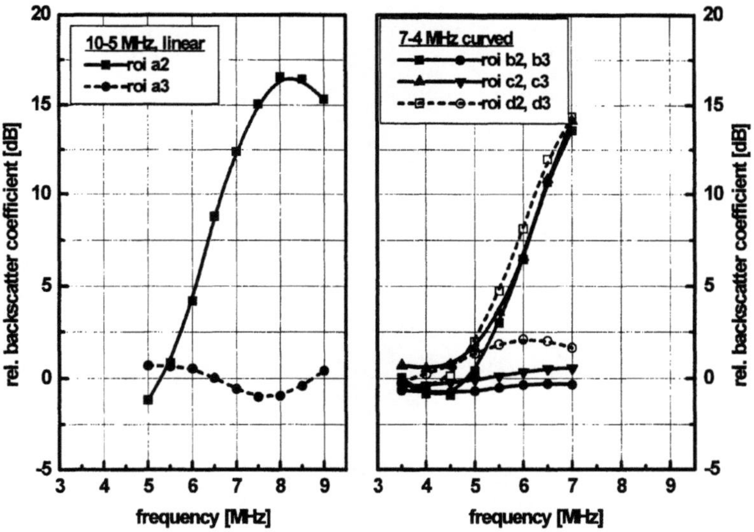

Figure 8 rel. backscatter coefficient (internal reference) for different trancducers and different intermediate layers

REFERENCES

1. Shung,K.K ; Thieme,G.A; (editors): *Ultrasonic Scattering in Biological Tissues*. CRC Press.; 1993
2. Waag,R.C.; Astheimer,J.P.: Characterization of measurement system effects in ultrasonic scattering experiments. *J Acoust Soc Am*; 88(5); 1990; 2418-2436
3. Waag;R.C.; Astheimer,J.P.; Smith,J.F.I.: Analysis and computations of measurement system effects in ultrasonic scattering experiments. *J Acoust Soc Am*; 91(3);1992; 1284-1297
4. Boote, E, Zagzebski, J. A, Madsen, E.L, : Backscatter coefficient imaging using a clinical scanner. *Med. Phys.*;19(5); 1992; 1145-1152
5. Feleppa,E.J. et al.: Typing of Prostate Tissue by Ultrasonic Spectrum Analysis. *IEEE Transactions on Ultrasound, Ferroelectrics, and Frequency Control*, 43(4); 1996; 609-619
6. Huisman: *In Vivo Ultrasonic Tissue Characterization of Liver Metastasis*. DTQP, Schiedam, 1998
7. Gärtner,T.; Jenderka,K.V.; Schneider,H.; Heynemann,H.:Tissue Characterization by Imaging of Acoustical Parameters. *Acoustical Imaging*; 22; 1996; 365-370

ULTRASOUND IMAGES OF SMALL TISSUE STRUCTURES FREE OF SPECKLE

William Tobocman,[1] Joseph A. Izatt,[2] and Nima Shokrollahi[3]

[1]Physics Department
Case Western Reserve University
10900 Euclid Ave.
Cleveland, OH 44106-7079

[2]Department of Medicine/Biomedical Engineering
University Hospitals of Cleveland, GI Division
11100 Euclid Ave.
Cleveland, OH 44106-5066

[3]MedSonics US, Inc.
455 West 23rd St.
New York, NY 10011-2156

We suggest an alternative to the conventional pulse-echo method for analyzing ultrasound reflections to image small tissue structures. The images captured by our method prove to have improved resolution and are free of speckle. The method applies to a tissue structure that can be regarded as a layered medium on the scale of a few millimeters and is not too weakly reflecting.

In both methods the incident pulse is made to be as nearly impulsive as possible. The pulse-echo method then rectifies the reflected pulse to create an image of the reflectivity distribution of the target convolved with the incident pulse. Instead of this, our method(1) analyzes the incident and reflected pulses to produce an image of the acoustic impedance distribution of the target.

A comparison of the two methods is shown in Fig. 1 where the target is three plastic films suspended in water. This phantom is insonified with 3.5 MHz ultrasound. The impedance image is seen to be much superior to the convolved reflectivity image. In Fig. 2 we compare a 20 MHz convolved reflectivity image with the 3.5 MHz impedance image. Although the 20 MHz reflectivity is much improved, it still does not match the 3.5 MHz impedance image.

Next in Fig. 3 we compare the images the two methods provide of a human aorta wall speciman from the reflection of 3.5 MHz ultrasound. The impedance image is free of the speckle that appears in the reflectivity image. In Fig. 4 the impedance image is

Acoustical Imaging, Volume 24.
Edited by Hua Lee, Kluwer Academic/Plenum Publishers, 2000.

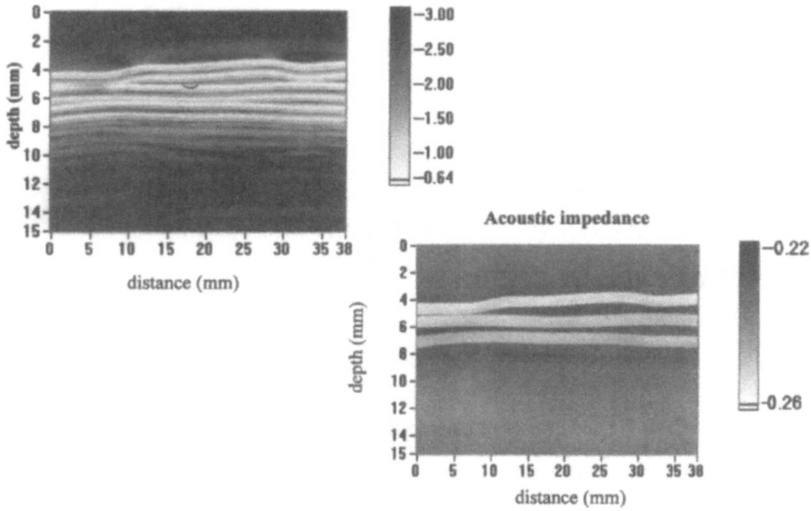

Figure 1. Convolved reflectivity and acoustic impedance images of three plastic films suspended in water. The images were captured from the reflection of 3.5 MHz ultrasound.

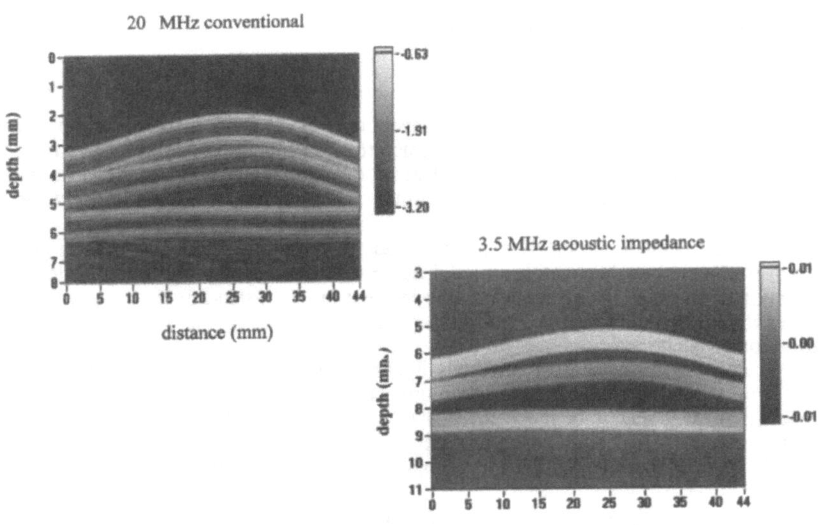

Figure 2. Images of three plastic films suspended in water. The convolved reflectivity image was captured from the reflection of 20 MHz ultrasound, and the acoustic impedance image was captured from the reflection of 3.5 MHz ultrasound.

compared with a magnified histology section from the same site on the aorta specimen. Both images display a thickening of the intimal layer. A 3.5 MHz aorta specimen impedance image is compared with a 20 MHz reflectivity image of the same site in Fig. 5. Note that the bright strip running along the top of the reflectivity image is an artifact arising from the convolution of the incident pulse with the reflectivity. In Fig. 6 we have 3.5 MHz reflectivity and impedance images from a plastic film which has three holes of diameter 3.56 mm drilled in it. The images of the holes are seen to be rather fuzzy. This reveals that the transverse resolution of our images is much inferior to the longitudinal resolution. The finite size (about 2mm) of the focussed ultrasound beam is responsible for this.

A schematic diagram of our ultrasound imaging system is presented in Fig. 7.

We will next outline how the incident and refelcted pulses are analyzed to capture the acoustic impedance distribution. For acoustic wave propagation in one dimension the excess pressure, $\psi_k(x)$, is a solution of[1]

$$(d^2/dx^2 + k^2)\,\psi_k(x) = (d/dx \ln z(x))d/dx\,\psi_k(x) \tag{1}$$

where x is the elapsed time multiplied by the speed of sound in water. $z(x) = Z(x)/Z$ (water) is the relative acoustic impedance, where Z is the speed of sound times the density.

We suppose the surface of the insonified specimen to be at $x = 0$ such that $z = 1$ for $x < 0$ and $z = z(x)$ for $x > 0$. Then for a plane wave incident on the specimen we have

$$\psi_k(x) = e^{ikx} + R(k)e^{-ikx} \quad \text{for } x < 0 \tag{2}$$

where R_k is called the reflection amplitude or the transfer function.

Now suppose the incident pulse is

$$\psi_{INC}(x) = \int_0^{\infty} dk\{a(k)e^{ikx} + \text{c.c.}\} \tag{3}$$

where c.c. denotes complex conjugate. Then the reflected pulse will be

$$\psi_{REF}(x) = \int_0^{\infty} dk\ \{a(k)\ R(k)\ e^{-ikx} + \text{c.c.}\} \tag{4}$$

If we denote the Fourier transform of $\psi(x)$ by $\Psi(k)$, then

$$\Psi_{INC}(k) = (2\pi)^{-1} \int_{-\infty}^{\infty} dx\ e^{-ikx}\ \psi_{INC}(x) = a(k) \tag{5}$$

and

$$\Psi_{REF}(k) = a(k)^{*}\ R(k)^{*} \tag{6}$$

It follows that the reflection coefficient spectrum can be deduced from the recorded incident and reflected pulses by means of

$$R(k) = \Psi_{REF}(k)^{*}/\Psi_{INC}(k) \tag{7}$$

The reflection coefficient is related to the acoustic impedance by the reflection formula

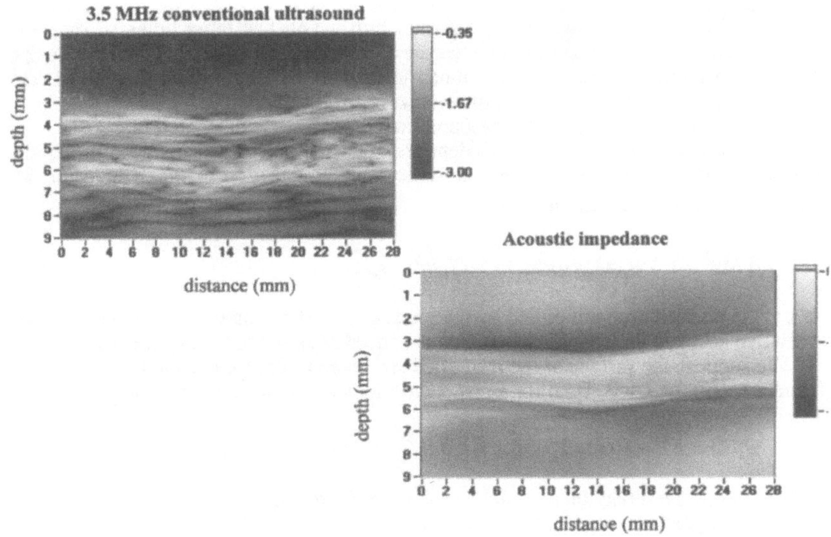

Figure 3. Images of a human aorta wall specimen in vitro captured from the reflection of 3.5 MHz ultrasound.

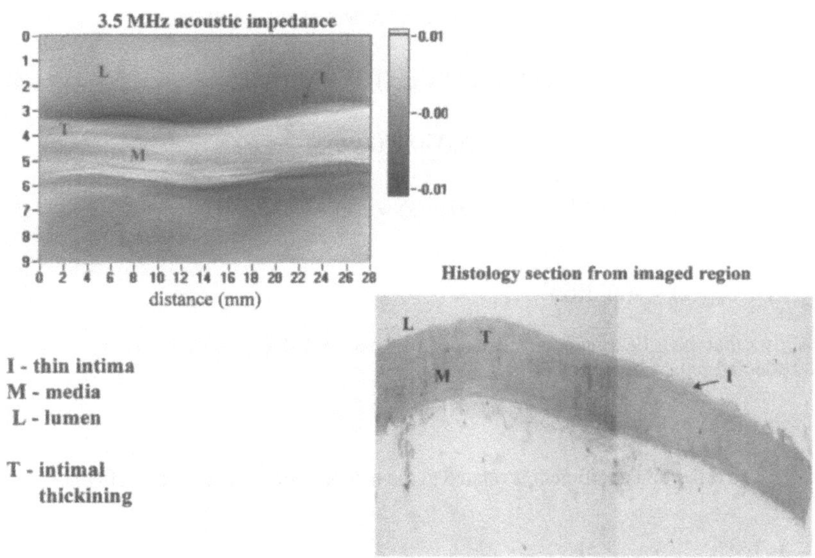

Figure 4. Image of a human aorta wall specimen in vitro captured from the reflection of 3.5 MHz ultrasound, and a magnified histology section from the same site.

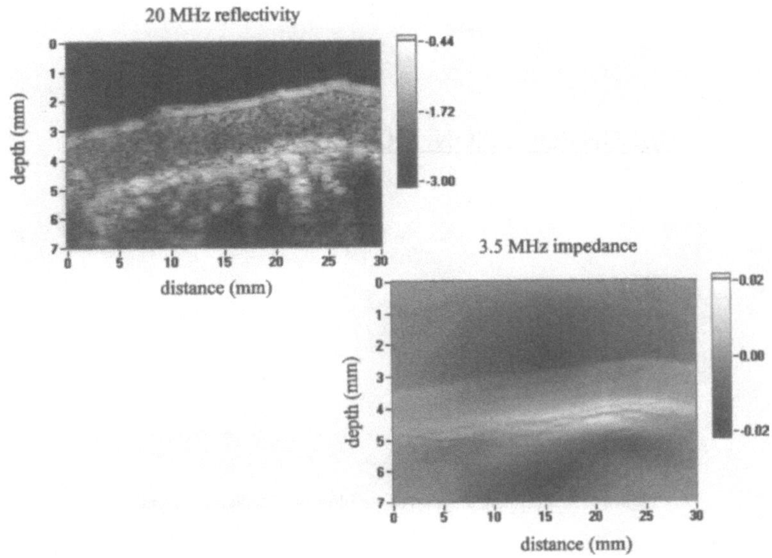

Figure 5. Images of a human aorta wall specimen in vitro. The convolved reflectivity image was captured from the reflection of 20MHz ultrasound. The acoustic impedance image was captured from the reflection of 3.5 MHz ultrasound.

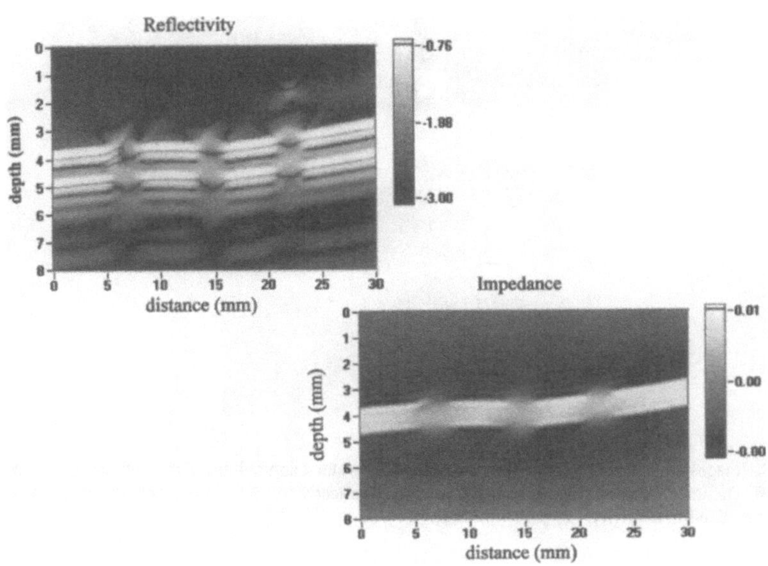

Figure 6. Images of a single plastic film suspended in water captured from the reflection of 3.5 MHz ultrasound. The film has three holes of diameter 3.56 mm drilled in it.

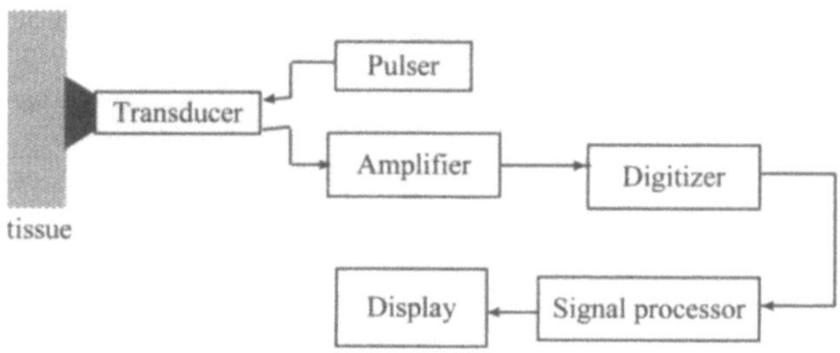

Figure 7. A schematic diagram of our ultrasound imaging system.

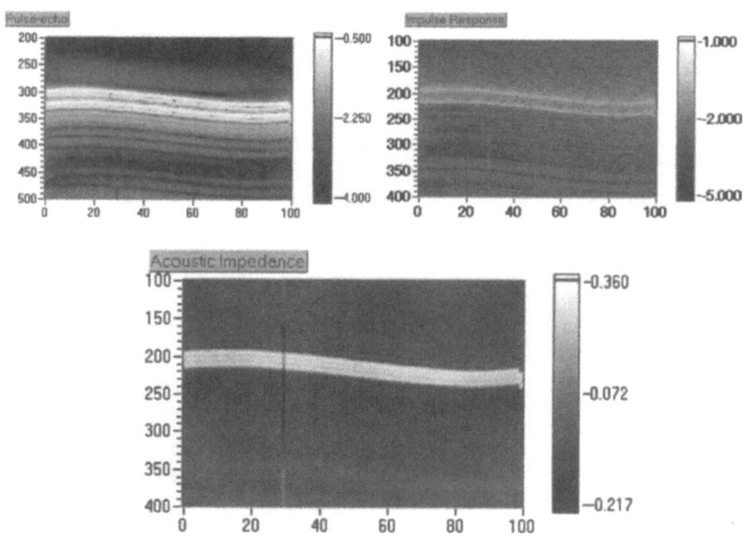

Figure 8. Images of a single plastic film suspended in water captured from the reflection of 3.5 MHz ultrasound. The image labeled pulse-echo is the convolved reflectivity. The image labeled impulse response is the (nonconvolved) reflectivity.

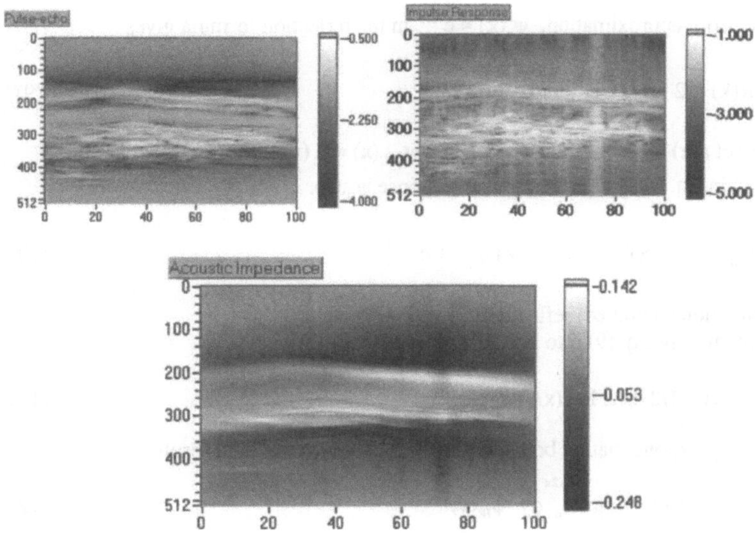

Figure 9. Images of a human aorta wall specimen in vitro captured with 3.5 MHz ultrasound reflection. The image labeled pulse-echo is the convolved reflectivity. The image labeled impulse response is the (nonconvolved) reflectivity.

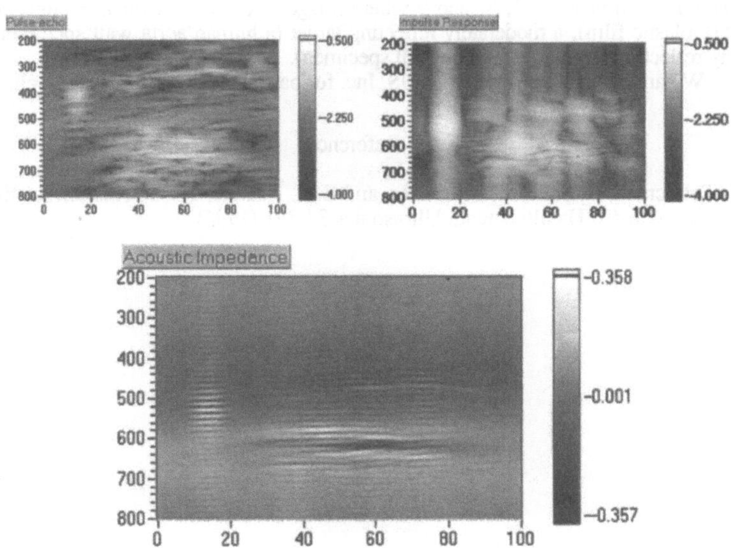

Figure 10. Images of a pig colon wall specimen in vitro captured with 3.5 MHz ultrasound reflection. The image labeled pulse-echo is the convolved reflectivity. The image labeled impulse response is the (nonconvolved) reflectivity.

$$R(k) = -\, i(2k)^{-1} \int_0^\infty dx\, e^{ikx} \{d/dx \ln\ z(x)\} d/dx\, \psi_k(x) \tag{8}$$

Making the Born approximation, $\psi_k(x) \approx e^{ikx}$, in the reflection formula gives

$$R(k) \approx 2^{-1} \int_0^\infty dx\, e^{i2kx}\, d/dx\ \ln z(x) \tag{9}$$

Now if we set $a(k) = (2\pi)^{-1}$ in Eq. (3) we get $\psi_{INC}(x) = \delta\,(x)$. It follows that setting $a(k) = (2\pi)^{-1}$ in Eq. (4) gives the impulse response $\psi_{IR}(x)$.

$$\psi_{IR}(x) = (2\pi)^{-1} \int_0^\infty dk\, \{R(k)\, e^{-ikx} + c.c.\} \tag{10}$$

$\psi_{IR}(x)$ is the (nonconvolved) reflectivity.
Finally, substituting Eq. (9) into Eq. (10) gives

$$\psi_{IR}(x) = 1/2\ d/dx \ln z(x/2) \tag{11}$$

Integrating and exponentiating both sides of Eq. (11) gives our final result

$$z(x) = \exp 2 \int_0^{2x} dy\ \ \psi_{IR}(y) \tag{12}$$

In practice, the ψ_{IR} in Eq. (12) is given by Eq. (10), the inverse Fourier transform of R(k), where R(k) is given by Eq. 7, the ratio of the Fourier transforms of the incident and reflected pulses.

The method requires that R(k) be known over a sufficiently wide band of frequencies. Due to the band-limited nature of the data and the presence of noise this condition is fulfilled only when the target is relatively strongly reflecting for ultrasound. This is illustrated in Figs. 7, 8, and 9 where images are captured for a strongly reflecting target (a plastic film), a moderately reflecting target (a human aorta wall specimen), and a weakly reflecting target (a pig colon wall specimen).

We are grateful to MedSonics US, Inc. for partial support of this research.

References

1. W. Tobocman, K. Santosh, J.R.Carter, and E.M. Haacke, Tissue characterization of arteries with 4 MHz ultrasound, Ultrasonics 33:331 (1995)

OPTIMIZATION OF NON-UNIFORM ARRAYS FOR FARFIELD BROADSIDE BEAMFORMING

Richard Y. Chiao

GE Corporate Research and Development
P. O. Box 8, Schenectady, NY 12301

INTRODUCTION

Non-uniform arrays have elements which vary in size as illustrated in Fig. 1(b). They form an intermediate step between the dense uniform array shown in Fig. 1(a) and the continuous aperture shown in Fig. 1(c). The elements in a non-uniform array are connected symmetrically around the center of the array since it is used only for broadside beamforming. Potential applications of non-uniform arrays include telecommunications, radar, sonar, and medical imaging [1].

The non-uniform array provides a useful balance between beam control and complexity. For broadside beamforming, the dense uniform array is an over-design such that the non-uniform array offers the advantage of significantly fewer elements and lower complexity with similar broadside beam control. Compared with a continuous aperture, the non-uniform array provides significantly better beam control (especially in the nearfield [1]) at the cost of moderately increased complexity.

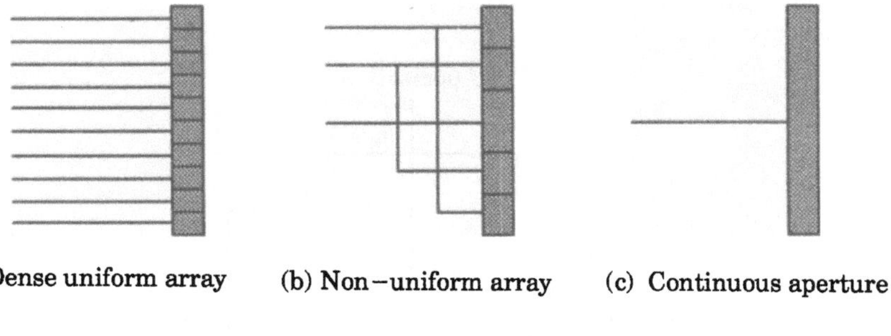

Figure 1. Comparison of the non-uniform array with the uniform array and the continuous aperture.

Acoustical Imaging, Volume 24.
Edited by Hua Lee. Kluwer Academic/Plenum Publishers, 2000.

In this paper, we present a method based on simulated annealing [2] to optimize non-uniform arrays for farfield broadside beamforming. For a given aperture size and number of elements, the element sizes and apodization values are selected to minimize the sidelobes relative to the mainlobe. This optimization problem is complicated by the nonlinear interaction between element sizes and apodization values such that standard gradient-descent methods are of limited usefulness. The simulated annealing method may be extended to include focusing delays and element turn-on for nearfield beam control [3].

OPTIMIZATION OF NON-UNIFORM ARRAYS

The farfield, narrow-band beam pattern $B(\omega)$ due to an aperture $a(x)$ of length W is given by its Fourier transform

$$B(\omega) = | \int_{-W/2}^{W/2} a(x)\ exp(-j\omega x)\ dx |, \tag{1}$$

where the spatial frequency ω is related to the beam angle θ and wavelength λ by

$$\omega = \frac{2\pi}{\lambda} sin(\theta). \tag{2}$$

Assuming the aperture is partitioned into five elements (three independent elements) located symmetrically around the origin as illustrated in Fig. 2, we have

$$B(\omega) = 2\ | \int_{0}^{W/2} a(x)\ cos(\omega x)\ dx\ | \tag{3}$$

$$= \frac{2}{\omega}\ | (1 - a_2)sin(\omega x_1)\ +\ (a_2 - a_3)sin(\omega x_2)\ +\ a_3 sin(\omega W/2)\ |,$$

where $\{x_1, x_2, x_3\}$ are the element endpoints and $\{a_1, a_2, a_3\}$ are the apodization values as depicted in Fig. 2. For N independent elements ($2N - 1$ total elements), Eq. (3) generalizes to

$$B(\omega) = \frac{1}{\omega}\ | \sum_{i=1}^{N} (a_i - a_{i+1})\ sin(\omega x_i)\ | \tag{4}$$

with $a_1 = 1$, $a_{N+1} = 0$, and $x_N = W/2$.

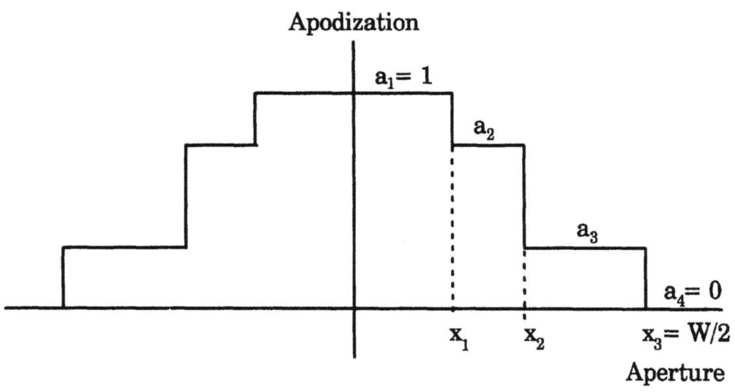

Figure 2. Depiction of aperture with $N = 3$ independent elements.

The problem is to determine the elements sizes and the apodization values to optimize the beam pattern given by Eq. (4). For the remainder of this paper we use the minimax criterion given by

$$MIN\{C_{mm}\} = MIN\{\ MAX\ B(\omega)/B(0)\ \}\ \ for\ \omega \in [\omega_1, \omega_2]. \tag{5}$$

Alternatively, the least squares criterion may be used

$$MIN\{C_{ls}\} = MIN\{\ \int_{\omega_1}^{\omega_2} (B(\omega)/B(0))^2 d\omega\ \}. \tag{6}$$

The user-specified parameter ω_1 (equivalently θ_1 using Eq. (2)) determines the mainlobe-sidelobe boundary. Typically ω_2 is set to be $2\pi/\lambda$ which corresponds to $\theta_2 = 90^o$.

This optimization problem is complicated by the nonlinear interaction between the element sizes and apodization values such that standard gradient-decent optimization methods cannot be used. We resort to the method of simulated annealing to search for solutions.

SIMULATED ANNEALING ALGORITHM

Simulated annealing is a method for global optimization using randomized searches [2]. Global optimization is necessary when the cost function has local minima. Its name derives from an analogy with physical annealing in statistical mechanics where a solid is placed in its lowest energy state through controlled cooling after initial heating. The thermal energy provides random perturbation to the lattice structure which naturally settles toward lower-energy, ordered states. However, higher-energy states may also be reached if sufficient thermal energy is present. If the cooling occurs too fast then parts of the solid may be frozen in high-energy states with irregular lattice structure which constitute flaws in the solid.

Simulated annealing uses the same random perturbation concept to perform global optimization as shown in Figure 3. The algorithm starts at an initial state vector $\mathbf{p}$ and temperature parameter T. In each iteration, the current state is perturbed by a random vector. If the perturbed state $\mathbf{p}^*$ has lower cost than the current state, then the current state moves downhill to the perturbed state; otherwise the Metropolis algorithm [2] is used to determine whether an uphill move occurs. The Metropolis algorithm permits an uphill transition with probability given by

$$Prob(transition) = exp(-\Delta C/T), \tag{7}$$

where ΔC is the increase in cost and T is the temperature. Thus, deeper valleys are more difficult to climb out of than shallower ones, and uphill moves are more difficult with lower temperature. As the temperature is gradually decreased, the state vector tends to become trapped in the deepest valley.

The simulated annealing algorithm for optimizing non-uniform arrays is shown in Fig. 4. The state vector $\mathbf{p}$ consists of the parameters to be optimized: element endpoints $x_1, x_2, ..., x_{N-1}$ and apodization values $a_2, ..., a_N$. The major addition in Fig. 4 is the lower triangular matrix $\mathbf{L}$ which contains adaptive search information [2]. The matrix $\mathbf{L}$ is obtained from the Cholesky decomposition of the covariance matrix $\mathbf{R}$ generated from previous moves. Initially, $\mathbf{L}$ is a diagonal matrix with each element set at a fraction of the search range for the corresponding parameter. Subsequently, it is updated using the computed covariance matrix after each temperature decrement. In this "terrain-learning" feature, the covariance matrix $\mathbf{R}$ is a quadratic estimate of the topology of the local cost function, and $\mathbf{L}$ directs the perturbations into regions that are more likely to have a lower cost, which speeds up the convergence.

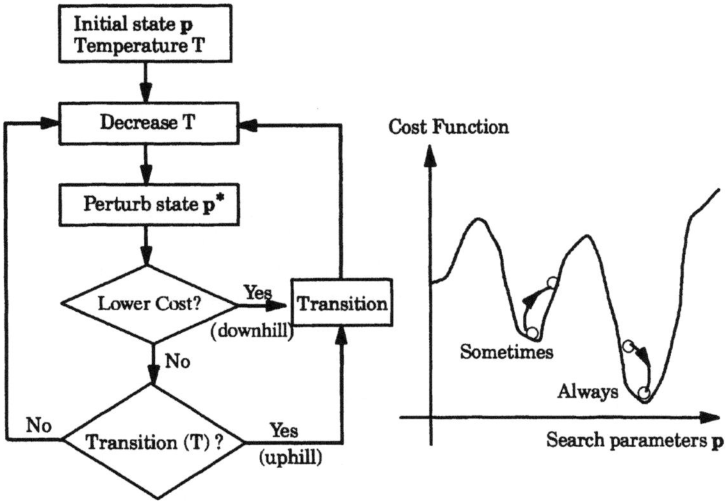

Figure 3. Global search using simulated annealing.

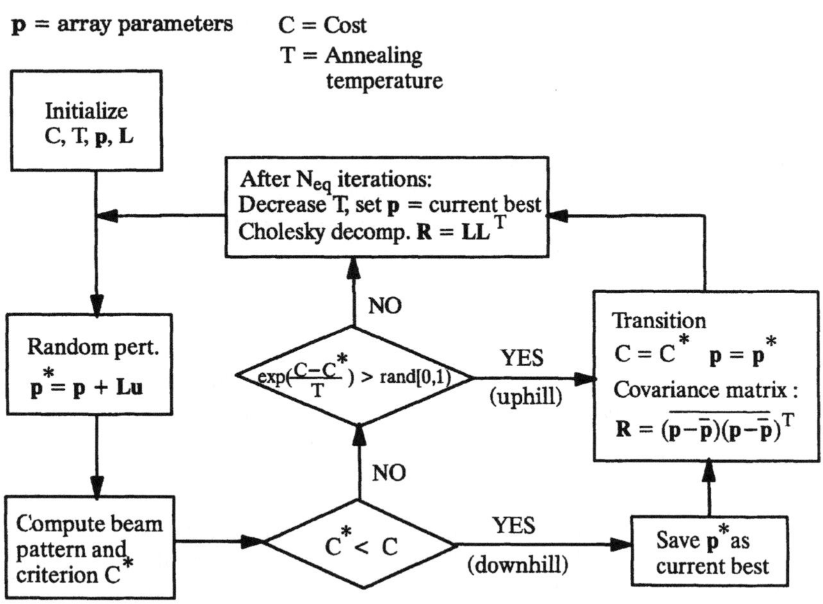

Figure 4. Simulated annealing algorithm for non-uniform array optimization.

At each temperature a fixed number of perturbations N_{eq} are tried (e.g. $N_{eq} = 25 \times StateVectorSize$). Each perturbation vector is generated using an independent, uniformly-distributed random vector multiplied by the matrix $\mathbf{L}$. After each set of the N_{eq} perturbations, the temperature is decreased according to a power series α^k (we used $\alpha = 0.96$). The algorithm terminates when the ratio of moves to perturbations falls below a certain threshold (e.g. 1 percent).

A major drawback of simulated annealing is that the final result is not guaranteed to be optimal, and more extensive searches are more likely to yield good results. However, searches may be performed in parallel by running the algorithm independently on multiple processors and selecting the best solution among the results [2].

RESULTS

This section illustrates the array optimization method described previously with a hypothetical 100 KHz, 30 cm array. For $N = 3$, the search space may be conveniently displayed as shown in Fig. 5. The left panel of Fig. 5 shows the element size search space where the horizontal axis is x_1 and the vertical axis is x_2, both ranging from 0 mm to 150 mm. Since $x_2 > x_1$, the search space is limited to the upper-left triangular region. The right panel of Fig 5 shows the apodization search space where the horizontal axis is a_2 and the vertical axis is a_3, both ranging from 0.0 to 5.0. The dots shown in Fig. 5 correspond to search locations (for this particular run) where the brightness of the dots increases with decreasing cost. This search diagram permits the search progress to be visualized when displayed in real-time. Initially when the temperature parameter is high, dim dots occur over large regions of the search space. Then, brighter regions corresponding to valleys are found as the search progresses. Finally, a small region of bright dots appear as the search converges to a solution.

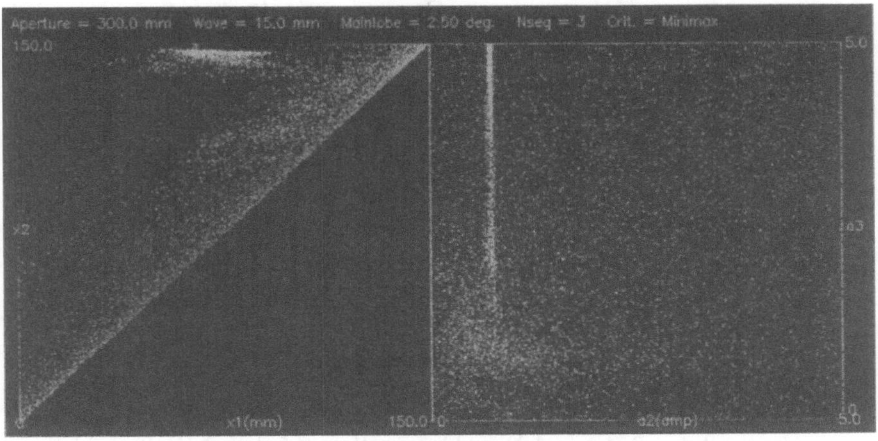

Figure 5. Example of search diagram for 5-element (N=3) array. The left panel shows the element size search space while the right panel shows the apodization search space. Dots correspond to search locations for this particular run of the simulated annealing algorithm where the brightness of the dots increases with decreasing cost.

Figure 6 shows the apodization aperture and beam pattern for the 100 KHz, 30 cm array optimized using $\theta_1 = 2.5^o$ corresponding to the search diagram shown in Fig. 5. The element edges are given by $x_1 = 6.5\ cm$ and $x_2 = 14.75\ cm$ while the apodization are given by $a_2 = 0.74$ and $a_3 = 4.99$. For comparison, the continuous Dolph-Chebyshev apodization is shown in the dotted line. As seen in the bottom panel of Fig. 6, the five-element array produces the same mainlobe width and 0.4 dB lower peak sidelobe than the Dolph-Chebyshev continuous apodization. Although this array is probably impractical due to low sensitivity, it does demonstrate the operation of the simulated annealing algorithm.

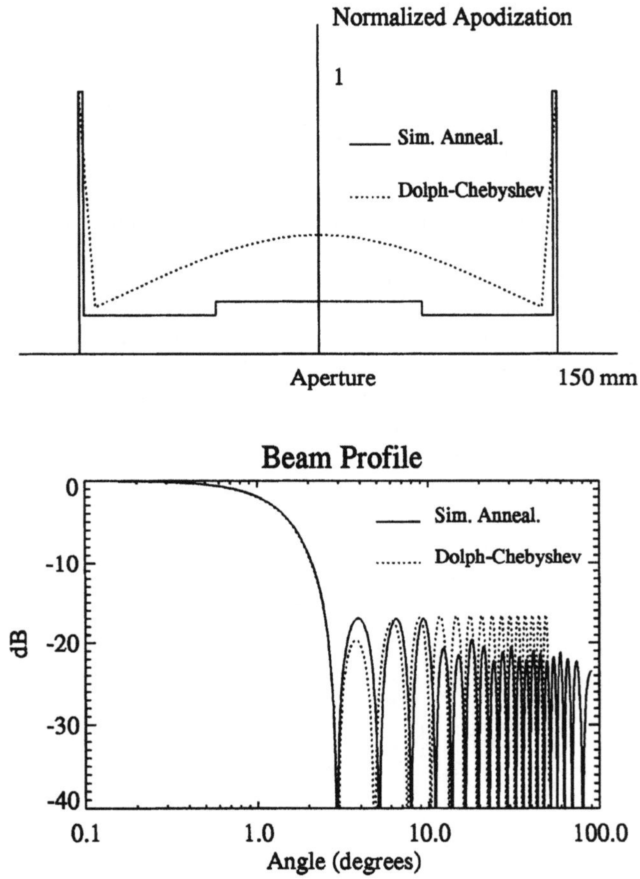

Figure 6. Array apodization ($x_1 = 6.5\ cm, x_2 = 14.75\ cm, x_3 = 15\ cm$ and $a_1 = 1.0, a_2 = 0.74, a_3 = 4.99$) and beam pattern of 5-element (N=3), 100 KHz, 30 cm array optimized using $\theta_1 = 2.5^o$ compared to Dolph-Chebyshev continuous apodization.

Figure 7 shows a 9-element (N=5) array and resulting beam pattern optimized using $\theta_1 = 5.5^o$. Here, the mainlobe is permitted to be wider to gain significantly lower sidelobes. Again, the mainlobe has the same width as the Dolph-Chebyshev continuous aperture, but the peak sidelobe is 1.4 dB lower than Dolph-Chebyshev.

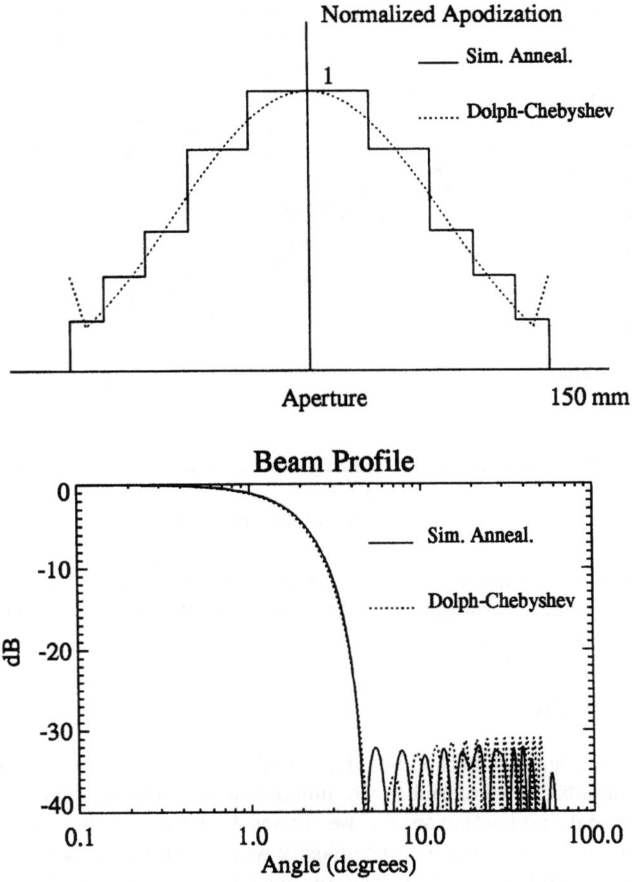

Figure 7. Array apodization ($x_1 = 3.8\ cm, x_2 = 7.6\ cm, x_3 = 10.3\ cm, x_4 = 12.9\ cm, x_5 = 15.0\ cm$ and $a_1 = 1.0, a_2 = 0.79, a_3 = 0.5, a_4 = 0.34, a_5 = 0.18$) and beam pattern of optimized 9-element (N=5), 100 KHz, 30 cm array compared with Dolph-Chebyshev continuous apodization.

Figure 8 shows the optimized mainlobe-sidelobe tradeoff for different number of independent elements. Mainlobe beamwidth is measured from beam center to the $-20\ dB$ point. The mainlobe-sidelobe tradeoff for Dolph-Chebyshev continuous apodization is shown in the solid line. As seen in Fig. 8, the peak sidelobe decreases monotonically with increasing mainlobe width as expected. However, the attainable mainlobe width and peak sidelobe depend on the number elements in the array. Increasing the number of elements is necessary for the mainlobe to get wider. For a given mainlobe width, increasing the number of elements generally decreases the sidelobe level, however there are instances (e.g. $\theta = 2.6^o$ and $\theta = 3.1^o$) where the 5-element and 9-element optimized arrays have almost identical performance. Interestingly, a few of the 5-element and most of the 9-element arrays optimized using simulated annealing outperform the Dolph-Chebyshev continuous apodization.

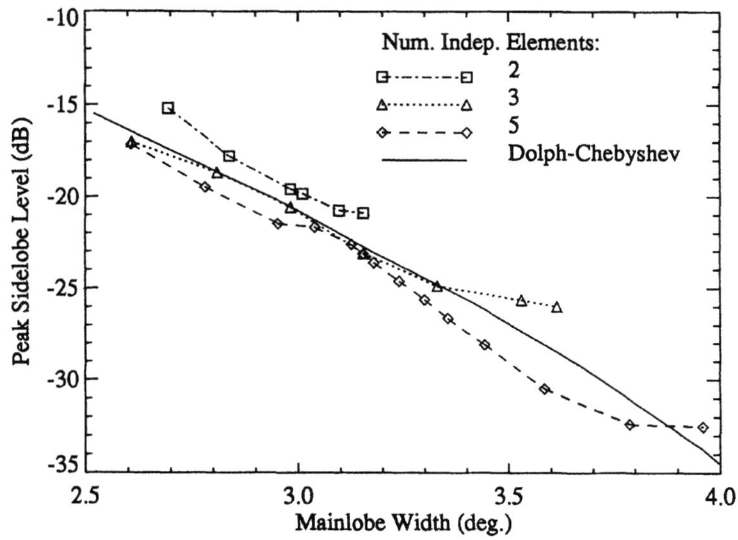

Figure 8. Mainlobe width (half-width at -20 dB) versus peak sidelobe level for different number of independent elements in a 5 MHz, 30 cm array and Dolph-Chebyshev continuous apodization.

CONCLUSIONS

We have presented a method based on simulated annealing to optimize non-uniform arrays for broadside beamforming. By minimizing the peak sidelobe level over the element sizes and apodization values, we showed that the non-uniform array can have good beam control with minimal complexity which makes it a useful compromise between the dense uniform array and the continuous aperture for broadside beamforming. Compared with mechanical apodization of a continuous aperture, this "segmented" apodization allows easier fabrication and more flexible apodization control to result in a potentially better beam pattern. In addition, design constraints such as minimal element size are easily incorporated into the optimization. Although we have demonstrated this method for farfield narrowband beamforming, it may be extended to nearfield broadband imaging [3].

REFERENCES

[1] D. G. Wildes, R. Y. Chiao, C. M. W. Daft, K. W. Rigby, L. S. Smith, and K. E. Thomenius, "Elevation Performance of 1.25D and 1.5D Transducer Arrays," *IEEE Trans. Ultrason. Ferroelec. Freq. Contr.*, vol. 44, no. 5, September 1997, pp. 1027-1037.

[2] P. J. M van Laarhoven and E. H. L. Aarts, *Simulated Annealing: Theory and Applications*, Boston: Dordrecht, 1987.

[3] R. Y. Chiao, K. W. Rigby, and D. G. Wildes, "Optimization of 1.5D Arrays", *1998 IEEE Ultrasonics Symposium.*

HYPERTHERMIA THERAPY USING ACOUSTIC PHASE CONJUGATION

Philippe Roux, Michael B. Porter, Hee C. Song and William A. Kuperman

Marine Physical Laboratory, Scripps Institution of Oceanography
University of California San Diego
9500 Gilman Drive, La Jolla, CA 92093

INTRODUCTION

One of the advantages of ultrasound phased-array treatments is the non-invasive manner in which ultrasonic energy, and then heat, can be delivered to the treatment volume. However, the heat generated in the medium induces variations in the sound speed, which affects the acoustic focusing and leads to strong effects such as self-focusing or self-defocusing of the sound beam. In this work, we use a numerical simulation to study coupled acoustic-heat propagation in tissue. By simultaneously modeling the propagation of acoustic waves and the diffusion of heat, we are able to quantitatively simulate acoustic refraction induced by heat in the treatment volume. Finally, using phase conjugation, we develop one procedure to control temperature and maintain the acoustic focusing at the initially desired focal point.

NUMERICAL SIMULATION

In order to explain the heat-ultrasound interaction, a non-linear formulation is classically introduced by coupling the wave and heat equations. This formulation allows for a change in the acoustic sound speed due to acoustic heating. More precisely, by assuming the acoustic streaming effect negligible compared to the acoustic heating, we obtain the coupled equations (1), (2) and (3).

Equation (1) describes the evolution of the pressure field $P=P(r,\omega)$ by a partial differential equation in first order with range known as the Parabolic Equation[1]. Equation (2) is the heat equation where $T=T(r,t)$ is the temperature field, κ is the heat diffusion coefficient, α the acoustic attenuation and $I=I(r,\omega)$ the acoustic intensity derived from Eq.

Acoustical Imaging, Volume 24,
Edited by Hua Lee. Kluwer Academic/Plenum Publishers, 2000.

(1). Last, Eq. (3) takes into account the dependence of the acoustic sound speed with the temperature, which will de discussed later.

$$\begin{cases} \dfrac{\partial P}{\partial r} = ik_0\left(\sqrt{n^2 + \dfrac{1}{k_0^2}\dfrac{\partial^2}{\partial z^2}} - 1\right)P & (1) \\ \dfrac{\partial T}{\partial t} - \kappa\Delta T = 2\alpha I & (2) \\ n = \dfrac{c_0}{c} = f(T) & (3) \end{cases}$$

To solve this set of coupled equations, we first compare the characteristic time scale of Eq. (1) and Eq. (2). In our simulation, the acoustic frequency is F=1.5Mhz, which corresponds to a wavelength λ=1 mm and a characteristic time scale $1/F \sim 10^{-6}$ s. The heat diffusion coefficient κ is on the order of 10^{-7} m^2/s in biological tissue[2,3]. So the dynamic time scale Δt of Eq. (2), which represents the time interval after which the temperature has spread over one wavelength λ due to heat diffusion, is $\Delta t \approx \dfrac{\lambda^2}{\kappa} \approx 10\text{s}$. We have $\Delta t >> 1/F$, which means that the heterogeneity of sound speed due to heat in the medium described by Eq. (3) is a slowly dependent function of time and space.

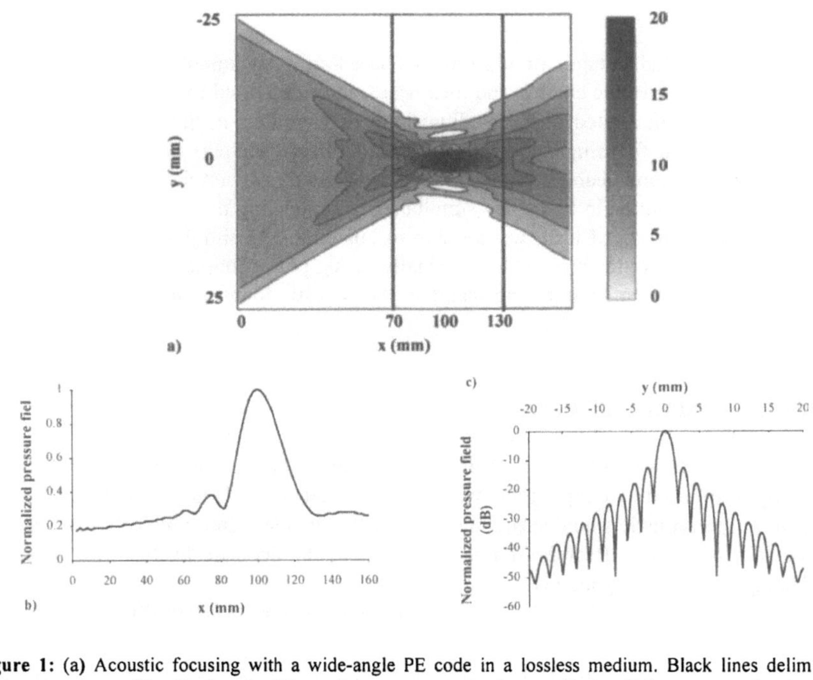

Figure 1: (a) Acoustic focusing with a wide-angle PE code in a lossless medium. Black lines delimit the treatment volume. The black-and-white scale corresponds to the amplitude of the pressure in atm. The intensity at the focal point is 185 W/cm^2; (b) Normalized pressure field along the x-axis at y=0 mm. (c) Normalized pressure field along the y-axis at x=100 mm. The size of the focal spot is 20 mm along the x-axis and 2.5 mm along the y-axis.

In this case, the acoustic heating can be considered as a steady state temperature field with respect to the rapidly oscillating sound field. Thus the heat equation and the acoustic Parabolic Equation can be treated separately: first, the solution of the PE gives the heat source term that is injected into the heat equation. This heat equation allows us then to compute a new temperature distribution that changes the sound speed field. Last, this new sound speed distribution is injected back into the PE equation. This algorithm has already been described by Le Floch et al[4], but they compute the acoustic field with a 2D finite-difference simulation. The advantage of the PE formulation used here lies in the fact that, for the same accuracy, PE code runs much faster.

Our simulations are carried out in two dimensions. The acoustic field is computed by a wide-angle Parabolic Equation method[1]. The acoustic source is a 50mm finite-aperture array outside the treatment volume in a water bath used as a lossless coupling medium. We consider the zone of interest, between x=70 mm and x=130 mm on Fig.1, as a fluid medium (we neglect shear waves) whose density, sound speed and temperature before treatment are taken equal to those of the coupling medium (c=1500 m/s, density=1000 kg/m^3). The emitted field is phased delayed to focus at x=100 mm in the zone of interest where attenuation[5] (α = 10 N/m) produces heat. The value of the acoustic intensity at the focal point, measured in a homogeneous lossless medium, is equal to 185 W/cm^2. The duration of insonification is set at 10 s in our simulations.

RESULTS

Roughly speaking, hyperthermia consists in locally heating the medium at the acoustic focal point. What is now the effect of the local increase of temperature on the acoustic field? To describe acoustic refraction induced by heat, two cases are to be taken into consideration: if the sound speed increases with the temperature ($\frac{dc}{dT} > 0$), we observe a *self-defocusing* of the acoustic field, whereas we see a *self-focusing* of the sound beam if the sound speed decreases with the temperature ($\frac{dc}{dT} < 0$). The physics behind these two cases are different because Snell's laws induce a different refraction depending on the sign of $\frac{dc}{dT}$ along an acoustic ray (Fig.2). Experimentally speaking, a fatty tissue corresponds to a negative value of $\frac{dc}{dT}$ and muscle fibers to a positive $\frac{dc}{dT}$[3].

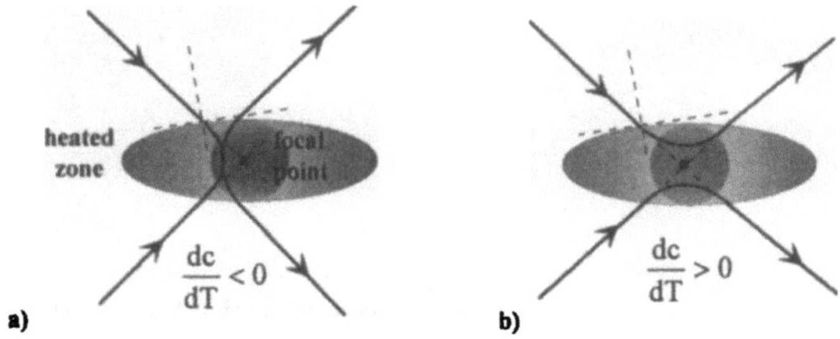

Figure 2: a) $\frac{dc}{dT} < 0$: self-focusing of the acoustic beam. b) $\frac{dc}{dT} > 0$: self-defocusing of the acoustic beam.

Figure 3 shows then the time-evolution of the acoustic field during a 10s insonification, respectively for $\frac{dc}{dT} = -4$ m/s/°C (Fig. 3a to Fig.3e) and for $\frac{dc}{dT} = 4$ m/s/°C (Fig. 3f to Fig. 3j).

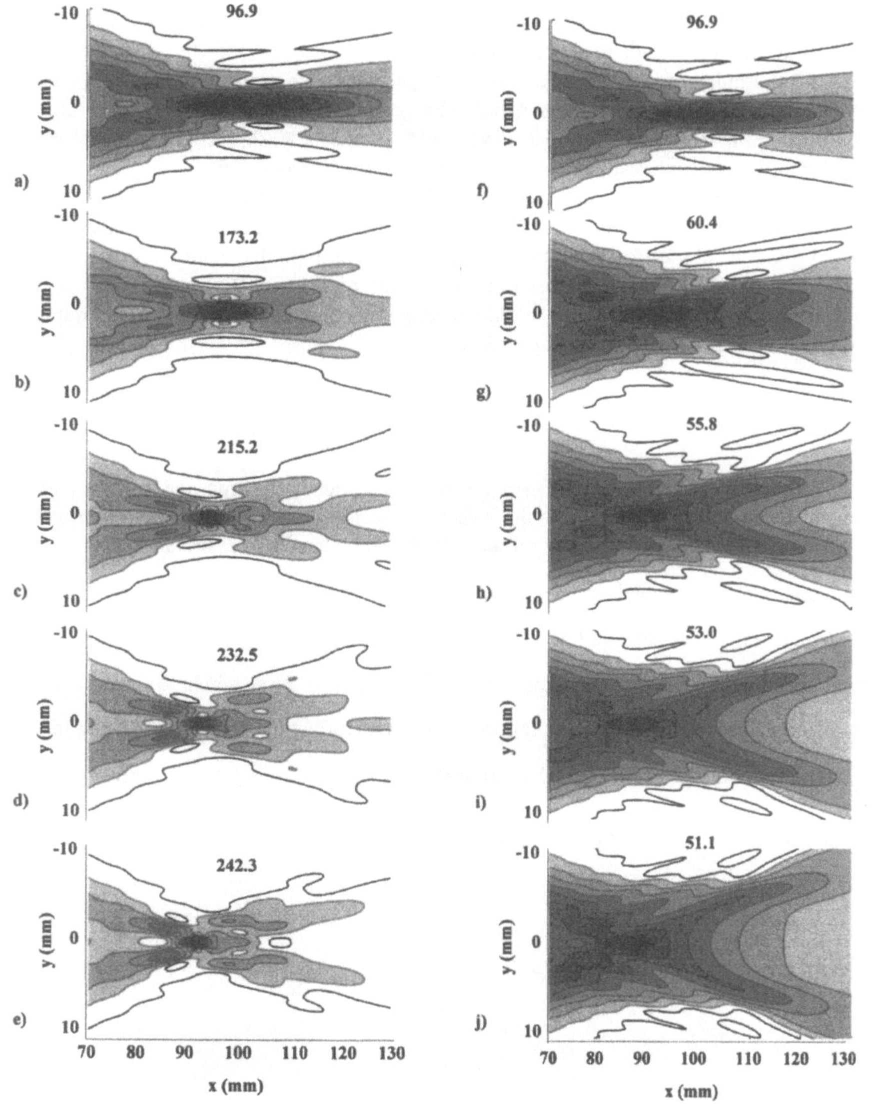

Figure 3: Pressure field in the treatment region for $\frac{dc}{dT} = -4$ m/s/°C at a) t=0, b) t=2.5 s, c) t=5 s, d) t=7.5 s, e) t=10 s. Pressure field in the treatment region for $\frac{dc}{dT} = 4$ m/s/°C at f) t=0, g) t=2.5 s, h) t=5 s, i) t=7.5 s, j) t=10 s. Each title corresponds to the maximum of the acoustic intensity in W/cm^2.

The time evolution of the pressure field shows clearly the self-focusing or self-defocusing of the acoustic field. If $\frac{dc}{dT} < 0$, the intensity increases dramatically at the focus, which means that the acoustic field is self-focused, whereas if $\frac{dc}{dT} > 0$, the intensity decreases which corresponds to a self-defocused effect. In both cases, the focal zone is shifted toward the array up to 10 mm, which corresponds to a ten-wavelength distance.

Figure 4 below represents the resulting temperature distributions in the treatment volume at the end of the 10s insonification.

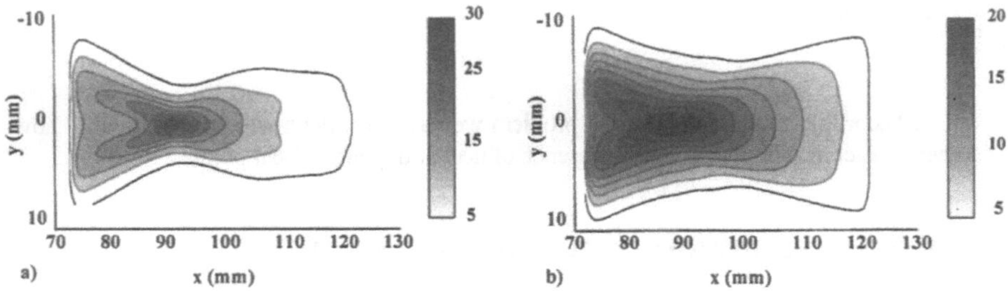

Figure 4: Temperature elevation (in degree Celsius) in the treatment volume after 10 s for a) $\frac{dc}{dT} = -4$ m/s/°C and b) $\frac{dc}{dT} = 4$ m/s/°C.

Two important points are: first, because of the self-focusing and self-defocusing effects, the temperature reached for $\frac{dc}{dT} < 0$ is much larger than the temperature obtained for $\frac{dc}{dT} > 0$. Second, the focal shift due to the heat-induced acoustic refraction leads to a misleading estimation of the temperature expected at the desired focal point if the acoustic-heat coupling is neglected. Actually, whatever the sign of $\frac{dc}{dT}$, the focus has moved out of the initial focal plan, which means that the damage done by heat in the zone of interest may not occur at the desired point. Fig. 4 shows for example that the highest temperature elevation after 10 s is located around 90 mm, which is 10 mm away from the initial focal point.

Finally, Fig. 5 represents the temperature elevation at the initial focal point (x=100 mm, y=0). Comparing Fig. 4 and Fig. 5, we clearly see that the temperature reached after 10 s is smaller than the maximum obtained in the whole treatment region. More precisely, if $\frac{dc}{dT} < 0$, we see a sudden saturation of the temperature rise due both to the self-focusing effect which concentrates the acoustic intensity and to the refraction-induced focal shift. If $\frac{dc}{dT} > 0$, the temperature rise is slower and seems less affected by the focal shift because of the self-defocusing effect.

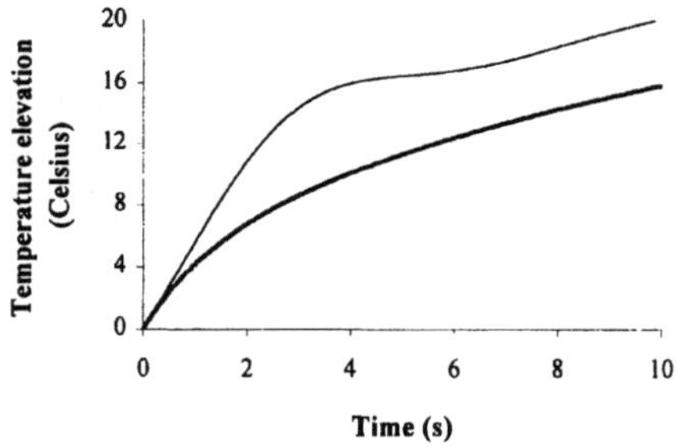

Figure 5: Temperature elevation at the initial focal point (x=100 mm, y=0) for $\frac{dc}{dT} = -4$ m/s/°C (thin line) and $\frac{dc}{dT} = 4$ m/s/C (bold line).

According to these results, the problem we have to solve now is the following: how to keep the acoustic focusing in the presence of heat at the initial focal point?

HOW TO MAINTAIN AND CONTROL THE FOCUSING?

To maintain focus, the idea is to take advantage of phase-conjugation invariance properties[6,7]. Indeed, phase conjugation invariance insures that a phase-conjugated field will focus back to its acoustical source whatever the sound speed distribution of the medium. So, if we were able to put an acoustic source at the desired focal point, a phase-conjugation hyperthermia treatment would consist of, first, phase-conjugating the field received from the source on the array and second, re-transmitting this phase-conjugated field with a high intensity in the medium. This would lead to a constant high-intensity focusing on the source whatever the heat distribution in the treatment volume. But, of course, it is not practical to place a source at the desired focus and such a phase-conjugation treatment is not workable.

To get round this problem, we propose a two-stage procedure based on phase-conjugation principles.

In the first stage, we compute a phase-conjugation hyperthermia treatment: we define "in the computer" a medium as closed as possible to the experimental one. Then we place an acoustic source at the desired focal point. Last, we phase-conjugate and retransmit the conjugated field back to the source with the same intensity as the actueal intensity available during the experimental hyperthermia treatment. The aim of this first stage is to realize an *in vitro* numerical experiment to record as often as needed during the treatment the time evolution of the phase-conjugated field, which focuses back to the source. We end up then with a temporal sequence of phase-conjugated vectors, which are adapted to the heat distribution at the focal point.

In a second stage, we use this phase-conjugated vectors sequence as the time-dependent emitted source field of the array during the *in vivo* hyperthermia treatment. The difference between this phase-conjugated hyperthermia treatment and a classical one lies in the fact that the emitted source field stays the same during a classical treatment and is not adapted to the heat distribution at the focal point, which leads to the previously discussed heat-induced acoustic refraction effects. In the case of the phase-conjugated hyperthermia treatment, the acoustic refraction still exists but is compensated by the shape of the phase-conjugated emitted field.

Figure 6 below represents the pressure field and the temperature distribution obtained after a 20s insonification phase-conjugation hyperthermia treatment for a medium with $\frac{dc}{dT} = -4$ m/s/°C. For the same temperature elevation as the one obtained classically after 7.5s (cf. Fig. 5), we see that the field is still in focus at the desired focal point.

During a phase-conjugation treatment, the intensity recorded on the array decreases because of the heat-induced refraction effect. This leads to a smaller intensity at the focus and then to a slower raise of the temperature. However, because some energy is still transmitted into the medium, the desired temperature is reached at the focus without burning neighboring tissues.

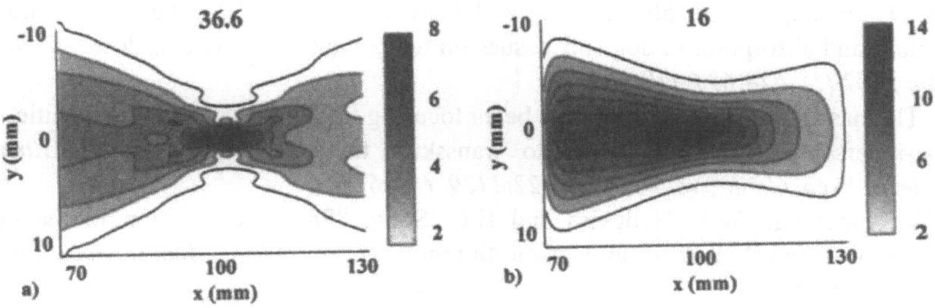

Figure 6: Phase-conjugated hyperthermia treatment after a 20 s insonification in a medium such as $\frac{dc}{dT} = -4$ m/s/°C : a) Pressure field – the black-and-white scale corresponds to the amplitude of the pressure in atm and the title to the maximum intensity in W/cm^2. b) Temperature field — the black-and-white scale corresponds to the temperature in degree Celsius and title to the maximum temperature.

The advantage of this phase-conjugation treatment is the ability to burn cells accurately. The limitations are principally that the quality of the treatment may depend dramatically on the quality of the numerical experiment made in the first stage.

In the future, we more carefully will study then the consequences of mismatch induced by an incorrect knowledge of the medium. For example, we must investigate what happens during the treatment if the experimental value of the attenuation is not equal to the value used for the numerical experiment?

CONCLUSION

We have shown numerically that, in a medium with attenuation, heat induced by acoustic focusing leads to refraction of acoustic waves and disturbs the focus. The effect observed at the focus is either a self-focusing or a self-defocusing of the acoustic beam depending on the sign of $\frac{dc}{dT}$ in the medium. We have proposed a procedure to maintain focus in the zone of interest using phase-conjugation invariance properties.

ACKNOWLEDGEMENT

This work is supported by DARPA under contract number N00014-97-D-0350.

REFERENCES

1. M.D. Collins, "Higher-order parabolic approximations for accurate and stable elastic parabolic equations with application to interface wave propagation", *J. Acoust. Soc. Am., 89 (3), 1050:1057, (1991).*

2. R.J. McGough, M.L. Kessler, E.S. Ebbini and C.A. Cain, "Treatment planning for hyperthermia with ultrasound phased arrays", *IEEE Trans. Ultrason., Ferroelect., Freq. Contr., vol 43, N6, 1074:1084, (1996).*

3. R. Seip, P. VanBaren, C.A. Cain and E.S. Ebbini, "Non-invasise real-time multipoint temperature control for ultrasound phased array treatments", *IEEE Trans. Ultrason., Ferroelect., Freq. Contr., vol 43, N6, 1063:1073, (1996).*

4. C. Le Floch, M. Tanter and M. Fink, *submitted to Applied Physics Letters, July 1998.*

5. C.A. Damianou, N.T. Sanghvi, F.J. Fry and R. Maass-Moreno, "Dependence of ultrasonic attenuation and absorption in dog soft tissues on temperature and thermal dose", *J. Acoust. Soc. Am., 102 (1), 628:634, (1997).*

6. J.L. Thomas and M. Fink, "Ultrasonic beam focusing through tissue inhomogeneities with a time-reversal mirror: application to transskull therapy", *IEEE Trans. Ultrason., Ferroelect., Freq. Contr., vol 43, N6, 1122:1129, (1996).*

7. W.A. Kuperman, W.S. Hodgkiss and H.C. Song, "Phase conjugation in the ocean: experimental demonstration of an acoustic time-reversal mirror", *J. Acoust. Soc. Am., 103 (1), 25:40, (1998)*

ULTRASONIC BROADBAND FIBER OPTIC SOURCE FOR NON DESTRUCTIVE EVALUATION AND CLINICAL DIAGNOSIS

E. Biagi, F. Margheri, L. Masotti, M. Pieraccini

Ultrasonic and Non Destructive Testing Laboratory
Department of Electronic Engineering
University of Florence
via Santa Marta, 3
50139 Firenze, Italy

INTRODUCTION

Most advanced ultrasonic diagnostic systems, both for Non Destructive Testing and for clinical applications, base themselves on spectral analysis of ultrasonic signals. The diagnostic power of these systems strongly depends on the available ultrasonic band. However, the present transducers make use of piezoelectric ceramics which exhibit a typical spectrum centred on a resonance frequency. The widening of the spectrum is obtained to detriment of efficiency, by mechanically damping the oscillation.

On the contrary, ultrasound generation by means of optical pulses conversion in mechanical waves through thermoelastic effect, it is a intrinsically wide band technique, and in certain conditions also flat band. Therefore, optoelectronic devices for ultrasonic generation through thermoelastic effect could be of great interest in the development of advanced systems for ultrasonic diagnostics.

The authors propose a fiber optic ultrasonic source. The working principle is based on thermoelastic conversion of laser light. A metallic layer deposed on fiber tip, absorbs and converts in heat the optical pulses guided by the fiber optic.

The authors developed a theoretical model for analyzing the proposed optoacustic source. Experimental results confirm theoretical predictions. Finally, it was experimentally demonstrated that a low cost diode laser can act as optical source for this kind of ultrasound transducer.

THEORETICAL ANALYSIS

The aim of this section is a theoretical analysis of the optoacoustic generation physics, in order to optimize the design and materials. Following theoretical formulation, although it takes the works of White [1], von Gutfeld [2] and Oksanen [3] as starting points, considers cases which are not previously analyzed, furthermore it removes same approximations of previous works.

Acoustical Imaging, Volume 24.
Edited by Hua Lee, Kluwer Academic/Plenum Publishers, 2000.

The device working principle is based on thermal conversion of laser pulses into a metallic film evaporated directly onto the tip of a fiber optic.

Let's consider two media, as in Figure 1, the first, indicated with subscript 1, is the laser light propagation medium (fiber optic); the second, subscript 2, is the surrounding medium.

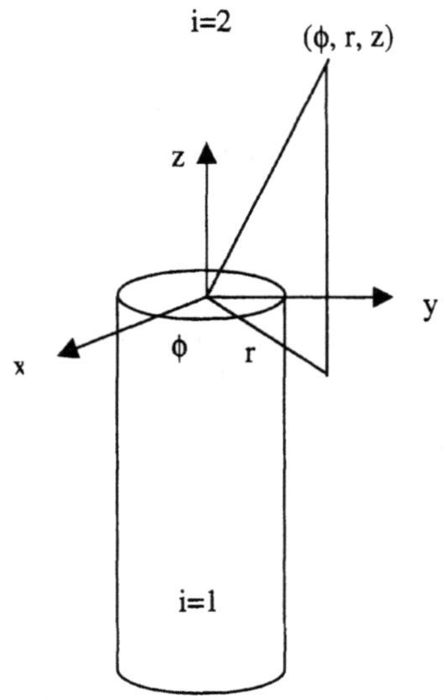

Figure 1. Transducer geometry

The heat diffusion and displacement wave equations are respectively:

$$\hat{k}_i \nabla^2 T_i = \frac{\partial T_i}{\partial t} \qquad (1)$$

$$\nabla^2 u_i - \frac{1}{v_i^2}\frac{\partial^2 u_i}{\partial t^2} = \beta_i \nabla \cdot T_i \qquad (2)$$

where $\hat{k}_i$ is the thermal diffusivity ($\hat{k}_i = K_i / (\rho_i c_i)$ with K_i thermal conductivity, ρ_i density, c_i specific heat), T_i is the temperature rise, u_i is the local displacement, v_i is the sound velocity, β_i is the expansion coefficient.

By supposing that thermal effects are limited in close proximity to the boundary surface z=0, the approximation $\nabla \cdot T_i \approx \frac{\partial T_i}{\partial x}$ is reasonable.

The analytical problem can be solved in transformed dominia of the Laplace and Hankel transforms. We remind the Laplacian operator ∇^2 in cylindrical coordinates can be separated into two parts:

$$\nabla^2 = \nabla_\perp^2 + \frac{\partial^2}{\partial z^2} \qquad (3)$$

where $\nabla_\perp^2 = \frac{1}{r}\frac{\partial}{\partial r}\left(r\frac{\partial}{\partial r}\right)$ and $\nabla_\perp^2 J_0(qr) = -q^2 J_0(qr)$. $J_0(x)$ is the zeroth order Bessel function, and q is the Hankel transform variable.

By applying the Laplace and Hankel transforms to (1) and (2), one obtains:

$$\frac{\partial^2 \breve{\tilde{T}}_i}{\partial z^2} - p_i^2 \breve{\tilde{T}}_i = 0 \tag{4}$$

$$\frac{\partial^2 \breve{\tilde{u}}_i}{\partial z^2} - k_i^2 \breve{\tilde{u}}_i = \beta_i \frac{\partial \breve{\tilde{T}}}{\partial z} \tag{5}$$

with $p_i = \sqrt{q^2 + s/\hat{k}_i}$ and $k_i = \sqrt{q^2 + s^2 / v_i^2}$. $\breve{\tilde{T}}_i$ and $\breve{\tilde{u}}_i$ are the Laplace and Hankel transforms of T_i and u_i.

The above linear equations can be solved by imposing appropriate boundary conditions, which are relative to following physical assumptions:

- thermal continuity equation

$$T_1(z=0) = T_2(z=0) \tag{6}$$

- energy conservation equation

$$K_1 \left.\frac{\partial T_1}{\partial x}\right|_{z=0} - K_2 \left.\frac{\partial T_2}{\partial x}\right|_{z=0} = \alpha I \tag{7}$$

with I optical intensity onto the film, α opto-thermal conversion coefficient of the film.

- displacement continuity equation

$$u_1(z=0) = u_2(z=0) \tag{8}$$

- pressure continuity equation

$$P_1(z=0) = P_2(z=0) \tag{9}$$

with

$$P_i = \rho_i v_i^2 \left(\frac{\partial u_i}{\partial x} - \beta_i T_i \right) \tag{10}$$

of note that temperature rise T_i can be considered zero, away from heat surface z=0.

Substituting transformed boundary conditions in (4) and (5), the following expression for displacement in medium 2 is obtained:

$$\breve{\tilde{u}}_2 = A e^{-k_2 x} + B e^{-p_2 x} \tag{11}$$

with A and B determined by boundary conditions. Only the first term in (11) represents an acoustic wave. Its amplitude is given by:

$$A = \frac{\alpha \breve{\tilde{I}}}{K_1 p_1 + K_2 p_2} \frac{k_1}{k_1 + r k_2} \left(\frac{\beta_1}{p_1} + \frac{\beta_2}{p_2} \right) \tag{12}$$

with $r = \rho_2 v_2^2 / \rho_1 v_1^2$.

In order to calculate the temporal pressure waveform, the optical forcing term has to be known. It is reasonable to suppose the optical intensity onto the film having the following expression:

$$I = I_0 rect\left(\frac{r - \frac{r_0}{2}}{r_0} \right) \left(\frac{t}{\tau} \right) e^{1 - \left(\frac{t}{\tau} \right)} \tag{13}$$

where r_0 is fiber optic core radius, τ is optical pulse duration.

The Laplace and Hankel transform of the (13) is:

$$\breve{\tilde{I}}(s,q) = I_s(s) \frac{r_0^2}{2} \sum_{n=0}^{\infty} \frac{(-1)^n}{n!(n+1)!} \left(\frac{r_0 q}{2} \right)^{2n} \tag{14}$$

with

$$I_s(s) = I_0 \frac{e\tau}{(1+\tau s)^2} \tag{15}$$

The Hankel antitrasform of derivative of $\hat{u}_2$ respect to z , is:

$$\frac{\partial \hat{u}_2}{\partial z} = I_s(s) \int_{q=0}^{q=\infty} \frac{\alpha}{K_1 p_1 + K_2 p_2} \frac{k_1 k_2}{k_1 + r k_2} \left(\frac{\beta_1}{p_1} + \frac{\beta_2}{p_2}\right) \frac{r_0^2}{2} \sum_{n=0}^{\infty} \frac{(-1)^n}{n!(n+1)!} \left(\frac{r_0 q}{2}\right)^{2n} J_0(qr) e^{-k_2 z} q\,dq \quad (16)$$

by substituting the (16) in (10), and imposing $s=j2\pi f$ with f frequency, one obtains acoustic pressure versus frequency .

Figure 2 shows the ratio between acoustic pressure spectrum and optical modulation spectrum ($I_s(s)$), at 2mm distance from the surface z=0, for different values of the fiber radius. Numerical values employed in calculation are in Table 1

Table 1. Numerical values for silice and water

	$Si\,O_2$	H_2O
ρ (Kg/m³)	2.3×10^3	1.0×10^3
c (J/Kg C)	740	4182
K (W/m C)	2.0	0.6
v (m/s)	5.66×10^3	1.48×10^3
β (C^{-1})	0.5×10^{-6}	0.67×10^{-4}

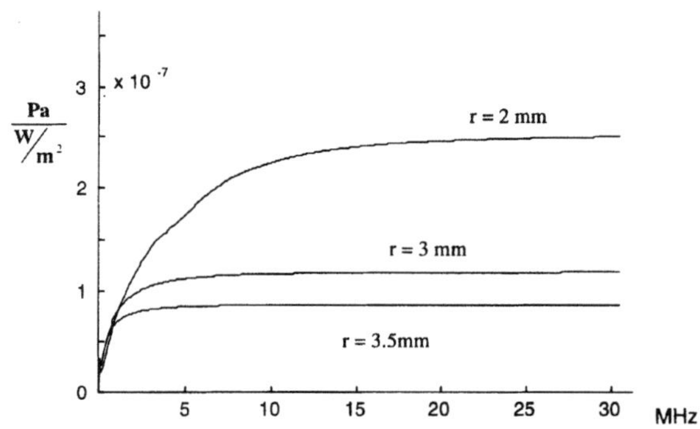

Figure 2 Ultrasonic spectra, varying fiber diameter. The distance from fiber is 2 mm.

Finally, the acoustic wave pressure $P(r,z,t)$ is given by

$$P(r,z,t) = \rho_2 v_2^2 \frac{\partial u_2}{\partial z} \quad (17)$$

since, by making all terms:

$$P(r,z,t) = \rho_2 v_2^2 I_0 L^{-1} \left\{ \int_{q=0}^{q=\infty} \frac{\alpha}{K_1 p_1 + K_2 p_2} \frac{k_1 k_2}{k_1 + r k_2} \left(\frac{\beta_1}{p_1} + \frac{\beta_2}{p_2}\right) \frac{r_0^2}{2} \sum_{n=0}^{\infty} \frac{(-1)^n}{n!(n+1)!} \left(\frac{r_0 q}{2}\right)^{2n} \frac{e\tau}{(1+\tau s)^2} J_0(qr) e^{-k_2 z} q\,dq \right\} \quad (18)$$

with L^{-1} Laplace antitrasform.

In order to numerically calculate Laplace antitrasform, the following expression is particularly useful [4]:

$$L^{-1}\{F(s)\}=\frac{e^{\Omega t}}{T}\left[\frac{F(\Omega)}{2}+\sum_{k=1}^{\infty}Re\left\{F\left(\Omega+\frac{jk\pi}{T}\right)\right\}cos\left(\frac{k\pi t}{T}\right)-Im\left\{F\left(\Omega+\frac{jk\pi}{T}\right)\right\}sin\left(\frac{k\pi t}{T}\right)\right] \quad (19)$$

where $F(s)$ is a generic complex function of complex variable. T is a time sufficiently longer than temporal length of $L^{-1}\{F(s)\}$. By writing $s=\Omega+j\omega$, Ω is an arbitrary value such as no poles exist in complex plane with real part smaller than Ω.

Figure 3 shows waveforms calculated for different fiber radius (light spot), at distance of 50 mm from the fiber tip. The data employed in the calculus are in Table I. Optical pulse duration was $\tau=130$ ns and peak power was 3.3 kW.

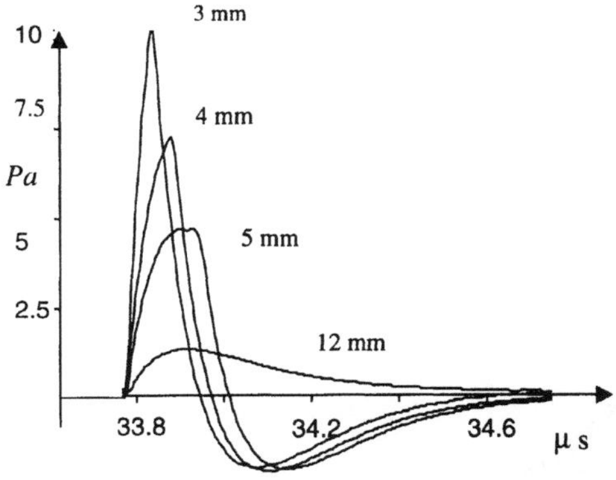

Figure 3. Waveforms calculated varying fiber diameter.

EXPERIMENTAL RESULTS

With the aim of experimentally testing the proposed optoacoustic devices, a aluminium metallic film was evaporated directly onto the tip face of a 600 μm core diameter PCS fiber. An experimental setup as in Figure 6 was devised.

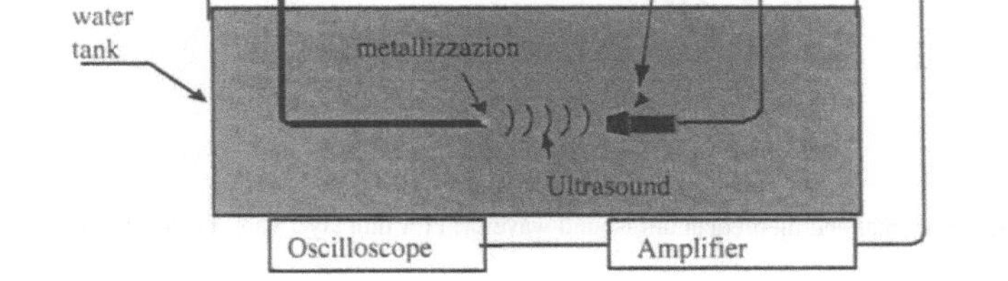

Figure 4 Experimental setup

The tests were conducted in a water tank where a receiving ultrasonic probe is positioned on axis with respect to the fiber. A 20 MHz V316 Panametric probe was employed as receiver. The signal was amplified by a commercial receiver (5052PR Panametrics) and acquired by a digital oscilloscope. In this measurement session pulses a Nd:YAG laser was been employed. Pulses of 1mJ energy and 1.36 kHz repetition frequency were delivered. By employing an appropriate laser pulse duration, about 350 ns, an ultrasonic signal with frequency content outside the bandwidth of the receiving traducer was generated. This experimental condition allows to make measurements not affected by the transducer frequency response. In fact, if the transducer is employed far from its resonance bandwidth, its response is practically flat versus frequency: in this working condition it behaves like a hydrophone.

Figure 5 compares experimental shape of the laser pulse with the shape supposed in theoretical model. This shape is given by the following expression

$$I \propto \left(\frac{t}{\tau}\right) e^{1-\left(\frac{t}{\tau}\right)} \qquad \text{with } \tau = 130\text{ns}$$

The agreement is very good.

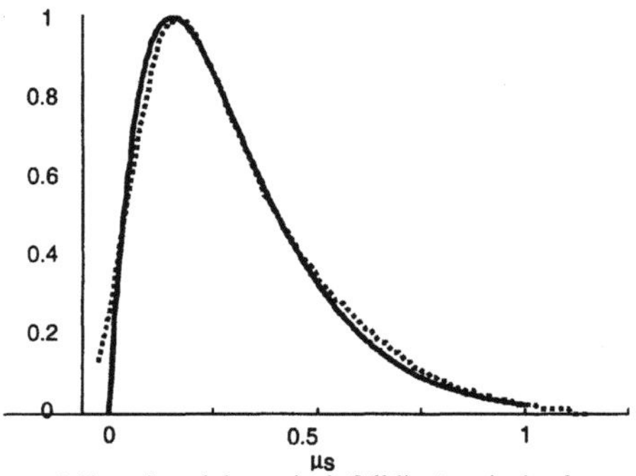

Figure 5. Experimental (dotted) and theoretical (full line) optical pulse

Figure 6 compares detected signal with pressure waveform calculated with the hypothesis of the metallic film case. Experiments is in very good accordance with the predicted waveform.

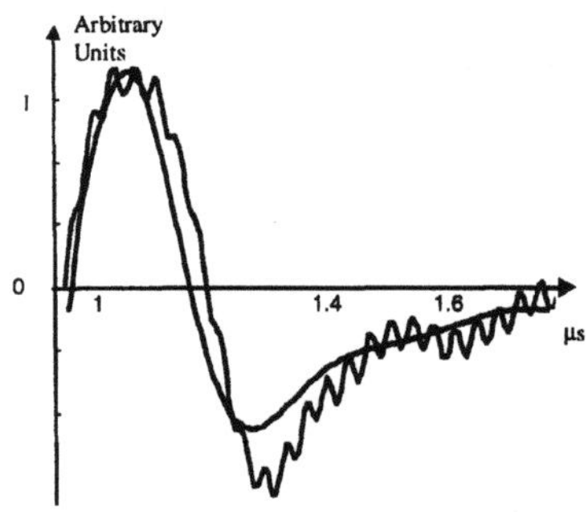

Figure 6. Experimental and theoretical ultrasound waveform for thin layer case. Distance from fiber tip: 1mm.

In previous measurement sessions high power lasers were employed, however optoacoustic technology could become effective only if it could make use of low cost semiconductor lasers. With this regard, measurements was performed using a very low cost diode laser of 0.8 W peak power. As absorber it was employed a glass slab with a cromium oxyde deposition. Optical absorption of the cromium oxyde is next to 1. In order to obtain a well detectable oscilloscope signal, receiver band was overlapping the band of the generated ultrasonics and consequently the detected signal waveform is strongly affected by receiver frequency response.

Figure 7 shows the obtained result. Signal is weak, but it demonstrate the feasibility, in terms of conversion efficiency, of optoacoustic sources employing laser diodes.

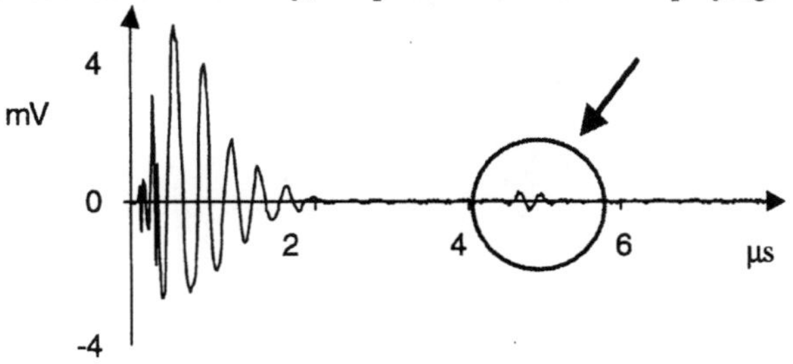

Figure 7. Experimental waveform obtained with a diode laser of 0.8 W peak power and a glass slab with a cromium oxyde deposition. The first signal is a electromagnetic interference due to laser driver.

A more effective solution could be to use a 100W peak power diode, today commercially available.

CONCLUSION

Proposed optoacoustic ultrasound source appears a promising device thanks to its intrinsic characteristics. This device acts as a tunable high frequency and broadband source. Most advanced ultrasonic ecographic equipment could take great advantage of this kind of source. The investigation of skin pathological morphology and intravascular ultrasound could be elective application because of device miniaturization and spectral characteristics of generated ultrasounds.

REFERENCES

[1] R.M. White, "Generation of Elastic Waves by Transient Surface Heating", Journal of Applied Physics, Vol. 34, No. 12, December 1963, pp. 3559-3567.

[2] R. J. von Gutfeld, H.F. Budd, "Laser-generated MHz elastic waves from metallic-liquid surfaces", Applied Physics Letters, Vol. 34, No. 10, May 1979, pp. 617-619.

[3] M. Oksanen, J. Wu, Prediction of the temporal shape of an ultrasonic pulse in a photoacustic sensing application, Ultrasonics, Vol. 32, No.1, 1994, pp. 43-46.

[4] K.S. Crump, "Numerical Inversion of Laplace Transforms Using a Fourier Series Approximation" Journal of the Association for Computing Machinery, Vol. 23, No.1, January 1976, pp. 89-96.

A STUDY OF SIGNAL-TO-NOISE RATIO OF THE FOURIER METHOD FOR CONSTRUCTION OF HIGH FRAME RATE IMAGES

Jian-yu Lu

Ultrasound Laboratory, Department of Bioengineering
The University of Toledo
Toledo, Ohio 43606, U.S.A.
Email: jilu@eng.utoledo.edu

INTRODUCTION

Limited diffraction beams were first developed in 1941 by stratton and was termed undistorted progressive waves (UPW)[1]. These beams have an infinite depth of field, i.e., they can propagate to an infinite distance without spreading, provided they are produced with an infinite large aperture. When produced with a finite aperture, they have a large depth of field[2–6]. Because of this property, limited diffraction beams have applications in medical imaging[7–10], tissue identification[11], Doppler blood flow velocity measurement[12–13], nondestructive evaluation (NDE) of materials[14], secure high-speed communications[15], and other areas such optics[6] and electromagnetics[16].

Based on the study of limited diffraction beams, recently, a new imaging method, called Fourier method, has been developed[17–20]. Unlike the conventional delay-and-sum method[21], where signals received with an array transducer are delayed and summed to focus beams on each point of an object, Fourier transform[22] is used to construct images with the new method. Because only one transmission is necessary to construct an image of an entire region of interests and the Fourier transform can be implemented with the fast Fourier transform, the new method has a potential to achieve a very high frame rate (up to 3750 m/s in biological soft tissues at a depth of 200 mm) with simplified electronics.

In this paper, we study the signal-to-noise ratio (SNR) of the new method and compare the results with the conventional delay-and-sum method. Results show that the SNR of the new method is almost identical to or a slightly better than the conventional method, provided that the same transmit beam is used.

Acoustical Imaging, Volume 24.
Edited by Hua Lee, Kluwer Academic/Plenum Publishers, 2000.

THEORETICAL PRELIMINARIES

Assuming a broadband transmission of a plane wave, from the X wave equation[3] one obtains receive signals from a linear array transducer weighted with a limited diffraction beam[17]

$$R_{k_x,k_y,k'_z}(t) = \frac{1}{2\pi} \int_{-\infty}^{\infty} \frac{A(k)T(k)H(k)}{c} F\left(k_x, k_y, k'_z\right) e^{-i\omega t} dk, \tag{1}$$

where t is time, $R_{k_x,k_y,k'_z}(t)$ is the receive signal with the parameters of limited diffraction beams of k_x, k_y, and $k'_z = k + k_z$, where $k = \omega/c$, ω is the angular frequency, c is the speed of sound, $k_z = \sqrt{k^2 - \left(k_x^2 + k_y^2\right)}$, $F(k_x, k_y, k'_z)$ is the Fourier transform of an object function representing the ultrasound reflectivity coefficient of the object, $A(k)$ and $B(k)$ are transmitting and receiving transfer functions of the transducer, $H(k)$ is the Heaviside step function[22]

$$H\left(\frac{\omega}{c}\right) = \begin{cases} 1 & , \quad \omega \geq 0 \\ 0 & , \quad \omega < 0 \end{cases}. \tag{2}$$

By Fourier transforming the receive signal in Eq. (1), $R_{k_x,k_y,k'_z}(t)$, one obtains a relationship between the Fourier transform of receive signals and the Fourier transform of the object to be constructed. With an inverse Fourier transform, the object function can be constructed.

SIGNAL-TO-NOISE RATIO OF THE FOURIER METHOD

From Eq. (1), one can construct images from receive signals. However, in any practical systems, there is always noise. Therefore, it is important to study the influence of noise on image construction methods. In the following, the influence of noise on both the Fourier and the conventional delay-and-sum methods is studied and compared with a computer simulation.

In the simulation, a linear array transducer of 48 elements, a dimension of 18.288 mm × 12.192 mm, 2.25 MHz central frequency, and a pulse-echo bandwidth of about 81% of the central frequency was assumed[17]. A broadband ultrasound plane wave was used to illuminate a wire target whose cross section was a point. Signals received with the array transducer were superposed with a band-pass random noise whose amplitude was about 50% of the peak of the receive signals. Images were constructed from these signals with both the Fourier and the conventional delay-and-sum methods and results are shown in Figure 1.

In Figure 1, the radio-frequency (RF) signals received with the linear array with and without noise are shown on the upper right and left hand sides, respectively. The horizontal dimension is time and the vertical direction represents the number of elements. Images constructed with and without noise are shown on the lower right and left hand sides, respectively. It is seen that the SNR of the images constructed with the Fourier and the conventional methods are very similar (Figure 1).

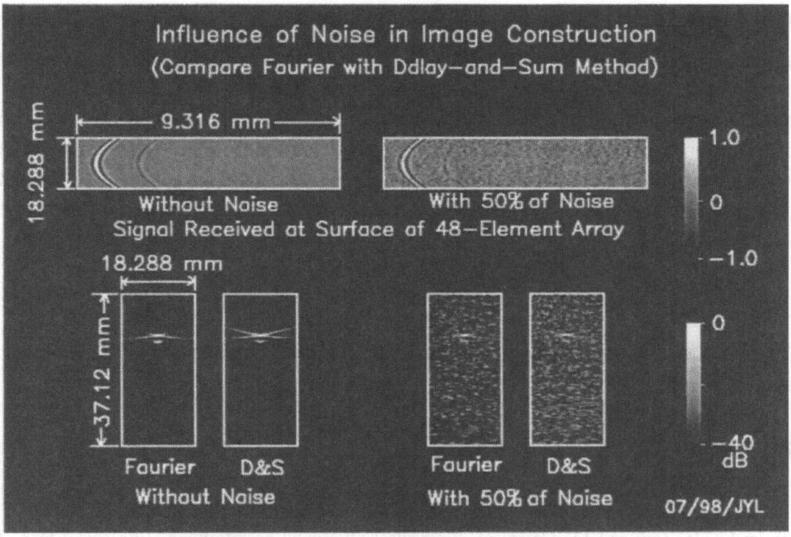

Figure 1. Influence of noise on the construction of images of a line target with the Fourier and the conventional (delay-and-sum) methods using a linear array transducer. Notice that the upper and lower grayscale bars are in linear and log scales, respectively. The speed of sound is assumed to be 1450 m/s.

DISCUSSION

The SNR is an important parameter of image quality. A higher SNR means a deeper penetration of ultrasound beams in biological soft tissues because signals from a deeper depth are weak due to tissue attenuation. From the study above, it is clear that both the Fourier and the conventional methods have a similar SNR on image constructions. Although the SNR is about the same for both the Fourier and the conventional methods, the Fourier method has the advantage that imaging systems can be greatly simplified, especially, for three-dimensional (3D) imaging where a two-dimensional (2D) array transducer is used. This is because the Fourier method can be implemented with the fast Fourier transform. As a comparison, the conventional delay-and-sum method requires to delay signals of each element and then sum the delayed signals to focus a beam at each point of an object. This process is very complex and is usually done by the so called digital beamformer.

In discussion above, a broad band plane wave is used in transmission to illuminate a large area of an object to construct high frame rate images. Because the plane wave does not diverge over the depth of interest, energy density is high as compared to a diverged beam[17]. Although a focused transmission beam may produce a higher energy density in its focal region when *f-number* is small, image frame rate is low and thus images of fast moving objects such as the heart and the blood flow within the heart will be distorted. Therefore, the Fourier method will be useful for high frame rate imaging at a high SNR.

CONCLUSION

A new image construction method, called the Fourier method, has been developed recently[17]. This method can be used to construct images at a high frame rate (up to 3750 frames/s for a depth of 200 mm in biological soft tissues) with simplified electronics because the fast Fourier transform can be used in image constructions. A computer simulation shows that the signal-to-noise ratio (SNR) of the new method is comparable to that of the conventional delay-and-sum (D&S) method when the same transmission beam is used for both methods.

ACKNOWLEDGMENTS

This work was supported in part by grant R01 HL60301 from the National Institutes of Health.

REFERENCES

1. J.A. Stratton. *Electromagnetic Theory.* McGraw-Hill Book Company, New York (1941), p. 356.
2. Jian-yu Lu, H. Zou, and J.F. Greenleaf, Biomedical ultrasound beamforming, *Ultrasound Med. Biol.* 20(5):403–428 (1994).
3. Jian-yu Lu and J.F. Greenleaf, Nondiffracting X waves — exact solutions to free-space scalar wave equation and their finite aperture realizations, *IEEE Trans. Ultrason. Ferroelec. Freq. Contr.* 39(1):19–31 (1992).
4. Jian-yu Lu and J.F. Greenleaf, Experimental verification of nondiffracting X waves, *IEEE Trans. Ultrason. Ferroelec. Freq. Contr.* 39(3):441–446 (1992).
5. J. Durnin, Exact solutions for nondiffracting beams. I. The scalar theory, *J. Opt. Soc. Am. A*, 4(4):651–654 (1987).
6. J. Durnin, J. J. Miceli, Jr., and J. H. Eberly, Diffraction-free beams, *Phys. Rev. Lett.*, 58(15):1499–1501, April 13 (1987).
7. Jian-yu Lu and J. F. Greenleaf, Ultrasonic nondiffracting transducer for medical imaging, *IEEE Trans. Ultrason., Ferroelect., Freq. Contr.*, 37(5):438–447, September (1990).
8. Jian-yu Lu and J. F. Greenleaf, Pulse-echo imaging using a nondiffracting beam transducer, *Ultrason. Med. Biol.*, 17(3), pp. 265–281, May (1991).
9. Jian-yu Lu, Bowtie limited diffraction beams for low-sidelobe and large depth of field imaging, *IEEE Trans. Ultrason. Ferroelec. Freq. Contr.* 42(6):1050–1063 (1995).
10. Jian-yu Lu, Producing bowtie limited diffraction beams with synthetic array experiment, *IEEE Trans. Ultrason. Ferroelec. Freq. Contr.* 43(5):893–900 (1996).
11. Jian-yu Lu and J.F. Greenleaf, Evaluation of a nondiffracting transducer for tissue characterization, in *IEEE 1990 Ultrason. Symp. Proc.* 90CH2938-9, 2:795–798 (1990).
12. Jian-yu Lu, Xiao-Liang Xu, Hehong Zou, and J. F. Greenleaf, Application of Bessel beam for Doppler velocity estimation, *IEEE Trans. Ultrason., Ferroelect., Freq. Contr.*, 42(4):649–662, July (1995).
13. Jian-yu Lu, Improving accuracy of transverse velocity measurement with a new limited diffraction beam, in *IEEE 1996 Ultrason. Symp. Proc.* 96CH35993, 2:1255–1260 (1996).
14. Jian-yu Lu and J.F. Greenleaf, Producing deep depth of field and depth-independent resolution in NDE with limited diffraction beams, *Ultrason. Imag.* 15(2):134–149 (1993).
15. Jian-yu Lu, High-speed transmissions of images with limited diffraction beams, in *Acoustical Imaging*, 23, S. Lees and L. A. Ferrari, Editors, pp. 249–254 (1997).
16. J. Ojeda-Castaneda and A. Noyola-Iglesias, Nondiffracting wavefields in grin and free-space, *Microwave and Optical Technology Letters* 3(12):430–433 (1990).
17. Jian-yu Lu, 2D and 3D high frame rate imaging with limited diffraction beams, *IEEE Transactions on Ultrasonics, Ferroelectrics, and Frequency Control*, 44(4):839–856, July (1997).
18. Jian-yu Lu, Experimental study of high frame rate imaging with limited diffraction beams, *IEEE Transactions on Ultrasonics, Ferroelectrics, and Frequency Control*, 45(1):84–97, January (1998).
19. Jian-yu Lu, Limited diffraction array beams, *Int. J. Imag. System and Tech.* 8(1):126–136 (1997).
20. Jian-yu Lu, Designing limited diffraction beams, *IEEE Trans. Ultrason. Ferroelec. Freq. Contr.* 44(1):181–193 (1997).

21. J. Shen, H. Wang, C. Cain, and E. S. Ebbini, A post-beamforming processing technique for enhancing conventional pulse-echo ultrasound imaging contrast resolution, in *IEEE 1995 Ultrason. Symp. Proc.*, 95CH35844, 2:1319–1322 (1995).
22. R. Bracewell, *The Fourier Transform and its Applications.* New York, NY: McGraw-Hill Book Company, chs. 4 and 6, (1965).
23. S. J. Norton and M. Linzer, Ultrasonic reflectivity imaging in three dimensions: exact inverse scattering solutions for plane, cylindrical, and spherical apertures, *IEEE Trans. Biomed. Eng.*, vol. BME-28(2):202–220, Feb. (1981).

INDEX

FSC
www.fsc.org
MIX
Papier aus verantwortungsvollen Quellen
Paper from responsible sources
FSC® C105338

If you have any concerns about our products,
you can contact us on
ProductSafety@springernature.com

In case Publisher is established outside the EU,
the EU authorized representative is:
Springer Nature Customer Service Center GmbH
Europaplatz 3, 69115 Heidelberg, Germany

Printed by Libri Plureos GmbH
in Hamburg, Germany